Chitosan-Based Edible Coatings in the Food Industry

This book provides comprehensive coverage of the application of naturally derived chitosan for improved food quality and shelf life. It describes the application of chitosan-based composite coatings in combination with antimicrobial agents, such as essential oils, organic acids, metal ions, and inorganic nanomaterials. It includes the sensory evaluation of chitosan-coated foods. It discusses the nutraceutical properties of the coatings. It also includes information about the marketing, technology transfer, commercialization, and regulatory guidelines associated with these products. This book is meant for researchers and industry experts in food processing and food engineering.

Key Features:

- Reviews the effects of chitosan coating on delaying loss in chlorophyll quality, malondialdehyde, weight loss, soluble solids content, pH, and firmness in various fruits and vegetables
- Discusses the nutraceutical properties and nutritional benefits of chitosan coating
- Highlights the combined effect of edible coatings during the storage of fruits and vegetables using evaporative coolant structures
- Discusses the application of techniques such as proteomics, metabolomics, bioinformatics, and chemometrics for the characterization of bioactive compounds present in these coatings
- Covers the various natural sources of chitosan

Current Developments in Agricultural Biotechnology and Food Security

Series Editor:
Charles Oluwaseun Adetunji
Professor and Deputy Vice Chancellor,

Federal University of Agriculture and Developmental Studies
Iragbiji

This series intends to provide comprehensive coverage of sustainable modern technologies, aimed at improving food production via agriculture and food biotechnological techniques. Proposed suite of books focuses on topics in agricultural microbiology, biotechnology, food science, crop production, and postharvest management, aimed at both basic and advanced food and agricultural biotechnology. The series seeks to discuss and provide foundational content from bench to bedside in food microbiology and food and agricultural biotechnology. The series' goal is to enhance knowledge and present updates on hot topics in the field of Agricultural Biotechnology; specifically, the series aims to translate results and recent findings of studies into enhanced food production. This series is primarily intended for researchers and students in Food and Agricultural Biotechnology at the graduate level and above, including those working in academic, corporate, or non-profit settings.

Phyto-Oxylipins: Metabolism, Physiological Roles, and Profiling Techniques
Edited by Sheikh Mansoor Shafi, Chukwuebuka Egbuna, and Charles Oluwaseun Adetunji

Agricultural Biotechnology: Food Security Hot Spots
Edited by Charles Oluwaseun Adetunji, Deepak Gopalrao Panpatte, and Yogeshvari Kishorsinh Jhala

Next Generation Probiotics: From Novel Drugs to Future Therapeutics
Edited By Shakira Ghazanfar, Charles Oluwaseun Adetunji, and Alessandro Di Cerbo, Ghulam Muhammad Ali, Wajya Ajmal, Muhammad Naeem Riaz

Chitosan-Based Edible Coatings in the Food Industry
Charles Oluwaseun Adetunji, Raghvendra Raman Mishra, Pragya Mishra, and Dr Juliana Bunmi Adetunji

Biotechnological Relevance of Cytochrome P450 in Boosting Agricultural Production, and Food Security
Wadzani Palnam Dauda and Charles Oluwaseun Adetunji

For more information about this series, please visit: Current Developments in Agricultural Biotechnology and Food Security - Book Series - Routledge & CRC Press

Chitosan-Based Edible Coatings in the Food Industry

Edited by
Charles Oluwaseun Adetunji, Raghvendra Raman Mishra, Pragya Mishra and Juliana Bunmi Adetunji

CRC Press
Taylor & Francis Group
Boca Raton London New York

CRC Press is an imprint of the
Taylor & Francis Group, an **informa** business

Designed cover image: Shutterstock

First edition published 2027
by CRC Press
2385 NW Executive Center Drive, Suite 320, Boca Raton FL 33431

and by CRC Press
4 Park Square, Milton Park, Abingdon, Oxon, OX14 4RN

CRC Press is an imprint of Taylor & Francis Group, LLC

ISBN: 9781032405384 (hbk)
ISBN: 9781032408484 (pbk)
ISBN: 9781003355458 (ebk)

DOI: 10.1201/9781003355458

Typeset in Times
by codeMantra

Contents

Preface viii
About the Editors x

Chapter 1 Application of Chitosan-Based Edible Coatings in the Food Industry 1

Muhammad Akram, Tsado Priscilla Yetu, Oluwafemi Adebayo Oyewole, Charles Oluwaseun Adetunji, Olowe Olugbenga Adekunle, Ikechukwu P. Ejidike, and Olaniyi Amos Fawole, O.O. Osemwegie, Saleh Alhaji Ado

Chapter 2 Utilization of Chitosan Derived from Natural Sources to Enhance Food Quality and Shelf Life in Agriculture, Seafood, and Poultry Sectors 10

Osarenkhoe Osarenkhoe Osemwegie, Adebowale Toba Odeyemi, Damilare Rotimi, Adeyemi Ayotunde Adeyanju, M.A. Acho, and Charles Oluwaseun Adetunji

Chapter 3 The Effect of the Edible Coatings and Their Combinative Effect during the Storage of Fruits and Vegetables Using Evaporative Coolant Structure 39

Muhammad Akram, Tsado Priscilla Yetu, Oluwafemi Adebayo Oyewole, Charles Oluwaseun Adetunji, Olowe Olugbenga Adekunle, Ikechukwu P. Ejidike, and Olaniyi Amos Fawole, O.O. Osemwegie, Saleh Alhaji Ado

Chapter 4 Recent Advances in Food Coating in the Management of Diseases 49

Samuel Tobi Farohunbi, Oluwafemi Emmanuel Ekun, Faith Jesupemi Olaitan, and Oluwadamilola Oyetayo Ezekiel

Chapter 5 Engineering Properties of Chitosan-Based Edible Films/Coatings 61

John Tsado Mathew, Jonathan Hussaini, Abel Inobeme, Charles Oluwaseun Adetunji, Elijah Yanda Shaba, Mary Jummai Jiya, Yakubu Azeh, Musah Monday, Muhammad Aishetu Ibrahim, Etsuyankpa Muhammad Bini, Musa Safiyanu Tanko, Ismaila Aliyu Oga, Tanko Ezekiel, Muhammed Jibrin Ndejiko, and Amos Mamman

Chapter 6 Application of Chitosan-Based Edible Coatings in the Prevention of the Growth of Spoilage Microorganisms 79

Abiola Folakemi Olaniran, Faith Oluwatoyosi Agaja, Oluwatobi Victoria Obayomi, and Clinton Emeka Okonkwo

Chapter 7 Integration of Chitosan Edible Coating and Modified Atmosphere Packaging....... 100

John Tsado Mathew, Jonathan Hussaini, Abel Inobeme, Elijah Yanda Shaba, Mary Jummai Jiya, Charles Oluwaseun Adetunji, Yakubu Azeh, Musah Monday, Muhammad Aishetu Ibrahim, Etsuyankpa Muhammad Bini, Musa Safiyanu Tanko, Ismaila Aliyu Oga, Tanko Ezekiel, and Amos Mamman

Chapter 8 Application of Chitosan for the Extension of the Shelf Life of Poultry, Meat, and Fishery Products... 116

John Tsado Mathew, Abel Inobeme, Yakubu Azeh, Musah Monday, Elijah Yanda Shaba, Mary Jumai Jiya, Etsuyankpa Muhammad Bini, Ismaila Aliyu Oga, Musa Safiyanu Tanko, Tanko Ezekiel, Jonathan Hussaini, Mohammed Jibrin Ndejiko, Adamu Isah Abubakar, Charles Oluwaseun Adetunji, and Amos Mamman

Chapter 9 Biodegradability Studies of Chitosan-Based Packaging... 134

John Tsado Mathew, Jonathan Hussaini, Abel Inobeme, Elijah Yanda Shaba, Charles Oluwaseun Adetunji, and Amos Mamman

Chapter 10 Application of Chitosan-Based Composite Coatings When Combined with Antimicrobial Agents with Nanocompounds.. 157

John Tsado Mathew, Jonathan Hussaini, Yakubu Azeh, Musah Monday, Abel Inobeme, Elijah Yanda Shaba, Mary Jumai Jiya, Muhammad Aishetu Ibrahim, Charles Oluwaseun Adetunji, Etsuyankpa Muhammad Bini, Ismaila Aliyu Oga, Musa Safiyanu Tanko, Tanko Ezekiel, Muhammed Jibrin Ndejiko, and Amos Mamman

Chapter 11 Chitosan and Its Derivatives as Potential Biomaterials for Medical and Food Technologies 176

Osarenkhoe Omorefosa Osemwegie, Damilare Rotimi, Charles Obiora Nwonuma, Marvellous Amarachi Acho, Adeyemi Adetunde Adeyanju, Bamidele Sunday Fakinle, and Charles Oluwaseun Adetunji

Chapter 12 Application of Chitosan-Based Composite Coatings Enhanced with Antimicrobial Agents, Including Essential Oils 205

Ekong Mercy and Ekpenyong Glory Maurice

Chapter 13 The Health, Nutraceutical, and Nutritional Benefits of Edible Coatings Based on Chitosan.. 221

John Tsado Mathew, Abel Inobeme, Charles Oluwaseun Adetunji, Elijah Yanda Shaba, Yakubu Azeh, Musah Monday, Mary Jumai Jiya, Muhammad Aishetu Ibrahim, Etsuyankpa Muhammad Bini, Ismaila Aliyu Oga, Musa Safiyanu Tanko, Jibrin Mohammed Ndaejiko, Tanko Ezekiel, Jonathan Hussaini, and Amos Mamman

Chapter 14 Antioxidant Prospects of Combining Chitosan and Essential Oil Biofilm Inhibitors in Fruits and Vegetables 237

Mayowa Jeremiah Adeniyi

Chapter 15 Chitosan and Its Modification: A Promising Biomaterial for Food and Medical Technology 253

John Tsado Mathew, Jonathan Hussaini, Abel Inobeme, Elijah Yanda Shaba, Mary Jummai Jiya, Tanko Ezekiel, Yakubu Azeh, Musah Monday, Muhammad Aishetu Ibrahim, Etsuyankpa Muhammad Bini, Ismaila Aliyu Oga, Musa Safiyanu Tanko, Muhammed Jibrin Ndejiko, Charles Oluwaseun Adetunji, and Amos Mamman

Index 265

Preface

It has been forecasted that the human population will increase tremendously to nine billion by 2050. In view of this, there is a need to search for a sustainable solution that will ensure an increase in food production and safety. Postharvest fruits and vegetables have been identified as living organisms that undergo the process of metabolism. It has been observed that their characteristics, such as flavor, appearance, and nutrition, could deteriorate, especially during postharvest storage and transportation, which might be linked to several factors such as decay, water loss, and browning. The methods used for the preservation of fruits and vegetables include irradiation, controlled atmospheric storage, and modified storage for slowing down respiration and delaying senescence. Unfortunately, these methods are expensive to install, costly to maintain, and impractical to use for small quantities of produce, especially for the farmers in developing countries. Some of the problems faced by these fruit farmers include lack of electricity in these rural areas, inaccessible roads, and high cost of synthetic pesticides used for the prevention of pests and pathogens affecting the postharvest of fruits and vegetables. The application of edible coatings to freshly harvested produce offers a less expensive alternative with potentially equal beneficial outcome, prevents moisture loss, and inhibits the oxygen penetration into the plant tissue or the microbial growth, apart from the fact that they are convenient and conform to food safety. Chitosan could be defined as an improved, natural biopolymer that is obtained through the deacetylation of chitin from the shells of crustaceans and from some beneficial microorganisms. It is widely accepted, especially in the food industry, due to numerous factors such as its functional features and distinctive biological activities. This portends them for a wider application in the food industry because of their enhanced film-forming property and antimicrobial activity, which guarantee their application in food preservation and for edible coatings.

Therefore, this book is intended to provide detailed information on the utilization of chitosan derived from natural sources, such as seafood, poultry, and agriculture, for the enhancement of quality and shelf life of various foods. Special information is provided on the application of chitosan-based composite coating when combined with antimicrobial agents, such as essential oils, organic acids, organic matter, metal ions, and inorganic nanomaterials, as well as their composite coating of chitosan and biological control agents. Moreover, the application of edible coating with non-coating treatment, integrated application of heat treatment and gas fumigation such as 1-methylcyclopropene, and integrated application of chitosan coating and modified atmosphere packaging are highlighted. The effects of chitosan on how it could delay a loss in chlorophyll quality, malondialdehyde, weight loss, soluble solids content, titratable acidity, ascorbic acid content, pH, and firmness in various fruits and vegetables are presented, while the influence of chitosan coating on superoxide dismutase, catalase, phenylalanine ammonia-lyase, peroxidase, and polyphenol oxidase, as well as the process involved in the inhibition of the generation of superoxide free radicals and prevention of the growth of spoilage microorganisms such as mesotrophic *and* psychrotrophic bacteria, yeasts, and molds, are also discussed. Detailed information on the sensory evaluation of foods and fruits coated with edible coating based on chitosan are provided in detail. This book also highlights the health, nutraceutical, and nutritional benefits, as well as the reported effects of these edible coating foods for the treatment of diseases such as in antibacterial, antifungal, antiulcer, antituberculosis, anti-inflammatory, anticancer, antidiabetic, antimalarial, and antiviral drugs. Relevant information on marketing, potential legal issues, intellectual properties, technology transfer, development, and commercialization of edible coating based on chitosan are highlighted. The effect of edible coatings and their combined effects during the storage of fruits and vegetables using evaporative coolant structures are highlighted. Relevant details regarding the promotion and adoption of edible coating from chitosan by farmers, food handlers, transporters, wholesalers, and retailers, along with the role of policymakers, government, and food scientists, are also highlighted. Relevant information on the application of multi-omics techniques such as proteomics, metabolomics, transcriptomics, genomics, bioinformatics, and chemometrics for the characterization of biologically active components present in this edible coating based on chitosan are also highlighted. The application of chitosan for the extension of the shelf life of poultry, meat, and fishery products are also highlighted.

This book is aimed at a diverse audience, including the food science and health sectors, federal and state ministries of health, policymakers, researchers, government officials, virologists, epidemiologists, computer scientists, chemical pathologists, computational biologists, biomedical engineers, and scientists in different fields, such as Biochemistry, Bioinformatics, Pharmacology, Biomedical, Microbiology, and Medical Healthcare. Furthermore, I want to thank my coeditors for their efforts and dedication during this project. Moreover, I wish to gratefully acknowledge the suggestions, help, and support of Dr Bhavik Sawhney under Taylor & Francis Books India Pvt Ltd.

Professor Charles Oluwaseun Adetunji (Ph.D., AAS affiliate, FRSB (UK) FNYA; FBSN; FNSM, MNBGN)
Deputy Vice-Chancellor of the Federal University of Agriculture and Developmental Studies, Iragbiji, Nigeria
March 2025

About the Editors

Charles Oluwaseun Adetunji is currently the Deputy Vice Chancellor of the Federal University of Agriculture and Developmental Studies, Iragbiji, and previously a faculty member in the Department of Microbiology within the Faculty of Science at Edo State University, Uzairue, Edo State, Nigeria. He employs biological techniques and microbial bioprocesses to advance sustainable development goals and foster agrarian revolution through quality teaching, research, and community development. He previously held roles as the Acting Director of Intellectual Property and Technology Transfer, the Head of the Department of Microbiology, Sub Dean of the Faculty of Science, and immediate Dean of the Faculty of Science, and he is presently serving as the Chairman of the Grant Committee and the Director of Research and Innovation at EDSU. He was the first in the Department of Microbiology at the Faculty of Science, Edo State University, Uzairue, to deliver the 11th Inaugural Lecture, titled "Microbial Biosphere: The Global Impact and Underexplored Currency". He was also elected by the members of the Senate as a Senate representative to search for a new Vice Chancellor in EDSU.

Professor Adetunji commenced his early career at the Nigerian Stored Product Research Institute (NSPRI) in Ilorin, Nigeria, as a Research Officer II and progressed to the position of Principal Scientist. During his tenure at NSPRI, he actively contributed to the development and testing of functional foods, the isolation and characterization of health-promoting components in foods, and the improvement of processes aimed at extending the shelf life of fruits and vegetables while controlling spoilage and foodborne pathogens. He then transitioned to Landmark University, Omu-Aran, where he served as a lecturer in various capacities, engaging in teaching and research. Additionally, he served as a visiting fellow at the Hebrew University of Jerusalem, Israel, specializing in Food Safety, and visited the Wageningen Centre for Development and Innovation in the Netherlands.

He is a Fellow of the Royal Society of Biology (UK), the Biotechnology Society of Nigeria, the Nigerian Young Academy, and the Nigerian Society of Microbiology. Additionally, he holds visiting professorships at the Center of Biotechnology, Precious Cornerstone University, Ibadan, and Trinity University, Lagos State. He is an affiliate member of the African Academy of Science. Presently, he serves as an external examiner for numerous academic institutions globally, particularly for Ph.D. and M.Sc. students, including the University of Johannesburg, South Africa, Covenant University, Ota Lagos State, Ahmadu Bello University, Zaira, Landmark University, Kwara State, as well as the Department of Food Hygiene and Environmental Health, Faculty of Veterinary Medicine, University of Helsinki, Finland. He is an external assessor for the Department of Biotechnology at Addis Ababa University to assist in establishing prima facie of candidates to be promoted to the rank of Full Professor, as well as at the University of Ilorin for the Department of Microbiology and many other universities worldwide. He was recently appointed as a Visiting Professor at the University of Johannesburg in South Africa; a Visiting Professor at the Center of Biotechnology, the Federal University of Technology in Abeokuta, Ogun State; a Visiting Professor at the National Open University of Nigeria; and an instructor at the University of the People, Pasadena, CA, USA. Moreover, he was recently appointed as a member of the Advisory Board for the College of Agriculture and Natural Sciences by the management of Joseph Ayo Babalola University in Nigeria.

He has received numerous scientific awards and grants from prestigious academic bodies such as the Council of Scientific and Industrial Research (CSIR) in India, the Department of Biotechnology (DBT) in India, The World Academy of Science (TWAS) in Italy, the Netherlands Fellowship Programme (NFP) in the Netherlands, the Agency for International Development Cooperation in Israel, the Royal Academy of Engineering in the UK, and the National Research Fund (TETFUND).

Due to his international recognition and significant contributions to knowledge and scientific impact, Professor Adetunji was recently listed in Stanford University's World's Top 2% Scientists Rankings for 2023, while he was also recognized in the prestigious 2024 Career-long List of the

World's Top 2% Scientists, as well as the single-year Most Impactful List of Distinguished Professors globally. He has authored numerous scientific journals and conference proceedings in refereed national and international journals, amounting to over 860 manuscripts, including more than 60 books, 547 book chapters, and numerous scientific patents, while he has also published many manuscripts in top-rated journals such as *Lancet*. Notably, he holds a Google Scholar i10-Index of 216. Additionally, he was recently ranked first among the top 500 prolific authors in Nigeria from 2020 to date by SciVal/SCOPUS in Elsevier, Netherlands. Moreover, he achieved the top ranking in two distinct fields, Biological Sciences and Agricultural Science, among 500 prolific authors from 58 different countries in Africa in 2023. Interestingly, he was also ranked number 1 in Agricultural Science and Biological Sciences among 500 prolific authors in 2024. Currently, he serves as a book series editor with Wiley and Sons, Elsevier, Springer, and Taylor & Francis, among others.

Professor Adetunji has participated in over 100 international scientific conferences and workshops, assuming various roles such as Invited Speaker, Plenary Speaker, Keynote Speaker, Chief Panelist, Lead Guest Speaker, Chairman of Scientific Sessions, and Judge of Posters, across several continents worldwide. Additionally, he serves as a reviewer for more than 100 scientific journals, including *Nature*, *Communications Earth & Environment* in *Nature Portfolio*, and *PLOS One*. His contributions to open science and the role his research plays in addressing global challenges were recently acknowledged by Elsevier, particularly after some of his scientific articles were associated with the United Nations (UN) Sustainable Development Goals (SDGs).

Currently, Professor Adetunji holds the positions of President and Chairman of the Governing Council of the Nigerian Bioinformatics and Genomics Network Society. Previously, he had served as the General Secretary of the Nigerian Young Academy, where he also chaired the Biological and Health Sciences Working Group. Furthermore, he holds the positions of Assistant General Secretary and National Coordinator for the South-West Nigerian Society for Microbiology. He was recently appointed to the Young Investigator Award Committee by the American Society of Tropical Medicine and Hygiene, USA. He also serves as the Vice President and Nigerian Country Director of the Genetics Society of West Africa, as well as the Director of the International Affiliation and Training Centre for Environmental and Public Health, Research, and Development, situated in Zaria. Moreover, he acted as the immediate past Business Manager and is currently the Vice President of the Biotechnology Society of Nigeria. He was appointed by the African Union Development Agency-NEPAD Centre of Excellence in Science, Technology, and Innovation (CoE STI) as a senior expert and consultant to facilitate the development of curricula in Biotechnology, Applied Microbiology, Microbiome, and Genome Editing in various African countries, including Nigeria, Ethiopia, Ghana, Zimbabwe, Mozambique, Malawi, and Kenya. He was recently appointed as the Vice Chairman of the Technical Expert Committee by the Standard Organization of Nigeria on Sugar, Confectionery, Biscuits, and Bakery Products.

In recognition of his expertise and extensive experience in the fields of Microbiology, Bioinformatics, and Biotechnology, Professor Adetunji holds the unique privilege of serving as a Consultant to the Government of the Republic of Serbia in Europe, specifically as a Project Peer Reviewer for the Science Fund in the areas of Microbiology, Nanotechnology, Food Science, and Biotechnology. Also, he was recently appointed as a senior expert by South Africa's National Research Foundation (NRF) in evaluating the quality, impact, and standing of the research portfolio of scientists in the area of Microbiology who are associated with a Higher Education/Research Institution (HEI) in South Africa (SA). Moreover, he was recently appointed to serve as a reviewer for the DSI/NRF South African Research Chair Initiative Programme (SARChI)—Performance of Grant Recipients received by distinguished professors in the area of Biotechnology Innovation & Engagement in South Africa. Recently, he was appointed as a member of the College of Experts from the UK to serve as a member of the selection committee for the Africa Research Excellence Fund to help in the timely delivery of its vision of an inspired, committed, and talented community of researchers in Africa leading world-class research and participating equitably in international research endeavors for health and well-being. He was appointed as the Project Peer Reviewer for the Mawazo Fellowship Programme by the Mawazo Institute, which is a women-led African organization based in Nairobi, Kenya, supporting early-career women researchers as they work to find solutions to local and global development challenges. Additionally, he was recently appointed as the general coordinator and supervisor for the 1000 scientific trials program by Harvest Harmonics in the USA. In this role, he oversees coordination across 58 countries in Africa, serving as the Country Representative for Nigeria. This appointment underscores the significance of his landmark publications in translational research, his diligent work ethic, and exceptional talents, positioning him as a renowned trailblazer in advancing academics and research in Africa.

On a global scale, Professor Adetunji has mentored numerous undergraduate, postgraduate, and postdoctoral students at both national and international universities, including Jawaharlal Nehru University, India. He has provided advice to numerous scientists and mentored young researchers on how they could design translational and innovative research. Based on his experience as the Director of Intellectual Property at EDSU, he has advised many scientists on how they could file their innovation as well as advised numerous foundations, specifically, on how to secure funding, grants, scholarships, and awards from non-profit organizations to assist qualified young Nigerians who aspire to pursue science-based disciplines at the advanced levels (Master/Ph.D.) and assisted them in obtaining tuition fees and research funds.

Raghvendra Raman Mishra is working as an Assistant Professor in the Medical Laboratory Technology, DDU KAUSHAL KENDRA, RGSC, and Warden of Shivalik Hostel (Received Best Hostel Award, 2016–2019) at the South Campus, Banaras Hindu University since the 2016 academic session. He received his master's degree in Biomedical Technology in 2005 from Jiwaji University, Gwalior. He completed his project at the Virology Laboratory, School of Life Sciences, Jawaharlal Nehru University, New Delhi. He earned his Ph.D. (2009) in Biomedical Technology from the Molecular Oncology Laboratory, Department of Surgical Oncology, Institute of Medical Sciences, Banaras Hindu University, Varanasi. He then went to Thyrocare Technologies Limited, Navi Mumbai, as Officer-Central Processing Lab (2010). In 2011, he joined as a postdoctoral research associate at the Bacteriophage Research Lab, Department of Microbiology, Institute of Medical Sciences, Banaras Hindu University, Varanasi. From 2012 to 2015, he served as an Assistant Professor in an affiliated institute at the Department of Biotechnology (Institute Code 641) and was appointed as Group Leader of the Controller of Examination of Dr. APJ Abdul Kalam Technical University in Lucknow. With these 15 years of research and teaching experiences, his areas of specialization include Biomedical Technology, Biotechnology, Bioinformatics, Cancer Biology, Microbiology, Molecular Biology, and Food Biotechnology. He has published more than 50 national and international peer-reviewed papers. He has received a grant from the Department of Science and Technology, Government of India funded project of the Rapid Grant for Young Investigator (2010). He completed the "General Course on Intellectual Property Rights" from the World Intellectual Property Organization (WIPO). He has received fellowships from the University Grant Commission, the Indian Council of Medical Research, and Council of Scientific and Industrial Research, New Delhi, at various stages of his research. He is a Lifetime Member of "The Cytometry Society" in Hyderabad, "All India Medical Laboratory Technologists Association" in New Delhi, and "Society of Bacteriophage" in Varanasi. He is an Associate Editor of the *Journal of Applied Pharmaceutical Science and Research* (ISSN 2581-5520) and the Editor of a special issue of the *International Journal of Academic Research and Development* (ISSN 2395-1737).

Pragya Mishra is presently working as an Assistant Professor of Food Processing and Management, DDU KAUSHAL Kendra, RGSC, Banaras Hindu University. She completed her master's degree in Food Technology in 2010 from the Centre of Food Technology, University of Allahabad, with distinction. She did her dissertation research on "Feasibility studies for enhancement of sensory attributes of pearl millet weaning food by fermentation" from the National Dairy Research Institute, Karnal, Haryana. She has been qualified University Grant Commission National Eligibility Test Examination. She earned her Doctor of Philosophy in Food Technology from the Centre of Food Technology, Institute of Professional Studies, University of Allahabad, India, in 2017.

Her dissertation title was "Isolation and molecular characterization of edible *Spirulina* species under different abiotic stress conditions and evaluation of their food supplement". She also worked as a Senior Research Fellow of Uttar Pradesh Council of Agricultural Research funded project and applied for a patent on a potato flakes energy bar (Application no. 201711014441). She has received an accession number (GenBank: KM407008.1) from the National Center of Biotechnology Information, USA, for new isolates of *Spirulina maxima* species from cyanobacterial isolates collected from the paddy fields of the Prayagraj (Soraon) region targeting specific 16S rRNA gene. She has published more than 20 international and national papers in peer-reviewed journals and has edited six books. She has contributed to several chapters in international books. She received the MFPA Technology, 2018, Young Scientist Award, organized by Banaras Hindu University. She is a Lifetime Member of the "All India Medical Laboratory Technologists Association" in New Delhi and the "Society of Bacteriophage" in Varanasi. She is a Member of the Association of Food Scientists & Technologists India. She was an Editor of a special issue in the *International Journal of Academic Research and Development* (ISSN 2395-1737).

Juliana Bunmi Adetunji, Ph.D., is a Faculty Member and the Head of the Department of Biochemistry, Faculty of Science, Osun State University, Osogbo, Nigeria. Her research interests over the years have focused on the nutritional safety of foods and the evaluation of medicinal plants in the management and maintenance of human health. Her work on phytomedicinal potentials of certain medicinal plants and on food safety has made valuable contributions to ongoing research in the use of alternative remedies for disease management in Africa, as well as sustainable food systems for the future. She is a recipient of the Tertiary Educational Fund in Nigeria. She has published several papers in scientific journals and conference proceedings in international and local refereed journals. She is a member of many scientific and professional bodies like the Organization for Women in Science in the Developing World, the Nigerian Society of Biochemistry and Molecular Biology, the Nigerian Society of Experimental Biology, and the Nigeria Bioinformatics and Genetics Network.

1 Application of Chitosan-Based Edible Coatings in the Food Industry

Muhammad Akram[1], *Tsado Priscilla Yetu*[2], *Oluwafemi Adebayo Oyewole*[3], *Charles Oluwaseun Adetunji*[3], *Olowe Olugbenga Adekunle*[4], *Ikechukwu P. Ejidike*[5], *and Olaniyi Amos Fawole*[6], *O.O. Osemwegie*[7], *Saleh Alhaji Ado*[8]

[1]Department of Eastern Medicine, Government College University Faisalabad, Faisalabad, Punjab, Pakistan

[2]Department of Microbiology, Federal University of Technology Minna, Niger State, Nigeria

[3]Department of Microbiology, Edo State University, Uzairue, Edo State, Nigeria

[4]Department of Chemical Sciences, Faculty of Science, Anchor University, Lagos, Nigeria

[5]Postharvest and Agroprocessing Research Centre, Department of Botany and Plant Biotechnology, University of Johannesburg, Auckland Park, Johannesburg, South Africa

[6]Department of Botany and Plant Biotechnology, South African Research Chairs Initiative in Sustainable Preservation and Agroprocessing Research, Faculty of Science, University of Johannesburg, Auckland Park, Johannesburg, South Africa

[7]Department of Food Science and Microbiology, Landmark University, Mu Aran, Kwara State, Nigeria

[8]Department of Microbiology, Ahmadu Bello University, Zaria

1.1 INTRODUCTION

The perishable nature of many foods, globalization, and growing customer demand have made it increasingly challenging for the food industry to ensure the safety, quality, and extended shelf life of food products. Conventional methods of food preservation, such as chemical preservatives and artificial packaging, can give rise to concerns regarding potential health risks and environmental sustainability. People are now searching for effective and safe alternatives as a result. One of the most potential applications of these breakthroughs is the use of chitosan-based edible coatings. Chitosan, a naturally occurring biopolymer derived from chitin, is abundant in crustacean shells, such as those of shrimp, crabs, and other marine animals. Its antimicrobial, biocompatible, and biodegradable properties make it an excellent choice for food preservation (Khan et al., 2020). Chitosan-based coatings offer thin, edible layers on food surfaces that act as barriers against moisture, oxygen,

DOI: 10.1201/9781003355458-1

and microbiological contaminants. Food products maintain their nutritional value and flavor for an extended period (Kumar et al., 2019). Apart from its inherent properties, chitosan is a versatile polymer that can be modified or combined with other beneficial substances to meet specific requirements. For example, the antibacterial and antioxidant qualities of chitosan coatings are enhanced by the addition of essential oils, plant extracts, and nanoparticles (Liu et al., 2021). These developments have led to an expansion of the usage of chitosan coatings in several food categories, including fresh fruit, meat, seafood, and dairy products. Edible coatings based on chitosan support global sustainability goals by reducing food waste and reliance on artificial additives. This is particularly relevant given the growing awareness of the environmental damage caused by plastic packaging and the desire for eco-friendly substitutes (Smith et al., 2022). These coatings are an appealing solution to today's food preservation issues since they also have the added benefit of being safe to consume and acting as a barrier. The purpose of this study is to provide a thorough review of the food industry's use of edible coatings based on chitosan. This study looks at how they are made, how they work, and how they affect the quality and safety of food. Recent developments, difficulties, and opportunities for expanding their industrial application in the future are also covered. The results show that chitosan-based coatings have the potential to be a sustainable and efficient way to improve food preservation while tackling urgent problems of environmental preservation and food safety.

1.2 CHITOSAN

Chitosan is a naturally occurring, biodegradable polymer derived from chitin, the second most common biopolymer on Earth after cellulose. Chitin is mostly found in the exoskeletons of crustaceans, such as shrimp, crabs, and lobsters, as well as certain fungi, according to Kumar et al. (2018). Chitosan is produced by the process of deacetylation, which eliminates acetyl groups from chitin to produce a polymer-rich amino groups. Chitosan's physicochemical and functional properties, such as its solubility, viscosity, and bioactivity, are greatly influenced by its molecular weight and level of deacetylation (Mourya & Inamdar, 2008). One of chitosan's most remarkable properties is its biocompatibility, which permits its safe use in food, pharmaceutical, and biomedical applications. Additionally, it exhibits exceptional antibacterial properties against a range of microorganisms, including bacteria, fungi, and yeasts. Chitosan also interacts with negatively charged microbial cell membranes, causing cell lysis and growth retardation; positively charged amino groups in chitosan are thought to have antibacterial properties (Kong et al., 2010). Chitosan's well-known ability to form films makes it suitable for use in food preservation films and edible coatings. These coatings effectively reduce oxidative deterioration and microbiological spoilage in perishable food products by serving as a barrier to oxygen and moisture (Rinaudo, 2006). Furthermore, by protecting food products from lipid oxidation, chitosan's remarkable antioxidant properties can help maintain their nutritional value and sensory appeal (No et al., 2007). Research on chitosan has broadened recently to examine how it might be combined with other materials and functionalized to improve its properties. It has been demonstrated that adding plant extracts, essential oils, or nanoparticles to chitosan significantly improves its mechanical and antibacterial properties, expanding its use in food packaging and preservation (Elsabee & Abdou, 2013). These modifications allow for the tailored delivery of bioactive compounds, making chitosan a versatile material for functional food applications (Zhang et al., 2015). Chitosan's versatility is demonstrated by the fact that it can be utilized both directly and as a base material to generate new composites. Customizing these composites can help achieve specific industrial objectives, such as enhanced biodegradability for ecologically friendly applications, controlled-release systems for active packaging, or better mechanical properties for flexible films (Aranaz et al., 2009). Chitosan is an environmentally beneficial substance that supports global initiatives to decrease plastic waste and promote green technology. Large-scale chitosan extraction and processing, as well as guaranteeing its constant quality and usefulness in a range of applications, remain difficult tasks. If chitosan is to realize its full potential as a basis for sustainable innovation in the food business and beyond, these problems must be resolved.

1.2.1 Edible Coatings

Edible coatings, produced from natural, biodegradable materials including proteins, lipids, and polysaccharides, have the potential to improve food preservation and address sustainability issues in the food industry. These thin, edible coatings are applied on the surfaces of food products to serve as a barrier against environmental factors such as moisture, oxygen, and microbial contamination (Bourtoom, 2008). In addition to extending the shelf life of perishable items, edible coatings are a sustainable substitute for traditional synthetic packaging. Polysaccharide-based coatings based on chitosan, alginate, and cellulose are particularly popular due to their excellent film-forming properties, transparency, and suitability for use on food products. Particularly, chitosan-based coatings have attracted attention for their antibacterial and antioxidant properties as they can reduce oxidative food degradation and inhibit the growth of bacteria that cause rotting (Shahidi & Arachchi, 1999). These coatings are especially effective in fresh fruit, seafood, and dairy products because microbial contamination is a significant concern in these products (Tiwari et al., 2009). In addition to polysaccharides, protein-based coatings, such as those derived from casein, whey, and soy proteins, have demonstrated exceptional mechanical and barrier properties. These coatings are often used in combination with other materials to enhance their functionality. In line with this, lipids improve the protein-based coatings' ability to withstand moisture, making them suitable for usage in items with high water content, such as fruits and vegetables (Krochta & De Mulder-Johnston, 1997). Edible coatings also frequently contain lipids, such as oils, waxes, and fatty acids. Their main purpose is to extend the shelf life of foods that contain a lot of moisture by decreasing the permeability of water vapor. To attain balanced mechanical properties, they must frequently be mixed with proteins or polysaccharides because of their restricted flexibility and brittleness (Debeaufort et al., 1998). One of the most significant advancements in edible coating technology is the application of functional ingredients such as bioactive compounds, plant extracts, and essential oils. These additives can enhance the coatings' antibacterial, antioxidant, and sensory properties in addition to providing further benefits like flavor enhancement or vitamin and mineral fortification (Dutta et al., 2009). Applying coatings with essential oils, such as thyme or oregano, which have shown strong antibacterial activity against foodborne pathogens, might extend the shelf life of meat and poultry products (Valencia-Chamorro et al., 2011). Edible coatings contribute to food preservation and worldwide environmental goals by reducing reliance on plastic packaging. Falguera et al. (2011) state that they are usually biodegradable, extend the shelf life of food, and are produced from renewable resources such as agricultural leftovers or waste from the food sector. Despite these advantages, problems with cost-effectiveness, scalability, and regulatory approval must be fixed to promote their widespread application in the food industry. Additionally, edible coatings are increasingly being created to satisfy certain industrial needs. Currently, research is focused on creating active coatings that release antioxidant or antibacterial substances during storage, as well as smart coatings that react to environmental factors like temperature or humidity (Mchugh & Krochta, 1994). These developments demonstrate how edible coatings have the potential to transform food preservation and packaging, while simultaneously encouraging the food industry's shift to sustainable business practices.

1.3 FOOD PRESERVATION

Food preservation is essential for human civilization because it minimizes waste and guarantees a supply of food during times of scarcity. In order to maintain food's safety, freshness, and nutritional value, food preservation attempts to stop microbial development, enzymatic reactions, and oxidative damage that leads to rotting. Traditional methods like drying, salting, and fermenting have been combined with modern technologies like refrigeration, freezing, canning, and packaging. However, new concerns like food safety, environmental sustainability, and the growing demand for minimally processed foods have led to changes in food preservation methods (Lisboa et al., 2024).

Microbial decomposition is one of the most common causes of food deterioration. Preservation methods like freezing and refrigeration prevent microbial proliferation by reducing the temperature to levels unsuitable for the metabolism of rotting bacteria (Karanth et al., 2023). However, these methods often have disadvantages, such as high energy consumption and temperature sensitivity. As a result, other tactics have been created, such as the use of natural antimicrobials and edible coatings. Chitosan-based coatings have demonstrated efficacy in inhibiting microbial development in fresh fruit, meat, and seafood due to their inherent antibacterial properties (No et al., 2007). Oxidation, which can degrade proteins, lipids, and vitamins and alter the meal's nutritional value and aesthetic appeal, is another significant cause of food rotting. Conventional antioxidants such as tocopherols and ascorbic acid have been utilized extensively to prevent oxidative degradation. Recently, there has been a rise in the usage of natural antioxidants made from plant extracts, essential oils, and bioactive components. As a result of their sustainability and preservation advantages, these natural ingredients are being used more and more in edible films and coatings (Liyanapathiranage et al., 2023; Ibrahim et al., 2026). Dehydration can cause fresh food to lose its texture and appearance, making moisture loss a serious problem. Edible coatings and modified atmosphere packaging (MAP) have been utilized to circumvent this issue by establishing a controlled microenvironment that lowers moisture loss and slows down respiration rates (Zdulski et al., 2024). Polysaccharide-based edible coatings, such as those made from chitosan or alginate, increase the shelf life of fruits and vegetables by forming a barrier that prevents the passage of gases and moisture (Alemu et al., 2025). In addition to physical and chemical methods, advanced technologies such as high-pressure processing (HPP), pulsed electric fields (PEF), and ultraviolet (UV) therapy are increasingly being used in food preservation. These non-thermal methods effectively inactivate viruses and decaying microbes while maintaining the food's flavor and nutritional content. For instance, without the use of chemical preservatives, high-pressure processing can extend the shelf life of meats, shellfish, and juices (Nabi et al., 2021). Food preservation is also crucial to addressing the problems of global sustainability. By reducing food waste and boosting resource efficiency, food preservation technologies help combat hunger and food insecurity. More biodegradable packaging and edible coatings are being developed to replace single-use plastics for environmental conservation aims (Falguera et al., 2011). Additionally, smart packaging technologies such as oxygen scavengers and freshness indicators enhance the effectiveness of food preservation by providing real-time information on product quality (Panou et al., 2025). Despite advancements, the food preservation sector continues to encounter several challenges, including consumer acceptability, regulatory constraints, and the scalability of new technology. Future studies should focus on developing sustainable, cost-effective, and multifunctional preservation methods that can be used with a range of food systems.

1.4 ANTIMICROBIAL PROPERTIES

Antimicrobial properties refer to a material's ability to prevent or eradicate microorganisms, such as bacteria, fungi, and viruses. Microbial activity management is crucial for quality and safety in a wide range of industries, but it is especially crucial for medicines, biomedical applications, and food preservation. Antimicrobial compounds are used in the food business to extend the shelf life and preserve sensory qualities of food products by avoiding rotting and foodborne infections. Antimicrobial drugs commonly interact with bacterial membranes, disrupt metabolic processes, or hinder DNA synthesis. For instance, natural biopolymers like chitosan have antibacterial properties due to their positively charged amino groups. The negatively charged bacterial cell membranes are impacted by these groups, which increases permeability and permits the release of internal contents, ultimately leading to cell death (Kong et al., 2010). Additionally, some antibiotics have the potential to interfere with enzyme systems that are essential for microbial metabolism or growth. The market for natural antibacterial agents has grown significantly as a result of consumer preferences for clean-label products and worries about synthetic chemicals. Natural antimicrobials include organic acids, bioactive peptides, and essential oils derived from plants. Essential oils such as those from oregano, thyme, and cinnamon include phenolic compounds like carvacrol and eugenol, which break down microbial membranes and inhibit enzyme activity (El-Sakhawy et al., 2026). Examples of organic acids that reduce the pH of the environment and stop microbes from developing are lactic and citric acids (Atasoy et al., 2024). As innovative methods of food preservation, edible films and coatings containing antimicrobials have garnered a lot of attention. Chitosan-based coatings provide broad-spectrum antibacterial properties that combat both Gram-positive and Gram-negative bacteria. According to No et al. (2007), these coatings work in concert with plant extracts or essential oils to enhance their efficacy. Valencia-Chamorro et al. (2011) state that these coatings are particularly effective in extending the shelf life of perishable items such as fresh vegetables, dairy products, and seafood. Recent advancements in antimicrobial materials, such as nanoparticles, have opened up new possibilities for microbial control. For instance, silver nanoparticles exhibit strong antibacterial activity by interacting with bacterial proteins and DNA to cause oxidative stress and cell damage (Khalifa et al., 2025). Bio-based materials, including functionalized polymers and nanocellulose, are also being researched for active and controlled-release packaging solutions (Silva, et al., 2020).

1.5 POSTHARVEST LOSSES

The phrase "postharvest losses" refers to the decline in the quantity and quality of agricultural products that occurs after harvest but prior to consumption. Global food security is seriously threatened by these losses, which impact both wealthy and developing nations. According to estimates from the Food and Agriculture Organization (FAO), almost one-third of the world's food production—or 1.3 billion tons—is lost or wasted annually (FAO, 2011). These losses have detrimental impacts on society, the economy, and the environment, in addition to the loss of vital food supplies. Harvesting, handling, storing, shipping, processing, and retailing are some of the supply chain stages where postharvest losses might happen. The main sources of these losses include environmental, biological, and physical factors. Produce is more vulnerable to microbial infection and deterioration due to physical problems such as mechanical damage from harvesting and handling, inappropriate storage, and insufficient packing. These losses, particularly in grains and cereals, are made worse by bacterial and fungal activity, insect and rat infestations, and other biological factors (Corina Fărcaş, 2024). Temperature, humidity, and light exposure are all important environmental elements since poor temperature control speeds up microbial growth and respiration rates, which causes spoiling quickly (Alemu et al., 2025). Low-income nations are particularly affected economically by postharvest losses, which are made worse by a lack of infrastructure and restricted access to technology. These losses result in lower profits for smallholder farmers and higher food prices for consumers. Additionally, the energy, labor, and water needed to produce the lost food are wasted, which has a major negative influence on the environment. Environmental deterioration and greenhouse gas emissions result from this (Read et al., 2020). Better practices, instruments, and laws are needed to address postharvest losses. Following the right harvesting and handling practices, such as using the right tools and containers and training staff on how to keep food from being physically damaged, can significantly lower losses (FAO, 2011). Storage and transportation systems like cold storage and controlled environment storage help preserve the quality and freshness of perishable goods by lowering respiration rates and inhibiting microbial proliferation (Zdulski et al., 2024). Innovations like edible coatings made of chitosan or alginate, which serve as barriers to prevent moisture loss and microbial contamination, extend the shelf life of fresh food (Bourtoom, 2008). Recent advancements in postharvest technology are promising. Smart packaging using sensors to monitor temperature, humidity, and gas composition provides real-time data for bettering storage and transit conditions. Non-thermal preservation methods like UV treatment and HPP have been employed to maintain food safety and quality without compromising nutritional value. To lower losses, policies that support farmer education, infrastructure investment, and market access are required. Initiatives to raise understanding of postharvest management give farmers the knowledge they need to adopt better practices, reducing waste and boosting food security (Alemu et al., 2025).

1.6 PACKAGING TECHNOLOGY

Packaging technology is crucial to the food industry because it ensures the safety, preservation, and protection of food items during storage, transportation, and consumption. Complex systems intended to increase food's shelf life, maintain its nutritional value, and reduce waste have replaced simple containers. Today's packaging technology is more practical and meets consumer demands for convenience, sustainability, and safety because of new materials, designs, and advancements. Protecting food products from contamination, microbiological development, and physical damage is one of the most important roles of food packaging. In addition to offering an efficient barrier against environmental elements such as moisture, air, and light that could degrade food quality, packaging materials must be strong enough to withstand damage during handling and transportation. To prolong the shelf life of fresh vegetables, meat, and seafood, for instance, MAP is a popular technology that modifies the gas composition surrounding a food product to slow down respiration and microbiological activity (Panou et al., 2025). Another crucial aspect of packing technology is preventing microbial contamination and degradation. Vacuum packaging and active packaging technologies are two examples of packaging methods intended to restrict the growth of bacteria, yeasts, and molds in food products. Vacuum packaging reduces the amount of oxygen available for microbial growth by removing air from the packaging environment. Active packaging extends the shelf life of food products by actively interacting with them through the use of additives such as moisture regulators, antioxidants, and antimicrobials. Chitosan-based coatings and films are increasingly being used in food packaging due to their antibacterial properties, which help to prevent bacterial development and rotting. Another factor propelling packaging technology developments is sustainability. Packaging materials that are recyclable, compostable, and biodegradable are becoming more and more popular as environmental worries about plastic waste increase (Fabiyi et al., 2026). As more ecologically friendly substitutes for conventional petroleum-based plastics, bioplastics derived from renewable resources such as cornstarch, sugarcane, and algae are being studied (Şomoghi et al., 2025). Another innovative idea that is gaining popularity is edible packaging, which can be eaten with the food. In addition to lowering packaging waste, edible coatings derived from natural substances like chitosan, gelatin, and proteins can also increase preservation (El-Sayed et al., 2024). Another recent advancement in package technology is smart packaging, which tracks the condition of the food and its surroundings using sensors and indicators. These smart sensors can help manufacturers and consumers ensure the safety and quality of the product by providing real-time data on variables like temperature, humidity, and gas composition (Panou et al., 2025). For instance, freshness indicators integrated into packaging can notify consumers when a product has exceeded its optimal storage conditions, preventing them from eating spoiled food. Similarly, time-temperature indicators (TTIs) can confirm that perishable items have been kept within the range of temperatures required to maintain their quality and safety by tracking their temperature history during storage and transit. Despite advancements in packaging technologies, problems with cost, scalability, and customer acceptance persist. The development of cost-effective and environmentally friendly packaging materials is a significant obstacle, particularly for small businesses. Furthermore, before novel packaging methods are extensively used, consumer preferences, food safety, and legal requirements must all be carefully considered (Silva et al., 2020). Ongoing research and development in packaging materials and technology provide promising solutions to these issues while also enhancing food safety, sustainability, and preservation.

1.7 PACKAGING TECHNOLOGIES

To maintain food products' safety, wholesomeness, and protection during storage, transportation, and consumption, modern food preservation significantly depends on packaging technology. Packaging technologies have evolved throughout time, moving from basic containers to more intricate methods that maintain nutritional value, extend food shelf life, and reduce food waste.

These technologies are essential for preserving the freshness and safety of food in addition to satisfying consumer demands for sustainability, convenience, and safety (Silva et al., 2020). Modern packaging technology uses a range of materials and advancements to meet various needs in the food industry. Packaging technology has advanced, but there are still several problems that need to be fixed. The development of cost-effective and efficient packaging materials remains a significant obstacle, particularly for small-scale food producers. Furthermore, before innovative packaging techniques are widely used, consumer preferences, food safety, and regulatory issues must all be considered (Silva et al., 2020). As the need for sustainable and safe packaging continues to grow, more research and innovation will be required to get beyond these barriers and ensure the ongoing advancement of packaging technology in the food industry.

1.8 CONCLUSION

To sum up, using edible coatings based on chitosan in the food industry is a practical way to improve food preservation and lower postharvest losses. Many perishable goods have longer shelf lives thanks to chitosan's inherent antibacterial and protective qualities, which operate as an efficient barrier against microbial contamination, moisture loss, and rotting. Chitosan-based coatings present a sustainable solution with substantial potential to improve food safety, quality, and sustainability in the food sector, as consumer demand for natural and sustainable substitutes for synthetic preservatives rises. A broader application of these coatings is anticipated as a result of ongoing research and development in this field, which will be advantageous to both manufacturers and consumers.

REFERENCES

Alemu, T. T., Intipunya, P. & Gebeyo, B.A. (2025). A comprehensive review of edible coatings for postharvest management of fruits and vegetables: enhancing food and nutrition security. *Discover Agriculture,* 3, 190. https://doi.org/10.1007/s44279-025-00348-8

Atasoy, M., Álvarez Ordóñez, A., Cenian, A., Djukić-Vuković, A., Lund, P. A., Ozogul, F., Trček, J., Ziv, C. & De Biase, D. (2024). Exploitation of microbial activities at low pH to enhance planetary health. *FEMS Microbiology Reviews*, 48(1), fuad062. https://doi.org/10.1093/femsre/fuad062

Aranaz, I., et al. (2009). Functional characterization of chitosan: Properties and applications. *Marine Drugs*, 7(2), 103–118.

Bourtoom, T. (2008). Edible films and coatings: Characteristics and properties. *International Food Research Journal*, 15(3), 237–248.

Corina Fărcaş, A. (2024). Food safety in cereal grains: Contaminants, legislation, and mitigation strategies. In *Worldwide Megatrends in Food Safety and Food Security* [Working Title]. IntechOpen. https://doi.org/10.5772/intechopen.1007523

Debeaufort, F., et al. (1998). Edible films and coatings: Tomorrow's packagings: A review. *Critical Reviews in Food Science and Nutrition*, 38(4), 299–313.

Dutta, P. K., et al. (2009). Perspectives for chitosan based antimicrobial films in food applications. *Food Chemistry*, 114(4), 1173–1182.

El-Sakhawy, M. A., El Raheim Mohammed Donia, A., Juraybi, T. N., El-Sherbiny Ateya, A. A., Osman, W., Abd-Elghany, A. A., Soliman, M. M., Kobisi, A. N. A. & Balah, M. A. (2026). Action mechanism of essential oils or their components against pathogenic bacteria: Literature review. *International Journal of Microbiology*, 2026, 7468450. https://doi.org/10.1155/ijm/7468450

Elsabee, M. Z., & Abdou, E. S. (2013). Chitosan-based edible films and coatings: A review on recent trends and innovations. *Journal of Carbohydrate Polymers*, 90(3), 1277–1291.

El-Sayed, S. M., Youssef, A. M. (2024). Emergence of cheese packaging by edible coatings for enhancing its shelf-life. *Food Measure,* 18, 5265–5280 https://doi.org/10.1007/s11694-024-02564-0

Fabiyi, O. A., Ogundele, A. V., Akinsipo, O. B. et al. (2026). Smart nano-packaging materials for sustainable food preservation and safety. *Discover Food,* 6, 209. https://doi.org/10.1007/s44187-026-00906-z

Falguera, V., et al. (2011). Edible films and coatings: Structures, active functions and trends in their use. *Trends in Food Science & Technology*, 22(6), 292–303.

FAO (2011). Global food losses and food waste. Study conducted for the International Congress SAVE FOOD! at Interpack2011 Düsseldorf, Germany. Retrieved from https://www.fao.org/4/mb060e/mb060e00.htm

Ibrahim, S., Hashim, A., El Said, A. et al. (2026). Utilization of plant derived natural antioxidants and nanofiber mats to improve oxidative stability and extend food shelf life. *Discovery Food,* 6, 116. https://doi.org/10.1007/s44187-026-00816-0

Karanth, S., Feng, S., Patra, D. & Pradhan, A. K. (2023). Linking microbial contamination to food spoilage and food waste: The role of smart packaging, spoilage risk assessments, and date labeling. *Frontiers in Microbiology,* 14, 1198124. https://doi.org/10.3389/fmicb.2023.1198124

Khalifa, H. O., Oreiby, A., Mohammed, T., Abdelhamid, M. A. A., Sholkamy, E. N., Hashem, H. & Fereig, R. M. (2025). Silver nanoparticles as next-generation antimicrobial agents: mechanisms, challenges, and innovations against multidrug-resistant bacteria. *Frontiers in Cellular and Infection Microbiology.* 15, 1599113. doi: 10.3389/fcimb.2025.1599113

Khan, M. A., et al. (2020). Applications of biopolymers in food preservation and sustainability. *Journal of Food Science and Technology*, 57(4), 1012–1024.

Kong, M., et al. (2010). Antimicrobial properties of chitosan and mode of action: A state of the art review. *Journal of Food Safety*, 30(3), 490–508.

Krochta, J. M., & De Mulder-Johnston, C. (1997). Edible and biodegradable polymer films: Challenges and opportunities. *Food Technology*, 51(2), 61–74.

Kumar, M. N. V. R., et al. (2018). Chitosan: A versatile biopolymer for food applications. *International Journal of Biological Macromolecules*, 104, 987–1000.

Kumar, R., et al. (2019). Chitosan-based edible films: Properties and applications in food packaging. *Food Packaging and Shelf Life*, 21, 100341.

Lisboa, H. M., Pasquali, M. B., dos Anjos, A. I., Sarinho, A. M., de Melo, E. D., Andrade, R., Batista, L., Lima, J., Diniz, Y. & Barros, A. (2024). Innovative and sustainable food preservation techniques: Enhancing food quality, safety, and environmental sustainability. *Sustainability*, 16(18), 8223. https://doi.org/10.3390/su16188223

Liu, Y., et al. (2021). Enhancement of antimicrobial activity in chitosan films with functional additives. *Trends in Food Science & Technology*, 112, 16–23.

Liyanapathiranage, A., Dassanayake, R. S., Gamage, A., Karri, R. R., Manamperi, A., Evon, P., Jayakodi, Y., Madhujith, T. & Merah, O. (2023). Recent developments in edible films and coatings for fruits and vegetables. *Coatings*, 13(7), 1177. https://doi.org/10.3390/coatings13071177

McHugh, T. H., & Krochta, J. M. (1994). Sorbitol- vs glycerol-plasticized whey protein edible films: Integrated oxygen permeability and tensile property evaluation. *Journal of Agricultural and Food Chemistry*, 42(4), 841–845.

Mourya, V. K., & Inamdar, N. N. (2008). Chitosan-modifications and applications: Opportunities galore. *Reactive and Functional Polymers*, 68(6), 1013–1051.

Nabi, B. G., Mukhtar, K., Arshad, R. N., Radicetti, E., Tedeschi, P., Shahbaz, M. U., Walayat, N., Nawaz, A., Inam-Ur-Raheem, M., & Aadil, R. M. (2021). High-Pressure Processing for Sustainable Food Supply. *Sustainability*, *13*(24), 13908. https://doi.org/10.3390/su132413908

No, H. K., et al. (2007). Antioxidant properties of chitosan and its derivatives. *Food Hydrocolloids*, 21(1), 143–149.

Panou, A., Lazaridis, D. G. & Karabagias, I. K. (2025). Application of smart packaging on the preservation of different types of perishable fruits. *Foods*, 14(11), 1878. https://doi.org/10.3390/foods14111878

Read, Q. D., Brown, S., Cuéllar, A. D., Finn, S. M., Gephart, J. A., Marston, L. T., Meyer, E., Weitz, K. A. & Muth, M. K. (2020). Assessing the environmental impacts of halving food loss and waste along the food supply chain. *The Science of the Total Environment*, 712, 136255. https://doi.org/10.1016/j.scitotenv.2019.136255

Rinaudo, M. (2006). Chitin and chitosan: Properties and applications. *Progress in Polymer Science*, 31(7), 603–632.

Shahidi, F., & Arachchi, J. K. V. (1999). Food applications of chitin and chitosan. *Trends in Food Science & Technology*, 10(2), 37–51.

Silva, F. A. G. S., Dourado, F., Gama, M. & Poças, F. (2020). Nanocellulose bio-based composites for food packaging. *Nanomaterials (Basel, Switzerland)*, 10(10), 2041. https://doi.org/10.3390/nano10102041

Smith, J. R., et al. (2022). Eco-friendly packaging alternatives in the food industry: A critical review. *Sustainability*, 14(6), 3218.

Şomoghi, R., Mihai, S., & Oancea, F. (2025). An overview of bio-based polymers with potential for food packaging applications. *Polymers*, 17(17), 2335. https://doi.org/10.3390/polym17172335

Tiwari, B. K., et al. (2009). Application of edible coatings in fresh and minimally processed fruits and vegetables: A review. *Journal of Food Science and Technology*, 46(6), 650–668.

Valencia-Chamorro, S. A., et al. (2011). Antimicrobial edible coatings for fresh fruit and vegetables: A review. *Food Science and Technology International*, 17(5), 459–476.

Zhang, H., et al. (2015). Enhancing functional properties of chitosan films by incorporation of bioactive compounds: A review. *Food Hydrocolloids*, 46, 93–100.

Zdulski, J. A., Rutkowski, K. P. & Konopacka, D. (2024). Strategies to extend the shelf life of fresh and minimally processed fruit and vegetables with edible coatings and modified atmosphere packaging. *Applied Sciences*, 14(23), 11074. https://doi.org/10.3390/app142311074

2 Utilization of Chitosan Derived from Natural Sources to Enhance Food Quality and Shelf Life in Agriculture, Seafood, and Poultry Sectors

Osarenkhoe Osarenkhoe Osemwegie[1], Adebowale Toba Odeyemi[1], Damilare Rotimi[2], Adeyemi Ayotunde Adeyanju, M.A. Acho[2], and Charles Oluwaseun Adetunji[3]

[1]Department of Food Science and Microbiology, Landmark University, Mu Aran, Kwara State, Nigeria

[2]Department of Biochemistry, Landmark University, Omu Aran, Kwara State, Nigeria

[3]Department of Microbiology, Edo University, Uzaire, Edo State, Nigeria

2.1 INTRODUCTION

Chitosan is a positively charged or cationic, linear, deacetylated, and abundant naturally occurring biopolymeric derivative of chitin, particularly α-chitin, comprising a structurally random arrangement of β-1,4-linked D-glucosamine moieties (Chatterjee et al., 2022; Iber et al., 2022). While it is commonly sourced from the exoskeleton of invertebrate animals like the crustaceans (crabs, crayfish, lobsters, and shrimps), shellfishes, squids, insects, mollusks, green algae, and fungi, or global seafood wastes, they rank along with starch, chitin, glycogen, and cellulose as one of the most biodegradable polycarbohydrates (polysaccharides) around the world (Kou et al., 2021; Chatterjee et al., 2022; Huq et al., 2022). Even though little is known about the biological phenomena underlying the diverse properties of chitosan polymer or the effect of its structural architecture on its end-use activity (biological, chemical, or physical), there has been growing research interest in their biocompatibility, biodegradation, physicochemical, versatility, and malleability (fabricability) properties, particularly their characterization (Wang et al., 2020). These rapidly growing research interests have created a new scientific paradigm exposing biopolymer's broad spectrum of applications, which has equally witnessed a rise in the number of utility patents (Aranaz et al., 2021; Wang and Zhuang, 2022). Amoli et al. (2022), and a few other authors involved the compound in computer design (3D) and biosensor studies as an extension of seeking to improve its end functions. For instance, fabricated chitosan complex (chitosan grafted with other natural or synthetic molecules) was utilized in the poultry industry to improve in vitro prebiotic effect, colon fermentation, and digestive stability while also having the potential to impact fecal microbiota (Tang et al., 2022). Despite these, the wider application of chitosan may be currently limited by the scope of the source distribution, and its high macromolecular weight makes it mechanically inadequate, vulnerable to acids, insoluble in water, and dosage-sensitive medically. While the presence

 DOI: 10.1201/9781003355458-2

of amino groups may be responsible for the properties of chitosan, as noted by Priyadarshi and Rhim (2020), these properties can be enhanced for better biotechnological applications using the 'hurdle technique'. This includes grafting the polymer with plasticizers, natural extracts (essential oil, phytochemicals, exudates, and enzymes), spices, crosslinkers, fibers, nanoparticles, whiskers, and other synthetic polymers. Also, the bioeconomic values transcend several sectors including medical, pharmaceutical, agriculture, food processing, wastewater treatments, and technology. In recent decades, this has been making a robust impact on the global natural product market. According to Riofrio et al. (2021), some of the impressive attractions to chitosan include its bioeconomic and financial viabilities coupled with its partial role in the environmental cleanup of seafood wastes generated, particularly in some coastal nations of the world. This suggests that the bioeconomic viability and circular utility of chitosan and its derivatives may rely on the sustainable valorization of their feedstock sources (Mahjoub and Domscheit 2020). Therefore, a larger number of chitosan research is required to trigger a transition toward the potential exploitation of more non-invasive chitosan sources and cross-sectoral collaborative efforts in optimization technology to meet its utility demands.

Organic and inorganic agricultural practices utilize chitosan in diverse ways that differ from the ones annotated in Figure 2.1. For example, Liu et al. (2021) noted that chitosan, when added to the chicken manure during soil fertilization, affects the soil microbial community by accelerating antibiotics and resistance genes removal while boosting crops' natural defenses against pathogens (Mukta et al., 2017). This is coupled with their potential capacity to immobilize microspheric soil heavy metals and act as an anti-transpirant coating to assist crops in managing drought, biomaterial for improving plants' resilience to climate eccentricities, and nutrients biocapture by crops as well as an eco-friendly graft in the management of crop pests/diseases (Turan et al., 2018; Chowdary et al., 2022).

Similarly, the seafood and poultry industries have witnessed diverse use of chitosan as a functional biomaterial, primarily to improve their product shelf life, sensory qualities, freshness, detoxification, and values (Baptista et al., 2020; Inanli et al., 2020). This is besides their application as dietary fiber, enzyme immobilizer, antioxidant, and health promoter. Studies have also shown that seafood and poultry biomass or waste can be viably valorized for biotechnological products or

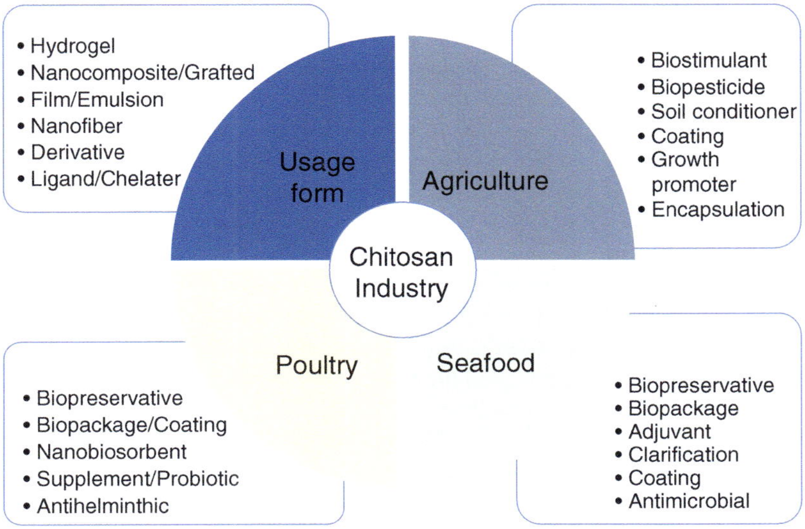

FIGURE 2.1 Schematic illustration of chitosan applications and application forms in the agriculture, seafood, and poultry industries.

beneficial products, including chitosan (Namboodiri and Pakshirajan, 2020; Singh et al., 2022; Tongwanichniyom et al., 2022).

The biocompatibility, malleability, biodegradability, antimicrobial, and antioxidant properties of chitosan have made it intriguing and versatile in the agriculture, seafood, and poultry industries (Figure 2.1). Numerous accounts on the use of chitosan, particularly in the agriculture, seafood, and poultry industries, have been documented in the body of literature even though more investigation may be required in determining more inherent capacities in their chitosan and its derivatives. Furthermore, studies into how benign fabrication (demineralization, deacetylation, deproteinization, and nanonization) and modification (cross-linking, scaffolding, etherification, graft copolymerization, alkylation, acylation, and esterification) technologies affect use functions, solubility, acid resistance, mechanical threshold, biodegradability, biocompatibility, hydrophobic interaction, antimicrobial, and thermoplasticity of chitosan complex properties is crucial to their rational use for human benefits (Abifarin et al., 2022). Fabrication and modification of chitosan have opened the utility of the polymer as a membrane and optimizer of its physicochemical and biological properties, making it amenable for coating, encapsulation, hydrogel, UV-shielding application, nanonization, coacervation, and regulation of beneficial principles (Huang et al., 2021; Amaregouda et al., 2022; Liu et al., 2022; Sochilina et al., 2022; Tu et al., 2022). Therefore, understanding the underlying biochemical tractability of chitosan as biopreservatives, biopackaging composites, antifouling, hydrogel, stimulant, and adjuvant requires more investigation (Priyadarshi and Rhim, 2020). Hence, this study reviews documented innovations in the utilization of chitosan polycarbohydrate in the agricultural, seafood, and poultry industries, and its commercial prospects.

2.2 PROPERTIES OF CHITOSAN

Chitosan has diverse and unique properties that have garnered growing biotechnological and bioeconomic interest over the last decade. An innovative combination of its physicochemical and biological properties has broadened its applications in different as well as several industrial contexts, including food processing, nutritional improvement, fabric and textiles, cosmetics, agriculture, and waste and water treatment. In addition, its bioactivity, biodegradability, and biocompatibility, together with its lack of allergenicity and toxicity, made it more appealing for a wide range of applications.

The structural uniqueness of chitosan underpins its physicochemical properties, which include the ability to create water-soluble salts with organic and inorganic acids like lactic, hydrochloric, glutamic, and acetic acids and become insoluble at alkaline and neutral pH (Mustafa, 2018). Its amino group is not protonated in alkaline or neutral media, but it becomes a protonated polysaccharide at an acidic pH, rendering it soluble in water. Several other types of acids, both inorganic and organic, may react with chitosan to produce water-soluble salts, such as citrate, acetate, lactate, glycolate, tartrate, pyruvate, glyoxylate, malate, malonate, and ascorbic acid. Chitosan dissolves in organic acids when the pH drops below 6.5, while its amino groups (NH_2) become protonated in an acidic environment, giving the molecule a positive charge. Therefore, the protonated amino groups in its polymer chain affect the sorption, solubility, biodegradability, and reactivity tendencies of chitosan (Ahmed and Ikram, 2015). Hence, the degrees of deacetylation and depolymerization have resulted in varying chitosan derivatives with exclusive physicochemical properties (Figure 2.2). Adjusting the conditions in chitosan formulation and fabrication processes results in desired chitosan properties like solubility and pK_a (Thanou et al., 2001). Deacetylation is a chemical process that affects the molecular weight (Mw), viscosity, chain elongation and breakage, and tensile strength of chitosan. Also, biodegradation by lysozyme and improved wound healing are only two examples of how viscosity may affect biological processes (Chatelet et al., 2001).

Since chitosan is hydrophilic and capable of retaining water, it may be polycationic in acidic settings with the potential to form gels. To achieve gelation, chitosan is acetylated in a hydroalcoholic medium, which allows for the selective alteration of the free amino groups (Madera-Santana et al., 2018).

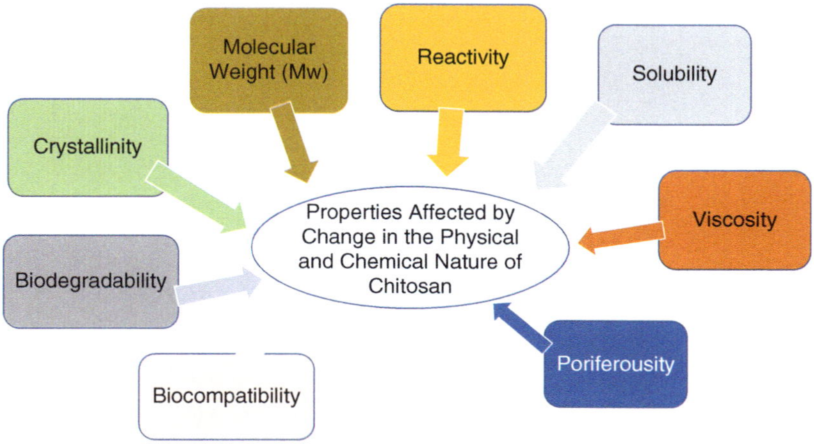

FIGURE 2.2 Graphic illustration of some chitosan properties affected by varying degrees of change in its physicochemical nature.

Deswelling and its reversibility are facilitated by conditions that decrease the charge density of the chain segments, which is a crucial characteristic for the application of gel as coatings due to its polyelectrolyte property (Şenel and McClure, 2004). Likewise, intermolecular cross-linking of chitosan with substances like glutaraldehyde greatly decreases the molecule's solubility. Increasing the quantity of cross-linking agents reduces film swelling, thereby affecting their amenability in biopreservation (Ahmadi et al., 2015).

Over the last two decades, chitosan has been employed as a benign excipient in pre-harvest and postharvest agriculture due to its mucoadhesive potential (Chopra et al., 2006). Chitosan is utilized in improving seeds, seedlings, and whole plant tolerance to adverse environmental or habitat stress-inducing factors basically because it can stick to hard and soft tissues. The biodegradation kinetics of chitosan are known to be affected by the polymer chain length and the distribution of acetyl groups. The cationic behavior of chitosan also showed its biological characteristics and antibacterial action. It is thought that chitosan's polycationic composition reacts with the anionic components, inducing alterations in permeability that ultimately result in cell death by promoting the leaking of internal components (Raafat and Sahl, 2009).

One of the primary motivations for the expanded use of naturally abundant chitosan is the adaptable biological, chemical, and physical characteristics (Morin-Crini et al., 2019). Although the effect of different chitosan-treatment methods used to adapt or functionalize chitosan molecules for broad-spectrum applicability or desired biotechnological end-use is unclear, a greater understanding of these phenomena is required for a paradigm of bioeconomic relevance of chitosan and its derivatives. Chitosan is relatively simpler compared to the vast majority of naturally occurring polysaccharides (cellulose, dextrin, pectin, agarose, agar, and carrageenans), most of which are acidic. In addition, the unique attributes of chitosan, such as high viscosity, solubility in a wide range of fluids, mucoadhesion, polyoxysalt creation, metal chelation, polyelectrolyte, and optical behaviors as well as film-forming potential, afforded it a relative advantage over other biopolymers (Azmana et al., 2021) (Figure 2.3).

Unlike cellulose, chitosan has an amino group at carbon-2 instead of a hydroxyl group, and it shares the (1,4) anhydroglucosidic bond with cellulose. The economic value of chitosan is based on its high nitrogen content (6.69%), which is far higher than that of synthetically replaced cellulose (1.25%). Hence, many polymers are not as biocompatible or biodegradable as natural cellulose, chitin, and chitosan (Bhatia, 2016). However, chitosan may be limited due to its poor solubility at

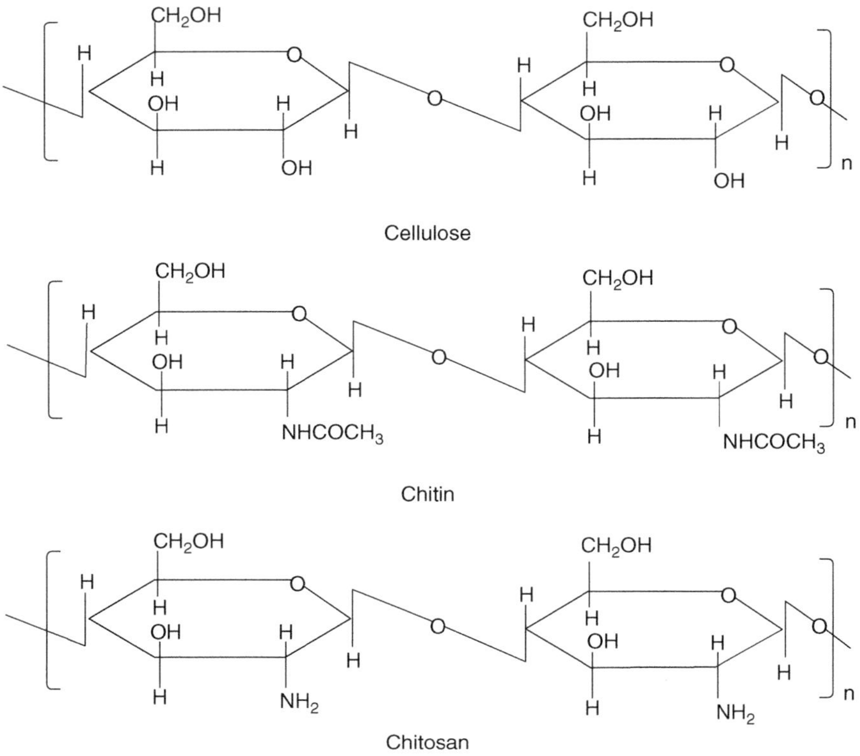

FIGURE 2.3 Structural relationships between cellulose, chitin, and chitosan (Ramya et al., 2012).

a physiological pH of 7.4, which affects its use as an absorption enhancer in systems. Another constraint of chitosan for the development of sustained release of bioactivity is its rapid adsorption of water and enhanced swelling degree in aqueous conditions. Several chemically modified chitosan derivatives have been developed to address these difficulties (Kayser et al., 2014).

2.3 CHITOSAN ANALOGS AND THEIR CHEMICAL DIFFERENCES

Navigating through the chitosan analog context is inexhaustive due to its versatility. However, the essence of this section of the chapter is to introduce the concepts underlying functionalization by chemical modification, fabrication, copolymerization, and hydrolysis function. The hydroxyl and amino groups in chitosan undergo a wide range of chemical reactions like acylation, carboxylation, hydroxylation, alkylation, and esterification. These reactions modify chitosan properties and functionalize their chemical architecture by introducing pendant groups through *N*-substitution, *O*-substitution, and free-radical graft copolymerization that can alter its crystal structure as well as biological properties (Figure 2.4).

2.4 BIOTECHNOLOGICAL USES OF CHITOSAN IN FOOD AND AGRICULTURE

Global health and environmental risks due to chronic applications of chemical-based synthetic materials in several spheres of human endeavors, including synthetic polymer applications in the food industry, are causing a revolution, shifting the focus to bio-based polymers, particularly chitosan, as a safer replacement or substitute.

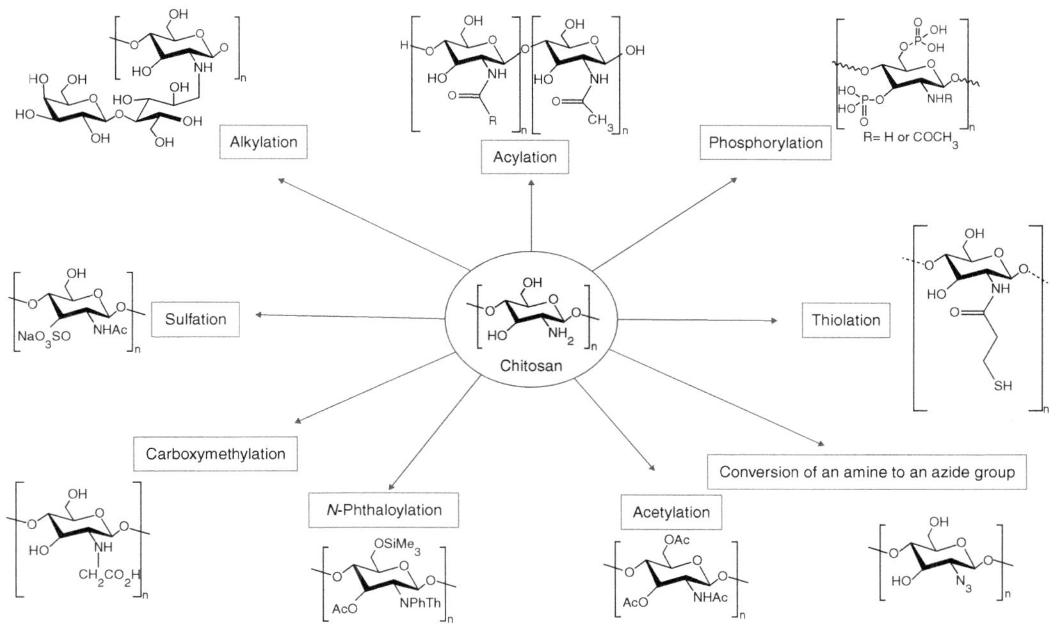

FIGURE 2.4 Illustrates changes in the structure of cellulose, chitin, and chitosan when exposed to different chemical reactions (Lima et al., 2022).

2.4.1 Chitosan in the Food Processing

The US Food and Drug Administration has regarded chitosan as a safe food ingredient, a valuable consumer component, and its purpose as dietary fiber (hypocholesterolemic impact). This has increased their applications as a nutritional ingredient, an antioxidant and antimicrobial agent, a coating for vegetables and fruits to prevent spoilage, a dietary supplement for lowering cholesterol, and more (Muxika et al., 2017). To combat harmful germs in food, scientists are looking at novel chitosan derivatives and oligomers as potential antimicrobials and nutraceuticals (Kardas et al., 2012). Chitosan is also applied in protecting seafood and poultry products from spoilage by increasing their shelf life or in creating biodegradable films for food coating, encapsulation, and packaging. The inhibition of the growth of two major waterborne food pathogens, *Escherichia coli* and *Staphylococcus aureus*, was seen when chitosan acetate was used (Li et al., 2007). Van den Broek et al. (2015) studied chitosan films and their blends in food packaging systems. They observed that their antibacterial and film-forming capabilities mitigate the dangers of using synthetic pesticides in food processing. The study also confirmed them as suitable replacements for chemical food preservatives or coating materials while lessening the reliance on potentially dangerous pesticides. Chitosan-based films may contain functional compounds like vitamins and act as carriers, releasing antibacterial chemicals. This showed excellent performance and good mechanical qualities (No et al., 2007; Morin-Crini et al., 2019). Its molecules may be damaged by hydrolysis and chain depolymerization in an acidic environment.

Improving food safety, bioactivity, and hygiene throughout the production and processing stages has received more attention in recent years. To improve the process of achieving food security, the food sector still relies on a range of synthetic, non-biodegradable components. However, the cost of resistant chemical, their non-biodegradability, and toxic gas emissions, along with the growing number of health and environmental hazards they pose, have raised serious concerns. As a result, there is an increasing need for food packaging materials that can improve food safety, extend shelf life, and reduce the usage of chemical preservatives and synthetic petroleum-based polymers in food applications. This gave impetus to the acceptance of biomaterials in the industry, making them relevant as preservatives, biopesticides, prebiotics, and flavoring and texturizing agents. It also functioned as a packaging material to store and transport food products. Chitosan is a polysaccharide-based, non-toxic biomaterial formed when all or part of chitin is deacetylated. The amino and hydroxyl groups in this polymer are highly reactive in conferring multifaceted bioactivities (antioxidant, antimicrobial, and chelating), which collectively enhance chitosan's ability to be biodegradable, biocompatible, non-toxic, and adaptable to desired end-use functions. These factors have collectively enhanced chitosan's effectiveness in culinary applications.

Numerous adverse responses occur inside and outside the food matrix, lowering the food's shelf life, value, and acceptance. Therefore, the use of food additives is intended to prevent or lessen these reactions and, more importantly, extend the food products' shelf life and general appeal. Chitosan's unique chemical makeup, low Mw, and resilience under extreme food processing conditions (heating, leaching) make it a suitable additive in different food products (Božič et al., 2012; Hashemi and Niakousari, 2017). Also, chitosan's structure contains two primary functional groups (NH_2 and OH), which give the polymer a strong hydrogen-donating capability. This helps the polymer bind metal ions and suppress free radicals (Yen et al., 2008; Vinsova and Vavrikova, 2011). The C-2 amino group and C-6 hydroxyl group of chitosan interact to chelate metal ions and stop lipid peroxidation (Yen et al., 2008; Inoue et al., 1988). Although chitosan's end-use functions depend heavily on its structurally linked properties, such as its Mw, degree of deacetylation, and origin, its antioxidant activity increases with the degree of deacetylation and decreases with increasing Mw (Luo and Wang, 2013). According to Yen et al. (2008), both effects correspond to a rise in solubility.

2.4.2 Chitosan as an Emulsifying Agent

Chitosan is a powerful emulsifier that may stabilize oil-in-water emulsions in different food items (de Souza Soares et al., 2017). The amount of chitosan's amino groups that are protonated in an acidic medium also correlates with this property. This protonation gives the molecule an amphiphilic characteristic, which subsequently permits adsorption at the oil/water interfaces and the creation of the emulsion by lowering the interfacial tension (Luo and Wang, 2013). Mw and degree of deacetylation of chitosan are key to determining the effectiveness of chitosan as an emulsifier (Luo and Wang, 2013). A study by de Souza Soares et al. (2017) investigated the efficiency of chitosan as a thickener and an emulsion stabilizer, confirming its prospects in food processing. They mixed 0.1% (w/v) chitosan dispersions with Tween 20 (1% v/v) and sunflower oil in lactic acid solutions (pH 3.0, 3.5, or 4.0) to create oil-in-water emulsions. Lactic acid solution emulsions containing Tween 20 and sunflower oil were used in the control group. The results showed that chitosan was an effective thickening agent, improving the size distribution and average size of droplets in emulsions, even at concentrations as low as 0.1% (m/v).

2.4.3 Chitosan as Packaging and Film Material

A few years back, the bulk of packing materials were made from synthetic and semi-synthetic polymers because they showed favorable application properties like low weight, transparency, thermolability, and acceptable mechanical capabilities. However, their xenobiotic nature, toxicity, and excessive use resulted in serious environmental and health consequences (Dilkes-Hoffman et al., 2018). However, chitosan's capacity to form films suggests that it is a critical and promising alternative to synthetic chemical-based polymers in biopackaging production. This creates opportunities to develop films and coatings with selective permeability to CO_2 and O_2 with comparatively higher mechanical strength and antimicrobial properties, which may be directly applied to improve food safety and shelf life. To improve the functional properties of chitosan films and increase their potential uses, researchers have tried changing the solvent type, pH, deacetylation level, and combining with plasticizers or components like nanoparticles, proteins, and polysaccharides (Cazón et al., 2017). Chitosan can be applied as an active food packaging agent or utilized as a thin edible film and coating in food. In a study by Li and Yu (2001), prochloraz-prepared peaches and peaches preserved with a chitosan covering were contrasted. The results showed that the presence of brown rot pathogens (*Rhizopus stolonifera* and *Monilinia fructicola*) was significantly reduced by chitosan. In comparison to peaches treated with prochloraz, chitosan-treated peaches also showed increased titratable acidity and vitamin C levels. In another study, increased concentrations of total soluble solids, titratable acid, and ascorbic acid, as well as decreased polyphenol oxidase and peroxidase activity, were found to be stable, and peeled litchi fruit had a low shelf life until it was coated with chitosan, which delayed weight loss and improved its sensory quality. Dong et al. (2004) found that the coating reduced malondialdehyde production, promoted superoxide dismutase activity, and preserved membrane integrity. The chitosan coating, which also enhances vitamin C content, soluble solid content, and titratable acidity while delaying water loss and sensory quality deterioration, successfully improved the qualitative features and shelf life of sliced mango fruit (Chien et al., 2007a,b). It has been shown that chitosan-based coatings can delay the microbiological breakdown of baby carrots without lowering their commercial, organoleptic, or nutritional qualities (Leceta et al., 2015). Consequently, chitosan films are perfect for stopping oxidation and microbiological degradation of meat fat because of their exceptional oxygen-barrier properties. Using chitosan coatings or films is an intriguing way to improve the preservation quality of fresh meat, fish, and poultry items. Therefore, chitosan can take the place of artificially produced antibacterial and antioxidant compounds, increasing the final product's value (Bonilla et al., 2014a, b). The capacity of chitosan films or coatings to perceptively respond to local climate change and self-regulate is enhanced by additional intriguing and creative uses. For example, chitosan films can now be used to create pH indicators that can track and notify consumers in real time about the condition of their food. These smart chitosan-based films have been enhanced with the extracts of blueberry and blackberry pomace. Between pH 2 and 12, these films display various colors, suggesting different levels of deterioration (Kurek et al., 2018). Another example is the use of chitosan-enriched maize starch as an indicator of fish spoilage (Pereira et al., 2015; Silva-Pereira et al., 2015)

2.4.4 Chitosan-Nanobased Material

Chitosan nanoparticles are considered excellent candidates for reinforcing films, coatings, and bioplastics due to their advantageous properties (Naskar et al., 2019; Ma et al., 2017; Sangeetha et al., 2016). While food packaging polymers can improve physical, biological, and chemical stability, they are not as good at improving food quality and safety. Conventional packaging materials are unable to actively control many food products' reactions, causing spoilage and propelling the emergence of a new market in the packaging industry. Chitosan nanoparticles are therefore great candidates for reinforcing films, coatings, and bioplastics (Naskar et al., 2019; Ma et al., 2017; Sangeetha et al., 2016). The biochemical mechanism underlying the bioactive effectiveness of chitosan-based packaging in maintaining food safety and quality is not fully understood (Khaneghah et al., 2018). Food pathogens can be reduced or eliminated by using chitosan nanocomposites as active carriers (Abdelgawad et al., 2014; Arkoun et al., 2018). Inhibiting pathogen proliferation using antimicrobial drugs becomes essential for improving food quality and shelf life (Aranaz et al., 2009; Llorens et al., 2012). Osheba et al. (2013) investigated the effects of different chitosan concentrations and chitosan nanoparticles as active composites with commercial edible coatings using a few quality measures from preserved fish at 18°C. Ceylan et al. (2018) used electrofusion to create chitosan/thymol/liquid smoke nanofibers to protect fish fillets from microbiological deterioration. Chitosan/thymol/liquid smoke electromelted nanofiber dispersions successfully decreased about 60% of mesophilic bacteria. The study also found that nanofibers were thermally stable up to 150°C, which is within the temperature range used in traditional food preservation techniques. The effects of chitosan coatings and nanoparticles on the sensory, physical-chemical, and in vitro antibacterial activity of grapes kept at 12°C and 25°C were documented by Melo et al. (2018). The maturation of the grapes was delayed by the coatings containing chitosan nanoparticles, which led to a decrease in weight loss, soluble solids, and sugar content, while other metrics like moisture retention and titratable acidity also improved.

2.4.5 Application of Chitosan in Encapsulation and Delivery of Bioactive Ingredients

Encapsulation constitutes an enclosure of either an active agent or beneficial microbe acting as the core in another material (wall) to primarily protect the core from external adversities and optimize its benefits (Dordevic et al., 2014). Despite being widely accepted to be stable in various settings, probiotic bacteria and bioactive chemicals (vitamins, phenolic compounds, and bioactive peptides) nevertheless show sensitive reactivity in the gastrointestinal system and under food processing conditions (Dordevic et al., 2014; Speranza et al., 2017). Therefore, materials suitable for encapsulation must be phenomenally able to tolerate extreme conditions, including toxicity, and adequately offer effective defense against a wide range of challenges limiting the core material. Because they create an encapsulating system for food ingredients and their regulated release of bioactivity, chitin and chitosan are commonly used in this context. Bioactive compounds such as vitamins and polyphenols are essential in nutraceutical encapsulation systems. Despite being beneficial to human health, polyphenols typically taste bad and become unstable during normal food processing (Fang and Bhandari, 2010). Therefore, encapsulation may serve as a useful strategy to maintain food quality stability and guarantee the optimal release of its bioavailability and bioactivity of its active compounds. Chitosan, due to its properties, is gaining attention in encapsulation processes as a surface coating material or matrix-forming polymer (Matshetshe et al., 2018). Vitamins, especially those with antioxidant potentials like vitamin C or vitamin E, are prone to deterioration and can be restricted by poor absorption and breakdown during delivery (Sabliov and Astete, 2008). Similarly, encapsulation systems preserve the viability of probiotics throughout food processing and storage as well as move the probiotic through the digestive tract (Varankovich et al., 2017).

Cholecalciferol (vitamin D) was encapsulated in tripolyphosphate (TPP)-chitosomes when chitosan nanoparticles containing a cross-linking agent, such as sodium TPP, were added to a liposomal suspension. The encapsulation efficiency (EE) of TPP-chitosomes was higher than that of liposomes, increasing from 5% to 97%. After 14 days of storage, cholecalciferol starts to degrade, while the vitamin D encapsulated in TPP-chitosomes held steady for 49 days (Iida et al., 2019). Additional recently published studies on encapsulated probiotics, vitamins, and polyphenols using chitosan-based systems are shown in Table 2.1. Considering the role chitosan is playing in encapsulation systems across many industries, it is unequivocally recognized as a compatible carrier-encapsulate of probiotics, vitamins, essential oils, antioxidants, and other functional materials, while the technology is being acknowledged as a successful strategy for promoting functional meals. To maximize chitosan encapsulation's effectiveness, functionality, and biological disposition, more research is necessary to decide how best to use it in the food sector.

Piran et al. (2020) investigated the antioxidant potential of green tea (GT) extract using chitosan encapsulation. Using the DPPH approach, it was discovered that the scavenging activity of green tea-loaded chitosan nanoparticles (CS-NP-GT) was greater than that of GT antioxidants. Additionally, because hydroxyl groups were formed and added to the aromatic ring of the phenolic compounds during the sonication stage, using encapsulation increases the total phenolic content. It has been shown that chitosan is a valued substance for encapsulating α-lipoic acid (Milašinović et al., 2016; Dutta et al., 2019).

TABLE 2.1
Encapsulation of Polyphenols, Vitamins, and Probiotic Microorganisms Using Chitosan-Based Systems

Encapsulated Substance	Materials Used for Encapsulation	Encapsulation Techniques	Major Findings	References
β-Carotene	Nucleic acid and polylactic acid combined with amphiphilic chitosan	Complex coacervation	Superior efficacy in maintaining the integrity of beta-structural carotene and increased antioxidant activity	Di Martino et al. (2018)
Catechin and quercetin	Chitosan and sodium tripolyphosphate, altered by genipin	The next step is cross-linking after ionic gelation	Sustained release. It is more or equally effective than native catechin and quercetin at scavenging DPPH, ABTS•+, •OH, and O2• radicals and preventing the growth of *Bacillus subtilis*, *Staphylococcus aureus*, and *Escherichia coli*	Li et al. (2018)
Eugenol	Chitosan, sodium tripolyphosphate, and Tween 20	Ionic gelation mediated by ultrasonic emulsification	Improved thermal stability of eugenol via encapsulation. The beads had a potent antibacterial effect	Shao et al. (2018)
Epigallocatechin gallate (EGCG)	Zein, chitosan hydrochloride, and carboxymethyl chitosan	Ionic gelation	Zein coating greatly increased the antioxidant activity and the regulated release of EGCG	Liang et al. (2017)
Lactobacillus plantarum (LAB 12)	Sodium alginate, xanthan gum, and chitosan β-cyclodextrin as a supplement	Coating comes after ionic gelation mediated by extrusion	Microorganisms that are encapsulated can survive at pH 1.8 and release more readily at pH 6.8. At 75°C and 90°C, as well as after four weeks of cold storage, a minimum level of bacterial viability loss was noted	Fareez et al. (2017)
Lactobacillus rhamnosus GGLMG 18243	Genipin, chitosan, and carboxymethyl cellulose	Coating and extrusion (syringe or nozzle sprayer) facilitate ionic gelation	The capsules were stable at the stomach pH settings (pH 2.4), but they expanded considerably at higher pHs that are near the intestinal pH (pH 7.4)	Singh et al. (2017)

2.5 APPLICATION OF CHITOSAN IN AGRICULTURE

Chitin can be extracted from a wide range of invertebrate animals, fungi, and a few species of algae and deacetylated into non-toxic chitosan for use in the agricultural value chain (Hahn et al., 2020). It is a bioresource that is commonly used in dietary supplements, food packaging, and wound care products (Matica et al., 2019). Chitosan is also implicated in broad-spectrum plant disease control due to its potential to biostimulate the plant's immune system. It has been applied in agriculture with amazing results in fertilizers, sprays, and coatings of seeds, leaves, fruit, and vegetables. Plant productivity is increased by chitosan, which also shields plants from dangerous microbes and serves as an eco-friendly replacement for chemical pesticides. Despite these chemical pesticides' efficiency in increasing crop yields, they still exacerbate environmental and health risks (Orzali et al., 2014). Therefore, their applications in crop protection since the 1940s by farmers have become less relevant in recent times due to the rapidly emerging biotechnologically invented plant growth and protection products of biological origins, which include chitosan biopolymer.

Chitosan has gained significant relevance in agriculture because of its unequivocal ability to increase plant yields through its positive influence on plant nutrition, growth, resistance to pathogenic microorganisms/pests, and resilience against a rapidly fluctuating climate. Therefore, the role of chitosan in crop health and productivity may be conceptualized as follows:

2.5.1 Plant Defense Mechanism

Plants' natural defense mechanisms are often triggered in response to biological and environmental stress. These responses, which may include the production of biochemical defenses like siderophores, enzymes, and phytohormones, may be stretched thin when they encounter more serious dangers. Chitosan is a fantastic polymer that triggers defense reactions in instances involving pests and pathogen attacks. According to Yan and Lin (2014), chitosan helps plants produce glucanases, chitinases, pathogenesis-related (PR) proteins, hydrogen peroxide, reactive oxygen species (ROS), phytoalexins, and protein inhibitors. Hence, chitosan is a bioelicitor or biostimulant that readily interacts with cellular DNA to induce a range of biochemical processes leading to a quick response to plant pathogen attacks. For example, in turmeric plants, chitosan bioactivity enhances their ability to withstand infection by *Pythium aphanidermatum* (Salachna and Zawadzinska, 2014), while it also causes the growth of callose in the cells of soybean and parsley, production of proteinase inhibitors in tomato leaves, and the synthesis of phytoalexins in pea. The unique role of chitosan as a plant elicitor is enhanced by its ability to mediate polycationic monomer interaction with negatively charged phospholipids and, by extension, ligand-protein receptor-like activity. Suffices to say chitosan exerts antimicrobial activity by damaging pathogens' transport including metal chelating and cell surface deposition and communication mechanisms, electrostatically interacting with the pathogen DNA and RNA crippling transcriptomic activity, or stimulating the processes involved in induced and acquired host plant defenses (Wang et al., 2021; Riseh et al., 2022a).

The susceptibility of many biological control agents and plant-growth promoters to environmental vagaries has increased the need for the introduction of polymers like chitosan as carriers in microbial cells' encapsulation/coating technique (bioencapsulation) to plant disease management. Riseh et al. (2022b) observed that encapsulating plant beneficial microbes or biocontrol agents of plant pathogens in chitosan gels or beads, which can range in size from a few nanometers to a few millimeters, improves the viability, resilience to fluctuating climate factors, and stability. In addition, irrespective of the method (spray, emulsification, capsules, lyophilization, extrusion, coacervation, gelation, and fluidized treatment) used for the encapsulation of a microbial inoculum, the chitosan polymer facilitates the controlled poriferous release of beneficial microbes or bioactive metabolites while equally protecting the inoculum from contamination or phagotrophic organisms (Chowdary et al., 2022). This application has also found attention for the controlled delivery of chemical pesticides and postharvest storage management (Chowdary et al., 2022; Elsherbiny et al., 2022; *Riseh et al., 2022*a).

2.5.2 Antimicrobial and Antinematode

The antifungal potential of chitosan in plant disease management was reported by Wei et al. (2021), and Sen et al. (2022). *Alternaria solani* radial and submerged growth were suppressed by chitosan at a concentration of 1 mg/mL, protecting tomato plants from blight infection. Furthermore,

Fusarium wilts and gray mold infections were prevented by adding chitosan to the soil (Rabea et al., 2003). Chitosan treatment has also been used successfully in controlling oomycete pathogens like *Phytophthora capsici* in peppers (Xu et al., 2007) and *Phytophthora infestans* in potatoes. Chitosan's disruption of the endomembrane system, particularly the integrity of the vacuoles, was hypothetically responsible for the control of *P. capsici* infection in peppers. It was also implicated in preventing the *Alternaria* blight disease of tomatoes (Raut et al., 2014), damping off disease (Trzcinska et al., 2015), *Fusarium oxysporum*, and plant growth morphological alteration by *Tracheiphilum* sp. (Berger et al., 2016; Khairy et al., 2022).

One of the most used biopesticides in agriculture is the bacteria *Bacillus subtilis*, which is a fungus pathogen. It is known that *B. subtilis* secretes chitinases into the media in which it is developing (Chen et al., 2010). By using chitosan in the carrier material, *B. subtilis* was able to multiply more readily, and prevent *Aspergillus niger* and *Fusarium*-caused pigeon pea wilt and peanut crown rot. The bacterium's ability to combat powdery mildew in strawberry plants was also enhanced by the addition of chitosan (Lowe et al., 2012). The Mw and functional groups of chitosan enhance its ability to inhibit bacterial and fungal growth in food preservation systems. For example, large-Mw oligomeric chitosan readily crosses the cell membrane of microorganisms and hinders the ability of the cell to replicate by blocking the synthesis of RNA (Goy et al., 2009; Hosseinnejad and Jafari, 2016). The Mw and the level of deacetylation of the chitosan sample were shown to be important factors in its destruction by bacterial enzymes.

Acetylation levels and differences in the Mw of chitosan affect its action on fungi, resulting in differential responses (Falcón-Rodríguez et al., 2011). Several phytopathogenic fungi, including *F. oxysporum* in tomatoes, *Botrytis cinerea* in cucumber plants, *A. solani, Penicillium* sp. in citrus fruit, and *Phytophthora infestans*, were reported to be inhibited by the antifungal property of chitosan (El Hadrami et al., 2010). Several defensive activities, including proteinase inhibitor production, chitinase accumulation, callus creation, and lignification responses, were stimulated upon the formation of a permeable chitosan layer on the surface of the crop (Cheung et al., 2015). Applications such as fungicidal agents may be possible with the help of silver nanoparticles that were dispersed both externally and internally in chitosan spheres, demonstrating a macroporous property (Muxika et al., 2017). Using chitosan, researchers increased the survival rate of *Galleria mellonella* larvae infected with *Candida albicans. Macrophomina phaseolina* was resistant to chitosan, a water-soluble polysaccharide. Infected seedlings were less likely to develop fungal infections after being exposed to water-soluble chitosan (Chatterjee et al., 2014). This was due to an increase in the activities of defense-related enzymes such as chitosanase and peroxidase. For instance, quince juice and cranberry in the formulation of whey protein-chitosan films increased their flexibility while decreasing their tensile strength. Apples preserved in chitosan or whey protein-chitosan films with additional quince and cranberry juice may have a longer shelf life (Brink et al., 2019). In general, as the Mw grew, so did the fungal inhibition indices of deacetylated chitosan. However, low hydrophobicity, high-Mw chitosan derivatives and high-Mw hydrophobicity, low-Mw chitosan derivatives showed the greatest effectiveness in inhibiting the growth of *Aspergillus flavus* in vitro (Viegas de Souza et al., 2013). The use of Cu-chitosan nanoparticles inhibited spore germination and mycelial proliferation in *A. solani* and *F. oxysporum* in vitro and managed the infection (Cheung et al., 2015).

Plant viral infections have been observed to be controlled by chitosan (Kulikov et al., 2006; Riseh et al., 2022a,b). Although it would seem odd given that viruses are not made of chitin or other comparable cellular polysaccharides, it is also unclear if chitosan decapacitates plant viruses directly. This may explain why relatively few literature reports exist on its antiviral application in plant viral disease management. Hence, it has been suggested that chitosan may be effective against plant viruses by stimulating the plant's response to the infection rather than by exerting direct harm on the viruses (Rendina et al., 2019). Therefore, more investigation is crucial to understanding the underlying mechanism by which chitosan inhibits plant viral infections and the veracity of its biocontrol role as a vehicle for viral particles into plants to induce hypersensitivity reactions.

Similar discoveries have also been made with the potato virus X, tobacco mosaic and necrosis viruses, alfalfa mosaic virus, peanut stunt virus, and cucumber mosaic virus.

Chitosan provokes chitinolytic microbes that kill nematode eggs and disintegrate the cuticle of immature nematodes, including insect pests, when introduced to the soil. Chitosan has a significant amount of nitrogen, which causes ammonia emissions to rise during degradation and may likely become harmful to worms (Mwaheb et al., 2017). Chitosan elicits local and systemic defenses in tomato plants against the root-knot nematode, *Meloidogyne incognita. M. incognita* is more effectively controlled by low-Mw chitosan (Khalil and Badawy, 2012). Chitosan also suppresses the pinewood nematode (*Bursaphelenchus xylophilus*), a stem nematode, from reproducing properly and increasing mortality by changing the bacterial population that helps the nematode colony thrive (Silva et al., 2014; Liang et al., 2018). Chitosan becomes an energy substrate for the development and invigoration of the nematophagous fungus, *Pochonia chlamydosporia*, to parasitize *Meloidogyne javanica* eggs for food (Escudero and Lopez-Llorca, 2016).

2.5.3 Postharvest Treatment

Chitosan coating has become a fundamental postharvest treatment technology. Its capacity to alter the internal tissue environment, coupled with its fungistatic properties, is the major attraction for its potential exploitation for postharvest storage shelf life and crop spoilage. Chitosan gives fruits and vegetables additional firmness and encourages a reduction in their topical microbiological load (Devlieghere et al., 2004), thereby prolonging the life of the product. Given that chitosan films are more selectively permeable to O_2 than to CO_2, chitosan coating is expected to alter the internal environment of the enclosure without triggering anaerobic respiration. It is effective in preserving the pulp's color, which shows that mango cv. Alphonso fruits ripen more slowly and can be stored for a longer period. The postharvest decay of Cherry Tomato Fruit potentially through the Mitogen-Activated Protein Kinase Signaling Pathway is also effectively controlled by chitosan coating (Zhang et al., 2015).

2.5.4 Plant Growth Stimulant and Abiotic Stress Mitigator

Chitosan formulations, composites, and nanonized chitosan have recently gained attraction as a plant growth promoter to a variety of crops, including Faba bean plants, radish, passion fruit, potatoes, gerbera, cabbage, and soybean. The rate of flowering, root, and shoot growth responds significantly to chitosan, which has been found to mitigate plant-water dynamics and boost the activity of some biological macromolecules, particularly hydrophilic compounds, thereby reducing stress-related cell damage in plant cells. Giglou et al. (2022) demonstrated the mitigating effect of Kitoplus and chitosan-coated Iron oxide nanoparticles (Fe-CTs NPS) on drought stress of peppermint plant growth, while studies of different chitosan nanoparticles and modified chitosan biomaterials have also demonstrated their effectiveness in improving plant growth parameters under salinity and plant toxicity stresses (Azimi et al., 2021; Balusamy et al., 2022). Via a gene expression network between the nucleus and chloroplast, chitosan promotes the growth of rice seedlings.

Most plants and agricultural crops in recent times are faced with drought stress due to rising global temperature or incidences of heat waves, water shortages, and other climate change eccentricities. Heat and drought stressors have become fundamental concerns to agriculturists and scientists who have continued to seek mitigation. Chitosan is one of the numerous biomaterials empirically affirmed as anti-transpirant (Mphande et al., 2020). Its use in coating plant leaves increases stomatal conductance and decreases transpiration without an obvious impact on the plant growth parameters like height, root length, leaf area, or biomass (Shinohara and Leskovar, 2014). Abscisic acid (ABA), which regulates stomata opening, is more abundant when chitosan is sprayed on plant leaves (Iriti et al., 2009). Chitosan has a promising prospect in the future of crop production as a non-toxic anti-transpirant as well as a partial solution to plants' battle against the escalating risks of global drought and heat waves. For instance, pepper plants under agricultural conditions showed a reduction in physiological water by 26%–43% due to increased water loss by transpiration. Therefore, chitosan-treated plants respond by increasing the concentrations of ABA and jasmonic acid, which are crucial hormones for the control of stomatal activities (opening and aperture gap).

Crop biofertilization techniques are slowly gaining acceptance and field application toward guaranteeing increased productivity and food safety. Chitosan-coated compound, chitosan-nano-based, and chitosan encapsulation of nutrients have been reported for some crops including maize (Chen et al., 2020; Kumaraswamy et al., 2021; Wang et al., 2022). However, the mechanism by which these compounds interact with rhizospheric plant exudates, nitrogen-fixing microbes, and other plant growth-promoting microbes for controlled delivery of physiological nutrients to the plants for growth is still unclear (Ashraf et al., 2022). Maize production increased by using biofertilizers that combine chitosan and rhizobacteria to invigorate plant growth and enhance phytochemical levels, enzymatic, and antioxidant activities of spinach (Kamari et al., 2011).

As a carrier substance, chitosan enhances the fertilizer's gradual release to plant roots while halting leaching and nutrient losses (Wu and Liu, 2008). The nutrient release mechanism from a chitosan-coated NPK fertilizer was discovered to be non-Fickian and controlled by a combination of the fertilizer's diffusion through the coating material and the rate of chitosan's degradation. Chitosan and citric acid were used to create a controlled-release phosphorus fertilizer, and it was discovered that adding chitosan to the coating causes the fertilizer to release phosphorus more slowly for plant use (Ahmad et al., 2015). Corradini et al. (2010) investigated the action of a chitosan nanoparticle suspension containing NPK fertilizer. According to the FTIR data, the elements N, P, and K found in the urea, calcium phosphate, and potassium chloride, respectively, interacted electrostatically with the $-COO^-$ and $-NH_3^+$ of the chitosan nanoparticles. The stability of the chitosan methacrylic acid colloidal suspension was better with the addition of potassium and nitrogen than with the addition of phosphorus because calcium phosphate has a higher anion charge than potassium chloride and urea. In another study, it was observed that urea encapsulated in chitosan microspheres through emulsification and then cross-linked with genipin was investigated. The results showed that increasing urea content improves the release rate, while increasing cross-linker and chitosan promotes delayed release behavior (Hussain et al., 2012). Using a coagulation technique, potassium-rich chitosan montmorillonite microspheres were made, and extremely high concentrations of fertilizer were sorbed onto the substance. Potassium-containing chitosan montmorillonite microspheres demonstrated two distinct times when fertilizer was released. Researchers looked at how adding a chitosan binder to urea-kaolinite fertilizer improved its ability to release nitrogen (Roshanravana et al., 2015). For the regulated release of water, urea, and micronutrients that will increase the effectiveness of fertilizer application, a super water-absorbing substance made of chitosan, EDTA, and urea has been discovered to be useful in agriculture (Narayanan and Dhamodharan, 2015).

Conversely, chitosan can be utilized to create hydrogels because of its gel-forming abilities. In vivo degradation of chitosan into harmless compounds suggests that it is superior in terms of biodegradability compared to some other polymers. It contains both reactive $-NH_2$ and $-OH$, which makes it simple to graft the polymerization of hydrophilic vinyl monomers onto it in mild reaction conditions. The resulting superabsorbent resin can absorb aqueous solution hundreds of times more efficiently than the original dry sample. Together with chitosan, sodium humate could increase water absorption, and a solution containing 10% sodium humate provided the best absorption (Liu et al., 2007).

2.5.5 Contaminated Soil Remediation

Chitosan may promote the activity of helpful microbes in the soil, such as *Bacillus* spp., luminescent *Pseudomonas* spp., Actinomycetes, Mycorrhiza, and Rhizobacteria, as a mechanism of bioremediation. By doing so, the rhizosphere's microbial balance becomes biased toward beneficial bacteria. Although *Elsholtzia splendens* can tolerate contaminated soil with a variety of heavy metals, the addition of chitosan and mycorrhizal inoculation expedited the bioremediation of the soil (Angelim et al., 2013). When a group of several plant growth-promoting rhizobacteria (PGPR) was encapsulated within chitosan, it facilitated soil bioaugmentation and biostimulation. Although this phenomenon has not been proven, soils treated with chitosan might be less susceptible to the leaching of fertilizers that contain anionic nutrients like nitrates and phosphates. It also enhances the water retention capacity of soil, which indirectly conditioned the soil (Pandey and De, 2017).

Chitosan and treated chitosan bind metal ions even when the dominant soil ions, such as K^+, Cl^-, and NO_3, were present. Soil additives, including chitosan and cross-linked treated chitosan, can be used to clean up heavy metal-contaminated soil to reclaim more arable space for agriculture. The primary binding site is nitrogen, and the production of covalent bonds, which could transfigurate the polymer, could be the cause of the metal complexation with chitosan (Kamari et al., 2011). It has been demonstrated that chitosan has the best chelating properties among other natural polymers. The functional group responsible for compound formation in chitosan is its amino groups. Also, the hydroxyl groups can participate in sorption, whereas nitrogen donates electron pairs to these groups. The mechanism of combining these reactive groups with metal ions differs greatly and can be affected by the type of ion, pH, and other major elements of the solution. The Lewis acid–base theory can also be used to explain the chitosan–metal complexes' formation: a metal ion, which is functioning as the acid, accepts a pair of electrons supplied by chitosan (acting as the base), with metal ion selectivity in mixtures following the order $Cu^{2+}>Hg^{2+}>Zn^{2+}>Cd^{2+}>Ni^{2+}>Co^{2+}>Ca^{2+}$ (Hu et al., 2021). The remediation of soil contaminated with Hg(II) and Cr(VI) in industrial applications could be improved by adding 1.15 g/kg of chitosan to the soil for 7 days (Youcefi et al., 2022). Also, chitosan/ceramic alumina composites function as an adsorbent to remove anionic and cationic heavy metals such as As(III), As(V), Cu(II), Ni(II), and Cr(VI). Chitosan composites aid in removing Cr from the leather industry (Thenmozhi et al., 2010). Chitosan hydrogel film that has been cross-linked with glyoxal (Fe_3O_4NPs/CS/glyoxal) is utilized as an adsorbent to remove hazardous Cr (VI) from water by 80%–90% (Mirabedini et al., 2016).

2.5.6 Seed Coating/Priming

Seed priming is a cost-effective technology innovated in crop production to overcome the range of potential oddities to seed germination processes and a seed's healthy development of an adult plant. Even though seed priming differs with geographic location, seed type, pathogenic, agronomic, and season challenges, it encompasses exogenous treatments (biological, chemical, or physical) that enhance a seed's tolerance to on-farm environmental stressors (salinity, temperature, drought, pH, and pollutants), germinability, resistance to pathogens as well as pest attacks, and regulation of physiological resources utilization (Thakur et al., 2022). Chitosan, among other bioactive compounds, has been identified as crucial in biomaterial for seed biopriming due to its malleable biological, physical, and chemical properties. It increases the quantities of sugars and proline, changes the seed plasma membrane's permeability, and improves the activity of the enzymes peroxidase (POD), catalase (CAT), phenylalanine ammonia-lyase (PAL), and tyrosine ammonia lyase (TAL) (Guan et al., 2009; Preetha et al., 2022). Seeds primed with chitosan material showed a rise in germination rates, seedlings' growth, and health. Chitosan-soaked peanut seeds showed improved germination energy, germination percentage, lipase activity, and high levels of gibberellic acid (GA) and indole acetic acid (IAA). Maize seed priming raises the resistance to cold. Less damage from low-temperature stress is caused by priming with chitosan, which decreased the relative permeability of maize plasma membranes under low temperatures (Guan et al., 2009). Chitosan has equally been found to be efficient as an exogenous carrier coating to transport seed-protecting or growth-promoting beneficial microorganisms with the seeds to the soil (Dalvi et al., 2022; Sood et al., 2021). This is also meant to decelerate the dependence on the more expensive, ecologically unfriendly, stereotypical chemical seed treatments and mitigate the associated health as well as environmental risks linked to these treatment methods.

2.6 HYPOTHETICAL CHITOSAN MECHANISMS OF ACTION IN FOOD SYSTEMS

Even though the understanding of biochemical processes leading to food spoilage is still sketchy, it is perceived as a multi-step process that includes microbial metabolism, enzymatic activity, chemical interactions, sensory degradation, and toxin build-up (Yu et al., 2021). Food spoilage caused by pathogenic microorganisms is one of the foremost challenges in the food sector. Additionally, residual oxygen trapped during food packaging can lead to lipid oxidation, which alters the physical characteristics and nutritional status of foods and introduces some hazardous chemicals into them. This may facilitate the process of food spoilage, even though a similar process may enable the intrinsic compounds engrafted into food packaging to combat microorganisms. It also involves the scavenging of free radicals and prevention of unwanted microorganisms while maintaining the oxidative stability of the food (Muxika et al., 2017). A positively charged linear chitosan has various applications in the food industry that include protecting food from the harmful effects of pathogenic microorganisms and improving shelf life. Due to the protonation of its amine moiety, chitosan is densely populated with positive charges in slightly acidic media. This made it a more favorable choice as a biodegradable film material to produce food packaging and an agent of food preservation. Chitosan is perhaps one of the most effective naturally occurring polymers that exhibits antioxidant, antiviral, antibacterial, antifungal, and other intriguing properties (Arkoun et al., 2017; Manigandan et al., 2018; Morin-Crini et al., 2019; Priyadarshi and Rhim, 2020). Chitosan delays food spoilage because of its high hydrogen-donating and radical scavenging activities (Hussain et al., 2021). The mechanism of chitosan's antibacterial action is very complex and not entirely clear despite a few hypothetical postulations in the body of literature. Hu and Gänzle (2018) posited that the mechanism of antimicrobial action involves modifications of the cell envelope and disorientation of the cytoplasmic membrane integrity. Some of the proposed mechanisms of chitosan antibacterial activity may include:

The electrostatic interactions between positive charges contained in low- and medium-Mw chitosan and the negative charges found on the peptidoglycan cell wall of bacteria (Hassan et al., 2018). This can result in the dissolution of the cell wall and damage to the cell membrane, resulting in rupture and even perforation of the cell membrane. Because chitosan is polycationic, it possesses amine groups of glucosamine ($R\text{-}NH_3^+$), which electrostatically interact with the negatively charged groups on bacteria's surface and disrupt their function (Hussain et al., 2021). This eventually leads to the leakage of internal components, bacterial lysis, and ultimately bacterial death (Arkoun et al., 2017; Irastorza et al., 2021; Melo et al., 2018; Priyadarshi and Rhim, 2020). The lipopolysaccharide and teichoic acid found in bacteria perform a crucial function in the binding of chitosan with the bacterial cell membrane, altering, destabilizing, and ultimately disrupting cell wall dynamics (Hussain et al., 2021). Chitosan also rivals divalent metals in its ability to bind polyanions and adheres to bacteria surfaces while blocking some of the transport channels used to acquire nutrients (Yu et al., 2021).

i. High-Mw chitosan forms a polymeric cover that wraps bacterial or fungal cells and blocks cell interactions. The wrap is an impenetrable coating, preventing the flow of oxygen and exchange of vital solutes between the enclosed interior and outside. Consequently, the cell becomes suffocated, its metabolic system fails, and it eventually dies off (Arkoun et al., 2017; Chakraborty et al., 2022; Duarte and Picone, 2022a; Priyadarshi and Rhim, 2020).
ii. Chitosan's antibacterial action is also underscored by its chelating property. Chitosan can capture oligoelements necessary for bacterial growth. In addition, this action destabilizes homeostasis and causes alteration of microbial metabolism (Arkoun et al., 2017; Priyadarshi and Rhim, 2020). Creating exterior barriers against microorganisms, chelating metals, and causing a reduction in the number of nutrients required for the growth

may plausibly underly chitosan's action as a biopreservative. The metal chelating action of chitosan is most effective when the surrounding pH is higher than pK_a, enabling positive metal ions to bind chitosan by the deprotonation of amino groups and donation to metal ions (Arkoun et al., 2017; Ferreira et al., 2020a). Chitosan action renders valuable growth compounds inaccessible to microbial growth (Sharma et al., 2016; Yu et al., 2021).

iii. Chitosan also attacks the intracellular constituents of bacteria and binds to DNA after entering the cell, where it influences the transcription of DNA, alters the expression and synthesis of proteins, and inhibits mRNA. This mechanism of action, though attributable to the low Mw, suggests that chitosan has good penetration properties due to the ease of infiltrating many layers of cross-linked murein to access the cell cytoplasm (Duarte and Picone, 2022a; Ferreira et al., 2020b; Yu et al., 2021).

iv. At neutral pH, chitosan is found to damage bacterial cell walls and cause cell lysis in Gram-negative bacteria. This is achieved by binding to lipopolysaccharide and the outer membrane protein A (OmpA) (Duarte and Picone, 2022a; Morin-Crini et al., 2019).

v. Chitosan interacts with components of the food matrix, creating a protective film or coating that envelops them and shields them physically from microbial attack as well as other deteriorating agents (Priyadarshi and Rhim, 2020; Romanazzi et al., 2017).

Some authors have reported enhanced antimicrobial activity in Gram-negative bacteria, while others have reported otherwise (Duan et al., 2019; Ferreira et al., 2020b; Melo et al., 2018; Vargas et al., 2012). The effectiveness of chitosan in inhibiting Gram-positive bacteria may be due to the absence of an outer membrane, which serves as an extra defense in Gram-negative bacteria (Melo et al., 2018; Duan et al., 2019; Ferreira et al., 2020a).

According to Muxika et al. (2017), the antibacterial activity of chitosan is dependent on some factors, which include the species of bacteria, type, growth phase, and age of bacterial cell; characteristics (charge, density, Mw , concentration, chitosan source, hydrophilic/hydrophobic properties, degree of acetylation, and chelating ability) of the chitosan preservative; physical state; and environmental conditions (ionic strength, pH, temperature, and time) (Muxika et al., 2017; Zhang et al., 2021). Some reports have shown that low-Mw chitosan and those with a low DD have a higher potential to reduce and prevent bacterial growth (Vargas et al., 2012; Li et al., 2016; Sharma et al., 2016; Zhang et al., 2021; Duarte and Picone, 2022b).

2.6.1 Mechanism of Antifungal Activities of Chitosan

It has been noted that chitosan's polycationic composition and the length of the polymer chain seem to be important factors in its antifungal properties. Chitosan can migrate into the hyphae, inhibiting the enzyme that catalyzes fungal growth (Ferreira et al., 2020a). The higher the DD of the chitosan, the better the antifungal activity, and vice versa. Chitosan has a similar antibacterial action mechanism to antifungal activity. Chitosan chelates metals, minerals, trace elements, or nutrients necessary for the growth of fungi and alters the morphogenesis of fungal cell walls. It also hinders spores from sprouting, germ tubes from elongating, and radial growth (Grande-Tovar et al., 2018). Additionally, it has been noted that chitosan activity is sensitive to fungi with plasma membranes rich in polyunsaturated free fatty acids (Lopez-Moya and Lopez-Llorca, 2016). It has been suggested that chitosan's ability to trigger the increased synthesis of defense-related enzymes like polyphenol oxidase and peroxidase in chitosan-coated fruits is a demonstration of its antifungal activity (Grande-Tovar et al., 2018). Chitosan's positively charged amino groups react with the phospholipids' negative charges on the cell membrane of fungi, causing lipid peroxidation, a change in conformation that permits pores to form in the plasma membrane, a reduction in H^+-ATPase activity, and a build-up of protons inside the cell (Qing et al., 2015).

2.6.2 Mechanism of Chitosan's Antiviral Activity

The antiviral activities of chitosan have been documented in literature, but not as much as its antifungal and antibacterial activities (Kumaraswamy et al., 2018; Riseh et al., 2022a). Chitosan's mechanism of antiviral action is still not clear, although some mechanisms have been proposed. The resistance to infection in plants treated with chitosan may be related to callose deposition, micro-oxidative bursts, and micro-hypersensitive reactions, which prevent the systemic transit of viruses from cell to cell (Kumaraswamy et al., 2018). Additionally, Iriti and Varoni (2015) report that chitosan upregulated the expression of an RNase connected with the resistance of potato virus X infection, which has an impact on viral replication. Also, it has been suggested that chitosan's ability to stimulate a plant's immune system is what makes it efficient against plant viruses (Jaber et al., 2022). According to reports, chitosan induces hypersensitive responses, programmed cell death, and systemic acquired resistance and promotes the production of callose while also enhancing ABA synthesis. Callose regulates how permeable plants are to substances, and as a result, viruses' capacity to spread from cell to cell is diminished (Jaber et al., 2022; Saharan and Pal, 2016). Consequently, it is most probable that the antiviral activity of chitosan is linked with its ability to induce programmed cell death, while on the other hand, building up callose defense to prevent the virus from spreading to neighboring cells (Saharan and Pal, 2016). The antifungal activity of chitosan is also a function of its Mw, DD, and the modification of its structure (Jaber et al., 2022; Riseh et al., 2022b).

2.6.3 Antioxidant Activity of Chitosan

The high hydrogen-donating capacity of chitosan makes it an effective antioxidant for food preservation purposes (Hussain et al., 2021). Oxidized lipids are usually degraded into harmful toxins that result in bad odors in foods. Chitosan's protective ability against oxidative damage in food is closely linked to its DD, which results in delayed spoilage (Pabast et al., 2018; Vieira et al., 2019). The following mechanism of chitosan's antioxidant property includes serving as an oxygen barrier, thereby preventing the interaction of the food with oxygen when used to layer fruits; and chelation of metal ions, particularly ferrous ions, found in muscle tissues, to become prooxidants, which are key to lipid oxidation. Studies have also shown that chitosan possesses the ability to scavenge hydroxyl radical through the ability of its amino group to interact with hydroxyl radical (Yu et al., 2021). The antioxidant activity of chitosan is also dependent on its Mw, as low-Mw chitosan has been observed to possess better antioxidant activity. This may be due to the strong intramolecular interaction and compact structure that high-Mw chitosan possesses, resulting in weakened amino and hydroxyl groups (Chang et al., 2018).

2.6.4 Mechanism of Chitosan's Enzyme Inhibitory Activity in Food Products

Enzymes' activity linked to softening, texture alteration, and browning of diverse food products plays a huge role in food spoilage. The browning of tissues characterized by damaged surfaces typically happens in fruits and vegetables owing to the activity of polyphenol oxidase. Chitosan has been reported to possess anti-browning properties through its ability to inhibit polyphenol oxidase activity (Lin et al., 2020). An enhanced texture has also been reported in food coated with chitosan (Yu et al., 2019). According to Yu et al. (2021), some of the anti-browning mechanisms of chitosan include its ability to hinder the permeation of oxygen, limiting the availability of the substrate to the enzyme responsible for the browning reaction in food and binding to the active site of polyphenol oxidase or chelating the metal ions present in the active site through electrostatic interactions, thereby preventing the substrate from binding to the enzyme. When employed to enhance plant defenses, chitosan elicits host defense responses in both monocotyledons and dicotyledons (Riseh et al., 2022b). Chitosan defense mechanisms include the activation of chitinase and glucanase, production of phytoalexin, biosynthesis of jasmonate, and the expression of specific early response and defense-associated genes (Riseh et al., 2022a,b).

The role of chitosan in sustaining food texture and preventing muscle softening may be attributed to its ability to inhibit the activity of cathepsin D and distort its protein structure (Chanphai and Tajmir-Riahi, 2016); the activation of the enzyme pathway by chelating calcium ion, as the enzymes involved in muscle softening such as calpain and matrix metalloproteases are calcium-ion dependent; to access the sarcoplasm and inhibit the enzyme by interacting with it; to bind to structural proteins, thereby altering its structural and functional properties; and to change the food sample's microenvironment to artificially change the post-murder biochemical characteristics of muscle (Yu et al., 2021).

2.6.5 Mechanism of Chitosan Action in Plant Growth

Recent research has revealed that chitosan activates plant defense genes via the octadecanoid pathway, promoting plant growth and development (Chakraborty et al., 2020). Chitosan has a large impact on the rate of shoot growth, root growth, flowering, and the number of flowers. Chitosan's high hydrophilicity reduces stress-related cell damage in plants by reducing the water content and enhancing the activities of many biological macromolecules (Chakraborty et al., 2020). Chitosan also helps to mobilize stored dietary components like protein and carbs and activate the hydrolytic enzymes required for their breakdown. It can stimulate auxin and cytokinin, which are hormones crucial to plant nutrient absorption, promoting root cell growth and plant resilience. Antioxidant enzymes that guard plant tissues against ROS damage during seed germination are activated by chitosan, which also enhances seed germination and seedling growth (Kumaraswamy et al., 2018, 2021). The ability of chitosan to protect plants from several abiotic stresses, such as drought, salt, temperature, and heavy metal toxicity, boosts plant growth. Plant's physiological features, in addition to the expression of genes that code for stress-resistance compounds, are found to be associated with chitosan activity (Sangwan et al., 2023).

2.7 CONCLUSION

This study has affirmed that chitosan is amenable to a wide range of applications in food manufacturing and agriculture and recognized that such uses are strategically gaining popularity. Chitosan may be used as an antibacterial agent, a plant growth stimulant, and an edible film for coating fresh fruits and vegetables. It is implicated as a suitable material for biofertilizer production, improving the ability of soil to retain water, and serving as an adsorbent for irrigation water treatment. However, using chitosan to facilitate soil fertility and biostimulate the proliferation of plant-growth-promoting rhizospheric microbes is vulnerable to unanticipated interferences posed by changing local factors (abiotic and biotic). This suggests that chitosan research in agriculture is at a nascent stage and requires much more attention, even though encouraging empirical results are emerging that show its suitability as a carrier in the delivery of agrochemicals and micronutrients. Chitosan uniquely promises to be a more effective biopolymer whose action is phenomenal in reducing the required dosage for efficacy and controlled release of bioactive agents of encapsulated materials while ensuring a controlled delivery. However, the sustainable application of chitosan in the agricultural and food sectors may be achieved if the skepticism associated with its large-scale production, commodification, production cost estimates, indirect impact on its sourced-animal diversity, and clearer knowledge of its toxicity nature are addressed in future studies. Poor elucidation of the underlying process of chitosan action in food matrix, plant surfaces, or soils, coupled with the nascent understanding of how to stabilize it for optimal benefit in the open field, also constitutes some forms of limitations. Notwithstanding, chitosan holds promise and remains crucial for bioeconomic and biotechnological development around the globe if only its technology can be perfected or improved to conquer its current application limitations. Although research on chitosan-based delivery systems and other dimensions of chitosan utility or value added is increasing, the current level of knowledge has failed to give a fair but comprehensive assessment of the pros and cons of the use of chitosan-based products or provide a long-term study, such as a decade of its use in food, to allay debates of its toxicity. The uniqueness of chitosan-based materials affords it a competitive advantage over other contemporary biopolymers, making it a more attractive, safer, and eco-friendlier choice in revolutionizing value adds to protecting growing crops from pathogenic, climate, and edaphic adversatives while guaranteeing the consumption of safe, quality foods that can be stored for a longer period. Consequently, exploring the promises inherent in chitosan-based materials will abrogate dependence on chemosynthetic compounds and mitigate the risks hitherto associated with their applications.

REFERENCES

Abdelgawad AM, Hudson SM, Rojas OJ. (2014). Antimicrobial wound dressing nanofibermats from multicomponent (chitosan/silver-NPs/polyvinyl alcohol) systems. *Carbohydrate Polymers*, 100, 166–178.

Abifarin JK, Prakash C, Singh S. (2022). Optimization and significance of fabrication parameters on the mechanical properties of 3D printed Chitosan/PLA scaffold. Materials Today*: Proceedings*, 50(5), 2018–2025, ISSN 2214-7853. https://doi.org/10.1016/j.matpr.2021.09.386.

Ahmad NNR, Fernando WJN, Uzir MH. (2015). Parametric evaluation using mechanistic model for release rate of phosphate ions from chitosan-coated phosphorus fertiliser pellets. *Biosystems Engineering*, 129, 78–86.

Ahmadi F, Oveisi Z, Samani SM, Amoozgar Z. (2015). Chitosan based hydrogels: Characteristics and pharmaceutical applications. *Research in Pharmaceutical Sciences*, 10(1), 1.

Ahmed S, Ikram S. (2015). Chitosan & its derivatives: A review in recent innovations. *International Journal of Pharmaceutical Sciences and Research*, 6(1), 14.

Amoli MS, Anand R, EzEldeen M, Amorim PA, Geris L, Jacobs R, Bloemen V. (2022). The development of a 3D printable chitosan-based copolymer with tunable properties for dentoalveolar regeneration. *Carbohydrate Polymers*, 289, 119441, ISSN 0144-8617. https://doi.org/10.1016/j.carbpol.2022.119441.

Amaregouda Y, Kamanna K, Gasti T. (2022). Fabrication of intelligent/active films based on chitosan/polyvinyl alcohol matrices containing Jacaranda cuspidifolia anthocyanin for real-time monitoring of fish freshness. *International Journal of Biological Macromolecules*, 218, 799–815, ISSN 0141-8130. https://doi.org/10.1016/j.ijbiomac.2022.07.174.

Angelim AL, Costa SP, Farias BC, Aquino LF, Melo VM. (2013). An innovative bioremediation strategy using a bacterial consortium entrapped in chitosan beads. *Journal of Environmental Management*, 127, 10–17.

Aranaz I, Alcántara AR, Civera MC, Arias C, Elorza B, Heras Caballero A, Acosta N. (2021). Chitosan: An overview of its properties and applications. *Polymers (Basel)*, 13(19), 3256. https://doi.org/10.3390/polym13193256.

Aranaz I, Mengíbar M, Harris R, Panos I, Miralles B, Acosta N, Galed G, Heras Á. (2009). Functional characterization of chitin and chitosan. *Current Chemical Biology*, 3, 203–230.

Arkoun M, Daigle F, Heuzey MC, Ajji A. (2017). Mechanism of action of electrospun chitosan-based nanofibers against meat spoilage and pathogenic bacteria. *Molecules*, 22(4), 585. https://doi.org/10.3390/molecules22040585.

Arkoun M, Daigle F, Holley RA, Heuzey MC, Ajji A. (2018). Chitosan-based nanofibers as bioactive meat packaging materials. *Packaging Technology and Science*, 3, 185–195.

Azimi F, Oraei M, Gohari G, Panahirad S, Farmarzi A. (2021). Chitosan-selenium nanoparticles (Cs–Se NPs) modulate the photosynthesis parameters, antioxidant enzymes activities and essential oils in Dracocephalum moldavica L. under cadmium toxicity stress. *Plant Physiology and Biochemistry*, 167, 257–268. https://doi.org/10.1016/j.plaphy.2021.08.013.

Azmana M, Mahmood S, Hilles AR, Rahman A, Arifin MAB, Ahmed S. (2021). A review on chitosan and chitosan-based bionanocomposites: Promising material for combatting global issues and its applications. *International Journal of Biological Macromolecules*, 185, 832–848.

Balusamy SR, Rahimi S, Sukweenadhi J, Sunderraj S, Shanmugam R, Thangavelu L, Mijakovic L, Perumalsamy H. (2022). Chitosan, chitosan nanoparticles and modified chitosan biomaterials, a potential tool to combat salinity stress in plants. *Carbohydrate Polymers*, 284, 119189. https://doi.org/10.1016/j.carbpol.2022.119189.

Baptista RC, Horita CN, Sant'Ana AS. (2020). Natural products with preservative properties for enhancing the microbiological safety and extending the shelf-life of seafood: A review. *Food Research International*, 127, 108762. https://doi.org/10.1016/j.foodres.2019.108762.

Berger LRR, Stanford NP, Walladino LG, Laranieira D, Barbosa, de-Lima MA, Malheiros SMM, Oliveria WJD, Stamford TCM. (2016). Cowpea resistance induced against Fusarium oxysporum f. sp. tracheiphilum by crustaceous chitosan and by biomass and chitosan obtained from Cunninghamellaelegans. *Biological Control*, 92, 45–54.

Bhatia S. (2016). Natural polymers vs synthetic polymer. In: *Natural Polymer Drug Delivery Systems* (pp. 95–118). Springer, Cham.

Bonilla J, Fortunati E, Atarés L, Chiralt A, Kenny JM. (2014a). Physical, structural and antimicrobial properties of poly vinyl alcohol-chitosan biodegradable films. *Food Hydrocolloids*, 35, 463–470.

Bonilla J, Vargas M, Atarés L, Chiralt A. (2014b). Effect of chitosan essential oil films on the storage-keeping quality of pork meat products. *Food and Bioprocess Technology*, 7, 2443–2450.

Božič M, Gorgieva S, Kokol V. (2012). Laccase-mediated functionalization of chitosan bycaffeicand gallic acids for modulating antioxidant and antimicrobial properties. *Carbohydrate Polymers*, 87(4), 2388–2398.

Brink I, Šipailienė A, Leskauskaitė D. (2019). Antimicrobial properties of chitosan and whey protein films applied on fresh cut turkey pieces. *International Journal of Biological Macromolecules*, 130, 810–817.

Cazón P, Velazquez G, Ramírez JA, Vázquez M. (2017). Polysaccharide-based films and coatings for food packaging: A review. *Food Hydrocolloids*, 68, 136–148.

Ceylan Z, Sengor GFU, Yilmaz MT. (2018). Nanoencapsulation of liquid smoke/thymolcombination in chitosan nanofibers to delay microbiological spoilage of sea bass (Dicentrarchus labrax) fillets. *Journal of Food Engineering*, 229, 43–49.

Chakraborty M, Hasanuzzaman M, Rahman M, Khan MAR, Bhowmik P, Mahmud NU, Islam T. (2020). Mechanism of plant growth promotion and disease suppression by chitosan biopolymer. *Agriculture*, 10(12), 624.

Chakraborty P, Nath D, Hoque M, Sarkar P, Hati S, Mishra BK. (2022). Biopolymer-based antimicrobial coatings for aquatic food products: A review. *Journal of Food Processing and Preservation*, 46(4), e16465.

Chang SH, Wu CH, Tsai GJ. (2018). Effects of chitosan molecular weight on its antioxidant and antimutagenic properties. *Carbohydrate Polymers*, 181, 1026–1032. https://doi.org/10.1016/j.carbpol.2017.11.047.

Chanphai PH, Tajmir-Riahi A. (2016). Chitosan nanoparticles conjugate with trypsin and trypsin inhibitor. *Carbohydrate Polymers*, 144, 346–352. https://doi.org/10.1016/j.carbpol.2016.02.066.

Chatelet C, Damour O, Domard A. (2001). Influence of the degree of acetylation on some biological properties of chitosan films. *Biomaterials*, 22(3), 261–268.

Chatterjee R, Maity M, Hasnain MS, Nayak AK. (2022). Chapter 1 - Chitosan: Source, chemistry, and properties. In: Md Saquib Hasnain, Sarwar Beg, Amit Kumar Nayak, Editor(s), *Chitosan in Drug Delivery* (pp. 1–22). Academic Press. ISBN 9780128193365. https://doi.org/10.1016/B978-0-12-819336-5.00001-7.

Chatterjee S, Chatterjee BP, Guha AK. (2014). A study on antifungal activity of water-soluble chitosan against Macrophomina phaseolina. *International Journal of Biological Macromolecules*, 67, 452–457.

Chen D, Bai R, Yong H, Zong S, Jin C, Liu J. (2022). Improving the digestive stability and prebiotic effect of carboxymethyl chitosan by grafting with gallic acid: In vitro gastrointestinal digestion and colonic fermentation evaluation. *International Journal of Biological Macromolecules*, 214, 685–696. https://doi.org/10.1016/j.ijbiomac.2022.06.170.

Chen F, Wang M, Zheng Y, Luo J, Yang X, Wang X. (2010). Quantitative changes of plant defense enzymes and phytohormone in biocontrol of cucumber Fusariumwilt by Bacillus subtilisB579. *World Journal of Microbiology & Biotechnology*, 26, 675–684.

Cheung RCF, Ng TB, Wong JH, Chan WY. (2015). Chitosan: An update on potential biomedical and pharmaceutical applications. *Marine Drugs*, 13(8), 5156–5186.

Chien P, Sheu F, Lin H. (2007a). Coating citrus (Murcott tangor) fruit with low molecular weight chitosan increases postharvest quality and shelf life. *Food Chemistry*, 100, 1160–1164.

Chien P, Sheu F, Yang F. (2007b). Effects of edible chitosan coating on quality and shelf life of sliced mango fruit. *Journal of Food Engineering*, 78, 225–229.

Chopra S, Mahdi S, Kaur J, Iqbal Z, Talegaonkar S, Ahmad FJ. (2006). Advances and potential applications of chitosan derivatives as mucoadhesive biomaterials in modern drug delivery. *Journal of Pharmacy and Pharmacology*, 58(8), 1021–1032.

Chowdary VT, Roshan Baba SK, Divya J, Varma PK. (2022). Chapter 10 - Role of chitosan in eco-friendly management of plant diseases for sustainable agriculture. In: Harikesh Bahadur Singh, Anukool Vaishnav, Editor(s), *New and Future Developments in Microbial Biotechnology and Bioengineering* (pp. 219–244). Elsevier. https://doi.org/10.1016/B978-0-323-85581-5.00019-7.

Corradini E, De-Moura MR, Mattoso LHC. (2010). A preliminary study of the in corparation of NPK fertilizer into chitosan nanoparticles. *Express Polymer Letters*, 4(8), 509–515.

Dalvi S, Wani K, Ithape D, Suprasanna P. (2022). Chapter 7 - Potential of biopriming with irradiated chitosan for sugarcane micropropagation. In: M. Naeem, Tariq Aftab, M., Editor(s), *Radiation-Processed Polysaccharides* (pp. 179–204). Academic Press.

de Souza Soares L, de Faria JT, Amorim ML, de Araujo JM, Minim LA, dos Reis Coimbra JS, de Carvalho Teixeira AVN, de Oliveira EB. (2017). Rheological and physicochemical studies on emulsions formulated with chitosan previously dispersed in aqueous solutions of lactic acid. *Food Biophysics*, 12(1), 109–118.

Devlieghere F, Vermeulen A, Debevere J. (2004). Chitosan: Antimicrobial activity, interactions with food components and applicability as a coating on fruit and vegetables. *Food Microbiology*, 21(6), 703–710.

Dilkes-Hoffman LS, Lane JL, Grant T, Pratt S, Lant PA, Laycock B. (2018). Environmental impact of biodegradable food packaging when considering food waste. *Journal of Cleaner Production*, 180, 325–334.

Di Martino A, Trusova ME, Postnikov PS, Sedlarik V. (2018). Enhancement of the antioxidant activity and stability of β-carotene using amphiphilic chitosan/nucleic acid polyplexes. *International Journal of Biological Macromolecules*, 117, 773–780.

Dong H, Cheng L, Tan J, Zheng K, Jiang Y. (2004). Effects of chitosan coating on quality and shelf life of peeled litchi fruit. *Journal of Food Engineering*, 64, 355–358.

Dordevic V, Balanc B, Belscak-Cvitanovic A, Levic S, Trifkovic K, Kalusevic A, Kostic I, Komes D, Bugarski B, Nedovic V. (2014). Trends in encapsulation technologies for delivery of food bioactive compounds. *Food Engineering Reviews*, 7(4), 452–490.

Duan C, Meng X, Meng J, Khan IH, Dai L, Khan A, An X. (2019). Chitosan as a preservative for fruits and vegetables: A review on chemistry and antimicrobial properties. *Journal of Bioresources and Bioproducts*, 4(1), 11–21. https://doi.org/10.21967/jbb.v4i1.189.

Duarte LGR, Picone CSF. (2022a). Antimicrobial activity of lactoferrin-chitosan-gellan nanoparticles and their influence on strawberry preservation. *Food Research International*, 159(April), 111586. https://doi.org/10.1016/j.foodres.2022.111586

Duarte LGR, Picone CSF. (2022b). Antimicrobial activity of lactoferrin-chitosan-gellan nanoparticles and their influence on strawberry preservation. *Food Research International*, 159, 111586. https://doi.org/10.1016/j.foodres.2022.111586.

Dutta S, Choudhary P, Moses JA, Anandharamakrishnan C. (2019). Liposomal encapsulation of α-lipoic acid as a food supplement. *International Journal of Food Engineering*, 5(2), 111–115.

El Hadrami A, Adam LR, El Hadrami I, Daayf F. (2010). Chitosan in plant protection. *Marine Dugs*, 8(4), 968–987.

Elsherbiny AS, Galal A, Ghoneem KM, Salahuddin NA. (2022). Novel chitosan-based nanocomposites as ecofriendly pesticide carriers: Synthesis, root rot inhibition and growth management of tomato plants. *Carbohydrate Polymers*, 282, 19111. https://doi.org/10.1016/j.carbpol.2022.119111.

Escudero R, Lopez-Llorca LV. (2016). Chitosan enhances parasitism of Meloidogynejavanica eggs by the nematophagous fungus Pochoniachlamydo sporia. *Fungal Biology*, https://doi.org/10.1016/j.funbio.2015.12.005.

Falcón-Rodríguez AB, Costales D, Cabrera JC, Martínez-Téllez MÁ. (2011). Chitosan physico–chemical properties modulate defense responses and resistance in tobacco plants against the oomycete Phytophthora nicotianae. *Pesticide Biochemistry and Physiology*, 100(3), 221–228.

Fang Z, Bhandari B. (2010). Encapsulation of polyphenols: A review. *Trends in Food Science and Technology*, 21(10), 510–523.

Fareez IM, Lim SM, Lim FT, Mishra RK, Ramasamy K. (2017). Microencapsulation of *Lactobacillus* sp. using chitosan-alginate-xanthan gum-β-cyclodextrin and characterization of its cholesterol reducing potential and resistance against pH, temperature and storage. *Journal of Food Process Engineering*, 40(3), e12458.

Ferreira DCM, de Souza AL, da Silveira JVW, Marim BM, Giraldo GAG, Mantovan J, Pelissari FM. (2020a). Chitosan nanocomposites for food packaging applications. In: *Multifunctional Hybrid Nanomaterials for Sustainable Agri-Food and Ecosystems* (Issue 1, pp. 393–435). Elsevier. https://doi.org/10.1016/B978-0-12-821354-4.00017-0.

Ferreira L, Rafael D, Lima C, De Andrade K., Maria D, Navarro F, Lafaiete J, Alves R, Nelson G. (2020b). Industrial Crops & Products Chemical composition and insecticidal effect of essential oils from Illicium verum and Eugenia caryophyllus on Callosobruchus maculatus in cowpea. *Industrial Crops & Products*, 145, 112088. https://doi.org/10.1016/j.indcrop.2020.112088.

Giglou MT, Giglou RH, Esmaeilpour B, Azarmi R, Padash A, Falakian M, Śliwka J, Gohari G, Lajayer HM. A new method in mitigation of drought stress by chitosan-coated iron oxide nanoparticles and growth stimulant in peppermint. *Industrial Crops and Products*, 187(Part A), 115286. https://doi.org/10.1016/j.indcrop.2022.115286.

Goy RC, Britto DD, Assis OB. (2009). A review of the antimicrobial activity of chitosan. *Polímeros*, 19, 241–247.

Grande-Tovar CD, Chaves-lopez C, Serio A, Rossi C. (2018). Chitosan coatings enriched with essential oils: Effects on fungi involved in fruit decay and mechanisms of action. *Trends in Food Science & Technology*, 78, 61–71. https://doi.org/10.1016/j.tifs.2018.05.019.

Guan YJ, Hu J, Wang X-J, Shao C-X. (2009). Seed priming with chitosan improves maize germination and seedling growth in relation to physiological changes under low temperature stress. *Journal of Zhejiang University Science*, 10(6), 427–433.

Hahn T, Roth A, Ji R, Schmitt E, Zibek S. (2020). Chitosan production with larval exoskeletons derived from the insect protein production. *Journal of Biotechnology*, 310, 62–67. https://doi.org/10.1016/j.jbiotec.2019.12.015.

Hashemi Gahruie H, Niakousari M. (2017). Antioxidant, antimicrobial, cell viability and enzymatic inhibitory of antioxidant polymers as biological macromolecules. *International Journal of Biological Macromolecules*, 104, 606–617.

Hassan MA, Omer AM, Abbas E, Baset W, Tamer TM. (2018). Preparation, physicochemical characterization and antimicrobial activities of novel two phenolic chitosan Schiff base derivatives. *Scientific Reports*, 8(1), 1–14.

Hosseinnejad M, Jafari SM. (2016). Evaluation of different factors affecting antimicrobial properties of chitosan. *International Journal of Biological Macromolecules*, 85, 467–475.

Hu W, Niu Y, Zhu H, Dong K, Wang D, Liu F. (2021). Remediation of zinc-contaminated soils by using the two-step washing with citric acid and water-soluble chitosan. *Chemosphere*, 282, 131092. https://doi.org/10.1016/j.chemosphere.2021.131092.

Hu Z, Gänzle MG. (2018). Challenges and opportunities related to the use of chitosan as a food preservative. *Journal of Applied Microbiology*, 126(5), 1318–1331. https://doi.org/10.1111/jam.14131.

Huang B-H, Shih-Yuan Li S-Y, Chiou Y-J, Chojniak D, Chou S-C, Wong VCM, Chen S-Y, Wu P-W. (2021). Electrophoretic fabrication of a robust chitosan/polyethylene glycol/polydopamine composite film for UV-shielding application. *Carbohydrate Polymers*, 273, 118560. https://doi.org/10.1016/j.carbpol.2021.118560.

Huq T, Khan A, Brown D, Dhayagude N, He Z, Ni Y. (2022). Sources, production and commercial applications of fungal chitosan: A review. *Journal of Bioresources and Bioproducts*, 7, 85–98.

Hussain A, Ahmed T, Kanti S, Mohammad M, Je W, Hamid M, Sharifuzzaman SM, Lyon C, Lee H, Lee E, Hasan T. (2021). Essential oils and chitosan as alternatives to chemical preservatives for fish and fisheries products: A review. *Food Control*, 129, 108244. https://doi.org/10.1016/j.foodcont.2021.108244.

Hussain MD, Rabiul D, Rashmi RM, Tarun K. (2012). Controlled release of urea from chitosan microspheres prepared by emulsification and cross-linking method. *Irarian Polymer Journal*, 21(8), 473–479.

Iber BT, Kasan NA, Torsabo D, Omuwa JW. (2022). A review of various sources of chitin and chitosan in nature. *Journal of Renewable Materials*, 10(4), 1097–1123.

Iida ASL, Luz KN, Barros-Alexandrino TT, Fávaro-Trindade CS, Pinho SCD, Assis OBG, Martelli-Tosi M. (2019). Investigation of TPP-Chitosomes particles structure and stability as encapsulating agent of cholecalciferol. *Polímeros,* 29(4), e2019049.

Inanli AG, Tümerkan ETA, El Abed N, Regenstein JM, Özogul F. (2020). The impact of chitosan on seafood quality and human health: A review. *Trends in Food Science & Technology*, 97, 404–416. https://doi.org/10.1016/j.tifs.2020.01.029.

Inoue K, Baba Y, Yoshizuka K, Noguchi H, Yoshizaki M. (1988). Selectivity series in the adsorption of metal ions on resin prepared by crosslinking copper (II)-complexed chitosan. *Chemistry Letters*, 17(8), 1281–1284.

Irastorza A, Zarandona I, Andonegi M, Guerrero P. (2021). The versatility of collagen and chitosan: From food to biomedical applications. *Food Hydrocolloids*, 116, 106633. https://doi.org/10.1016/j.foodhyd.2021.106633.

Iriti M, Picchi V, Rossoni M, Gomarasca S, Ludwig N, Gargano M, Faoro F. (2009). Chitosan antitranspirant activity is due to abscisic acid-dependent stomatal closure. *Environmental and Experimental Botany*, 66, 493–500.

Iriti M, Varoni EM. (2015). Chitosan-induced antiviral activity and innate immunity in plants. *Environmental Science and Pollution Research*, 22(4), 2935–2944. https://doi.org/10.1007/s11356-014-3571-7.

Jaber N, Al-Remawi M, Al-Akayleh F, Al-Muhtaseb N, Al-Adham IS, Collier PJ. (2022). A review of the antiviral activity of chitosan, including patented applications and its potential use against COVID-19. *Journal of Applied Microbiology*, 132(1), 41–58. https://doi.org/10.1111/jam.15202.

Kamari A, Pulford ID, Hargreaves JSJ. (2011). Chitosan as a potential amendment to remediate metal contaminated soil: A characterization study. *Colloids and Surfaces B: Biointerfaces*, 82, 71–80.

Kardas I, Struszczyk MH, Kucharska M, van den Broek LA, van Dam JE, Ciechańska D. (2012). Chitin and chitosan as functional biopolymers for industrial applications. In: Navard, P. Editor, *The European Polysaccharide Network of Excellence (EPNOE)*. Springer, Vienna. (pp. 329–373).

Kayser H, Pienkoß F, de María PD. (2014). Chitosan-catalyzed biodiesel synthesis: Proof-of-concept and limitations. *Fuel*, 116, 267–272.

Khairy AM, Tohamy MRA, Zayed MA, Mahmoud SF, El-Tahan AM, El-Saadony MT, Mesiha PK. (2022). Eco-friendly application of nano-chitosan for controlling potato and tomato bacterial wilt. *Saudi Journal of Biological Sciences*, 29(4), 2199–2209. https://doi.org/10.1016/j.sjbs.2021.11.041.

Khalil MS, Badawy ME. (2012). Nematicidal activity of a biopolymer chitosan at different molecular weights against root-knot nematode, Meloidogyne incognita. *Plant Protection Science*, 48, 170–178.

Khaneghah AM, Hashemi SMB, Limbo S. (2018). Antimicrobial agents and packaging systems in antimicrobial active food packaging: An overview of approaches and interactions. *Food and Bioproducts Processing*, 111, 1–19.

Kou SG, Peters LM, Mucalo MR. (2021). Chitosan: A review of sources and preparation methods. *International Journal of Biological Macromolecules*, 169, 85–94, ISSN 0141-8130. https://doi.org/10.1016/j.ijbiomac.2020.12.005.

Kulikov SN, Chirkov SN, Ilina AV, Lopatin SA, Varlamov VP. (2006). Effect of the molecular weight of chitosan on its antiviral activity in plants. *Prikladnaya Biokhimiya i Mikrobiologiya*, 42, 224–228.

Kumaraswamy RV, Kumari S, Chandra R, Pal A, Raliya R, Biswas P, Saharan V. (2018). Engineered chitosan based nanomaterials: Bioactivities, mechanisms and perspectives in plant protection and growth. *International Journal of Biological Macromolecules*, 113, 494–506. https://doi.org/10.1016/j.ijbiomac.2018.02.130.

Kumaraswamy RV, Saharan V, Kumari S, Choudhary RC, Pal A, Sharma SS, Rakshit S, Raliya R, Biswas P. (2021). Chitosan-silicon nanofertilizer to enhance plant growth and yield in maize (Zea mays L.). *Plant Physiology and Biochemistry*, 159, 53–66. https://doi.org/10.1016/j.plaphy.2020.11.054.

Kurek M, Garofulić I, Bakić M, Ščetar M, Uzelac V, Galić K. (2018). Development and evaluation of a novel antioxidant and pH indicator film based on chitosan and food waste sources of antioxidants. *Food Hydrocolloid*, 84, 238–246.

Leceta I, Molinaro S, Guerrero P, Kerry JP, De la Caba K. (2015). Quality attributes of map packaged ready-to-eat baby carrots by using chitosan-based coatings. *Postharvest Biology and Technology*, 100, 142–150.

Li F, Jin H, Xiao J, Yin X, Liu X, Li D, Huang Q. (2018). The simultaneous loading of catechin and quercetin on chitosan-based nanoparticles as effective antioxidant and antibacterial agent. *Food Research International*, 111, 351–360.

Li H, Yu T. (2001). Effect of chitosan on incidence of brown rot, quality and physiologicalattributes of postharvest peach fruit. *Journal of the Science of Food and Agriculture*, 81, 269–274.

Li JH, Wu YG, Zhao LQ. (2016). Antibacterial activity and mechanism of chitosan with ultra high molecular weight. *Carbohydrate Polymers*, 148, 200–205. https://doi.org/10.1016/j.carbpol.2016.04.025.

Li L, Chen D, Zhang Y, Deng Z, Ren X, Meng X, Tang F, Ren J, Zhang L. (2007). Magnetic and fluorescent multifunctional chitosan nanoparticles as a smart drug delivery system. *Nanotechnology*, 18(40), 405102.

Liang J, Yan H, Wang X, Zhou Y, Gao X, Puligundla P, Wan X. (2017). Encapsulation of epigallocatechin gallate in zein/chitosan nanoparticles for controlled applications in food systems. *Food Chemistry*, 231, 19–24.

Liang W, Yu A, Wang G, Zheng F, Jia J, Xu H. (2018). Chitosan-based nanoparticles of avermectin to control pine wood nematodes. *International Journal of Biological Macromolecules*, 112, 258–263. https://doi.org/10.1016/j.ijbiomac.2018.01.174.

Lima R, Fernandes C, Pinto MM. (2022). Molecular modifications, biological activities, and applications of chitosan and derivatives: A recent update. *Chirality*, 34, 1166–1190.

Lin Y, Li N, Lin H, Lin M, Chen Y, Wang H, Lin Y. (2020). Effects of chitosan treatment on the storability and quality properties of longan fruit during storage. *Food Chemistry*, 306, 125627.

Liu H, Ye X, Chen S, Sun A, Duan X, Zhang Y, Zou H, Zhang Y. (2021). Chitosan as additive affects the bacterial community, accelerates the removals of antibiotics and related resistance genes during chicken manure composting. *Science of the Total Environment*, 792, 148381, ISSN 0048-9697. https://doi.org/10.1016/j.scitotenv.2021.148381.

Liu J, Qin W, Aiqin, W. (2007). Synthesis and characterization of chitosan-g-poly (acrylic acid)/ sodium humate superabsorbent. *Carbohydrate Polymers*, 70, 166–173.

Liu Y-W, Zhou Y, Huang G-Q, Guo L-P, Li X-D, Xiao J-X. (2022). Fabrication of lipase-loaded particles by coacervation with chitosan. *Food Chemistry*, 385, 132689. https://doi.org/10.1016/j.foodchem.2022.132689.

Llorens A, Lloret E, Picouet PA, Trbojevich R, Fernandez A. (2012). Metallic based micro and nanocomposites in food contact materials and active food packaging. *Trends in Food Science and Technology*, 24, 19–29.

Lopez-Moya F, Lopez-Llorca LV. (2016). Omics for investigating chitosan as an antifungal and gene modulator. *Journal of Fungi*, 2(11). https://doi.org/10.3390/jof2010011.

Lowe A, Rafferty-McArdle SM, Cassells AC. (2012). Effects of AMF and PGPR-root inoculation and a foliar chitosan spray in single and combined treatments on powdery mildew disease in strawberry. *Agricultural and Food Science*, 21, 28–38.

Luo Y, Wang Q. (2013). Recent advances of chitosan and its derivatives for novel applications in food science. *Journal of Food Processing & Beverages*, 1(1), 1–13.

Ma Z, Garrido-Maestu A, Jeong KC. (2017). Application, mode of action, and in vivo activity of chitosan and its micro– and nanoparticles as antimicrobial agents: A review. *Carbohydrate Polymers*, 176, 257–265.

Madera-Santana TJ, Herrera-Méndez CH, Rodríguez-Núñez JR. (2018). An overview of the chemical modifications of chitosan and their advantages. *Green Materials* 6(4), 131–142. https://doi.org/10.1680/jgrma.18.00053.

Mahjoub B, Domscheit E. (2020). Chances and challenges of an organic waste–based bioeconomy. *Current Opinion in Green and Sustainable Chemistry*, 25, 100388, ISSN 2452-236. https://doi.org/10.1016/j.cogsc.2020.100388.

Manigandan V, Karthik R, Ramachandran S, Rajagopal S. (2018). Chitosan applications in food industry. In: *Biopolymers for Food Design* (pp. 469–491). Academic Press.

Matica MA, Aachmann FL, Tøndervik A, Sletta H, Ostafe V. (2019). Chitosan as a wound dressing starting material: Antimicrobial properties and mode of action. *International Journal of Molecular Sciences*, 20(23), 5889. https://doi.org/10.3390/ijms20235889.

Matshetshe KI, Parani S, Manki SM, Oluwafemi OS. (2018). Preparation, characterization and in vitro release study of β-cyclodextrin/chitosan nanoparticles loaded Cinnamomumzeylanicum essential oil. *International Journal of Biological Macromolecules*, 118, 676–682.

Melo NFCB, de MendonçaSoares BL, Diniz KM, Leal CF, Canto D, Flores MA, Tavares-Filho JHC, Galembecj A, Stamford TCM, Stamford-Arnaud TM, Stamford TCM. (2018). Effects of fungal chitosan nanoparticles as eco-friendly edible coatings on the quality of postharvest table grapes. *Postharvest Biology and Technology*, 139, 56–66. https://doi.org/10.1016/j.postharvbio.2018.01.014.

Milašinović N, Čalija B, Vidović B, Sakač MC, Vujić Z, Knežević-Jugović Z. (2016). Sustained release of α-lipoic acid from chitosan microbeads synthetized by inverse emulsion method. *Journal of the Taiwan Institute of Chemical Engineers*, 60, 106–112.

Mirabedini M, Kassaee MZ, Poorsadeghi S. (2016). Novel magnetic chitosan hydrogel film, cross-linked with glyoxal as an efficient adsorbent for removal of toxic Cr (VI) from water. *Arabian Journal for Science and Engineering*, 42, 1–10.

Morin-Crini N, Lichtfouse E, Torri G, Crini G. (2019). Applications of chitosan in food, pharmaceuticals, medicine, cosmetics, agriculture, textiles, pulp and paper, biotechnology, and environmental chemistry. *Environmental Chemistry Letters*, 17(4), 1667–1692.

Mphande W, Kettlewell PS, Grove IG, Farrell AD. (2020). The potential of antitranspirants in drought management of arable crops: A review. *Agricultural Water Management*, 236, 106143. https://doi.org/10.1016/j.agwat.2020.106143.

Mukta JA, Rahman M, Sabir AA, Gupta DR, Surovy MZ, Rahman M, Islam MT. (2017). Chitosan and plant probiotics application enhance growth and yield of strawberry. *Biocatalysis and Agricultural Biotechnology*, 11, 9–18. https://doi.org/10.1016/j.bcab.2017.05.005.

Mustafa A. (2018). Pharmaceutical uses of chitosan in the medical field. *European Journal of Medicine and Natural Sciences*, 1(1), 1–6.

Muxika A, Etxabide A, Uranga J, Guerrero P, De La Caba K. (2017). Chitosan as a bioactive polymer: Processing, properties and applications. *International Journal of Biological Macromolecules*, 105, 1358–1368.

Mwaheb MAMA, Hussain M, Tian J, Zhang X, Imran Hamid M, El-Kassim NA, Hassan GM, Xiang M, Liu X. (2017). Synergetic suppression of soybean cyst nematodes by chitosan and Hirsutella minnesotensis via the assembly of the soybean rhizosphere microbial communities. *Biological Control*, 115, 85–94. https://doi.org/10.1016/j.biocontrol.2017.09.011.

Namboodiri MMT, Pakshirajan K. (2020). Chapter 10 - Valorization of waste biomass for chitin and chitosan production. In: Thallada Bhaskar, Ashok Pandey, Eldon R. Rene, Daniel C. W. Tsang, Editor(s), *Waste Biorefinery* (pp. 241–266). Elsevier. https://doi.org/10.1016/B978-0-12-818228-4.00010-1.

Narayanan A, Dhamodharan R. (2015). Super water-absorbing new material from chitosan, EDTA and urea. *Carbohydrate Polymers*, 134, 337–343.

Naskar S, Sharma S, Kuotsu K. (2019). Chitosan-based nanoparticles: An overview of biomedical applications and its preparation. *Journal of Drug Delivery Science and Technology*, 49, 66–81.

No HK, Meyers SP, Prinyawiwatkul W, Xu Z. (2007). Applications of chitosan for improvement of quality and shelf life of foods: A review. *Journal of Food Science*, 72(5), R87–R100.

Orzali L, Forni C, Riccioni L. (2014). Effect of chitosan seed treatment as elicitor of resistance to Fusarium graminearum in wheat. *Seed Science and Technology*, 42, 132–149.

Osheba AS, Sorour MA, Abdou ES. (2013). Effect of chitosan nanoparticles as active coating on chemical quality and oil uptake of fish fingers. *Journal of Agriculture and Environmental Sciences*, 2, 1–14.

Pabast M, Shariatifar N, Beikzadeh S, Jahed G. (2018). Effects of chitosan coatings incorporating with free or nano-encapsulated Satureja plant essential oil on quality characteristics of lamb meat. *Food Control*, 91, 185–92. https://doi.org/10.1016/j.foodcont.2018.03.047.

Pandey P, De N. (2017). Effect of chitosan based superabsorbent on water retention behaviour of soil and seedling growth of Alfalfa (Medicago sativa L.). *Indian Journal of Ecology*, 44(5), 456–460.

Pereira V, de Arruda I, Stefani R. (2015). Active chitosan/PVA films with anthocyanins from Brassica oleraceae (red cabbage) as time–temperature indicators for application in intelligent food packaging. *Food Hydrocolloids*, 43, 180–188.

Piran F, Khoshkhoo Z, Hosseini SE, Azizi MH. (2020). Controlling the antioxidant activity of green tea extract through encapsulation in chitosan-citrate nanogel. *Journal of Food Quality*. https://doi.org/10.1155/2020/7935420.

Preetha RT, Pagare V, Joseph BV, Praveen N. (2022). Chapter 10 - Impact of chitosan and chitosan nanoparticles on seed germination: probabilities and prospects. In: Santosh Kumar, Sundararajan V. Madihally, Editor(s), *Nanomaterial-Plant Interactions, Role of Chitosan and Chitosan-Based Nanomaterials in Plant Sciences* (pp. 213–237). Academic Press. https://doi.org/10.1016/B978-0-323-85391-0.00010-1.

Priyadarshi R, Rhim J-W. (2020). Chitosan-based biodegradable functional films for food packaging applications. *Innovative Food Science & Emerging Technologies*, 62, 102346, ISSN 1466-8564. https://doi.org/10.1016/j.ifset.2020.102346.

Qing W, Jin-Hua Z, Qian W, Yang N, Li-Pu G. (2015). Inhibitory effect of chitosan on growth of the fungal phytopathogen, Sclerotinia sclerotiorum, and sclerotinia rot of carrot. *Journal of Integrative Agriculture*, 14(4), 691–697.

Raafat D, Sahl HG. (2009). Chitosan and its antimicrobial potential: A critical literature survey. *Microbial Biotechnology*, 2(2), 186–201.

Rabea EI, Badawy MET, Stevens CV, Smagghe G, Steurbaut W. (2003). Chitosan as antimicrobial agent: Applications and mode of action. *Biomacromolecules* 4, 1457–1465.

Ramya R, Venkatesan J, Kim SK, Sudha PN. (2012). Biomedical applications of chitosan: An overview. *Journal of Biomaterials and Tissue Engineering*, 2(2), 100–111.

Raut SA, Borkar SG, Nagraled T. (2014). Effect of disease (alternaria leaf blight) resistance elicitors on growth parameters of tomato plant. *The Bioscan*, 9(3), 1157–1159.

Rendina N, Nuzzaci M, Scopa A, Cuypers A, Sofo A. (2019). Chitosan-elicited defense responses in Cucumber mosaic virus (CMV)-infected tomato plants. *Journal of Plant Physiology*, 234–235, 9–17. https://doi.org/10.1016/j.jplph.2019.01.003.

Riofrio A, Alcivar T, Baykara H. (2021). Environmental and economic viability of chitosan production in Guayas-Ecuador: A robust investment and life cycle analysis. *ACS Omega*, 6(36), 23038–23051.

Riseh RS, Hassanisaadi M, Vatankhah M, Babaki SA, Barka EA. (2022a). Chitosan as a potential natural compound to manage plant diseases. *International Journal of Biological Macromolecules*, 220, 998–1009. https://doi.org/10.1016/j.ijbiomac.2022.08.109.

Riseh RS, Tamanadar E, Hajabdollahi N, Vatankhah M, Thakur VK, Skorik YA. (2022b). Chitosan microencapsulation of rhizobacteria for biological control of plant pests and diseases: Recent advances and applications. *Rhizosphere*, 23, 100565. https://doi.org/10.1016/j.rhisph.2022.100565.

Romanazzi G, Feliziani E, Baños SB, Sivakumar D. (2017). Shelf life extension of fresh fruit and vegetables by chitosan treatment. *Critical Reviews in Food Science and Nutrition*, 57(3), 579–601.

Roshanravana SMS, Suraya ARC, Fariba M, Mohd Khanif Y. (2015). Enhancement of nitrogen release properties of urea–kaolinite fertilizer with chitosan binder. *Chemical Speciation & Bioavailability*, 27(1), 44–51.

Sabliov CM, Astete CE. (2008). Encapsulation and controlled release of antioxidants and vitamins. In: *Delivery and Controlled Release of Bioactives in Foods and Nutraceuticals* (pp. 297–330). Elsevier.

Saharan V, Pal A. (2016). *Chitosan Based Nanomaterials in Plant Growth and Protection*. Springer.

Salachna P, Zawadzinska A. (2014). Effect of chitosan on plant growth, flowering and corms yield of potted freesia. *Journal of Ecological Engineering*, 15(3), 93–102.

Sangeetha V, Sudha PN, Gomathi T. (2016). Experimental analysis of binary and ternary polymer blends of nano chitosan. *Materials Today: Proceedings*, 3, 2169–2177.

Sangwan S, Sharma P, Wati L, Mehta S. (2023). Effect of chitosan nanoparticles on growth and physiology of crop plants. In: Azamal Husen, Editor, *Engineered Nanomaterials for Sustainable Agricultural Production, Soil Improvement and Stress Management* (pp. 99–123). Elsevier Inc. Mexico. https://doi.org/10.1016/B978-0-323-91933-3.00011-8.

Sen SK, Tripathi R, Mandal P, Choudhuri C. (2022). Assessment of antifungal activity of nano-chitosan against mung bean seed borne pathogen Aspergillus flavus through solid matrix priming. *South African Journal of Botany*, 150, 372–386. https://doi.org/10.1016/j.sajb.2022.06.016.

Şenel S, McClure SJ. (2004). Potential applications of chitosan in veterinary medicine. *Advanced Drug Delivery Reviews*, 56(10), 1467–1480.

Shao Y, Wu C, Wu T, Li Y, Chen S, Yuan C, Hu Y. (2018). Eugenol-chitosan nanoemulsions by ultrasound-mediated emulsification: Formulation, characterization and antimicrobial activity. *Carbohydrate Polymers*, 193, 144–152.

Sharma S, Barman K, Siddiqui MW. (2016). Chitosan: Properties and roles in postharvest quality preservation of horticultural crops. In: *Eco-Friendly Technology for Postharvest Produce Quality* (pp. 269–296). Elsevier Inc. https://doi.org/10.1016/B978-0-12-804313-4.00013-6.

Shinohara T, Leskovar DI. (2014). Effects of ABA, antitranspirants, heat and drought stress on plant growth, physiology and water status of artichoke transplants. *Scientia Horticulturae*, 165, 225–234. https://doi.org/10.1016/j.scienta.2013.10.045.

Silva M, Nunes D, Cardosol AR, Ferreiral D, Britol M, Pintadol ME, Vasconcelos MW. (2014). Chitosan as a biocontrol agent against the pinewood nematode (Bursaphelenchus xylophilus). *Forest Pathology*. https://doi.org/10.1111/efp.12136.

Silva-Pereira M, Teixeira J, Pereira-Júnior V, Stefani R. (2015). Chitosan/corn starch blend films with extract from Brassica oleraceae (red cabbage) as a visual indicator of fish deterioration. *LWT-Food Science and Technology*, 61(1), 258–262.

Singh P, Medronho B, Alves L, da Silva GJ, Miguel MG, Lindman B. (2017). Development of carboxymethyl cellulose-chitosan hybrid micro- and macroparticles for encapsulation of probiotic bacteria. *Carbohydrate Polymers*, 175, 87–95.

Singh S, Negi T, Sagar NA, Kumar Y, Tarafdar A, Sirohi R, Sindhu R, Pandey A. (2022). Sustainable processes for treatment and management of seafood solid waste. *Science of the Total Environment*, 817, 152951. https://doi.org/10.1016/j.scitotenv.2022.152951.

Sochilina AV, Akasov RA, Arkharova NA, Klechkovskaya VV, Mironov AV, Prostyakova AI, Sholina NV, Zubov VP, Generalova AN, Vikhrov AA. (2022). Fabrication of moldable chitosan gels via thermally induced phase separation in aqueous alcohol solutions. *International Journal of Biological Macromolecules*, 215, 501–511. https://doi.org/10.1016/j.ijbiomac.2022.06.094.

Sood M, Kumar V, Rawal R. (2021). 8 - Seed biopriming a novel method to control seed borne diseases of crops. In: Sudisha Jogaiah, Editor, *Biocontrol Agents and Secondary Metabolites* (pp. 181–223). Woodhead Publishing. https://doi.org/10.1016/B978-0-12-822919-4.00008-9.

Speranza B, Petruzzi L, Bevilacqua A, Gallo M, Campaniello D, Sinigaglia M, and Corbo MR. (2017). Encapsulation of active compounds in fruit and vegetable juice processing: Current state and perspectives. *Journal of Food Science*, 82(6), 1291–1301.

Tang R, Yu H, Ruan Z, Zhang L, Xue Y, Yuan X, Qi M. (2022). Yao Y. Effects of food matrix elements (dietary fibres) on grapefruit peel flavanone profile and on faecal microbiota during in vitro fermentation. *Food Chemistry*, 371, 131065, ISSN 0308-8146. https://doi.org/10.1016/j.foodchem.2021.131065.

Thakur M, Tiwari S, Kataria S, Anand A. (2022). Recent advances in seed priming strategies for enhancing planting value of vegetable seeds. *Scientia Horticulturae*, 305, 111355. https://doi.org/10.1016/j.scienta.2022.111355.

Thanou M, Verhoef JC, Junginger HE. (2001). Oral drug absorption enhancement by chitosan and its derivatives. *Advanced Drug Delivery Reviews*, 52(2), 117–126.

Thenmozhi N, Narayanaswamy PY, Sudha PN. (2010). Efficacy of polymer blend in the treatment of tannery waste water. *The Ecoscan*, 4(4), 301–303.

Tongwanichniyom S, Kitjaruwankul S, Phornphisutthimas S. (2022). Production of biomaterials from seafood waste for application as vegetable wash disinfectant. *Heliyon*, 8(5), e09357. https://doi.org/10.1016/j.heliyon.2022.e09357.

Trzcinska A, Bogusiewic A, Szkop M, Drozdowski S. (2015). Effect of chitosan on disease control and growth of scots pine (Pinussylvestris L.) in a forest nursery. *Forest*, 6, 3165–3176.

Tu G, Li S, Han Y, Li Z, Liu J, Liu X, Li W. (2022). Fabrication of chitosan membranes via aqueous phase separation: Comparing the use of acidic and alkaline dope solutions. *Journal of Membrane Science*, 646, 120256. https://doi.org/10.1016/j.memsci.2022.120256.

Turan V, Khan SA, Mahmood-ur-Rahman, Iqbal M, Ramzani PMA, Fatima M. (2018). Promoting the productivity and quality of brinjal aligned with heavy metals immobilization in a wastewater irrigated heavy metal polluted soil with biochar and chitosan. *Ecotoxicology and Environmental Safety*, 161, 409–419. https://doi.org/10.1016/j.ecoenv.2018.05.082.

van den Broek LA, Knoop RJ, Kappen FH, Boeriu CG. (2015). Chitosan films and blends for packaging material. *Carbohydrate Polymers*, 116, 237–242.

Varankovich N, Martinez MF, Nickerson MT, Korber DR. (2017). Survival of probiotics inpea protein-alginate microcapsules with or without chitosan coating during storage and in a simulatedgastrointestinal environment. *Food Science and Biotechnology*, 26(1), 189–194.

Vargas M, Cháfer M, Chiralt A, De València UP. (2012). Edible chitosan coatings for fresh and minimally processed foods. In: *Emerging Food Packaging Technologies: Principles and Practice*. Woodhead Publishing Limited. https://doi.org/10.1533/9780857095664.1.66.

Viegas de Souza RHF, Takaki M, de Oliveira Pedro R, dos Santos Gabriel J, Tiera MJ, de Oliveira Tiera VA. (2013). Hydrophobic effect of amphiphilic derivatives of chitosan on the antifungal activity against Aspergillus flavus and Aspergillus parasiticus. *Molecules*, 18(4), 4437–4450.

Vieira BB, Mafra JF, Bispo ASdR, Ferreira MA, Silva FdL, Rodrigues AVN, Evangelista-Barreto NS. (2019). Combination of chitosan coating and clove essential oil reduces lipid oxidation and microbial growth in frozen stored tambaqui (Colossoma macropomum) fillets. *LWT-Food Science and Technology*, 116, 108546. https://doi.org/10.1016/j.lwt.2019.108546.

Vinsova J, Vavrikova E. (2011). Chitosan derivatives with antimicrobial, antitumour and antioxidant activities: A review. *Current Pharmaceutical Design*, 17(32), 3596–3607.

Wang J, Tian T, Wang H, Cui J, Shi X, Song J, Li T, Li W, Zhong M, Zhang W. (2022). Chitosan-coated compound fertilizer application and crop rotation alleviate continuous cotton cropping obstacles by modulating root exudates. *Rhizosphere*, 23, 10058. https://doi.org/10.1016/j.rhisph.2022.100581.

Wang J, Zhuang S. (2022). Chitosan-based materials: Preparation, modification and application. *Journal of Cleaner Production*, 355, 131825. https://doi.org/10.1016/j.jclepro.2022.131825.

Wang W, Xue C, Mao X. (2020). Chitosan: Structural modification, biological activity and application. *International Journal of Biological Macromolecules*, 164, 4532-4546, ISSN 0141-8130. https://doi.org/10.1016/j.ijbiomac.2020.09.042.

Wang X, Zheng K, Cheng W, Li J, Liang X, Shen J, Dou D, Yin M, Yan S. (2021). Field application of star polymer-delivered chitosan to amplify plant defense against potato late blight. *Chemical Engineering Journal*, 417, 129327. https://doi.org/10.1016/j.cej.2021.129327.

Wei L, Zhang J, Tan W, Wang G, Li Q, Dong F, Guo Z. (2021). Antifungal activity of double Schiff bases of chitosan derivatives bearing active halogeno-benzenes. *International Journal of Biological Macromolecules*, 179, 292–298. https://doi.org/10.1016/j.ijbiomac.2021.02.184.

Wu L, Liu M. (2008). Preparation and properties of chitosan-coated NPK compound fertilizer with controlledrelease and water-retention. *Carbohydrate Polymers*, 72, 240–247.

Xu J, Zhao X, Han X, Du Y. (2007). Antifungal activity of oligochitosan against Phytophthora capsici and other plant pathogenic fungi in vitro. *Pesticide Biochemistry and Physiology*, 87, 220–228.

Yan H, Lin G. (2014). Usage of chitosan on the complexation of heavy metal contents and vertical distribution of Hg (II) and Cr (VI) in different textural artificially contaminated soils. *Environmental Earth Sciences*, 73(5), 2483–2488.

Yen M-T, Yang J-H, Mau J-L. (2008). Antioxidant properties of chitosan from crab shells. *Carbohydrate Polymers*, 74(4), 840–844.

Youcefi F, Ouahab LW, Borsali L, Bengherbi SEI. (2022). Heavy metal removal efficiency and antibacterial activity of chitosan beads prepared from crustacean waste. *Materials Today: Proceedings*, 53(Part 1), 265–268. https://doi.org/10.1016/j.matpr.2022.01.089.

Yu D, Regenstein JM, Xia W. (2019). Bio-based edible coatings for the preservation of fishery products: A review. *Critical Reviews in Food Science and Nutrition*, 59(15), 2481–2493. https://doi.org/10.1080/10408398.2018.1457623.

Yu D, Yu Z, Zhao W, Regenstein JM, Xia W. (2021). Advances in the application of chitosan as a sustainable bioactive material in food preservation. *Critical Reviews in Food Science and Nutrition*, 62(14), 3782–3797. https://doi.org/10.1080/10408398.2020.1869920.

Zhang DW, Wang H, Hu Y, Liu Y. (2015). Chitosan controls postharvest decay on cherry tomato fruit possibly via the mitogen-activated protein kinase signaling pathway. *Journal of Agricultural and Food Chemistry*, 63(33), 7399–7404.

Zhang X, Ismail BB, Cheng H, Jin TZ, Qian M, Ali S, Liu D, Guo M. (2021). Emerging chitosan-essential oil films and coatings for food preservation: A review of advances and applications. *Carbohydrate Polymers*, 273, 118616. https://doi.org/10.1016/j.carbpol.2021.118616.

3 The Effect of the Edible Coatings and Their Combinative Effect during the Storage of Fruits and Vegetables Using Evaporative Coolant Structure

Muhammad Akram[1], Tsado Priscilla Yetu[2], Oluwafemi Adebayo Oyewole[3], Charles Oluwaseun Adetunji[3], Olowe Olugbenga Adekunle[4], Ikechukwu P. Ejidike[5], and Olaniyi Amos Fawole[6], O.O. Osemwegie[7], Saleh Alhaji Ado[8]

[1]Department of Eastern Medicine, Government College University Faisalabad, Faisalabad, Punjab, Pakistan

[2]Department of Microbiology, Federal University of Technology Minna, Niger State, Nigeria

[3]Department of Microbiology, Edo State University, Uzairue, Edo State, Nigeria

[4]Department of Chemical Sciences, Faculty of Science, Anchor University, Lagos, Nigeria

[5]Postharvest and Agroprocessing Research Centre, Department of Botany and Plant Biotechnology, University of Johannesburg, Auckland Park, Johannesburg, South Africa

[6]Department of Botany and Plant Biotechnology, South African Research Chairs Initiative in Sustainable Preservation and Agroprocessing Research, Faculty of Science, University of Johannesburg, Auckland Park, Johannesburg, South Africa

[7]Department of Food Science and Microbiology, Landmark University, Mu Aran, Kwara State, Nigeria

[8]Department of Microbiology, Ahmadu Bello University, Zaria

3.1 INTRODUCTION

Fresh product preservation is one of the main problems facing the food and agriculture industries. These perishable items can lose quality due to a variety of causes, such as enzymatic activity, moisture loss, microbial deterioration, and physical damage during postharvest storage (Urugo et al., 2024). Extending the shelf life of fresh produce while maintaining its nutritional and sensory qualities is necessary to reduce food waste and meet consumer demands for fresh products. Numerous strategies, including edible coatings and evaporative cooling systems, have been researched to get

DOI: 10.1201/9781003355458-3

around this. Creating a thin, biodegradable layer on the surface of fruits and vegetables that serves as a semi-permeable barrier to moisture, oxygen, and carbon dioxide, edible coatings, which use natural biopolymers like polysaccharides, proteins, and lipids offer an economical and sustainable method for extending the shelf life of fresh produce (Lohita & Srijaya, 2024). Edible coatings prevent microbial development, slow down ripening, and restrict weight loss by regulating gas exchange and reducing water loss. To further enhance these coatings' preservation and protection qualities, functional additives such as minerals, antioxidants, or antimicrobials may be applied (Perez-Vazquez et al., 2023). The use of evaporative cooling devices is another eco-friendly way to keep fresh fruit fresh while it's being stored. Absorbing heat through water evaporation, these devices help to create a humid yet cool environment. Fruits and vegetables respire more slowly in these conditions, delaying senescence and reducing the generation of ethylene (Ebrahimi et al., 2022). As a result of low energy consumption and affordability, evaporative cooling is a sensible option for small-scale farmers and locations with limited access to refrigeration facilities. The combination of edible coatings and evaporative cooling structures demonstrates a synergistic approach to fruit and vegetable storage. Combining the advantages of both technologies, the coating lowers moisture loss and microbial activity while the cooling system maintains optimal temperature and humidity levels (Misra et al., 2022). Together, they provide a comprehensive approach to preserving fruit freshness following harvest. This study aims to investigate the effects of edible coatings and evaporative cooling, both independently and in combination, on the storage of fruits and vegetables. Important factors like weight reduction, hardness, microbiological development, and nutritional content will be investigated in order to evaluate their efficacy. The findings will aid in the development of cost-effective and environmentally responsible storage methods, which will help address the issues of food preservation in both developed and developing countries.

3.2 EDIBLE COATINGS

Edible coatings, which are composed of natural materials such as proteins, lipids, and polysaccharides, are applied in thin layers on fruits and vegetables. They act as a semi-permeable barrier that controls the passage of gases like oxygen and carbon dioxide while reducing moisture loss and microbial contamination. Edible coatings, in contrast to synthetic packaging materials, are environmentally safe and biodegradable, which aligns with the growing emphasis on sustainability in food preservation. As an inventive and sustainable method of enhancing the freshness of fresh vegetables after harvest, edible coatings have attracted a lot of attention. Because of their capacity to form films and hold onto moisture, polysaccharide-based coatings—such as those made from cellulose, chitosan, and alginates—are frequently utilized. By preventing veggies from drying out, these coatings are especially good in preserving their texture and firmness (Ungureanu et al., 2023). Chitosan is a dual-purpose coating for disease management and preservation due to its proven antibacterial qualities (Egorov et al., 2023). Because they prevent oxidative browning and nutritional loss, protein-based coatings—like those derived from soy, casein, or whey protein—offer remarkable oxygen barrier qualities (Kandasamy et al., 2021). The main purpose of lipid-based coatings, such as fatty acids and waxes, is to provide water-resistant barriers that stop moisture loss in fruits with high transpiration rates. The addition of bioactive components, such as essential oils, antioxidants, or antimicrobials, can enhance the performance of edible coatings. These components give customers further health advantages in addition to extending the products' shelf life (Muñoz-Tebar et al., 2023). Coatings containing essential oils such as clove or cinnamon have shown substantial antibacterial activity against common microbes that cause rotting (Sarengaowa et al., 2022). Similarly, coatings that include antioxidants like ascorbic acid can stop oxidative damage without compromising the product's nutritional value. The adaptability of edible coatings is one of their main benefits. Depending on the type of produce and the desired result, they can be applied by brushing, spraying, or dipping procedures (Leya et al., 2024). Additionally, coatings can be customized to fulfill certain requirements, such as improving green vegetables' color retention or regulating ethylene production of climacteric fruits. Edible coatings are a useful tool in the contemporary food supply chain because of their effectiveness and adaptability. Edible coatings have several drawbacks in spite of their great advantages. The type of goods, storage conditions, and environmental considerations can all affect how successful they are. Moreover, it is still difficult to develop coatings with robust barrier qualities and market acceptance in terms of flavor and texture (Gali et al., 2024). By creating composite coatings and investigating cutting-edge materials like nanocomposites, ongoing research attempts to overcome these problems and enhance the overall performance of edible coatings. In conclusion, edible coatings are a practical method of extending the freshness and shelf life of fruits and vegetables. Because they decrease moisture loss, stop microbial development, and preserve nutrients, they are crucial to sustainable food preservation. Further advancements in formulation and application techniques are expected to enhance their ability to address the global issue of food waste.

3.3 EVAPORATIVE COOLING

Fresh fruit is frequently preserved with evaporative cooling since it uses less energy and is better for the environment. It is especially useful in locations where traditional refrigeration is costly or logistically difficult. This chilling method creates a cold, moist microenvironment around the food being stored by utilizing the principle of latent heat absorption during water evaporation. Evaporative cooling extends the shelf life of perishable goods by reducing respiration rates, delaying ripening, and reducing water loss while maintaining ideal temperature and relative humidity (Lufu et al., 2025). Two of the main advantages of evaporative cooling systems are their affordability and simplicity of usage. These systems are often constructed from porous materials, such as clay, bricks, or jute sacks, which are kept moist to encourage evaporation. The interior of the storage chamber becomes cooler and more humid due to heat loss from air passing through these moist surfaces. This passive cooling device is ideal for small-scale farmers and locations with erratic power supplies since, when handled manually, it consumes little to no electricity (Sibanda & Workneh, 2020). The high humidity levels of evaporative cooling systems are crucial for reducing moisture loss from fresh goods. As a result of transpiration, fruits and vegetables naturally lose water, which makes them smaller, less aesthetically pleasing, and heavier. By keeping the air moist and reducing the vapor pressure differential between the produce and its surroundings, evaporative cooling reduces water loss and maintains the freshness and texture of the food being preserved (Adeniyi et al., 2023). Senescence and spoiling are further postponed by the lower temperatures, which slow down physiological functions like respiration and ethylene production. Studies have shown that produce stored in evaporative cooling systems has superior quality attributes, such as color retention, hardness, and nutritional value, compared to fruit stored at room temperature (Patel et al., 2022). The technology is also versatile, extensible, flexible, and scalable enough to accommodate a variety of storage requirements, from small home units to large public storage facilities. Evaporative cooling is especially effective for crops that thrive in high humidity conditions, such as leafy greens, tomatoes, and cucumbers. Improving the efficiency and usability of evaporative cooling systems has been one of the primary objectives. Examples of contemporary designs include improved insulation, water-retentive materials, and electrical control systems that regulate temperature and humidity. Evaporative cooling has also been studied in conjunction with other preservation techniques, such as edible coatings or modified environment packing, to maximize its benefits and minimize its disadvantages (Adeniyi et al., 2023. In conclusion, evaporative cooling is a cost-effective and sustainable way to preserve fresh fruit. Its ability to slow down spoiling processes and provide a favorable microenvironment makes it an essential tool for postharvest management, particularly in rural and developing countries. Initiatives for global food security might significantly reduce postharvest losses.

3.4 FRUITS AND VEGETABLES

Fruits and vegetables are essential components of a balanced diet because they are rich in vitamins, minerals, dietary fiber, and bioactive compounds that promote overall health and wellness. According to Kaparapu et al. (2020), they are crucial in preventing chronic illnesses like diabetes, heart disease, and several forms of cancer. Due to their high water content and low calorie content, most fruits and vegetables are also great for promoting hydration and weight management. However, because of their perishability, they can provide significant challenges to the food supply chain, and poor handling and storage can lead to significant losses after harvest (Etefa et al., 2022). Since fresh fruits and vegetables are living tissues, their metabolic processes continue after harvest. These processes, which include respiration and transpiration, lead to a decline in quality over time. Respiration burns up stored carbohydrates to produce heat and reduce the nutritional quality of the produce, whereas transpiration causes moisture loss that leads to withering, shrinkage, and a loss of freshness (Rana et al., 2021). These factors, along with their susceptibility to mechanical damage and microbial spoilage, make effective postharvest handling and storage procedures crucial to maintaining their quality and prolonging their shelf life. Since the texture, size, and composition of fruits and vegetables vary greatly, storage and preservation techniques must be tailored. For instance, during ripening, the synthesis and respiration rate of ethylene increase in climacteric fruits such as tomatoes, bananas, and mangoes, necessitating careful regulation of temperature, humidity, and ethylene exposure (Alonso-Salinas et al., 2024). On the other hand, non-climacteric vegetables, such as cucumbers, lettuce, and strawberries, need to be stored at constant conditions to avoid rapid quality degradation. Storage practices must also take into account the sensitivity of certain objects to chilling damage, which can occur when they are exposed to temperatures below their recommended storage range (Rodrigues et al., 2022). To overcome these obstacles, novel preservation methods have been created, such as evaporative cooling, packing in a modified environment, and edible coatings. While preserving the food's texture and appearance, edible coatings offer a layer of defense against moisture loss and microbial contamination (Nunes et al., 2023). Similarly, evaporative cooling systems efficiently reduce the respiration and transpiration rates of fruits and vegetables by generating a cool, humid microenvironment (Alonso-Salinas et al., 2024). These methods lessen dependency on artificial packaging materials and energy-intensive refrigeration systems, increase product shelf life, and encourage sustainable and eco-friendly behaviors. A variety of factors, such as harvest time, storage conditions, and maturity, can affect the nutritional value and bioactive components of fruits and vegetables. Leafy greens like spinach and kale are rich in vitamins A, C, and K, while citrus fruits are a fantastic source of vitamin C and flavonoids (Etefa et al., 2022). Storage practices need to be modified to preserve vital nutrients because improper handling could lead to significant losses. Vitamin C, an important antioxidant, is particularly susceptible to heat, light, and oxygen. This emphasizes how important it is to keep the right storage conditions in order to sustain its levels (Nunes et al., 2023). In addition to being nutrient-dense, fruits and vegetables are essential to the world economy, particularly in developing countries where smallholder farmers depend mostly on them for their livelihood. However, postharvest losses—which can range from 30% to 40% in some regions—are a major barrier that reduces both food availability and economic returns (Olakiumide, 2021). Investments in infrastructure, technology, and training are necessary for the efficient delivery of fresh food and the mitigation of these losses. In summary, fruits and vegetables are vital for human nutrition and health, but they need to be handled carefully after harvest to maintain their quality and reduce losses. Advances in sustainable storage practices and preservation technology could address these problems and ensure that consumers always have access to high-quality, fresh produce while also increasing the profits of farmers and other supply chain operators.

3.5 FOOD PRESERVATION

Because it prolongs the shelf life of perishable goods while preserving their nutritional value and visual appeal, food preservation is crucial to the global food supply chain. It is essential for maintaining food security, lowering postharvest losses, and avoiding waste in areas with poor distribution and storage facilities (Bianchi and Venturi, 2025). Over time, preservation techniques have changed dramatically, combining more contemporary technology like vacuum packaging, refrigeration, and chemical treatments with more conventional techniques like drying and salting. The three main drivers of food deterioration—oxidation, enzymatic activity, and microbiological development—are the emphasis of each technique (Bilal & Iqbal, 2020). Refrigeration is one of the most often used contemporary techniques for food preservation. Low temperatures successfully extend the freshness of fruits, vegetables, dairy products, and meats by lowering microbial activity and metabolic processes (Agriopoulou et al., 2020). For the transportation and storage of perishable commodities, refrigeration has become essential. However, due to its high energy consumption, it is more expensive and less accessible in places with little resources. Modern innovations like evaporative cooling systems, which use the principle of water evaporation to create a cool, moist atmosphere for preserving fresh fruit, especially in tropical climates, provide more environmentally friendly solutions. Pasteurization and canning are two common thermal processing methods that employ heat to destroy bacteria and inhibit enzymes in order to preserve both liquid and solid goods. Milk, juices, and other liquids benefit greatly from pasteurization since it preserves quality and guarantees safety. However, by sealing them in airtight containers, canning increases the shelf life of a wider variety of products, including fruits, vegetables, and prepared meals (Rajput et al., 2022). Despite their efficacy, these treatments may result in textural alterations and nutrient losses, therefore, research into reducing these effects is still continuing. For high-value products like coffee, herbs, and some fruits, freeze-drying is the best approach since it preserves the food's original structure, flavor, and nutritional value, even if it is more expensive. By eliminating moisture, which is essential for microbial growth, modern technologies like vacuum drying and spray drying have enhanced dehydration, one of the oldest processes, and increased efficiency and product quality (Fathi et al., 2022). The use of chemical agents and natural additives to prevent oxidation and microbial growth is another important advancement in food preservation. To extend the shelf life of processed foods, preservatives including ascorbic acid, citric acid, and sodium benzoate are frequently employed (Acar et al., 2022). Due to customer desire for clean-label products, natural alternatives such as essential oils, plant extracts, and bioactive substances are becoming more and more well-liked. These materials not only guarantee quality and safety but also support the increasing emphasis on environmentally friendly and sustainable preservation techniques. Emerging technologies such as high-pressure processing (HPP), pulsed electric fields (PEF), and radiation are altering the way food is preserved. These non-thermal methods effectively eradicate viruses and decomposing organisms while maintaining the food's flavor and nutritional content. HPP eliminates bacteria without significantly altering the food's natural state, which makes it particularly helpful for meats, juices, and prepared foods (Bilal & Iqbal, 2020). Similarly, PEF has shown promise in the preservation of liquid foods like milk and juices by employing electrical pulses to degrade microbial cell membranes (Fathi et al., 2022). Food preservation technology has advanced, however issues still exist. The type of food, the intended market, and the resources at hand all influence the preservation method selection. Research and innovation are always focused on finding a way to satisfy the need for a longer shelf life while maintaining nutritional and sensory qualities. Furthermore, the development of sustainable solutions like edible coatings, biodegradable packaging, and devices powered by renewable energy is being driven by environmental concerns including energy consumption and packaging waste. To sum up, food preservation is crucial to contemporary food systems because it ensures safe and efficient food transit, satisfying the demands of a world population that is expanding. The discipline keeps developing while tackling issues with food security, waste reduction, and environmental sustainability by fusing state-of-the-art technologies with conventional knowledge. Future developments in this field have the potential to drastically change how we prepare and eat food, guaranteeing its quality and availability for future generations.

3.6 MOISTURE LOSS

Moisture loss is a significant concern when handling and storing fresh produce after harvest since it directly impacts the fruits and vegetables' quality, weight, and marketability. This phenomenon is caused by the natural process of transpiration, in which water vapor escapes the surface of the produce and enters the surrounding environment. Temperature, relative humidity, air circulation, and the physical properties of the produce itself all have a significant impact on the rate of moisture loss (Kays & Paull, 2004). Excessive moisture loss reduces the fresh food's visual attractiveness and consumer acceptance through wilting, shrinkage, and texture changes. Transpiration is driven by the vapor pressure differential (VPD) between the product and its environment. Higher VPD values, which signify drier ambient air, accelerate moisture loss. Leafy plants, such as lettuce and spinach, are particularly at risk due to their high water content and vast surface area. However, fruits with thick or waxy skins, such as apples and oranges, flow more slowly due to their natural defenses (Ben-Yehoshua et al., 1985). The storage environment must be carefully regulated, which includes lowering the temperature and maintaining a high relative humidity, in order to stop moisture loss and preserve the product's freshness. One of the most important factors affecting moisture loss is temperature. Higher temperatures cause water to evaporate more quickly, which causes the product to lose weight and decay more quickly. By reducing respiration and transpiration rates, lowering storage temperatures has been demonstrated to postpone senescence and spoiling (Hertog et al., 2001). Chilling damage, which can happen when tropical fruits and vegetables are kept at temperatures lower than their tolerance levels, must be avoided, though. Packaging techniques have been developed to prevent moisture loss by reducing water vapor exchange and establishing a microclimate. Vacuum packing and MAP are commonly used to maintain humidity and regulate oxygen and carbon dioxide levels. This prolongs shelf life and lessens weight loss (Sandhya, 2010). Furthermore, edible coatings derived from natural polymers like chitosan and alginate provide a semi-permeable barrier that lowers transpiration and maintains fruit freshness. Produce's nutritional value and visual appeal are impacted by moisture loss. Vitamin C and other B vitamins are among the water-soluble nutrients that are most susceptible to deterioration during storage, particularly when moisture loss takes place. To preserve these nutrients and guarantee the nutritious value of fruits and vegetables, optimal storage conditions must be maintained. Researchers are looking into creative ways to prevent fresh food from drying out. For instance, by providing greater control over water vapor permeability, coatings and films based on nanotechnology have demonstrated potential in improving the barrier qualities of packing materials (Emamifar et al., 2010). In order to limit moisture loss in a sustainable and economical way, evaporative cooling chambers and other controlled environment storage systems also lower temperature while preserving high humidity levels. Moisture loss needs to be addressed for the sake of world food security. In both rich and developing nations, postharvest losses—including those brought on by water loss—are a serious problem. Simple measures like using shade, humidifiers, and water-retentive materials can greatly reduce moisture loss and improve the quality of stored commodities in resource-constrained areas with limited access to refrigeration and sophisticated packaging (Kitinoja et al., 2011). In summary, one of the main factors influencing the quality of fresh fruit after harvest is moisture loss. Its effects can be lessened and consumers can be guaranteed access to wholesome, premium fruits and vegetables by comprehending the underlying concepts and utilizing the relevant technologies. To advance this field, more investigation and ingenuity are required.

3.7 MICROBIAL CONTROL

In order to stop or lessen the growth of harmful and rotting microorganisms in food products, microbial management is a crucial component of food preservation. These microorganisms, which include bacteria, fungus, and yeasts, can cause foodborne diseases, spoiling, and financial losses if they are not controlled. A range of techniques, including biological, chemical, and thermal procedures, as well as cutting-edge technologies designed to guarantee food safety and prolong shelf life, can be used to achieve microbial control (Fathi et al., 2022). In the food business, thermal techniques including pasteurization, sterilization, and blanching continue to be some of the most effective means of bacterial control. These procedures guarantee the stability and safety of food products by using heat to destroy or inactivate germs. In the dairy and juice industries, for example, pasteurization is frequently employed to get rid of bacteria like *Salmonella* spp. and *Listeria monocytogenes* without drastically changing the flavor of the product (Lewis & Heppell, 2000). Heat treatments work well, but they might damage sensitive nutrients, which is why non-thermal solutions are becoming more and more popular. In processed foods, chemical preservatives have been widely utilized to stop bacteria from growing. Chemicals such as sodium benzoate, potassium sorbate, and nitrates are commonly used to prevent the metabolism and growth of microorganisms. According to Lück and Jager (1997), these substances work especially well in acidic foods like fruit juices, sauces, and dressings. Research into natural antibacterial agents, such as plant extracts, essential oils, and bioactive peptides, has been prompted by consumer concerns about synthetic chemicals. These compounds offer a more "clean-label" alternative without sacrificing efficacy (Burt, 2004). Using naturally occurring bacteria and their metabolites to prevent harmful pathogens from decomposing, biological approaches to microbial management have grown in popularity recently. Lactic acid bacteria (LAB), for instance, generate organic acids, hydrogen peroxide, and bacteriocins. According to Delves-Broughton et al. (1996), bacteriocins such as nisin are permitted to be used as natural preservatives in meat, dairy, and other perishable foods. Infections in fresh fruit and foods originating from animals can also be managed with probiotics and competitive exclusion methods. New technologies including HPP, PEF, and ultraviolet (UV) irradiation are revolutionizing the field of microbial control in food preservation. For example, HPP dissolves cell membranes and enzyme systems under high pressure, rendering bacteria inactive without generating a lot of heat (Rastogi et al., 2007). Similarly, because PEF includes the short-term permeabilization of the membranes of microbiological cells, it is effective with liquid foods such as milk and juices. Another important component of germ control is packaging techniques. MAP and vacuum packing change the gaseous environment around food to prevent the growth of microorganisms. Church (1994) asserts that while declining oxygen levels inhibit aerobic microorganisms, rising carbon dioxide levels inhibit the growth of fungi and decomposing bacteria. Using antimicrobial films and coatings with substances like natural extracts or silver nanoparticles adds an extra layer of protection, especially for fresh and minimally processed foods (Appendini & Hotchkiss, 2002). In order to address the problems caused by antimicrobial resistance (AMR), a developing global concern, microbial control is especially important. The misuse of chemical preservatives and antibiotics has led to the creation of resistant bacteria, necessitating the development of alternative approaches. To efficiently tackle resistant microbes while reducing the danger of resistance development, researchers are investigating the use of natural antimicrobials, biopreservation techniques, and combination treatments (Martínez et al., 2015). To sum up, maintaining food quality, ensuring food safety, and reducing waste in the food supply chain all depend on microbial control. Technological developments and an increasing emphasis on natural and sustainable solutions are influencing the direction of this profession. More research and innovation will be required to address emerging problems like antibiotic resistance and satisfy consumer demands for minimally processed, preservative-free foods.

3.8 SUSTAINABLE STORAGE SOLUTIONS

Particularly in areas with limited resources, sustainable storage options are crucial for lowering postharvest losses, limiting environmental impact, and fostering food security. Despite being in use for a millennium, traditional storage systems frequently fail to meet the demands of contemporary urban and agricultural consumption patterns, which need comprehensive, energy-efficient solutions. Modern developments concentrate on fusing cutting-edge technologies with eco-friendly methods. Evaporative cooling systems are an economical and energy-efficient substitute for refrigeration in the preservation of fresh food because they employ the natural process of water evaporation to create a cold and humid atmosphere. In tropical and arid areas with inadequate refrigeration infrastructure, these devices are extremely beneficial (Kitinoja et al., 2011). Storage systems that run on renewable energy are becoming more popular as an environmentally responsible substitute for traditional storage options. For instance, by using solar energy to maintain low temperatures, solar-powered cold storage facilities lessen their reliance on fossil fuels and the electrical grid. These devices are particularly helpful in remote locations with limited electrical supplies yet a need for effective food preservation. Furthermore, hybrid strategies that combine wind and biomass energy improve the feasibility of sustainable storage even further (Deshmukh & Thombre, 2013). To increase energy efficiency, innovations like phase-change materials (PCMs) are also being used. PCMs offer reliable temperature control with little energy input by storing thermal energy during off-peak hours and releasing it during high demand (Sharma et al., 2009). Using biodegradable materials and cutting down on packaging waste are also essential elements of sustainable storage solutions. Advances in intelligent and active packaging technologies, including bio-based sheets reinforced with ethylene absorbers or antibacterial chemicals, offer the dual benefits of increasing shelf life and reducing environmental impact. Traditional methods, such as clay-pot refrigeration and underground storage pits, are also being reevaluated and enhanced using modern engineering approaches to boost efficiency while preserving their environmental benefits (Singh et al., 2014). Sustainable storage solutions combine traditional techniques with cutting-edge innovations to address both immediate and long-term problems in the food supply chain. These developments guarantee that everyone on the planet has equitable access to wholesome food while lowering greenhouse gas emissions and food waste.

3.9 CONCLUSION

To sum up, using edible coatings—either by themselves or in conjunction with evaporative cooling systems—offers a useful way to prolong the shelf life and quality of fruits and vegetables while they are being stored. While the evaporative cooling structure maintains ideal temperature and humidity levels, hence reducing postharvest losses, edible coatings aid in preventing moisture loss, delaying rotting, and preserving nutritional content. When combined, these methods provide a cost-effective, sustainable, and energy-efficient way to preserve fresh fruit while improving customer quality, cutting waste, and boosting food security.

REFERENCES

Acar, C., Dincer, I., & Mujumdar, A. (2022). A comprehensive review of recent advances in renewable-based drying technologies for a sustainable future. *Drying* Technology, 40(6), 1029–1050.

Adeniyi, M. J., Adetunji, C. O., Olaniyan, O. T., Adetuyi, B. O., Adetunji, J. B., Omosigho, O. P., ... & Dauda, W. P. (2023). Effect of an evaporative cooling structure as a storage technique in the preservation of fresh produce/horticulture to benefit human health. In Charles Oluwaseun Adetunji, Daniel Ingo Hefft, Duncan Onyango Mbuge, & Tilahun Seyoum Workneh (eds), *Evaporative Coolers for the Postharvest Management of Fruits and Vegetables* (pp. 249–272). Elsevier Science.

Agriopoulou, S., Stamatelopoulou, E., Sachadyn-Król, M., & Varzakas, T. (2020). Lactic acid bacteria as antibacterial agents to extend the shelf life of fresh and minimally processed fruits and vegetables: Quality and safety aspects. Microorganisms, 8(6), 952.

Alonso-Salinas, R., López-Miranda, S., Pérez-López, A. J., & Acosta-Motos, J. R. (2024). Strategies to delay ethylene-mediated ripening in climacteric fruits: Implications for shelf life extension and postharvest quality. Horticulturae, 10(8), 840.

Appendini, P., & Hotchkiss, J. H. (2002). Review of antimicrobial food packaging. *Innovative Food Science & Emerging Technologies*, 3(2), 113–126.

Ben-Yehoshua, S., Burg, S. P., & Young, R. (1985). Resistance of citrus fruit to mass transport of water vapor and other gases. *Plant Physiology*, 77(1), 9–16.

Bianchi, A., & Venturi, F. (2025). Enhancing the Shelf Life of Food Products: Strategies, Challenges, and Innovations. *Foods* (Basel, Switzerland), 14(23), 4034. https://doi.org/10.3390/foods14234034

Bilal, M., & Iqbal, H. M. (2020). State-of-the-art strategies and applied perspectives of enzyme biocatalysis in food sector—current status and future trends. *Critical Reviews in Food Science and Nutrition*, 60(12), 2052–2066.

Burt, S. (2004). Essential oils: Their antibacterial properties and potential applications in foods—: A review. *International Journal of Food Microbiology*, 94(3), 223–253.

Church, N. (1994). Developments in modified atmosphere packaging and related technologies. *Trends in Food Science & Technology*, 5(11), 345–352.

Delves-Broughton, J., Blackburn, P., Evans, R. J., & Hugenholtz, J. (1996). Applications of the bacteriocin, nisin. *Antonie van Leeuwenhoek*, 69(2), 193–202.

Deshmukh, S. S., & Thombre, S. B. (2013). Solar energy assisted low-temperature grain storage: A review. *Renewable and Sustainable Energy Reviews*, 16(6), 4583–4590.

Ebrahimi, A., Zabihzadeh Khajavi, M., Ahmadi, S., Mortazavian, A. M., Abdolshahi, A., Rafiee, S., & Farhoodi, M. (2022). Novel strategies to control ethylene in fruit and vegetables for extending their shelf life: A review. *International Journal of Environmental Science and* Technology, 19(5), 4599–4610.

Egorov, A. R., Kirichuk, A. A., Rubanik, V. V., Rubanik Jr, V. V., Tskhovrebov, A. G., & Kritchenkov, A. S. (2023). Chitosan and its derivatives: Preparation and antibacterial properties. Materials, 16(18), 6076.

Emamifar, A., Kadivar, M., Shahedi, M., & Soleimanian-Zad, S. (2010). Evaluation of nanocomposite packaging containing Ag and ZnO on shelf life of fresh orange juice. *Innovative Food Science & Emerging Technologies*, 11(4), 742–748.

Etefa, O. F., Forsido, S. F., & Kebede, M. T. (2022). Postharvest loss, causes, and handling practices of fruits and vegetables in Ethiopia: Scoping review. *Journal of Horticultural Research*, 30(1), 1–10.

Fathi, F., Ebrahimi, S. N., Matos, L. C., Oliveira, M. B. P. P., & Alves, R. C. (2022). Emerging drying techniques for food safety and quality: A review. *Comprehensive Reviews in Food Science and Food* Safety, 21(2), 1125–1160.

Gali, S. P., Reshma, R. G., Krishnamoorthy, S., & Meenatchi, R. (2024). Coat the flavor, preserve the quality: Elevating food stability with cutting-edge coating materials. In Mousumi Sen (ed.), *Food Coatings and Preservation Technologies* (pp. 211–258). John Wiley & Sons.

Hertog, M. L. A. T. M., Nicholson, S. E., & Whitmore, K. (2001). The effect of modified atmosphere packaging on the quality of fruits and vegetables. *Postharvest Biology and Technology*, 21(3), 287–297.

Kandasamy, S., Yoo, J., Yun, J., Kang, H. B., Seol, K. H., Kim, H. W., & Ham, J. S. (2021). Application of whey protein-based edible films and coatings in food industries: An updated overview. Coatings, 11(9), 1056.

Kaparapu, J., Pragada, P. M., & Geddada, M. N. R. (2020). Fruits and vegetables and its nutritional benefits. In Chukwuebuka Egbuna, & Genevieve Dable Tupas (eds.), *Functional Foods and Nutraceuticals: Bioactive Components, Formulations and Innovations* (pp. 241–260). Springer International Publishing.

Kays, S. J., & Paull, R. E. (2004). *Postharvest BiologyBiology*. Exon Press.

Kitinoja, L., Saran, S., Roy, S. K., & Kader, A. A. (2011). Postharvest technology for developing countries: Challenges and opportunities in research, outreach, and advocacy. *Journal of the Science of Food and Agriculture*, 91(4), 597–603.

Lewis, M. J., & Heppell, N. J. (2000). *Continuous Thermal Processing of Foods: Pasteurization and UHT Sterilization*. Springer.

Leya, B., Nivetha, T. U., Freeda Blessie, R., & Pragalyaashree, M. M. (2024). Edible coating deposition methods: Dipping, spraying, fluidized bed, and panning. In Mousumi Sen (ed.), *Food Coatings and Preservation Technologies* (pp. 485–514). John Wiley & Sons.

Lohita, B., & Srijaya, M. (2024). Novel technologies for shelf-life extension of food products as a competitive advantage: A review. In Rakhi Chakraborty, Piyush Mathur, & Swarnendu Roy (eds.), *Food Production, Diversity, and Safety under Climate Change* (pp. 285–306). Springer Nature.

Lück, E., & Jager, M. (1997). *Antimicrobial Food Additives: Characteristics, Uses, Effects*. Springer.

Lufu, R., Ambaw, A., & Opara, U. L. (2025). Evaporative cooling systems for perishables in Sub-Saharan Africa: A review. *Journal of Food Process Engineering*, 48(5), e70127.

Martínez, J. L., Coque, T. M., & Baquero, F. (2015). What is a resistance gene? Ranking risk in resistomes. *Nature Reviews Microbiology*, 13(2), 116–123.

Misra, S., Pandey, P., Dalbhagat, C. G., & Mishra, H. N. (2022). Emerging technologies and coating materials for improved probiotication in food products: A review. *Food and Bioprocess Technology*, 15(5), 998–1039.

Muñoz-Tebar, N., Pérez-Álvarez, J. A., Fernández-López, J., & Viuda-Martos, M. (2023). Chitosan edible films and coatings with added bioactive compounds: antibacterial and antioxidant properties and their application to food products: A review. Polymers, 15(2), 396.

Nunes, C., Silva, M., Farinha, D., Sales, H., Pontes, R., & Nunes, J. (2023). Edible coatings and future trends in active food packaging–fruits' and traditional sausages' shelf life increasing. Foods, 12(17), 3308.

Olakiumide, O. (2021). Post-harvest loss reduction: Enhancing food security and economic sustainability. *Journal Siplieria Sciences*, 2(2), 7–17.

Patel, D. P., Jain, S. K., Lakhawat, S. S., & Wadhawan, N. (2022). A low-cost storage for horticulture commodities for enhancing farmer's income: An overview on evaporative cooling. *Journal of Food Process Engineering*, 45(10), e14134.

Perez-Vazquez, A., Barciela, P., Carpena, M., & Prieto, M. A. (2023). Edible coatings as a natural packaging system to improve fruit and vegetable shelf life and quality. Foods, 12(19), 3570.

Rajput, H., Goswami, D., Arya, M., & Randhawa, A. (2022). Technology for canning. *Global Hi-Tech Horticulture*, 6(November), 135–151.

Rana, S. S., Pradhan, R. C., & Mishra, S. (2021). Physicochemical and physiological changes during storage. In Tanweer Alam (ed.), *Packaging and Storage of Fruits and Vegetables* (pp. 1–24). CRC Press.

Rastogi, N. K., Raghavarao, K. S. M. S., Balasubramaniam, V. M., Niranjan, K., & Knorr, D. (2007). Opportunities and challenges in high-pressure processing of foods. *Critical Reviews in Food Science and Nutrition*, 47(1), 69–112.

Rodrigues, C., Gaspar, P. D., Simões, M. P., Silva, P. D., & Andrade, L. P. (2022). Review on techniques and treatments toward the mitigation of the chilling injury of peaches. *Journal of Food Processing and Preservation*, 46(8), e14358.

Sandhya, S. (2010). Modified atmosphere packaging of fresh produce: Current status and future needs. *LWT – Food Science and Technology*, 43(3), 381–392.

Sarengaowa, Wang, L., Liu, Y., Yang, C., Feng, K., & Hu, W. (2022). Screening of essential oils and effect of a chitosan-based edible coating containing cinnamon oil on the quality and microbial safety of fresh-cut potatoes. *Coatings*, 12(10), 1492.

Sharma, A., Tyagi, V. V., Chen, C. R., & Buddhi, D. (2009). Review on thermal energy storage with phase change materials and applications. *Renewable and Sustainable Energy Reviews*, 13(2), 318–345.

Sibanda, S., & Workneh, T. S. (2020). Potential causes of postharvest losses, low-cost cooling technology for fresh produce farmers in Sub-Sahara Africa. *African Journal of Agricultural* Research, 16(5), 553–566.

Singh, S. P., Mahawar, M. K., & Reddy, B. S. (2014). Evaporative cooling: An energy-saving and sustainable cooling option for fruits and vegetables storage. *Renewable and Sustainable Energy Reviews*, 39, 515–525.

Ungureanu, C., Tihan, G., Zgârian, R., & Pandelea, G. (2023). Bio-coatings for preservation of fresh fruits and vegetables. Coatings, 13(8), 1420.

Urugo, M. M., Yohannis, E., Teka, T. A., Gemede, H. F., Tola, Y. B., Forsido, S. F., & Abdu, J. (2024). Addressing post-harvest losses through agro-processing for sustainable development in Ethiopia. *Journal of Agriculture and Food Research*, 18, 101316.

4 Recent Advances in Food Coating in the Management of Diseases

Samuel Tobi Farohunbi[1], *Oluwafemi Emmanuel Ekun*[2], *Faith Jesupemi Olaitan*[3], *and Oluwadamilola Oyetayo Ezekiel*[3]

[1]Enzymology and Drug Discovery Laboratory, Department of Biological Sciences, Thomas Adewumi University, Oko-Irese, Kwara State, Nigeria

[2]Protein Chemistry and Enzymology Unit, Department of Biochemistry, Adekunle Ajasin University, Akungba-Akoko, Ondo State, Nigeria

[3]Microbiology and Parasitology Laboratory, Department of Biological Sciences, Thomas Adewumi University, Oko-Irese, Kwara State, Nigeria

4.1 INTRODUCTION

Vegetables and fruits have up to 90% water by weight and are an important component of a healthy diet highly demanded globally. Because of the presence of high-water composition in fruits and some foods, spoilage and contamination have become a recurring decimal in the food industry (Dhall, 2013). Food coatings are a class of thin-layered materials that are generally edible and are applied to the surface of food items to serve specific functions. Edible coatings have also been applied in food processing to provide an eco-friendly alternative to harmful, non-biodegradable packaging. The surface of food is coated with tiny films of edible materials to extend shelf life, improve appearance, and sometimes deliver active compounds such as antimicrobial agents. Food coatings have helped to preserve the quality of coated foods, reduce bacterial contamination, and control respiration and oxidation rates (Priya et al., 2023).

There are different kinds of materials used in food coatings, including complex carbohydrates (polysaccharides), lipids, proteins and certain synthetic substances such as biodegradable polymers (McClements, 2020). The applications of food coatings are many and varied, depending on the purpose. One important use of food coatings is their deployment as delivery agents for a myriad of health-promoting substances such as essential minerals, vitamins, antioxidants, and probiotics, among other bioactive substances. These substances are integrated into the coating material, and when ingested, they may be released by partial gut digestion to elicit their therapeutic effects (McClements, 2020). Cutting-edge technologies can be employed in the formation of food coatings for therapeutic purposes, including encapsulation and nanoemulsion. In encapsulation, the bioactive materials are entrapped in a matrix, which serves the function of preventing the deterioration of these active ingredients and controlling their eventual release. Nanoemulsion, on the other hand, is important for increasing the bioavailability of lipophilic substances, thereby promoting absorption by body cells and tissues (Sánchez-Ortega et al., 2016).

DOI: 10.1201/9781003355458-4

4.2 PROPERTIES OF EDIBLE COATINGS

Edible Coatings:

1. They are efficient as barriers to moisture, water, oxygen, carbon dioxide, and ethylene.
2. They enhance the appearance and mechanical handling of fruits and vegetables to preserve their color and structure.
3. They contain active ingredients that improve the nutritional makeup of fruits and vegetables without compromising their quality, such as vitamins and antioxidants.
4. They provide a protective coating to fruits and vegetables to increase their shelf life (Sanjay et al., 2022).

4.3 BIOMATERIALS UTILIZED IN FOOD COATINGS

In the global chain of food supply and distribution, food preservation is vital to limit waste, maintain its original quality, and consequently extend its shelf life. Edible and therapeutic food coatings have become game changers in this regard owing to their multidirectional properties they possess in addition to their ease of biodegradation and safety. Food products are preserved because of the barriers these coatings put up against microbial invasion, oxygen, and moisture. Biomaterials in common use are derived from polysaccharides, lipids, and proteins.

4.3.1 Polysaccharides Employed in Food Coatings

Commonly utilized polysaccharides in food coatings include cellulose, alginate, starch, and chitosan. Others include Pectin and Aloe Vera. They are widely used in coating food products largely because of their compatibility with biological systems, bioavailability, and ease of forming stable films (Priya et al., 2023).

4.3.1.1 Cellulose

One of the most prevalent sources of carbon on Earth is cellulose. This glucose homopolymer is the key structural element that gives plant cell walls their rigidity and form. With up to 15,000 D-glucose units, the linear polymer is connected by β-(1–4) glycosidic bonds (Koolman and Roehm, 2005) (Figure 4.1).

In the cellulose molecule, there are interchain hydrogen bonds, contributed by parallel chains, along with Van der Waals forces that keep the molecule stable. The hydrogen bonds form crosslinks, which gives cellulose high tensile strength and forms a compact, cohesive structure that is insoluble in water, despite being hydrophilic (Koolman and Roehm, 2005). These desired properties make cellulose a widely used biomaterial for food coating. Fruits and vegetables are commonly coated

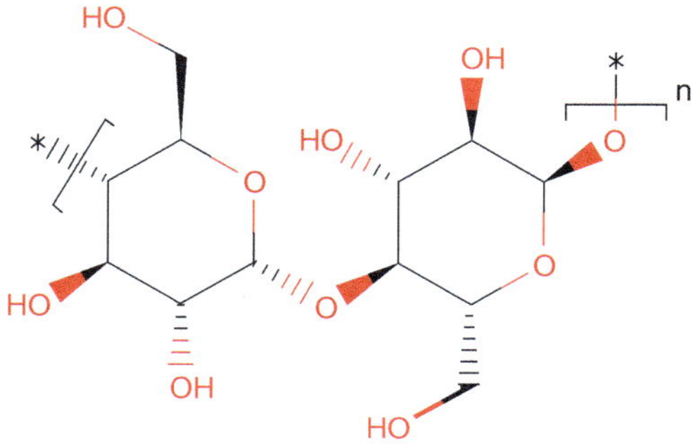

FIGURE 4.1 Chemical structure of cellulose (MarvinSketch 18.8, 2018) (http://www.chemaxon.com).

using cellulose, because they impede both the movement of moisture and the exchange of gases like O_2 and CO_2. Derivatives of cellulose, which are in industrial use as edible food coatings, include methylcellulose, carboxymethyl cellulose, hydroxypropyl cellulose, and hydroxypropyl methylcellulose (Kumar, 2019). A study by Panahirad et al. (2020) revealed that carboxymethyl cellulose (1% concentration) used to coat plum fruits caused an increase in its phenolic and flavonoid content, which in turn enhances the antioxidant capacity of the fruits when consumed. In similar fashion, hydroxyethyl cellulose was used to coat strawberries, which conferred antimicrobial potential on these fruits against the fungus *Penicillium italicum* (Liu et al., 2021). Hydroxyethyl cellulose derived from asparagus waste was used to coat strawberries, controlling color changes, preserving phenolic and flavonoid content, and increasing the shelf life to 8 days at 80% relative humidity and 25°C (Liu et al., 2021).

Wang et al. (2023) have also reported the application of cellulose-based edible coatings in the safety of deep-fried foods. Cellulose-based coatings have good foaming properties and are easily used to incorporate active ingredients in food and drugs for controlled release in the body (Bahmid et al., 2024).

4.3.1.2 Alginates

Alginates are water-soluble polysaccharides primarily obtained from several brown colored species of seaweed like *Laminaria hyperborean, Ascophyllum nodosum, Macrocystis pyrifera*, etc. Structurally, alginates are composed of α-L-guluronic acid and β-D-mannuronic acid residues with linear structures joined by β-(1–4) glycoside linkages (Hay et al., 2013) (Figure 4.2).

The functional properties of alginate are attributed to the amount of L-guluronic acid (G) and D-mannuronic acid (M). A lower M/G ratio favors a stronger alginate gel, whereas a more flexible gel is determined by a higher mannuronic acid content in the polymer (Parreidt et al., 2018). They have the capacity to complex with divalent metal ions to synthesize stable hydrogels at 25°C without the requirement of heating or addition of chemicals. This exceptional property makes alginates valuable materials for use as food coatings (Coles, 2003; Parreidt et al., 2018). Alginate-based coatings have been explored in the preservation of several food items including cheese, fruits, vegetables, meat, seafood, and poultry products (Parreidt et al., 2018). Alginate coatings in meat and

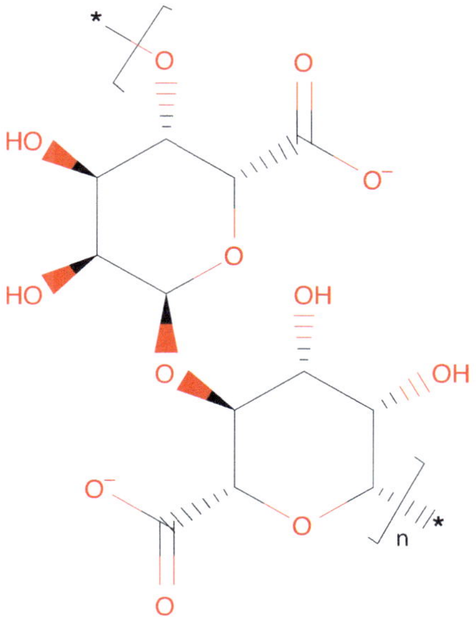

FIGURE 4.2 Alginate structure (MarvinSketch 18.8, 2018) (http://www.chemaxon.com).

fish products serve as oxygen barriers, reducing lipid oxidation and microbial spoiling; for example, alginate coatings and essential oils have been shown to work in concert to delay fish fillet spoilage, preserving its nutritional value and sensory appeal (Chidanandaiah et al., 2009; Ehsani et al., 2017). Alginates are also used in bakery and confectionery products to stop moisture migration, enhancing texture and prolonging freshness (Hambleton et al., 2009).

4.3.1.3 Starch

Starch is one of the most readily available and biodegradable biopolymers in nature, and it finds immense applications as in agriculture and other bioprocesses as an important raw material (Versino et al., 2016). It is a polysaccharide composed of two macromolecular structures, the linear polymer amylose and the highly branched polymer amylopectin. Amylose is made up of α-D-glucose units joined by α-(1–4) glycosidic bonds, while amylopectin has α-1,6-glycosidic connections at the branch points and α-(1–4) glycosidic bonds at the linear chain (Koolman and Roehm, 2005). The physicochemical characteristics of starch, such as its capacity to form films, mechanical strength, and water vapor permeability, are greatly influenced by the ratio of amylopectin to amylose (Pérez and Bertoft, 2010). Granules of starch are insoluble in cold water, but when heated in water, they gelatinize, breaking down their structure and enabling the polysaccharides to form a gel or viscous solution. Because it makes it easier to create continuous films with the right mechanical and barrier qualities, this gelatinization technique is essential for creating starch-based edible coatings (Jane, 2004; Versino et al., 2016). Starch-based coatings are biodegradable, non-toxic, odorless, and colorless, and they are commonly used to extend the shelf life and enhance the quality of various foods and products (Guo, 2019). One primary application of starch coatings is their use in preserving fresh vegetables and fruits to prevent moisture loss, exchange of gas, and microbiological contamination (Arvanitoyannis et al., 1994).

4.3.1.4 Chitosan

Chitosan is an organic polymer primarily obtained from the shells of crustaceans, especially crabs and shrimps. It consists of alternatively repeating units of *N*-acetyl-D-glucosamine and D-glucosamine connected by β-(1–4) glycoside linkages (Harugade et al., 2023) (Figure 4.3).

Chitosan has natural antimicrobial, semipermeable and mucoadhesive properties, which are attributed to the presence of functional groups like amino and hydroxyl groups. These groups enable chitosan to participate in reactions, including ether formation, amination, and esterification reactions, which contribute to the functional properties of the polysaccharide (Shariatinia, 2019). Chitosan has been demonstrated in several studies to improve the shelf life of a number of perishable fruits (such as strawberries), in part by limiting the exchanges of gases such as carbon (IV) oxide and oxygen (Xu et al., 2019).

4.4 LIPID-BASED FOOD COATINGS

The lipids are generally hydrophobic molecules that are naturally suited to serve as food coatings, as they exclude moisture from the food product (Rhim and Shellhammer, 2005). In practice, they are commonly used in combination with proteins and polymeric carbohydrates because lipids lack high tensile strength when compared to other biomolecules used in edible coating (Yousuf and Qadri, 2020). However, lipids as edible coatings prevent moisture and add to the attractiveness of the food product. Fatty acids, petroleum waxes, resins, and mineral oils are some of the lipids used to coat food products (Rhim and Shellhammer, 2005; Yousuf et al., 2021). Fruits such as tomatoes and cherries are some of the several examples of food products, which have been preserved using lipid-based coatings (Yaman and Bayoındırlı, 2002; Rhim and Shellhammer, 2005; Ruiz-Martínez et al., 2020).

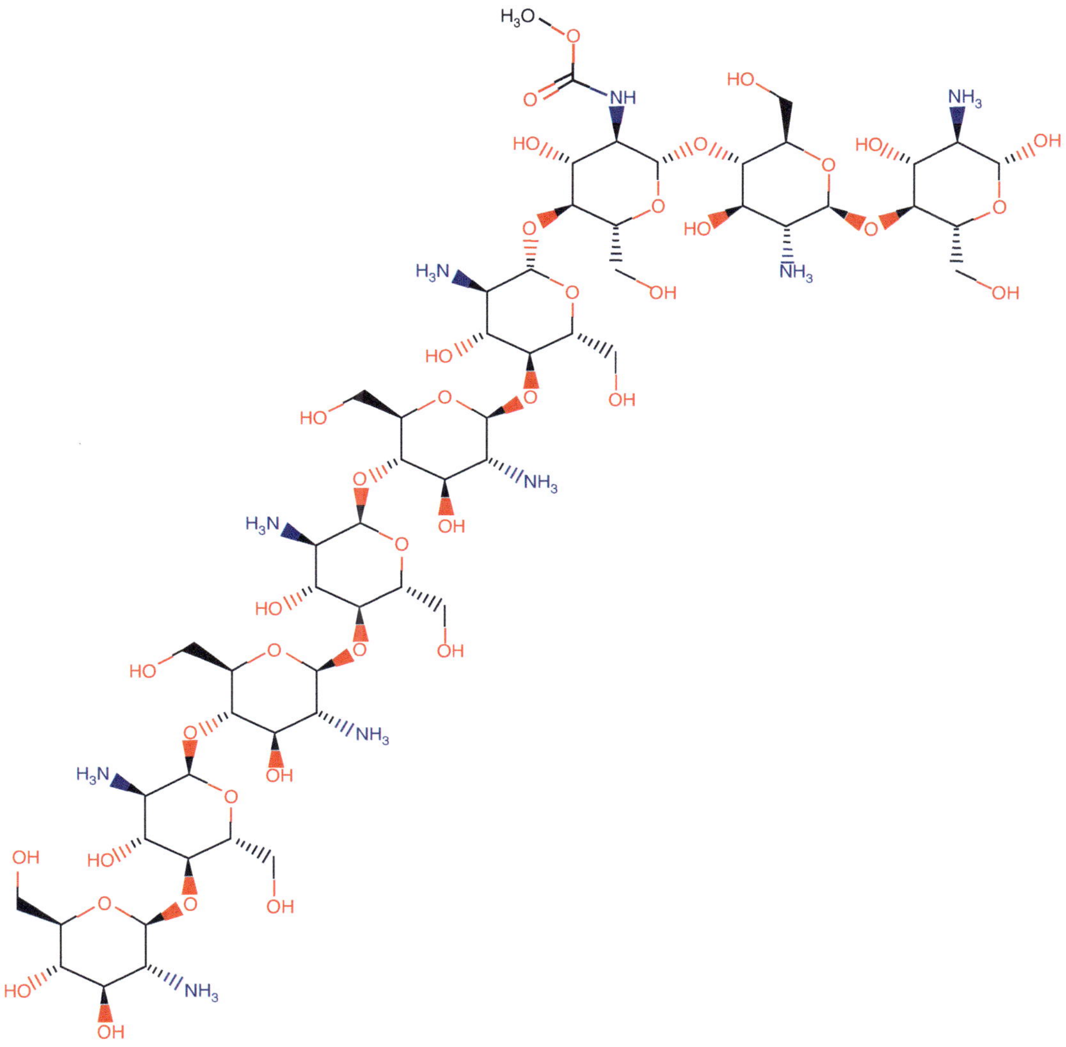

FIGURE 4.3 Chitosan structure (MarvinSketch 18.8, 2018) (http://www.chemaxon.com).

4.5 PROTEIN FOOD COATINGS

Protein-based coatings such as gelatin, casein, whey, and soy, as well as certain plant proteins, have been shown to increase food quality and prolong food products' shelf life, while reducing oxidative damage (Hassan et al., 2018). They have the ease of being used in combination with other materials such as lipids, as well as bioactive components such as antimicrobial agents or antioxidant, for value-added, health promoting benefits (Rhim et al., 2013). Corn gluten flour is the source of zein protein. Because corn-zein protein has a high O_2 barrier, it can help fruits and vegetables last longer on the shelf by reducing color changes, hardness, and weight loss (Sanjay et al., 2022). Pérez-Mateos et al. (2009) found that antimicrobial protein coatings on meat products and seafood help to limit microbial growth and prevent lipid oxidation. This provided improved texture and extended shelf life of these products.

4.6 CURRENT UNDERSTANDING OF THE FOOD COATING PROCESSES

The food coating process is a function of the coating type, the food product itself, and the expected properties of the food to be preserved. There are several methods employed in food coating, including the dipping method, spraying, electrostatic coating, and pan coating methods.

The *dipping method* is both the simplest and the most common method used in coating most food products. It is particularly the method of choice in the coating of fruits and vegetables as well as snacks that require a thick and uniform coating (Krochta et al., 1995; Ju et al., 2019). This method ensures that it has one of its greatest advantages, the ability to ensure the complete and homogenous coating of food products. In the dipping method, it involves the complete immersion of the food substance in a solution of the coating material within a given time frame, after which the product is allowed to dry and then the coating is set (Ju et al., 2019). The dipping method may be summarized in three steps. First, the sample prepared for coating is first submerged in a coating solution. Next, the contact made with the coating mixture installs a layer of edible coating on the food product. Lastly, excess coating solution is removed from the food product by means of drying or gentle heating (Priya et al., 2023).

The major factor affecting the coating method by dipping is the length of time during which the product makes contact with the coating solution. This factor is immediately dependent upon properties like density of the solution, surface tension, and viscosity of coating solution (Suhag et al., 2020). Longer contact times means that the product may also absorb more moisture, which is detrimental to the shelf life of the food. Too short an exposure time will cause an uneven coating of the food product. It is also in the coating solution that antioxidant materials and antimicrobial agents are introduced, which ultimately form the coating material (Priya et al., 2023).

In the *spraying method*, nozzles are used to spray the food item with the coating solution at increased pressures. This method is efficient for large-scale production and is used for the application of lipid coatings to snacks and baked food products, so as to improve uniformity in appearance and texture while extending their shelf life. Spray coating is noted for its precision and ability to reduce waste (Suhag et al., 2020). Factors affecting the efficiency of spray methods include the physicochemical characteristics of the spray solution in use (viscosity, temperature), nozzle design, and duration of spray and drying method, among other conditions (Ju et al., 2019).

The *Electrostatic coating method* involves the application of charged coating particles to oppositely charged food surfaces. An electrostatic field is applied to charge the coating material, which is then applied to the food product, having an opposite or neutral charge. This causes the food product to attract the coating material, ensuring a uniform application in the process (Tan et al., 2018). This method provides equal distribution and good adherence, making it suitable for applying seasonings, powders, or fine coatings. Electrostatic coating is often used in the snack food industry to add spice blends to chips or nuts. This method's main advantage is the enhanced adhesion of the coating particles to the food surface as its mechanism takes advantage of the opposite polarities between them (Ubbink and Krüger, 2006; Tan et al., 2018).

4.7 THERAPEUTIC PROSPECTS OF EDIBLE FRUIT AND VEGETABLE FOOD COATINGS

There is a growing interest in the fortification of food coatings with active ingredients, with proven health promoting benefits. Recent innovations in food coating technology have tended toward the incorporation of plant extracts, which may have beneficial effects such as probiotic potential (Massoud et al., 2023), antimicrobial (Valdés et al., 2017), and antioxidant effects (Ganiari et al., 2017). Edible coatings are also applied to fresh and processed meat to extend shelf life. These coatings minimize moisture loss, reduce lipid oxidation and discoloration, and protect volatile flavors. They also function as carriers for antimicrobial and antioxidant agents, safeguarding meat from physical, chemical, and microbiological damage (Shendurse et al., 2018) (Figure 4.4).

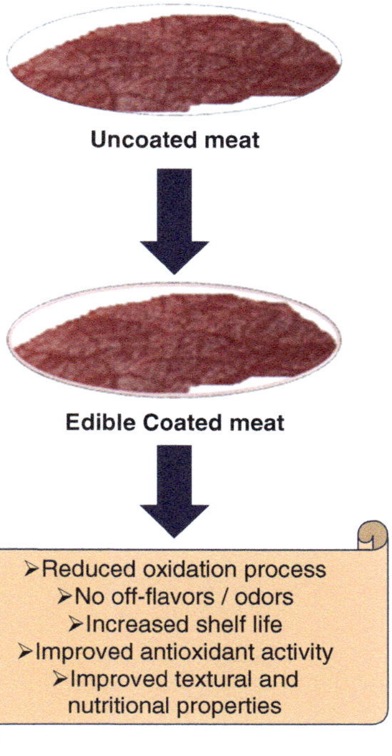

FIGURE 4.4 Merits of edible coating on meat (Priya et al., 2023).

Phenolics and flavonoids, which have antioxidant potentials, are being added to edible food coatings to prevent oxidation of the food material and to aid in scavenging free radicals when ingested, combating cellular oxidative stress in the process (Eça et al., 2014; Ganiari et al., 2017). In similar fashion, plant-derived antimicrobial agents such as essential oils, Nisin, carvacrol, and Trans-cinnamaldehyde have been added to coatings applied on fruits, which protect them from microbial attack (Valdés et al., 2017). Experimentally, probiotic strains such as *Lactobacillus* have been used in the preservation of fruit slices such as apple, lettuce, and pear. And the results showed that they reduced the growth of pathogenic bacteria such as *Listeria* species (Massoud et al., 2023). The prospects of edible food coatings should leverage on their ability to facilitate controlled release of bioactive agents using cutting edge technologies such as encapsulation and nanoemulsion. Areas needed to be explored in the value-added potentials of edible food coatings should include the aspects of the incorporation of plant-derived agents with cardioprotective activities, antidiabetic properties, and anti-neoplastic effects, among other therapeutic potentials. Some edible coatings contain compounds with known anti-inflammatory properties, such as curcumin (from turmeric), ginger, or certain polyphenols. These compounds act by inhibiting the activity of pro-inflammatory cytokines, enzymes (e.g., COX-2), and signaling pathways (such as NF-kB), which play central roles in the inflammatory response (Khare et al., 2020).

4.8 PROSPECTS OF EDIBLE COATING FOODS IN THE TREATMENT OF MALARIA

Plasmodium parasites produce malaria (WHO, 2023b), which is spread by the bites of infected Anopheles mosquitoes and continues to pose a major global health concern, particularly in tropical and subtropical areas (WHO, 2023c). Traditional treatments include antimalarial drugs such as artemisinin and its derivatives (Chinaeke et al., 2014). While antimalarial drugs have proven effective, issues like drug resistance, poor bioavailability, and non-compliance necessitate innovative delivery strategies. One such emerging solution is the application of edible coatings on food for the delivery of antimalarial agents (Mbassi et al., 2023).

In medicinal applications, these coatings can act as carriers for bioactive compounds or drugs, making food a vehicle for therapeutic delivery (Li et al., 2023; Sorrentino et al., 2018). By incorporating active pharmaceutical ingredients (APIs) into consumable food coatings, this approach offers a promising, patient-friendly pathway for malaria treatment. However, challenges like drug resistance and limited accessibility call for innovative solutions. One such emerging approach is the use of edible coatings for drug delivery, offering potential advantages in improving the efficacy and bioavailability of antimalarial drugs. Active antimalarial drugs, such as artemisinin, chloroquine, or quinine, can be incorporated into edible coatings. These coatings allow controlled release of drugs when consumed. It enhances the stability of drugs by protecting them from environmental degradation (e.g., heat and moisture). It also provides a palatable way to administer bitter antimalarial drugs (Zambrano-Zaragoza et al., 2018; Avramescu et al., 2020).

Functional foods like fruits, nuts, or snacks can be coated with layers containing natural antimalarial agents, such as:

- Curcumin: Found in turmeric and known for its anti-inflammatory and antimalarial properties.
- *Artemisia annua* Extracts: A natural source of artemisinin.
- Quinine: A plant-derived antimalarial compound (Ghosh and Banerjee, 2021; Patel and Singh, 2022b).

These foods serve as dual-purpose products, offering nutrition alongside malaria treatment. Children and pregnant women are among the most affected by malaria and often face difficulties with conventional medication. Edible-coated foods offer a non-invasive delivery system that can be tailored to provide age-appropriate dosages and improve adherence by integrating treatment into regular dietary habits (Berbudi et al., 2020; Shah et al., 2020).

The delivery of antimalarial drugs in a palatable and accessible manner, particularly for susceptible populations such as infant and expectant mothers, is an innovation (Ghosh and Banerjee, 2021). Edible coatings can enhance malaria treatment by controlling release mechanisms, reducing peak drug concentrations, minimizing adverse reactions, and masking bitter drug taste (e.g. quinine), making them palatable and accessible, particularly for vulnerable populations such as infant and expectant mothers (Ghosh and Banerjee, 2021).

The application of edible coatings in malaria treatment offers a novel approach to improving the efficacy, stability, and accessibility of antimalarial drugs. However, comprehensive studies on drug interactions and stability are essential to ensure the compatibility of antimalarial agents with food-based components. This method, which integrates treatment within food systems, has the potential to enhance patient compliance and address key challenges in malaria management, particularly among vulnerable populations.

4.9 EFFECTS OF EDIBLE COATING FOODS IN THE TREATMENT OF TUBERCULOSIS

On an international scale, tuberculosis (TB), a fatal respiratory illness caused by *Mycobacterium tuberculosis*, causes a significant amount of morbidity and mortality. Globally, a projected 10.6 million people received a TB diagnosis in 2021, with 46% of cases recorded occurring in South-East Asia. In a year, one active TB patient may infect 5–15 people. By 2035, the World Health Organization's (WHO) End TB Strategy targets to reduce TB-related mortality by 95% and TB incidence by 90% (Bourguignon et al., 2023; Usharani et al., 2023).

With 450,000 new cases of rifampicin-resistant TB (RR-TB) reported in 2021, drug-resistant TB (DR-TB) remains a serious problem (Muthukrishnan, 2021). Patients with isoniazid-susceptible TB ought to undergo a six-month regimen of rifampicin (R), ethambutol (E), pyrazinamide (Z), and isoniazid (H), according to the WHO's recommended standardized treatment strategy. Current anti-tubercular treatments are lengthy, complicated, and often associated with significant side effects, which contribute to low patient adherence (Bourguignon et al., 2023).

However, recent advancements in drug delivery systems, including edible coatings, offer innovative solutions to enhance the efficacy and accessibility of TB treatment. Edible coatings are thin, consumable layers made from natural or synthetic materials applied to foods, drugs, or bioactive compounds. Commonly used materials include Polysaccharides (Alginate, chitosan, and cellulose derivatives), Proteins (Gelatin, whey protein, and soy protein), and Lipids (Beeswax and stearic acid).

These coatings can encapsulate drugs, enhancing their stability, bioavailability, and controlled release while minimizing side effects. Examples include:

- Chitosan-based nanoparticles coated with rifampicin, which allow for sustained drug release, reducing the frequency of doses (Patel and Singh, 2022a,b).
- Alginate-encapsulated pyrazinamide, which improves drug solubility and absorption in the gastrointestinal tract.

Anti-TB drugs are often sensitive to environmental conditions such as heat and light. Edible coatings act as protective barriers, preserving the drugs' integrity during storage and transit. The bitter taste of TB medications can reduce adherence to the regimen. Edible coatings can mask unpleasant flavors, making treatment more palatable, particularly for children and elderly patients (Khan and Ahmed, 2021; Muthukrishnan, 2021).

Advancements in edible coatings integrated with nanotechnology enable drugs to target infection sites more precisely, reducing systemic side effects. pH-sensitive coatings, for instance, release drugs in specific regions of the gastrointestinal tract, enhancing absorption (Lin and Zhao, 2007). By incorporating TB medications or natural anti-TB agents, edible coatings transform functional foods into therapeutic tools. Anti-TB drugs like rifampicin and isoniazid can be encapsulated within edible coatings for controlled release. These coatings improve the palatability of medications when integrated with foods like fruits or snacks (Khan and Ahmed, 2021; Patel and Singh, 2022b). Edible coatings can also incorporate natural compounds with anti-TB properties, such as:

- Curcumin: Known for its anti-inflammatory and antimicrobial effects.
- Garlic-Derived Allicin: Shown to inhibit *M. tuberculosis* growth.
- Green Tea Polyphenols: Potential adjuncts in TB therapy (WHO, 2023a).

Foods coated with these bioactive agents can offer adjunctive support to conventional treatments. Nutrient-dense foods, such as nuts or energy bars, can be coated with anti-TB agents to address malnutrition commonly associated with TB patients. This dual-action approach combines nutritional supplementation with therapeutic delivery, protects drugs from gastric acid, enhances absorption in the gastrointestinal tract, ensures consistent plasma drug levels, reduces dosing frequency, and minimizes systemic exposure and side effects (WHO, 2023a; Bourguignon et al., 2023; Lin and Zhao, 2007).

4.10 CONCLUSION

The food packaging industry is a rapidly evolving one, and modifications to the methods and contents of food coatings are inevitable. It is imperative that processing industries should shift attention to more health-enhancing bioactives that can be incorporated into edible food coatings, which will have beneficial effects, not just on the food product to be preserved, but also possess therapeutic potentials as consumers ingest foods fortified in this fashion.

REFERENCES

Arvanitoyannis, I., Kalichevsky, M., Blanshard, J. M. and Psomiadou, E. (1994). Study of diffusion and permeation of gases in undrawn and uniaxially drawn films made from potato and rice starch conditioned at different relative humidities. *Carbohydrate Polymers*, *24*(1), 1–15.

Avramescu, S. M., Butean, C., Popa, C. V., Ortan, A., Moraru, I. and Temocico, G. (2020). Edible and functionalized films/coatings: Performances and perspectives. *Coatings*, *10*(7), 687. https://doi.org/10.3390/coatings10070687

Berbudi, A., Fatih, M. Q. and Memvanga, A. (2020). The role of curcumin as an antimalarial agent. *Systematic Reviews in Pharmacy*, *11*(7), 20–37 https://doi.org/10.31838/srp.2020.7.04.

Bahmid, N. A., Siddiqui, S. A., Ariyanto, H. D., Sasmitaloka, K. S., Rathod, N. B., Wahono, S. K., and Indrianingsih, A. W. (2024). Cellulose-based coating for tropical fruits: method, characteristic and functionality. *Food Reviews International, 40*(4), 1069–1092.

Bourguignon, T., Godinez-Leon, J. A. and Gref, R. (2023). Nanosized drug delivery systems to fight tuberculosis. *Pharmaceutics*, *15*(2), 393. https://doi.org/10.3390/pharmaceutics15020393.

Chidanandaiah, Keshri, R. C. and Sanyal, M. K. (2009). Effect of sodium alginate coating with preservatives on the quality of meat patties during refrigerated (4±1c) storage. *Journal of Muscle Foods*, *20*, 275–292.

Chinaeke, E. E., Chime, S. A., Onyishi, V. I., Attama, A. A. and Okore, V. C. (2014). Formulation development and evaluation of the anti-malaria properties of sustained release artesunate-loaded solid lipid microparticles based on phytolipids. *Drug Delivery*, *22*(5), 652–665. https://doi.org/10.3109/10717544.2014.881633

Coles, R. (2003). Introduction. In: *Food Packaging Technology*; Coles, R., McDowell, D., Kirwan, M. J., Eds.; Blackwell Publishing: London, pp. 1–31.

Dhall, R. K. (2013). Advances in edible coatings for fresh fruits and vegetables: A review. *Critical Reviews in Food Science and Nutrition*, *53*(5), 435–450.

Eça, K. S., Sartori, T. and Menegalli, F. C. (2014). Films and edible coatings containing antioxidants: A review. *Brazilian Journal of Food Technology*, *17*, 98–112.

Ehsani, A., Paktarmani, M. and Yousefi, M. (2017). Efficiency of dietary sodium alginate coating incorporated with lycopene in preserving rainbow trout. *Food Science and Biotechnology*, 26, 557–562.

Ganiari, S., Choulitoudi, E. and Oreopoulou, V. (2017). Edible and active films and coatings as carriers of natural antioxidants for lipid food. *Trends in Food Science & Technology*, *68*, 70–82.

Ghosh, A. and Banerjee, S. (2021). Antimalarial drug delivery using food-based edible coatings: A novel strategy. *Journal of Drug Delivery Science and Technology*, *63*, 102489. https://doi.org/10.1016/j.jddst.2021.102489

Guo, M. (Ed.). (2019). *Whey Protein Production, Chemistry, Functionality, and Applications*. John Wiley & Sons, Hoboken, NJ.

Hambleton, A., Debeaufort, F., Bonnotte, A. and Voilley, A. (2009). Influence of alginate emulsion-based films structure on its barrier properties and on the protection of microencapsulated aroma compound. *Food Hydrocolloids,* 23, 2116–2124.

Harugade, A., Sherje, A. P. and Pethe, A. (2023). Chitosan: A review on properties, biological activities and recent progress in biomedical applications. *Reactive and Functional Polymers*, *191*, 105634.

Hassan, B., Chatha, S. A. S., Hussain, A. I., Zia, K. M. and Akhtar, N. (2018). Recent advances on polysaccharides, lipids and protein based edible films and coatings: A review. *International Journal of Biological Macromolecules, 109*, 1095–1107.

Hay, I. D., Rehman, Z. U., Moradali, M. F., Wang, Y. and Rehm, B. H. (2013). Microbial alginate production, modification and its applications. *Microbial Biotechnology*, *6*, 637–650.

Jane, J. L. (2004). *Starch: Structure and Properties*. CRC Press LLC, Boca Raton, FL.

Ju, J., Xie, Y., Guo, Y., Cheng, Y., Qian, H. and Yao, W. (2019). Application of edible coating with essential oil in food preservation. *Critical Reviews in Food Science and Nutrition, 59*, 2467–2480.

Khan, S. and Ahmed, R. (2021). Alginate-based edible films for controlled delivery of anti-TB drugs. *International Journal of Biological Macromolecules*, *178*, 341–353.

Khare, T., Palakurthi, S. S., Shah, B. M., Palakurthi, S. and Khare, S. (2020). Natural product-based nanomedicine in treatment of inflammatory bowel disease. *International Journal of Molecular Sciences*, *21*(11), 3956.

Koolman, J. and Roehm, K. (2005). *Color Atlas of Biochemistry* (2nd edition). Geog Thieme, New York, p. 42.

Krochta, J. M., Baldwin, E. A. and Nisperos-Carriedo, M. O. (1995). *Edible Coatings and Films to Improve Food Quality*. CRC Press, Boca Raton, FL.

Kumar, N. (2019). Polysaccharide-based component and their relevance in edible film/coating: A review. *Nutrition & Food Science*, *49*(5), 793–823.

Li, T., Chen, Y. and Chen, Q. (2023). Applications of prolamin-based edible coatings in food preservation: A review. *MDPI Foods*, *12*(15), 23121. https://doi.org/10.3390/foods13152321

Lin, D., and Zhao, Y. (2007). Innovations in the development and application of edible coatings for fresh and minimally processed fruits and vegetables. *Comprehensive Reviews in Food Science and Food Safety*, *6*(3), 60–75.

Liu, C., Jin, T., Liu, W., Hao, W., Yan, L. and Zheng, L. (2021). Effects of hydroxyethyl cellulose and sodium alginate edible coating containing asparagus waste extract on postharvest quality of strawberry fruit. *LWT*, *148*(2021), Article 111770. https://doi.org/10.1016/j.lwt.2021.111770

MarvinSketch (Version 18.8). (2018). Software for drawing and displaying chemical structures, developed by ChemAxon. https://www.chemaxon.com/products/marvin/marvinsketch/

Massoud, R., Khodaeii, D., Hamidi-Esfahani, Z. and Khosravi-Darani, K. (2023). The effect of edible probiotic coating on quality of fresh fruits and vegetables: Fresh strawberries as a case study. *Biomass Conversion and Biorefinery*, *13*(4), 2517–2526.

Mbassi, D. E., Pfaffendorf, C. and Mombo-Ngoma, G. (2023). Real-life effectiveness of anti-malarial treatment regimens: What are we aiming for? *Malaria Journal*, *22*, 189. https://doi.org/10.1186/s12936-023-04606-2

McClements, D. J. (2020). Nanotechnology approaches for improving the health benefits of functional foods. *Annual Review of Food Science and Technology*, *11*(1), 53–73.

Muthukrishnan, L. (2021). Multidrug resistant tuberculosis: Diagnostic challenges and its conquering by nanotechnology approach: An overview. *Chemico-Biological Interactions*, *337*, 109397. https://doi.org/10.1016/j.cbi.2021.109397

Panahirad, S., Naghshiband-Hassani, R., Bergin, S., Katam, R. and Mahna, N. (2020). Improvement of postharvest quality of plum (Prunus domestica L.) using polysaccharide-based edible coatings, *Plants*, *9*, 1–16.

Parreidt, S. T., Müller, K. and Schmid, M. (2018). Alginate-based edible films and coatings for food packaging applications. *Foods*, *7*(10), 170.

Patel, R. and Singh, A. (2022a). Edible coatings for enhanced drug delivery: Applications in tuberculosis therapy. *Journal of Drug Delivery Science and Technology*, *65*, 102701.

Patel, R. and Singh, V. (2022b). Applications of polysaccharide-based edible coatings in pharmaceutical foods. *International Journal of Biological Macromolecules*, *178*, 45–58.

Pérez, S. and Bertoft, E. (2010). The molecular structures of starch components and their contribution to the architecture of starch granules: A comprehensive review. *Starch-Stärke*, *62*(8), 389–420.

Pérez-Mateos, M., Montero, P. and Gómez-Guillén, M. C. (2009). Formulation and stability of biodegradable films made from cod gelatin and sunflower oil blends. *Food Hydrocolloids*, *23*(1), 53–61.

Priya, K., Thirunavookarasu, N. and Chidanand, D. V. (2023). Recent advances in edible coating of food products and its legislations: A review. *Journal of Agriculture and Food Research*, *12*, 100623.

Rhim, J. W., Park, H. M. and Ha, C. S. (2013). Bio-nanocomposites for food packaging applications. *Progress in Polymer Science*, *38*(10–11), 1629–1652.

Rhim, J. W. and Shellhammer, T. H. (2005). Lipid-based edible films and coatings. In: *Innovations in Food Packaging*. Seoul: Academic Press, pp. 362–383.

Ruiz-Martínez, J., Aguirre-Joya, J. A., Rojas, R., Vicente, A., Aguilar-Gonzalez, M. A., Rodríguez-Herrera, R., Alvarez-Perez, O. B., Torres-Leon, C. and Aguilar, C. N. (2020). Candelilla wax edible coating with *Flourensia cernua* bioactives to prolong the quality of tomato fruits. *Foods*, *9*, 1–12.

Sánchez-Ortega, I., García-Almendárez, B. E., Santos-López, E. M., Reyes-González, L. R. and Regalado, C. (2016). Characterization and antimicrobial effect of starch-based edible coating suspensions. *Food Hydrocolloids, 52*, 906–913.

Sanjay, P., Saxena, D., and Kazimi, R. (2022). Enhancing shelf life of fresh fruits by the application of different edible coatings. *International Journal of Innovation Science*, *11*, 626–632.

Shah, A., Zhang, C. and Li, J. (2020). Edible bioactive film with curcumin: A potential functional packaging? *MDPI Foods*, *13*(7), 7513. https://doi.org/10.3390/foods13223613

Shariatinia, Z. (2019). Pharmaceutical applications of chitosan. *Advances in Colloid and Interface Science*, *263*, 131–194.

Shendurse, A., Gopikrishna, G., Patel, A. C., and Pandya, A. J. (2018). Milk protein based edible films and coatings–preparation, properties and food applications. *Journal of Nutritional Health Food Engineering*, *8*(2), 219–226.

Sorrentino, A., Gorrasi, G. and Vittoria V. (2018). Nanosystems in edible coatings: A novel strategy for food preservation. *International Journal of Molecular Sciences, 19*(3), 705. https://doi.org/10.3390/ijms19030705

Suhag, R., Kumar, N., Petkoska, A. T. and Upadhyay, A. (2020). Film formation and deposition methods of edible coating on food products: A review. *Food Research International*, *136*, 109582.

Tan, A. W. Y., Sun, W., Bhowmik, A., Lek, J. Y., Marinescu, I., Li, F. and Liu, E. (2018). Effect of coating thickness on microstructure, mechanical properties and fracture behaviour of cold sprayed Ti6Al4V coatings on Ti6Al4V substrates. *Surface and Coatings Technology*, *349*, 303–317.

Ubbink, J. and Krüger, J. (2006). Physical approaches for the delivery of active ingredients in foods. *Trends in Food Science & Technology*, *17*(5), 244–254.

Usharani, N., Kanth, S. V. and Saravanan, N. (2023). Current nanotechnological strategies using lipids, carbohydrates, proteins and metal conjugates-based carrier systems for diagnosis and treatment of tuberculosis: A review. *International Journal of Biological Macromolecules*, *227*, 262–272. https://doi.org/10.1016/j.ijbiomac.2022.12.087

Valdés, A., Ramos, M., Beltrán, A., Jiménez, A. and Garrigós, M. C. (2017). State of the art of antimicrobial edible coatings for food packaging applications. *Coatings*, *7*(4), 56.

Versino, F., Lopez, O. V., Garcia, M. A. and Zaritzky, N. E. (2016). Starch-based films and food coatings: An overview. *Starch/Staerke*, *68*, 1026–1037.

Wang, Z., Ng, K., Warner, R. D., Stockmann, R., and Fang, Z. (2023). Application of cellulose-and chitosan-based edible coatings for quality and safety of deep-fried foods. *Comprehensive Reviews in Food Science and Food Safety*, *22*(2), 1418–1437.

World Health Organization. (2023a). Global Tuberculosis Report. Retrieved from WHO website https://www.who.int.

World Health Organization. (2023b). Malaria Fact Sheet. Retrieved from WHO website https://www.who.int.

World Health Organization. (2023c). *Malaria*. World Health Organisation, Geneva. https://www.who.int/news-room/fact-sheets/detail/malaria.

Xu, Q., Ji, Y., Sun, Q., Fu, Y., Xu, Y. and Jin, L. (2019). Fabrication of cellulose nanocrystal/chitosan hydrogel for controlled drug release. *Nanomaterials*, *9*(2), 253.

Yaman, Ö. and Bayoındırlı, L. (2002). Effects of an edible coating and cold storage on shelf-life and quality of cherries. *LWT-Food Science and Technology*, *35*(2), 146–150.

Yousuf, B. and Qadri, O. S. (2020). Preservation of fresh-cut fruits and vegetables by edible coatings. In: *Fresh-Cut Fruits and Vegetables*. United States: Academic Press, pp. 225–242.

Yousuf, B., Sun, Y. and Wu, S. (2021). Lipid and lipid-containing composite edible coatings and films. *Food Reviews International*, *38*, 1–24.

Zambrano-Zaragoza, M. L., González-Reza, R., Mendoza-Muñoz, N., Miranda-Linares, V., Bernal-Couoh, T. F., Mendoza-Elvira, S., and Quintanar-Guerrero, D. (2018). Nanosystems in edible coatings: A novel strategy for food preservation. *International Journal of Molecular Sciences*, *19*(3), 705.

5 Engineering Properties of Chitosan-Based Edible Films/Coatings

John Tsado Mathew[1], Jonathan Hussaini[1], Abel Inobeme[2], Charles Oluwaseun Adetunji[3], Elijah Yanda Shaba[4], Mary Jummai Jiya[5], Yakubu Azeh[1], Musah Monday[1], Muhammad Aishetu Ibrahim[1], Etsuyankpa Muhammad Bini[6], Musa Safiyanu Tanko[6], Ismaila Aliyu Oga[6], Tanko Ezekiel[4], Muhammed Jibrin Ndejiko[7], and Amos Mamman[1]

[1]Department of Chemistry, Ibrahim Badamasi Babangida University Lapai, Niger State, Nigeria

[2]Department of Chemistry, Edo University Uzairue, Auchi, Edo State, Nigeria

[3]Department of Microbiology, Applied Microbiology, Biotechnology and Nanotechnology Laboratory, Edo University Uzairue, Auchi, Edo State, Nigeria

[4]Department of Chemistry, Federal University of Technology, Minna, Niger State, Nigeria

[5]Department of Food Science and Technology, Federal University of Technology, Minna, Niger State, Nigeria

[6]Department of Chemistry, Federal University of Lafia, Nasarawa State, Nigeria

[7]Department of Microbiology, Ibrahim Badamasi Babangida University Lapai, Niger State, Nigeria

5.1 INTRODUCTION

In recent years, the quest for sustainable and eco-friendly packaging solutions in the food industry has intensified, prompting researchers and industries to explore novel materials with remarkable properties. Among these materials, chitosan, a derivative of chitin, has emerged as a promising candidate for the development of edible films and coatings. Chitosan, obtained from the exoskeletons of crustaceans and insects or through the deacetylation of chitin, exhibits unique engineering properties that render it suitable for a wide range of applications in the realm of food packaging (Egorov et al. 2023).

Chitosan, a polysaccharide composed of β-(1–4)-linked D-glucosamine and *N*-acetyl-D-glucosamine units, inherits its distinct properties from its chitin precursor. As an abundant biopolymer found in the shells of crustaceans, chitin has long been recognized for its potential in various applications, but its insolubility in most solvents limited its practical utility. The deacetylation of chitin results in chitosan, a more versatile material due to its improved solubility and film-forming abilities. This transformation has paved the way for the exploration of chitosan in the development of edible films

DOI: 10.1201/9781003355458-5

and coatings, presenting an opportunity to address environmental concerns associated with conventional packaging materials (Jiménez-Gómez & Cecilia 2020).

The significance of chitosan in the context of edible films and coatings lies in its multifaceted engineering properties that contribute to the enhancement of food preservation, safety, and sustainability. As a biodegradable and biocompatible material, chitosan aligns with the growing demand for environmentally friendly alternatives to traditional packaging. Its film-forming ability, attributed to the unique molecular structure, provides a natural barrier against moisture, gases, and lipids, thereby extending the shelf life of perishable food products. This inherent property positions chitosan as a key player in addressing the global issue of food waste by minimizing spoilage and maintaining the quality of packaged goods. Moreover, the mechanical properties of chitosan-based films can be tailored to suit the specific requirements of different food items. The concentration of chitosan in the formulation plays a crucial role in determining the mechanical strength of the films, allowing for versatility in applications. From delicate fruits to robust meat products, chitosan-based coatings offer a customizable solution for various packaging needs (Stefanowska et al. 2023).

In addition to its physical attributes, chitosan exhibits notable antimicrobial properties. The ability of chitosan to inhibit the growth of pathogenic microorganisms adds an extra layer of food safety to the packaged products. This feature is particularly relevant in the context of minimizing the risk of contamination and ensuring the well-being of consumers. The optical and sensory attributes of chitosan-based films also contribute to their effectiveness in food packaging. Transparency, color, and flexibility are essential factors in consumer perception and acceptance. Chitosan films demonstrate favorable optical and mechanical properties, striking a balance between functionality and aesthetic appeal. This ensures that the packaging not only serves its primary purpose but also aligns with the visual and tactile expectations of consumers (Bose et al. 2023).

Furthermore, the solubility and water vapor permeability of chitosan-based films can be fine-tuned to meet the specific needs of diverse food items. This adaptability makes chitosan an attractive material for designing customized packaging solutions that cater to the unique requirements of different products. In the broader context of environmental sustainability, chitosan stands out as an eco-friendly material. Chitosan, derived from abundant marine resources, embodies the principles of a circular economy by utilizing waste from the seafood industry. Its biodegradability ensures that the end-of-life disposal of chitosan-based packaging aligns with environmental stewardship (Khubiev et al. 2023).

In addition, the exploration of chitosan-based edible films and coatings represents a significant advancement in the quest for sustainable and effective food packaging solutions. The multifaceted engineering properties of chitosan, encompassing film-forming ability, mechanical strength, antimicrobial activity, and environmental sustainability, position it as a versatile and promising material for the future of food packaging. As we delve deeper into the specific aspects of chitosan's engineering properties, a comprehensive understanding will emerge, guiding further research and development in this dynamic field (Purewal et al. 2024).

5.2 MOLECULAR STRUCTURE AND FILM-FORMING ABILITY

The unique molecular structure of chitosan is fundamental to its exceptional film-forming ability, making it a standout material in the development of edible films and coatings. Chitosan, derived from the deacetylation of chitin, is a polysaccharide composed of β-(1–4)-linked D-glucosamine and *N*-acetyl-D-glucosamine units. This distinctive arrangement imparts chitosan with properties that distinguish it from its precursor and other biopolymers (El-Araby et al. 2024).

The deacetylation process involves the removal of acetyl groups from chitin, resulting in a polymer that is more soluble in common solvents. This increased solubility is a pivotal characteristic for the film-forming capability of chitosan. Unlike chitin, which is insoluble in most solvents, chitosan can be dissolved in acidic solutions, particularly in the presence of acetic acid. This solubility opens up avenues for the fabrication of films and coatings through various processing techniques.

The film-forming process involves the creation of a thin, continuous layer of chitosan on a substrate, such as the surface of a food product. The solubility of chitosan allows for the preparation of casting solutions or suspensions, which can be subsequently applied to the desired surface and then dried or cured to form a coherent film. The resulting films exhibit a range of thicknesses and morphologies, depending on the processing parameters and the intended application (Mathaba & Daramola 2020).

Chitosan's film-forming ability is attributed to the presence of amino and hydroxyl functional groups in its molecular structure. These groups enable the formation of intermolecular and intramolecular hydrogen bonds during film formation. The hydrogen bonding contributes to the cohesion of chitosan molecules, leading to the development of a stable film. Additionally, the presence of positively charged amino groups in the chitosan structure facilitates interactions with negatively charged components in food matrices, enhancing adhesion and film stability. The molecular weight of chitosan also plays a crucial role in its film-forming properties. Higher-molecular-weight chitosan generally results in films with better mechanical strength and barrier properties. However, the solubility of chitosan decreases with increasing molecular weight, necessitating a balance in choosing the appropriate molecular weight for a given application (Aranaz et al. 2021).

Furthermore, the film-forming ability of chitosan can be influenced by the degree of deacetylation, which refers to the extent of acetyl group removal from chitin. As the degree of deacetylation increases, the solubility of chitosan improves, affecting the film's properties such as transparency, flexibility, and water vapor permeability. In practical terms, the film-forming ability of chitosan allows for the creation of thin coatings that serve as protective barriers for food products. The films can be tailored to meet specific requirements, providing a customizable solution for various applications in the food industry. Whether applied to fruits, vegetables, meat products, or other perishable goods, chitosan-based films contribute to extending shelf life, preventing spoilage, and maintaining the quality of the packaged items (Piekarska et al. 2023).

More so, the molecular structure of chitosan, characterized by its unique arrangement of glucosamine and *N*-acetylglucosamine units, underlies its exceptional film-forming ability. The solubility of chitosan, achieved through the deacetylation of chitin, enables the fabrication of films and coatings with diverse applications in the food industry. The interplay of molecular weight, degree of deacetylation, and functional groups within the chitosan structure influences the film's properties, making it a versatile material for creating protective and sustainable packaging solutions. As we delve deeper into the engineering properties of chitosan, the exploration of its molecular structure and film-forming ability sets the stage for understanding the broader implications of this biopolymer in the realm of edible films and coatings (Li et al. 2020).

5.3 MOLECULAR STRUCTURE AND ITS IMPACT ON FILM FORMATION

The molecular structure of chitosan, a derivative of chitin, plays a pivotal role in governing its film-forming ability, influencing various aspects of film properties, including structure, stability, and functionality. As we explore the impact of chitosan's molecular structure on film formation, it becomes evident that the arrangement of glucosamine and *N*-acetylglucosamine units is a key determinant of the unique properties exhibited by chitosan-based films. The chitosan molecule is composed of repeating units of β-(1–4)-linked D-glucosamine and *N*-acetyl-D-glucosamine. The deacetylation process transforms chitin, an insoluble biopolymer, into chitosan, a more soluble and versatile material. This structural modification enhances the film-forming ability of chitosan, making it amenable to various processing techniques in the creation of edible films and coatings (Egorov et al. 2023).

The impact of chitosan's molecular structure on film formation is particularly pronounced in the solubility of the polymer. The deacetylation of chitin introduces amino groups, making chitosan soluble in acidic solutions, especially in the presence of acetic acid. This solubility is a crucial prerequisite for preparing casting solutions or suspensions that can be applied to different substrates, facilitating the formation of thin films upon drying or curing. The presence of amino and hydroxyl functional groups within the chitosan structure significantly influences the intermolecular and intramolecular interactions during film formation. Hydrogen bonding, a prevalent force in the formation of chitosan-based films, contributes to the cohesion and stability of the films. The establishment of hydrogen bonds between adjacent chitosan molecules results in the creation of a continuous and uniform film, enhancing its structural integrity (Freitas et al. 2020).

Furthermore, the positively charged amino groups in the chitosan structure interact with negatively charged components in food matrices, promoting adhesion and compatibility. This interaction is particularly relevant when chitosan-based films are applied directly to the surfaces of food products, creating a protective barrier that adheres effectively and helps maintain the quality of the packaged items. The molecular weight of chitosan is another crucial factor that influences its impact on film formation. Higher-molecular-weight chitosan tends to result in films with superior mechanical strength and barrier properties. However, a trade-off exists, as higher-molecular-weight chitosan is often less soluble. Therefore, selecting an optimal molecular weight is essential, balancing the need for film performance with the practicality of processing and application (Thambiliyagodage et al. 2023).

The degree of deacetylation, representing the extent of acetyl group removal from chitin, also affects the film-forming properties of chitosan. As the degree of deacetylation increases, so does the solubility of chitosan, allowing for the preparation of casting solutions with a higher concentration of chitosan. This influences the transparency, flexibility, and water vapor permeability of the resulting films, tailoring them to meet specific requirements in diverse applications. In practical terms, the impact of chitosan's molecular structure on film formation extends to the creation of coatings with tailored properties for different food products. Whether applied as a protective layer on fruits, vegetables, or meat, the molecular characteristics of chitosan contribute to the development of films that address specific needs in terms of mechanical strength, barrier properties, and adhesion (Rehman et al. 2023).

However, the molecular structure of chitosan serves as the foundation for its remarkable film-forming ability, influencing every facet of the film's properties. The interplay of amino and hydroxyl functional groups, positively charged amino groups, molecular weight, and degree of deacetylation collectively determine the unique attributes of chitosan-based films. As a result, chitosan emerges as a versatile material with the potential to revolutionize the field of edible films and coatings, offering sustainable and effective solutions for food packaging and preservation. Understanding the intricate relationship between molecular structure and film formation is essential for unlocking the full potential of chitosan in the realm of food packaging technologies (Loo et al. 2022).

5.4 TAILORING MECHANICAL STRENGTH THROUGH CHITOSAN CONCENTRATION

The mechanical strength of chitosan-based films and coatings is a critical parameter that determines their applicability across a spectrum of food products with varying structural requirements. One of the key avenues for tailoring the mechanical strength of these films lies in adjusting the concentration of chitosan in the formulation. This parameter not only influences the strength of the films but also plays a crucial role in shaping their barrier properties, thus making it a pivotal factor in designing packaging solutions for different applications (Muñoz-Tebar et al. 2023).

Chitosan concentration refers to the percentage of chitosan present in the film or coating formulation. This concentration can be manipulated to achieve films with varying degrees of mechanical strength, catering to the specific needs of different food items. A nuanced understanding of the relationship between chitosan concentration and mechanical strength allows for the customization of films, ensuring optimal performance in diverse packaging scenarios. At lower chitosan concentrations (around 1%), the resulting films tend to exhibit lower mechanical strength. These films are suitable for applications involving delicate fruits and leafy greens, where the packaging must provide a protective layer without subjecting the produce to excessive pressure. The flexibility of these films makes them ideal for items that are susceptible to bruising or damage from external forces during handling and transportation (Xu et al. 2022).

As the chitosan concentration is increased to around 2%, a medium level of mechanical strength is achieved. This concentration range strikes a balance, making the films suitable for a broader range of applications. Bakery products and confectionery items, for instance, benefit from films with moderate mechanical strength. These films offer adequate protection without being overly rigid, allowing for the preservation of the product's structural integrity while still providing a reliable barrier against environmental factors. Stepping up the chitosan concentration to approximately 3% results in films with high mechanical strength. This range is particularly beneficial for packaging meat products and cheeses, where a robust barrier is essential to prevent spoilage and maintain product quality. The increased mechanical strength ensures that the films can withstand the handling and storage demands associated with these items, which often have higher weight and density compared to fruits and vegetables (Dybka-Stępień et al. 2021).

For applications demanding even greater mechanical strength, a chitosan concentration of 4% or higher may be employed. Films produced in this range exhibit very high mechanical strength, making them suitable for durable packaging solutions, especially in industrial settings. Products that undergo extensive shipping, storage, and handling, such as bulk goods or industrial components, benefit from the structural integrity provided by these high-chitosan-concentration films. Beyond mechanical strength, adjusting chitosan concentration also influences the barrier properties of the films. Higher chitosan concentrations often correlate with improved barrier performance against moisture, gases, and other environmental factors. This is particularly advantageous for extending the shelf life of packaged goods and minimizing the impact of external conditions on product quality (Perera et al. 2023; Inobeme et al. 2024a).

In addition, tailoring mechanical strength through chitosan concentration offers a versatile approach to designing edible films and coatings for diverse applications in the food industry. The ability to customize films based on the specific requirements of different products underscores the adaptability of chitosan as a packaging material. Whether it's providing a delicate shield for fruits or a robust barrier for industrial products, the manipulation of chitosan concentration empowers researchers and industries to craft packaging solutions that align with the mechanical demands of varied food items, contributing to enhanced food preservation and safety. As we delve deeper into the engineering properties of chitosan, the tailoring of mechanical strength stands out as a crucial aspect in the quest for effective and sustainable food packaging solutions (Souza et al. 2020).

5.5 MEETING FUNCTIONAL AND AESTHETIC REQUIREMENTS IN FOOD PACKAGING

Chitosan-based edible films seamlessly blend functional and aesthetic elements, making them versatile materials for food packaging. Transparency, a fundamental property, ensures clear visibility of the packaged product, meeting the functional requirement of showcasing the contents while enhancing visual appeal. Flexibility, another key attribute, enables the film to conform to the product's shape, meeting practical needs in handling and storage, while providing a pleasant tactile experience for consumers. The mechanical strength, inherent in chitosan films, fulfills the functional requirement of protecting the product during handling, while conveying durability for a consistent appearance. Barrier properties, crucial for preserving freshness and preventing spoilage, also contribute to a clean and neat presentation, meeting both functional and aesthetic demands. Examples of applications range from flexible packaging for delicate bakery items to extended shelf life for meats, cheeses, and fresh produce. Chitosan-based films exemplify a harmonious integration of functionality and aesthetics, offering a sustainable and effective solution in the dynamic landscape of food packaging (Table 5.1) (Muñoz-Tebar et al. 2023).

TABLE 5.1
Meeting Functional and Aesthetic Requirements in Food Packaging

Requirement	Functional Features	Aesthetic Features	Examples	References
Preservation and protection	Barrier against moisture, oxygen, light, and contaminants	Appealing colors and graphics to attract consumers	Vacuum-sealed pouches, airtight containers	Thompson-Witrick et al. (2021)
Convenience	Easy-to-open packaging	User-friendly design, visually pleasing	Resealable bags, easy-tear openings	Oloyede and Lignou (2021)
Information and communication	Clear nutritional labels, expiration dates, and usage instructions	Eye-catching graphics, branding, and product information	Colorful product imagery, brand logos	Bopape et al. (2021) and Bodunrinde et al. (2025)
Sustainability	Use of eco-friendly materials, recyclability, reduced waste	Nature-inspired designs, eco-friendly color schemes	Biodegradable packaging, recycled materials	Ibrahim et al. (2022)
Visibility and display	Transparent windows for product visibility	Attractive product presentation, unique shapes	Clear plastic containers, unique box shapes	Ma et al. (2020)
Durability	Sturdy materials that protect the product during handling	Durable graphics and designs that resist wear and tear	Thick cardboard boxes, reinforced edges	Willskytt and Brambila-Macias (2020)
Stackability and storage	Stackable designs for efficient storage	Consistent branding and design for shelf appeal	Square or rectangular shapes for easy stacking	Srivastava et al. (2022)
Brand identity	Consistent with the overall brand image and messaging	Distinctive brand colors, logos, and fonts	Coca-Cola's iconic red and white packaging	
Regulatory compliance	Adherence to food safety and labeling regulations	Clear and compliant text, symbols, and warnings	Allergen information, FDA-approved labels	Chang et al. (2023)

5.6 WATER VAPOR PERMEABILITY AND ITS IMPACT ON FOOD PRESERVATION

In the realm of food preservation, the control of moisture transfer is a critical factor influencing the quality, safety, and shelf life of various products. Water vapor permeability, a measure of a material's ability to allow the passage of water vapor, plays a pivotal role in determining the effectiveness of packaging solutions in preserving the freshness and integrity of food items (Deshwal et al. 2019; Mathew et al. 2025a).

Understanding Water Vapor Permeability: Water vapor permeability is a material property that quantifies the rate at which water vapor can move through a substance. In the context of food packaging, it refers to how easily water vapor can permeate the packaging material and interact with the enclosed product. This property is crucial because water vapor can lead to undesirable consequences, such as changes in texture, taste, and microbial spoilage (Buchwalder et al. 2023).

The Impact of Water Vapor on Food Products: Food products have unique moisture content and sensitivity levels, and exposure to excessive or insufficient moisture can significantly impact their quality. High water vapor permeability can lead to moisture loss, resulting in undesirable outcomes such as the drying out of fresh produce or the formation of ice crystals in frozen goods. Conversely, low permeability may trap moisture, leading to a loss of crispness in snacks or fostering conditions conducive to mold growth (Shi et al. 2020; Adetunji et al. 2022a).

Chitosan-Based Films and Water Vapor Permeability: Chitosan, derived from chitin, has gained prominence as a material for edible films and coatings, offering a sustainable and effective solution to mitigate the impact of water vapor on food preservation. The unique properties of chitosan allow for the modulation of water vapor permeability, making it a versatile choice for packaging applications. Chitosan's film-forming ability allows the creation of thin, protective layers that can be tailored to specific water vapor permeability requirements. By adjusting factors such as chitosan concentration, molecular weight, and processing conditions, researchers and manufacturers can fine-tune the permeability of chitosan-based films to align with the moisture needs of diverse food products (Kaur et al. 2024).

Low Water Vapor Permeability for Dry Products: Chitosan-based films can be engineered to exhibit low water vapor permeability, making them highly effective for products that require minimal moisture exposure, such as dry snacks or powdered goods. By incorporating chitosan into the packaging, the films create a robust barrier against external moisture, thereby preserving the product's texture, flavor, and shelf life. The low water vapor permeability of chitosan films helps prevent moisture ingress, which is crucial for maintaining the quality and stability of moisture-sensitive products. This protective feature ensures that goods remain unaffected by environmental humidity, reducing the risk of degradation due to moisture-induced changes. Consequently, chitosan-based films provide a valuable solution for extending the shelf life of products while maintaining their desired attributes (Prakoso et al. 2023).

Moderate Water Vapor Permeability for Fresh Produce: In the case of fresh produce, maintaining an optimal level of moisture is crucial to prevent wilting and maintain crispness. Chitosan films can be designed with a moderate water vapor permeability, allowing for controlled exchanges of moisture between the product and its environment. This controlled permeability helps strike a balance, preventing excessive moisture loss while still allowing for respiration processes in fruits and vegetables, contributing to extended freshness (Guivier et al. 2024).

High Water Vapor Permeability for Perishable Goods: Certain food items, such as bakery products or some types of cheeses, benefit from controlled moisture release. Chitosan-based films can be engineered with higher water vapor permeability to facilitate the escape of excess moisture, preventing the formation of a humid environment that could lead to textural deterioration or mold growth. This controlled permeability ensures that the packaging actively contributes to preserving the quality and safety of the enclosed perishable goods (Munteanu & Vasile 2021; Ogundolie et al. 2023).

Antimicrobial Properties Enhancing Preservation: In addition to regulating water vapor permeability, chitosan offers inherent antimicrobial properties that enhance its effectiveness in food packaging. These antimicrobial characteristics inhibit the growth of spoilage microorganisms and pathogens, providing an extra layer of protection to food products. By combining controlled moisture levels with its natural ability to prevent microbial growth, chitosan-based packaging effectively extends the shelf life of a wide range of food items. The controlled moisture exchange helps maintain the desired texture and quality of the food, while the antimicrobial action minimizes the risk of contamination and spoilage. This dual functionality makes chitosan an excellent choice for preserving the freshness and safety of various food products, ensuring that they remain in optimal condition for longer periods (Fadiji et al. 2023).

Water vapor permeability stands out as a crucial parameter in the design of effective food packaging solutions, influencing the preservation of products across various categories. Chitosan-based films, with their tunable properties and inherent advantages, offer a versatile means of addressing the challenges associated with moisture control in food packaging. From low permeability for dry goods to moderate and high permeability for fresh produce and perishable items, chitosan provides a customizable solution that aligns with the unique preservation needs of different food products.

As we continue to explore innovative approaches in food packaging, the impact of water vapor permeability and the role of materials like chitosan remain central to achieving optimal preservation outcomes (Long et al. 2023).

5.7 CHITOSAN AS AN ECO-FRIENDLY MATERIAL

In the quest for sustainable and environmentally friendly materials, chitosan has emerged as a frontrunner, showcasing its versatility and ecological benefits. Chitosan, derived from chitin—a naturally abundant biopolymer found in the exoskeletons of crustaceans and insects—offers a compelling alternative in the realm of food packaging, aligning with the global shift toward sustainable practices (Maliki et al. 2022).

Renewable Source and Biodegradability: Chitosan's eco-friendliness stems from its renewable source—byproducts of the seafood industry. Shells from shrimp, crab, and other crustaceans, which are otherwise considered waste, become the raw material for chitosan production. This repurposing not only reduces the environmental impact of seafood waste but also aligns with the principles of a circular economy. Furthermore, chitosan is inherently biodegradable. Unlike conventional petroleum-based plastics, which persist in the environment for centuries, chitosan-based materials break down into natural components, minimizing their environmental footprint. This characteristic addresses the global concern of plastic pollution and contributes to the reduction of long-lasting waste in landfills and oceans (Zhao et al. 2022; Mathew et al. 2025b).

Reduced Dependency on Petrochemicals: The production of traditional plastics is largely dependent on petrochemicals, leading to the depletion of fossil fuel resources and increased greenhouse gas emissions. In contrast, chitosan is a bio-based material derived from natural sources, such as chitin found in crustacean shells. Utilizing chitosan as an alternative to conventional plastics helps reduce reliance on non-renewable resources and lowers the environmental impact associated with plastic production. By incorporating chitosan into food packaging, industries can significantly cut their carbon footprint and contribute to a more sustainable future. Chitosan's biodegradability further enhances its environmental benefits, as it breaks down naturally without leaving harmful residues, aligning with the growing demand for eco-friendly packaging solutions and promoting a circular economy (Trotta et al. 2023).

Energy-Efficient Processing: The extraction and processing of chitosan are generally more energy-efficient compared to the production of synthetic polymers. The conversion of chitin into chitosan involves a deacetylation process that is typically conducted under mild conditions, reducing the need for high temperatures and extensive energy input. This lower energy requirement not only makes the production process more sustainable but also contributes to the overall eco-friendliness of chitosan-based materials. By opting for chitosan, industries can significantly reduce their energy consumption and environmental footprint, aligning with sustainability goals. The energy-efficient processing of chitosan underscores its attractiveness as a greener alternative to conventional plastics, catering to eco-conscious industries seeking to minimize their environmental impact while maintaining functional performance (Kozma et al. 2022).

Customizable Biopolymer Properties: Chitosan's versatility extends to its customizable properties, allowing it to be tailored for specific applications in food packaging. This adaptability ensures that chitosan-based materials meet the functional requirements of packaging without compromising on eco-friendliness. Whether designing films with specific barrier properties or coatings with targeted adhesion, chitosan provides a platform for creating sustainable packaging solutions that cater to diverse needs (Díaz-Montes & Castro-Muñoz 2021; Adetunji et al. 2022b).

Edible Films Reducing Food Waste: Chitosan's use in edible films presents a promising solution for combating food waste. These films are designed to be both biodegradable and antimicrobial, providing multiple benefits to perishable goods. By extending the shelf life of food products, chitosan-based edible films help prevent spoilage and maintain product quality. Their biodegradable nature ensures that, even after their useful life, the films break down naturally without contributing to environmental pollution. This combination of extended freshness and environmental responsibility helps reduce the amount of food that is discarded due to deterioration. Consequently, chitosan-based edible films not only support sustainability but also play a crucial role in minimizing food waste, aligning with efforts to enhance food security and reduce overall waste in the supply chain (Moura-Alves et al. 2023).

Chitosan's Antimicrobial Advantage: Chitosan's antimicrobial properties offer dual benefits for both food preservation and environmental impact. By effectively inhibiting the growth of spoilage microorganisms and pathogens, chitosan reduces the need for chemical preservatives in food products. This aligns with the growing consumer preference for clean-label products, which are free from synthetic additives. Additionally, the use of chitosan minimizes the release of synthetic preservatives into the environment, contributing to greater environmental sustainability. Reducing the reliance on chemical preservatives not only lessens the potential for chemical contamination but also supports more sustainable and eco-friendly food packaging solutions. Thus, chitosan plays a vital role in both enhancing food safety and reducing the ecological footprint associated with conventional food preservation methods (Ke et al. 2021; Mathew et al. 2023a).

Promoting Circular Economy Practices: Chitosan's eco-friendly profile supports the principles of a circular economy, which emphasizes the reduction, reuse, and recycling of materials. Chitosan-based packaging materials can be designed with recyclability in mind, facilitating their reuse and minimizing waste generation. This approach aligns with the circular economy's goal of managing materials in a closed-loop system, where products are designed for longevity, easy recycling, or composting. By incorporating chitosan into packaging solutions, industries can contribute to a more sustainable lifecycle for materials, reducing overall environmental impact. Chitosan's biodegradability ensures that, when not reused, the material breaks down naturally, further supporting the reduction of waste and promoting environmental stewardship. This makes chitosan an integral component in advancing circular economy practices and sustainable resource management (Ezeudu & Ezeudu 2019; Adetunji et al. 2023a).

Challenges and Future Perspectives: Despite its many eco-friendly advantages, chitosan faces challenges including cost, scalability, and specific application limitations. The relatively high cost of raw materials and processing can limit its widespread adoption, while scalability issues arise in producing chitosan-based products at a large scale. Additionally, certain application-specific limitations, such as performance under varied environmental conditions, need to be addressed. Ongoing research aims to tackle these challenges by developing more cost-effective processing techniques, optimizing sourcing strategies, and exploring functional modifications to enhance performance. Innovations in these areas are crucial for making chitosan a more economically viable and mainstream solution for sustainable food packaging. As these advancements continue, chitosan's potential as a key player in eco-friendly packaging solutions becomes increasingly feasible (Vieira et al. 2023).

Chitosan stands out as a beacon of eco-friendliness in the landscape of food packaging materials. Its renewable source, biodegradability, reduced dependency on petrochemicals, and energy-efficient processing collectively position chitosan as a sustainable alternative. The customizable properties of chitosan enable the creation of functional and environmentally conscious packaging solutions, contributing to the reduction of plastic pollution and fostering a circular economy. As industries increasingly prioritize sustainability, the versatile and eco-friendly nature of chitosan is poised to play a pivotal role in shaping the future of environmentally responsible food packaging (Fazeli *et al.* 2023).

5.8 ALIGNING WITH THE DEMAND FOR SUSTAINABLE PACKAGING SOLUTIONS

In an era where environmental consciousness is at the forefront of consumer preferences and global initiatives, the demand for sustainable packaging solutions has surged. The packaging industry, a significant contributor to environmental challenges, is now undergoing a transformative shift toward eco-friendly alternatives. Chitosan, derived from chitin, has emerged as a notable player in this sustainability paradigm, aligning seamlessly with the growing demand for packaging solutions that minimize environmental impact (Mahmoud et al. 2022; Mathew et al. 2024a).

Addressing Plastic Pollution: Plastic pollution has become a global crisis, with vast amounts of plastic waste inundating landfills and oceans. The persistence of traditional plastics poses threats to ecosystems, wildlife, and human health. Chitosan, being a biodegradable and naturally occurring material, offers a promising avenue to address the environmental repercussions of plastic pollution. Its capacity to break down into harmless components alleviates concerns about long-lasting plastic waste, contributing to the mitigation of pollution (Bidashimwa et al. 2023; Mathew et al. 2024b).

Circular Economy Principles: Chitosan naturally aligns with the principles of a circular economy by adopting a regenerative approach to material use. It is derived from chitin, which is extracted from the exoskeletons of crustaceans and insects, often considered waste products from the seafood industry. By repurposing these byproducts, chitosan transforms waste into a valuable resource, exemplifying circular economy practices that emphasize the reduction of waste and the efficient use of resources. This process not only minimizes the ecological footprint of production but also supports a more sustainable production cycle. By integrating waste materials into valuable products, chitosan contributes to a circular economy by promoting resource efficiency and reducing overall environmental impact (De Aguiar et al. 2021; Dawud et al. 2025).

Renewable Resource and Reduced Dependency on Fossil Fuels: Chitosan's renewable nature, derived from abundant marine resources, distinguishes it from traditional packaging materials that rely on fossil fuels. Sourced from chitin in seafood waste, chitosan provides an alternative to petrochemical-based materials, reducing the demand for non-renewable resources. The extraction and processing of chitosan utilize byproducts from the seafood industry, minimizing the need for raw materials that contribute to fossil fuel depletion. This reduced dependency on fossil fuels helps mitigate the environmental impact associated with the extraction, transportation, and processing of petrochemicals. As a result, chitosan emerges as an environmentally responsible choice, aligning with sustainability goals and offering a more eco-friendly solution in the realm of packaging materials (Teixeira-Costa & Andrade 2021; Inobeme et al. 2023a).

Energy-Efficient Production: Chitosan production is notably energy-efficient compared to the energy-intensive processes required for manufacturing synthetic polymers. The key step in converting chitin into chitosan is deacetylation, which typically occurs under mild conditions, thus requiring significantly less energy input. This energy-efficient process aligns well with sustainable manufacturing practices, focusing on resource efficiency and minimizing the carbon footprint associated with material production. By reducing energy consumption during production, chitosan supports a more environmentally friendly approach to material manufacturing, contributing to lower greenhouse gas emissions and a reduced overall environmental impact. This makes chitosan an attractive option for industries seeking to enhance sustainability and decrease their carbon footprint in material production (Kaczmarek et al. 2019; Adetunji et al. 2023b).

Customizable Properties for Diverse Applications: Chitosan's versatility is evident in its customizable properties, making it suitable for a wide range of packaging applications. Its film-forming capability allows for the creation of thin, protective layers that can be adjusted to meet specific needs, whether for fresh produce, perishable goods, or industrial packaging. This adaptability enables chitosan to cater to the diverse requirements of various industries. For example, chitosan films can be engineered to control moisture levels, enhance antimicrobial protection, and provide barrier properties tailored to the product's needs. This flexibility ensures that chitosan not only meets functional demands but also supports environmental goals by offering a sustainable alternative to conventional packaging materials. Its ability to address both practical and ecological concerns positions chitosan as a valuable asset in the packaging industry (Desai et al. 2023; Muhammad et al. 2025).

Edible Films and Reduced Food Waste: Chitosan's application in creating edible films offers a unique advantage in the battle against food waste. These films, with their biodegradability and antimicrobial properties, contribute to extending the shelf life of perishable goods. By preventing spoilage and maintaining product quality, chitosan-based edible films play a vital role in reducing the amount of food discarded due to deterioration. This aligns with global efforts to address the significant issue of food waste, contributing to a more sustainable and resource-efficient food supply chain (Martins et al. 2024).

Consumer Preferences and Clean Labeling: The growing demand for sustainable packaging is driven by shifting consumer preferences toward ethically sourced and environmentally friendly products. Chitosan, with its natural and biodegradable properties, aligns perfectly with these preferences and supports the trend toward clean labeling. As consumers increasingly seek products that reflect their commitment to environmental stewardship, chitosan-based packaging offers a compelling option. Its biodegradability and eco-friendly profile resonate with consumers who prioritize sustainability and minimal environmental impact. By using chitosan for packaging, companies can cater to this evolving market demand, enhancing their brand's appeal to eco-conscious consumers and demonstrating a commitment to sustainable practices. This alignment with consumer values positions chitosan as a key player in the future of sustainable packaging (Cook et al. 2023; Mathew et al. 2024c).

Challenges and Collaborative Innovation: While chitosan presents numerous advantages in the pursuit of sustainability, challenges such as cost, scalability, and certain application-specific limitations persist. Collaborative efforts between researchers, industries, and policymakers are essential to address these challenges and drive innovations that enhance the overall sustainability of chitosan-based materials. This collaborative approach ensures that the benefits of chitosan are maximized and that it continues to play a pivotal role in meeting the demand for sustainable packaging solutions (Lam et al. 2023; Malik et al. 2025).

Chitosan's role in aligning with the demand for sustainable packaging solutions is multifaceted, addressing issues ranging from plastic pollution to renewable resource utilization. Its biodegradability, reduced dependency on fossil fuels, energy-efficient production, and customizable properties make it a powerful contender in the pursuit of eco-friendly packaging. As industries and consumers increasingly prioritize sustainability, chitosan's versatility and ecological benefits position it as a key player in shaping the future of packaging materials that harmonize with both environmental responsibility and functional requirements. The continued exploration and integration of chitosan into diverse packaging applications signify a positive step toward a more sustainable and resilient packaging landscape (Hameed et al. 2022; Inobeme et al. 2024b).

5.9 FUTURE PROSPECTS AND DEVELOPMENTS IN CHITOSAN-BASED EDIBLE FILMS AND COATINGS

The future prospects of chitosan-based edible films and coatings are promising, driven by ongoing developments in sustainable packaging solutions. Researchers and industries are actively exploring innovative formulations, processing techniques, and applications for chitosan. With its eco-friendly profile, biodegradability, and antimicrobial properties, chitosan is poised to play a pivotal role in addressing environmental concerns associated with conventional packaging materials. The continuous refinement of chitosan-based formulations, considering factors like mechanical strength, barrier properties, and scalability, suggests a trajectory toward broader adoption in diverse industries. As consumer preferences increasingly gravitate toward sustainable choices, chitosan's potential to revolutionize food packaging while minimizing environmental impact positions it as a frontrunner in the evolution of packaging materials for a more sustainable future (Table 5.2) (Tanpichai et al. 2023; Chimbekujwo et al. 2025).

TABLE 5.2
Future Prospects and Developments in Chitosan-Based Edible Films and Coatings

Aspect	Potential Developments	Benefits	References
Enhanced barrier properties	Incorporation of nanotechnology for improved barrier against gases and water	Prolonged shelf life, reduced food spoilage	Biswas et al. (2022) and Adetunji et al. (2023b)
Active packaging	Integration of active compounds like antioxidants or antimicrobials	Inhibition of microbial growth, preservation of freshness	Popa et al. (2022) and Adetunji et al. (2023c)
Tailored mechanical properties	Engineering films with adjustable strength, elasticity, and flexibility	Customizable packaging suitable for various food types	Shah et al. (2023)
Smart packaging technologies	Incorporation of sensors for real-time monitoring of food quality	Early detection of spoilage, improved safety assurance	Osmólska et al. (2022)
Biodegradability enhancements	Developing chitosan blends for faster degradation without environmental harm	Reduced environmental impact, aligning with sustainability	Rafiqah et al. (2021) and Inobeme et al. (2023b)
Improved printability	Advancements in printing technologies for better quality and detail	Enhanced branding opportunities, attractive packaging	Iftekar et al. (2023)
Nanocomposite films	Incorporation of nanomaterials to enhance mechanical and barrier properties	Improved film strength, extended shelf life	Khezerlou et al. (2021) and Mathew et al. (2023b)
Regulatory compliance	Establishment of clear guidelines and standards for chitosan-based packaging	Facilitates wider adoption, ensures consumer safety	Wallelign et al. (2022)
Market expansion	Increasing applications beyond food, such as pharmaceuticals and cosmetics	Diversification of product offerings and market reach	Pawlak and Kołodziejczak (2020) and Abimbola et al. (2023)
Cost-effective production methods	Development of more efficient and cost-effective production processes	Reduces overall production costs, making it more accessible	Gold et al. (2022)

5.10 CONCLUSION

The engineering properties of chitosan-based edible films and coatings present a compelling and sustainable solution for the food packaging industry. With its innate biodegradability, biocompatibility, and antimicrobial attributes, chitosan serves as a versatile material, offering enhanced barrier properties against moisture, gases, and contaminants. The tunable mechanical properties enable the customization of film strength, elasticity, and flexibility, making it suitable for diverse food products. The concept of active packaging, achieved through the incorporation of antioxidants or antimicrobials, enhances the preservation of freshness and inhibits microbial growth. The integration of smart packaging technologies, such as sensors, allows real-time monitoring, ensuring early detection of spoilage and elevated safety standards. Chitosan's biodegradability and compatibility with nanocomposite materials position these films as environmentally friendly alternatives, aligning with the increasing demand for sustainable packaging. Improved printability not only contributes to the aesthetic appeal but also provides enhanced branding opportunities. As the industry moves toward eco-conscious practices, chitosan-based edible films and coatings are poised to play a pivotal role. Continued research and development efforts are crucial for optimizing production methods, establishing regulatory standards, and expanding applications beyond food packaging. In this rapidly evolving landscape, chitosan stands out as a promising material, offering a harmonious balance between functional efficacy and environmental responsibility, shaping the future of innovative packaging technologies.

REFERENCES

Abimbola, O. F., Adetunji, C. O., Dare, B. J., Mathew, J. T., Inobeme, A., Ghazanfar, S., Olaniyan, O. T, Ajiboye, M. D., & Dauda, W. (2023). Packaging of animal-based food products: The need for intelligent freshness sensors. In: *Sensing and Artificial Intelligence Solutions for Food Manufacturing*. CRC Press, pp. 89–102. https://doi.org/10.1201/9781003207955-6

Adetunji, C. O., Egbuna, C., Ajiboye, A. A. A., Ajayi, O. O., Dauda, W. P., Ghazanfar, S., Ogundolie, F. A., Mathew, J. T., Adetunji, J. B., Inobeme, A., Olotu, T., Olaniyan, O. T., Ijabadeniyi, O. A., Aborode, A. T., Ficai, A., Akram, M., Bodunrinde, R. E., Laila, U., Olaniyan, O. P., Micheal, O. S., Aliyu, A., Akinbo, O., & Adetuyi, B. O. (2023c). Biogenic nanomaterials with diverse biological activities in the food and biomedical industries. In: *Evaporative Coolers for the Postharvest Management of Fruits and Vegetables*: C. O. Adetunji, D. I. Hefft, D. O. Mbuge, T. S. Workneh (Eds.). Academic Press, pp. 395–420. https://doi.org/10.1016/B978-0-323-89864-5.00001-1

Adetunji, C. O., Mathew, J. T., Inobeme, A., Olaniyan, O., Singh, K. R. B., Ogundolie, F. A., Vanya, N., Singh, J., & Singh, R. P. (2022b). Microbial and plant cell biosensors for environmental monitoring. In: *Nanobiosensors for Environmental Monitoring*. Springer, Cham, pp. 175–190. https://doi.org/10.1007/978-3-031-16106-3_9

Adetunji, C. O., Mathew, J. T., Singh, K. R. B., Inobeme, A., Olaniyan, O., & Singh, J. (2022a). Antibiofilm potential of zero-dimensional carbon nanomaterials. In: *Zero-dimensional Carbon Nanomaterials: Fundamentals and Applications*. IOP Publishing. https://doi.org/10.1088/978-0-7503-4048-9ch12

Adetunji, C. O., Shariati, M. A., Olugbenga, S. M., Osarenkhoe, O. O, Odoh, U. E., Olaniyan, O. T., Rebezov, M., Ajayi, O. O., Baibalinova, G., Bodunrinde, R. E., Adetunji, J. B., Adeniyi, M. J., Ugwu, P. N., Inobeme, A., Mathew, J. T., Elufisan, T. O., & Omotayo, O. O. (2023a). Pineapple [*Ananas comosus* (L.) Merr.]: A biological and pharmacological active medicinal plant. In: *Phytochemical Composition and Pharmacy of Medicinal Plants*: T. Pullaiah (Ed.). Apple Academic Press, Inc. Co-published with CRC Press (Taylor & Francis), Palm Bay, FL, Vol. 1, pp. 537–567.

Adetunji, C. O., Olaniyan, O. T., Olugbenga, S. M., Shariati, M. A., Osarenkhoe, O. O., Odoh, U. E., Oyedara, O. O., Rebezov, M., Ajayi, O. O., Babaeva, M., Inobeme, A., Mathew, J. T., Adetunji, J. B., Ogechukwu, H. U., Bodunrinde, R. E., & Adeniyi, M. J. (2023b). Pharmacology, photochemical, and healing effectiveness of *Chromolaena odorata* (L.) R. M. King and H. Robinson. In: *Phytochemical Composition and Pharmacy of Medicinal Plants*: T. Pullaiah (Ed.). Apple Academic Press, Inc. Co-published with CRC Press (Taylor & Francis), Palm Bay, FL, Vol. 2, pp. 461–482.

Aranaz, I., Alcántara, A. R., Civera, M. C., Arias, C., Elorza, B., Heras Caballero, A., & Acosta, N. (2021). Chitosan: An overview of its properties and applications. *Polymers*, 13(19), 3256. https://doi.org/10.3390/polym13193256

Bidashimwa, D., Hoke, T., Huynh, T. B., Narkpitaks, N., Priyonugroho, K., Ha, T. T., Burns, A., & Weissman, A. (2023). Plastic pollution: How can the global health community fight the growing problem? *BMJ Global Health*, 8(Suppl 3), e012140. https://doi.org/10.1136/bmjgh-2023-012140

Biswas, R., Alam, M., Sarkar, A., Haque, M. I., Hasan, M. M., & Hoque, M. (2022). Application of nanotechnology in food: Processing, preservation, packaging and safety assessment. *Heliyon*, 8(11), e11795. https://doi.org/10.1016/j.heliyon.2022.e11795

Bodunrinde, R. E., Chimbekujwo, K. I., Oyewole, O. A., Adetunji, C. O., Inobeme, A., Eniola, K. I. T., Yerima, M. B., & Mathew, J. T. (2025). The use of nanochitosan for enhancement in the quality and yield of fruit crops. In: *Nanochitosan Applications for Enhanced Crop Production and Food Security*. John Wiley & Sons, pp. 111–122. https://doi.org/10.1002/9781394214945.ch5

Bopape, M., Taillie, L. S., Frank, T., Murukutla, N., Cotter, T., Majija, L., & Swart, R. (2021). South African consumers' perceptions of front-of-package warning labels on unhealthy foods and drinks. *PloS One*, 16(9), e0257626. https://doi.org/10.1371/journal.pone.0257626

Bose, I., Roy, S., Pandey, V. K., & Singh, R. A. (2023). Comprehensive review on significance and advancements of antimicrobial agents in biodegradable food packaging. *Antibiotics*, 12, 968. https://doi.org/10.3390/antibiotics12060968

Buchwalder, S., Nicolier, C., Hersberger, M., Bourgeois, F., Hogg, A., & Burger, J. (2023). Development of a water transmission rate (WTR) measurement system for implantable barrier coatings. *Polymers*, 15(11), 2557. https://doi.org/10.3390/polym15112557

Chang, F., Eng, L., & Chang, C. (2023). Food allergy labeling laws: International guidelines for residents and travelers. *Clinical Reviews in Allergy & Immunology*, 65(2), 148–165. https://doi.org/10.1007/s12016-023-08960-6

Chimbekujwo, K. I., Yakubu, J. G., Omoregie, I. P., Oyewole, O. A., Muhammad, W. M., Adetunji, C. O., Eniola, K. I. T., Yerima, M. B., & Mathew, J. T. (2025). Synthesis of nano-chitosan using agricultural waste. In: *Nanochitosan Applications for Enhanced Crop Production and Food Security*. John Wiley & Sons, Inc., pp. 39–68. https://doi.org/10.1002/9781394214945.ch2

Cook, B., Costa Leite, J., Rayner, M., Stoffel, S., van Rijn, E., & Wollgast, J. (2023). Consumer interaction with sustainability labelling on food products: A narrative literature review. *Nutrients*, 15(17), 3837. https://doi.org/10.3390/nu15173837

Dawud, I., Bodunrinde, R. E., Oyewole, O. A., Adetunji, C. O., Inobeme, A. Eniola, K. I. T. Yerima, M. B., & Mathew, J. T. (2025). Application of nanochitosan for effective fruit production. In: *Nanochitosan Applications for Enhanced Crop Production and Food Security*. John Wiley & Sons. https://doi.org/10.1002/9781394214945.ch6

De Aguiar Saldanha Pinheiro, A. C., Martí-Quijal, F. J., Barba, F. J., Tappi, S., & Rocculi, P. (2021). Innovative non-thermal technologies for recovery and valorization of value-added products from crustacean processing by-products: An opportunity for a circular economy approach. *Foods*, 10, 2030. https://doi.org/10.3390/foods10092030

Desai, N., Rana, D., Salave, S., Gupta, R., Patel, P., Karunakaran, B., Sharma, A., Giri, J., Benival, D., & Kommineni, N. (2023). Chitosan: A potential biopolymer in drug delivery and biomedical applications. *Pharmaceutics*, 15(4), 1313. https://doi.org/10.3390/pharmaceutics15041313

Deshwal, G. K., Panjagari, N. R., & Alam, T. (2019). An overview of paper and paper based food packaging materials: health safety and environmental concerns. Journal of food science and technology, 56(10), 4391–4403. https://doi.org/10.1007/s13197-019-03950-z

Díaz-Montes, E., & Castro-Muñoz, R. (2021). Trends in chitosan as a primary biopolymer for functional films and coatings manufacture for food and natural products. *Polymers*, 13(5), 767. https://doi.org/10.3390/polym13050767

Dybka-Stępień, K., Antolak, H., Kmiotek, M., Piechota, D., & Koziróg, A. (2021). Disposable food packaging and serving materials: Trends and biodegradability. *Polymers*, 13, 3606. https://doi.org/10.3390/polym13203606

Egorov, A. R., Kirichuk, A. A., Rubanik, V. V., Rubanik Jr., V. V., Tskhovrebov, A. G., & Kritchenkov, A. S. (2023). Chitosan and its derivatives: Preparation and antibacterial properties. *Materials*, 16, 6076. https://doi.org/10.3390/ma16186076

El-Araby, A., Janati, W., Ullah, R., Ercisli, S., & Errachidi, F. (2024). Chitosan, chitosan derivatives, and chitosan-based nanocomposites: Eco-friendly materials for advanced applications (a review). *Frontiers in Chemistry*, 11, 1327426. https://doi.org/10.3389/fchem.2023.1327426

Ezeudu, O. B., & Ezeudu, T. S. (2019). Implementation of circular economy principles in industrial solid waste management: Case studies from a Developing Economy (Nigeria). *Recycling*, 4, 42. https://doi.org/10.3390/recycling4040042

Fadiji, T., Rashvand, M., Daramola, M. O., & Iwarere, S. A. (2023). A review on antimicrobial packaging for extending the shelf life of food. *Processes*, 11, 590. https://doi.org/10.3390/pr11020590

Fazeli, M., Mukherjee, S., Baniasadi, H., Abidnejad, R., Mujtaba, M., Lipponen, J., Seppälä, J., & Rojas, O. J. (2023). Lignin beyond the status quo: recent and emerging composite applications. *Green Chemistry: An International Journal and Green Chemistry Resource: GC*, 26(2), 593–630. https://doi.org/10.1039/d3gc03154c

Freitas, E. D., Moura Jr., C. F., Kerwald, J., & Beppu, M. M. (2020). An overview of current knowledge on the properties, synthesis and applications of quaternary chitosan derivatives. *Polymers*, 12, 2878. https://doi.org/10.3390/polym12122878

Gold, H. T., McDermott, C., Hoomans, T., & Wagner, T. H. (2022). Cost data in implementation science: Categories and approaches to costing. *Implementation Science*, 17, 11. https://doi.org/10.1186/s13012-021-01172-6

Guivier, M., Chevigny, C., Domenek, S., Casalinho, J., Perré, P., & Almeida, G. (2024). Water vapor transport properties of bio-based multilayer materials determined by original and complementary methods. *Scientific Reports*, 14(1), 50. https://doi.org/10.1038/s41598-023-50298-8

Hameed, A. Z., Raj, S. A., Kandasamy, J., Baghdadi, M. A., & Shahzad, M. A. (2022). Chitosan: A sustainable material for multifarious applications. *Polymers*, 14, 2335. https://doi.org/10.3390/polym14122335

Ibrahim, I. D., Hamam, Y., Sadiku, E. R., Ndambuki, J. M., Kupolati, W. K., Jamiru, T., Eze, A. A., & Snyman, J. (2022). Need for sustainable packaging: An overview. *Polymers*, 14(20), 4430. https://doi.org/10.3390/polym14204430

Iftekar, S. F., Aabid, A., Amir, A., & Baig, M. (2023). Advancements and limitations in 3D printing materials and technologies: A critical review. *Polymers*, 15, 2519. https://doi.org/10.3390/polym15112519

Inobeme, A., Adetunji, C. O., Ajai, A. I., Inobeme, J., Mathew, J. T., Obar, A., Maliki, M., Nwakife, N., & Eziukwu, C. (2023b). Chemical nanosensors for monitoring environmental pollution. In: *Advanced Application of Nanotechnology to Industrial Wastewater*. Springer Nature, Singapore, pp. 93–103. https://doi.org/10.1007/978-981-99-3292-4_6

Inobeme, A., Adetunji, C. O., Mathew, J. T., Ajai, A. I., Inobeme, J., Bamigboye, M. O., Onyeaku, S., Maliki, M., Eziukwu, C., & Tawa, K. (2023a). Nanotechnology for bioremediation of heavy metals. In: *Microbial Technologies in Industrial Wastewater Treatment*. Springer Nature, Singapore, pp. 19–30. https://doi.org/10.1007/978-981-99-2435-6_2

Inobeme, A., Mathew, J. T., Ajai, I. A., Adetunji, C. O., Inobeme, J., Maliki, M., Adekoya, M. A., Shaba, E. Y., Eric, O., Akhor, S., & Eziukwu, C. (2024a). Common techniques in food processing technologies. In: *Nanobiotechnology for Food Processing and Packaging*. Elsevier Inc., pp. 223–234. https://doi.org/10.1016/B978-0-323-91749-0.00002-2

Inobeme, A., Shahnawaz, M., Adetunji, C. O., Mathew, J. T., Adetuyi, B. O., Popoola, O. A., Olaitan, F. Y., Akinbo, O., Kolawole, O. M., Oyewole, O. A., Eniola, K. I. T., & Yerima, B. M. (2024b). Microplastic as a pathogenic vector to cause diseases in marine biota. In: *Microplastic Pollution*. Springer, Singapore, Vol. 1, pp. 321–349. https://doi.org/10.1007/978-981-99-8357-5_19

Jiménez-Gómez, C. P., & Cecilia, J. A. (2020). Chitosan: A natural biopolymer with a wide and varied range of applications. *Molecules (*Basel, Switzerland*)*, 25(17), 3981. https://doi.org/10.3390/molecules25173981

Kaczmarek, M. B., Struszczyk-Swita, K., Li, X., Szczęsna-Antczak, M., & Daroch, M. (2019). Enzymatic modifications of chitin, chitosan, and chitooligosaccharides. *Frontiers in Bioengineering and Biotechnology*, 7, 243. https://doi.org/10.3389/fbioe.2019.00243

Kaur, N., Somasundram, C., Razali, Z., Mourad, A. I., Hamed, F., & Ahmed, Z. F. R. (2024). Aloe vera/chitosan-based edible film with enhanced antioxidant, antimicrobial, thermal, and barrier properties for sustainable food preservation. *Polymers*, 16(2), 242. https://doi.org/10.3390/polym16020242

Ke, C.-L., Deng, F.-S., Chuang, C.-Y., & Lin, C.-H. (2021). Antimicrobial actions and applications of chitosan. *Polymers*, 13, 904. https://doi.org/10.3390/polym13060904

Khezerlou, A., Tavassoli, M., Alizadeh Sani, M., Mohammadi, K., Ehsani, A., & McClements, D. J. (2021). Application of nanotechnology to improve the performance of biodegradable biopolymer-based packaging materials. *Polymers*, 13(24), 4399. https://doi.org/10.3390/polym13244399

Khubiev, O. M., Egorov, A. R., Kirichuk, A. A., Khrustalev, V. N., Tskhovrebov, A. G., & Kritchenkov, A. S. (2023). Chitosan-based antibacterial films for biomedical and food applications. *International Journal of Molecular Sciences*, 24(13), 10738. https://doi.org/10.3390/ijms241310738

Kozma, M., Acharya, B., & Bissessur, R. (2022). Chitin, chitosan, and nanochitin: Extraction, synthesis, and applications. *Polymers*, 14, 3989. https://doi.org/10.3390/polym14193989

Lam, W. S., Lam, W. H., & Lee, P. F. (2023). The studies on chitosan for sustainable development: A bibliometric analysis. *Materials (*Basel, Switzerland*)*, 16(7), 2857. https://doi.org/10.3390/ma16072857

Li, B., Elango, J., & Wu, W. (2020). Recent advancement of molecular structure and biomaterial function of chitosan from marine organisms for pharmaceutical and nutraceutical application. *Applied Sciences*, 10, 4719. https://doi.org/10.3390/app10144719

Long, J., Zhang, W., Zhao, M., & Ruan, C. Q. (2023). The reduce of water vapor permeability of polysaccharide-based films in food packaging: A comprehensive review. *Carbohydrate Polymers*, 321, 121267. https://doi.org/10.1016/j.carbpol.2023.121267

Loo, H. L., Goh, B. H., Lee, L. H., & Chuah, L. H. (2022). Application of chitosan-based nanoparticles in skin wound healing. *Asian Journal of Pharmaceutical Sciences*, 17(3), 299–332. https://doi.org/10.1016/j.ajps.2022.04.001

Ma, X., Zhuang, X., & Ma, G. (2020). Transparent windows on food packaging do not always capture attention and increase purchase intention. *Frontiers in Psychology*, 11, 593690. https://doi.org/10.3389/fpsyg.2020.593690

Mahmoud, M. A., Tsetse, E. K. K., Tulasi, E. E., & Muddey, D. K. (2022). Green packaging, environmental awareness, willingness to pay and consumers' purchase decisions. *Sustainability*, 14, 16091. https://doi.org/10.3390/su142316091

Malik, S. I., Chimbekujwo, K. I., Maku, A. M., Oyewole, O. A., Omoregie, I. P., Adetunji, C. O., Eniola, K. I. T., Yerima, M. B., & Mathew, J. T. (2025). The application of nanochitosan biopesticides as a replacement to synthetic pesticides. In: *Nanochitosan Applications for Enhanced Crop Production and Food Security*. John Wiley & Sons, pp. 89–110. https://doi.org/10.1002/9781394214945.ch4.

Maliki, S., Sharma, G., Kumar, A., Moral-Zamorano, M., Moradi, O., Baselga, J., Stadler, F. J., & García-Peñas, A. (2022). Chitosan as a tool for sustainable development: A mini review. *Polymers*, 14, 1475. https://doi.org/10.3390/polym14071475

Martins, V. F. R., Pintado, M. E., Morais, R. M. S. C., & Morais, A. M. M. B. (2024). Recent highlights in sustainable bio-based edible films and coatings for fruit and vegetable applications. Foods (Basel, Switzerland*)*, 13(2), 318. https://doi.org/10.3390/foods13020318

Mathaba, M., & Daramola, M. O. (2020). Effect of chitosan's degree of deacetylation on the performance of PES membrane infused with chitosan during AMD treatment. *Membranes*, 10(3), 52. https://doi.org/10.3390/membranes10030052

Mathew, J. T., Adetunji, C. O., Inobeme, A., Musah, M., Azeh, Y., Etsuyankpa, M. B., Otori, A. A., Chado, I. M., Shaba, E. Y., Akoji, J. N., Mamman, A., & Oyewole, O. A. (2025b). Application of nanochitosan for the biodegradation of agricultural wastes. In: *Nanochitosan Applications for Enhanced Crop Production and Food Security*. John Wiley & Sons, pp. 361–382. https://doi.org/10.1002/9781394214945.ch16

Mathew, J. T., Adetunji, C. O., Inobeme, A., Otori, A. A., Musah, M., Azeh, Y., Shaba, E. Y., Muhammad, I. A., Mamman, A., & Oyewole, O. A. (2025a). Relevance of nanochitosan in food sector and food packaging sectors: Current trends. In: *Nanochitosan Applications for Enhanced Crop Production and Food Security*. John Wiley & Sons, pp. 345–360. https://doi.org/10.1002/9781394214945.ch15.

Mathew, J. T., Inobeme, A., Etsuyankpa, M. B., Adetunji, C. O., Shaba, E. Y., Musah, M., Azeh, Y., Jiya, M. J., Tanko, E., Muhammad, A. I., Mamman, A., & Maurice, J. (2024b). Applications of diatom for treatment of industrial effluents: Waste to wealth approach. In: *Marine Biomass: Biorefinery, Bioproducts and Environmental Bioremediation*. Walter de Gruyter GmbH & Co KG, pp. 1–20. https://doi.org/10.1515/9783111353951-001

Mathew, J. T., Inobeme, A., Hussaini, J., Shaba, E. Y., Azeh, Y., Musah, M., Muhammad, E. B., Tanko, M. S., Ismail, H., Adetunji, C. O., Kanwa, A. M., & Mamman, A. (2024a). Properties of nanomaterials for corrosion control. In: *Innovations in Nanomaterials-Based Corrosion Inhibitors*: A. Thakur, A. Kumar (Eds.). IGI Global, pp. 74–99. https://doi.org/10.4018/979-8-3693-3088-3.ch004

Mathew, J. T., Inobeme, A., Musah, M., Azeh, Y., Etsuyankpa, M. B., Adetunji, C. O., Shaba, E. Y., Jiya, M. J., Josiah, J. G., Tanko, E., Muhammad, A. I., Mamman, A., & Jemkur, M. (2024c). Utilization of marine green algae for development of bioplastics. In: *Marine Biomass: Biorefinery, Bioproducts and Environmental Bioremediation*. Walter de Gruyter GmbH & Co KG, pp. 39–55. https://doi.org/10.1515/9783111353951-003

Mathew, J. T., Musah, M., Azeh, Y., & Muhammed, M. (2023a). Adsorptive removal of selected toxic metals from pharmaceutical wastewater using Fe_3O_4/ZnO nanocomposite. *Dutse Journal of Pure and Applied Sciences*, 9(4a), 236–248. https://doi.org/10.4314/dujopas.v9i4a.22

Mathew, J. T., Musah, M., Azeh, Y., & Muhammed, M. (2023b). Kinetic study of heavy metals removal from pharmaceutical wastewater using geopolymer/Fe_3O_4 nanocomposite. *Bima Journal of Science and Technology*, 7(4), 152–163. https://doi.org/ 10.56892/bima.v7i4.539

Moura-Alves, M., Esteves, A., Ciríaco, M., Silva, J. A., & Saraiva, C. (2023). Antimicrobial and antioxidant edible films and coatings in the shelf-life improvement of chicken meat. Foods (Basel, Switzerland*)*, 12(12), 2308. https://doi.org/10.3390/foods12122308

Muhammad, W. M., Ibrahim, A. B., Samuel, J. O., Khalifa, A. S., Oyewole, O. A., Adetunji, C. O., Eniola, K. I. T., Yerima, M. B., & Mathew, J. T. (2025). The role of nanomaterials in agriculture as nanofertilizers. In: *Nanochitosan Applications for Enhanced Crop Production and Food Security*. John Wiley & Sons, Inc., pp. 1–38. https://doi.org/10.1002/9781394214945.ch1

Muñoz-Tebar, N., Pérez-Álvarez, J. A., Fernández-López, J., & Viuda-Martos, M. (2023). Chitosan edible films and coatings with added bioactive compounds: Antibacterial and antioxidant properties and their application to food products: A review. *Polymers*, 15, 396. https://doi.org/10.3390/polym15020396

Munteanu, B. S., & Vasile, C. (2021). Encapsulation of natural bioactive compounds by electrospinning-applications in food storage and safety. *Polymers*, 13(21), 3771. https://doi.org/10.3390/polym13213771

Ogundolie, F. A., Adetunji, C. O., Mathew, J. T., Inobeme, A., Ghazanfar, S., Olaniyan, O. T., Ajiboye, M. D., Dauda, W. P., Adetunji, J. B., Akinbo, O., & Adetuyi, B. O. (2023). Intelligent freshness indicators in packaging of plant-based food products. In: *Evaporative Coolers for the Postharvest Management of Fruits and Vegetables*: C. O. Adetunji, D. I. Hefft, D. O. Mbuge, T. S. Workneh (Eds.). Academic Press, 323–331, https://doi.org/10.1016/B978-0-323-89864-5.00008-4

Oloyede, O. O., & Lignou, S. (2021). Sustainable paper-based packaging: A consumer's perspective. *Foods*, 10, 1035. https://doi.org/10.3390/foods10051035

Osmólska, E., Stoma, M., & Starek-Wójcicka, A. (2022). Application of biosensors, sensors, and tags in intelligent packaging used for food products: A review. *Sensors*, 22, 9956. https://doi.org/10.3390/s22249956

Pawlak, K., & Kołodziejczak, M. (2020). The role of agriculture in ensuring food security in developing countries: Considerations in the context of the problem of sustainable food production. *Sustainability*, 12, 5488. https://doi.org/10.3390/su12135488

Perera, K. Y., Jaiswal, A. K., & Jaiswal, S. (2023). Biopolymer-based sustainable food packaging materials: Challenges, solutions, and applications. *Foods*, 12, 2422. https://doi.org/10.3390/foods12122422

Piekarska, K., Sikora, M., Owczarek, M., Jóźwik-Pruska, J., & Wiśniewska-Wrona, M. (2023). Chitin and chitosan as polymers of the future: Obtaining, modification, life cycle assessment and main directions of application. *Polymers*, 15, 793. https://doi.org/10.3390/polym15040793

Popa, E. E., Miteluț, A. C., Râpă, M., Popescu, P. A., Drăghici, M. C., Geicu-Cristea, M., & Popa, M. E. (2022). Antimicrobial active packaging containing nisin for preservation of products of animal origin: An overview. *Foods (*Basel, Switzerland*)*, 11(23), 3820. https://doi.org/10.3390/foods11233820

Prakoso, F. A. H., Indiarto, R., & Utama, G. L. (2023). Edible film casting techniques and materials and their utilization for meat-based product packaging. *Polymers*, 15(13), 2800. https://doi.org/10.3390/polym15132800

Purewal, S. S., Kaur, A., Bangar, S. P., Singh, P., & Singh, H. (2024). Protein-based films and coatings: An innovative approach. *Coatings*, 14, 32. https://doi.org/10.3390/coatings14010032

Rafiqah, S. A., Khalina, A., Harmaen, A. S., Tawakkal, I. A., Zaman, K., Asim, M., Nurrazi, M. N., & Lee, C. H. (2021). A review on properties and application of bio-based poly(butylene succinate). *Polymers*, 13(9), 1436. https://doi.org/10.3390/polym13091436

Rehman, K. U., Hollah, C., Wiesotzki, K., Heinz, V., Aganovic, K., Rehman, R. U., Petrusan, J.-I., Zheng, L., Zhang, J., Sohail, S., Mansoor, M. K., Rumbos, C. I., Athanassiou, C., & Cai, M. (2023). Insect-derived chitin and chitosan: A still unexploited resource for the edible insect sector. *Sustainability*, 15, 4864. https://doi.org/10.3390/su15064864

Shah, Y. A., Bhatia, S., Al-Harrasi, A., Afzaal, M., Saeed, F., Anwer, M. K., Khan, M. R., Jawad, M., Akram, N., & Faisal, Z. (2023). Mechanical properties of protein-based food packaging materials. *Polymers*, 15(7), 1724. https://doi.org/10.3390/polym15071724

Shi, Y., Zhang, W., & Zhou, G. (2020). Effects of different moisture-permeable packaging on the quality of aging beef compared with wet aging and dry aging. Foods (Basel, Switzerland*)*, 9(5), 649. https://doi.org/10.3390/foods9050649

Souza, V. G. L., Pires, J. R. A., Rodrigues, C., Coelhoso, I. M., & Fernando, A. L. (2020). Chitosan composites in packaging industry-current trends and future challenges. *Polymers*, 12(2), 417. https://doi.org/10.3390/polym12020417

Srivastava, P., Ramakanth, D., Akhila, K., & Gaikwad, K. K. (2022). Package design as a branding tool in the cosmetic industry: Consumers' perception vs. reality. *SN Business & Economics*, 2(6), 58. https://doi.org/10.1007/s43546-022-00222-5

Stefanowska, K., Woźniak, M., Dobrucka, R., & Ratajczak, I. (2023). Chitosan with natural additives as a potential food packaging. Materials (Basel, Switzerland*)*, 16(4), 1579. https://doi.org/10.3390/ma16041579

Tanpichai, S., Pumpuang, L., Srimarut, Y., Woraprayote, W., & Malila, Y. (2023). Development of chitin nanofiber coatings for prolonging shelf life and inhibiting bacterial growth on fresh cucumbers. *Scientific Reports*, 13, 13195. https://doi.org/10.1038/s41598-023-39739-6

Teixeira-Costa, B. E., & Andrade, C. T. (2021). Chitosan as a valuable biomolecule from seafood industry waste in the design of green food packaging. *Biomolecules*, 11, 1599. https://doi.org/10.3390/biom11111599

Thambiliyagodage, C., Jayanetti, M., Mendis, A., Ekanayake, G., Liyanaarachchi, H., & Vigneswaran, S. (2023). Recent advances in chitosan-based applications: A review. Materials (Basel, Switzerland*)*, 16(5), 2073. https://doi.org/10.3390/ma16052073

Thompson-Witrick, K. A., Pitts, E. R., Nemenyi, J. L., & Budner, D. (2021). The impact packaging type has on the flavor of wine. *Beverages*, 7, 36. https://doi.org/10.3390/beverages7020036

Trotta, F., Da Silva, S., Massironi, A., Mirpoor, S. F., Lignou, S., & Ghawi, S. K. (2023). Charalampopoulos, D. Silver bionanocomposites as active food packaging: Recent advances & future trends tackling the food waste crisis. *Polymers*, 15, 4243. https://doi.org/10.3390/polym15214243

Vieira, H., Lestre, G. M., Solstad, R. G., Cabral, A. E., Botelho, A., Helbig, C., Coppola, D., de Pascale, D., Robbens, J., Raes, K., Lian, K., Tsirtsidou, K., Leal, M. C., Scheers, N., Calado, R., Corticeiro, S., Rasche, S., Altintzoglou, T., Zou, Y., & Lillebø, A. I. (2023). Current and expected trends for the marine chitin/chitosan and collagen value chains. *Marine Drugs*, 21(12), 605. https://doi.org/10.3390/md21120605

Wallelign, T. M., Selam, M. N., Wondie, G. B., & Habte, B. M. (2022). Assessment of compliance to packaging and labeling regulatory requirements of locally manufactured alcohol-based hand sanitizers marketed in Addis Ababa, Ethiopia. *Journal of Pharmaceutical Policy and Practice*, 15, 60. https://doi.org/10.1186/s40545-022-00456-6

Willskytt, S., & Brambila-Macias, S. A. (2020). Design guidelines developed from environmental assessments: A design tool for resource-efficient products. *Sustainability*, 12, 4953. https://doi.org/10.3390/su12124953

Xu, J., Liu, K., Chang, W., Chiou, B. S., Chen, M., & Liu, F. (2022). Regulating the physicochemical properties of chitosan films through concentration and neutralization. Foods (Basel, Switzerland*)*, 11(11), 1657. https://doi.org/10.3390/foods11111657

Zhao, Z., Li, Y., & Du, Z. (2022). Seafood waste-based materials for sustainable food packing: From waste to wealth. *Sustainability*, 14, 16579. https://doi.org/10.3390/su142416579

6 Application of Chitosan-Based Edible Coatings in the Prevention of the Growth of Spoilage Microorganisms

Abiola Folakemi Olaniran[1,2,3], Faith Oluwatoyosi Agaja[2,3], Oluwatobi Victoria Obayomi[2,3], and Clinton Emeka Okonkwo[4]

[1]Landmark University SDG 12 (Responsible Consumption and Production Research Group), Omu-Aran, Kwara State, Nigeria

[2]Department of Food Science and Nutrition, College of Pure and Applied Sciences, Landmark University, Omu-Aran, Kwara State, Nigeria

[3]Department of Food Science and Microbiology, College of Pure and Applied Sciences, Landmark University, Omu-Aran, Kwara State, Nigeria

[4]Department of Food Science, United Arab Emirates University, Al Ain, United Arab Emirates

6.1 INTRODUCTION

Food loss comprises fresh produce and fluid milk each contributing to 20%, grain products at about 15%, caloric sweeteners at almost 12%, processed fruits and vegetables at 8.6%, poultry, meat, and fish at 8.5%, and fats and oil at 7.1%. Of this food, some were still fit for human consumption but were thrown away as it was expired, way past the sell-by date, or exceeded requirements (Olaniran et al., 2020). Food spoilage and loss are also accompanied by environmental and resource expenses. Food spoilage processes involve deterioration factors such as microbial metabolism by enzymes, chemical changes, and microbial spoilage involving bacterial, yeast, and mold growth (Figure 6.1). These microorganisms are involved in influencing the quality, safety, and shelf life of foods that are on the market (Shaltout, 2024). Food spoilage has become a threat to sustainability and security due to what happens across the world on food preservation. Approximately more than one-third of the global output of food is said to be wasted, lost, or thrown away, as reported by the Food and Agriculture Organization (FAO) of the United Nations. This leads to financial loss, degradation of the environment, and decreased food security (McCarthy et al., 2018). It is estimated that every year during postharvest activities, over one-third of the world's food production estimated 1.5 billion tons, worth about one trillion dollars, is lost. The highest losses are observed in fruits and vegetables, especially in developing countries, where they contribute to 42% of the overall loss and waste (Porat et al., 2018). This issue of food waste and loss is significant and increasingly challenging, particularly in developing nations like Nigeria. In these regions, food loss leads to a reduction in income by at least 15% for the 470 million smallholder farmers and other participants in the downstream value chain, many of whom are part of the 1.2 billion people experiencing food insecurity (Dooshima et al., 2022). Bacteria are one of the primary causes of food spoilage. Certain bacteria can provoke infections and foodborne illnesses, whereas others can produce

DOI: 10.1201/9781003355458-6

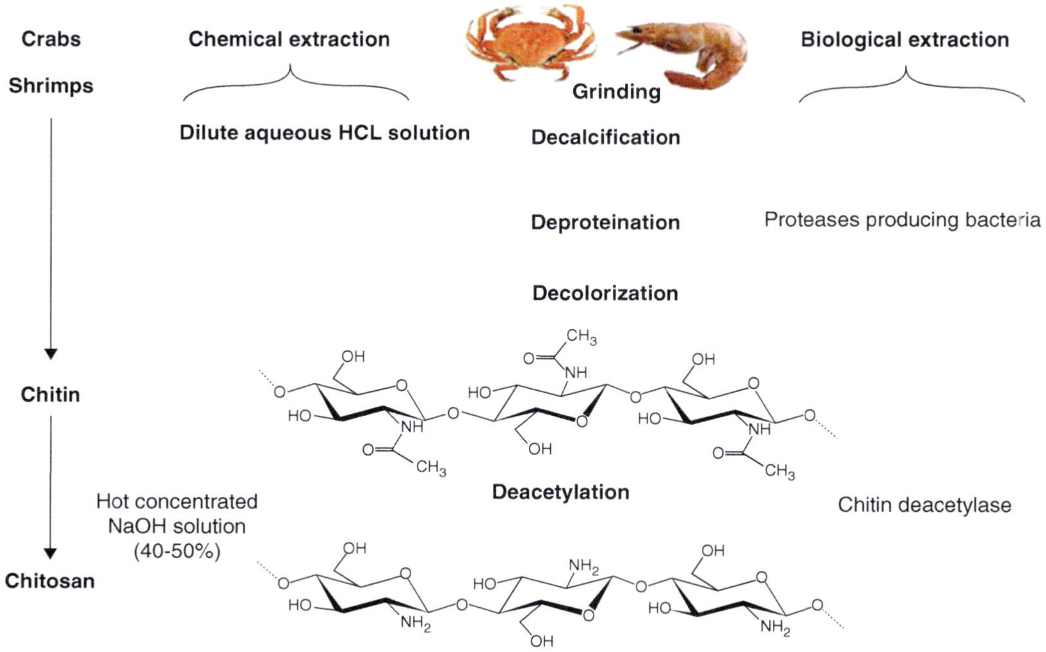

FIGURE 6.1 Picture showcasing the biological and chemical mechanisms of chitosan and chitin.

spores that are more resistant to preservation methods. Bacteria, such as *Proteus*, *Pseudomonas*, and *Clostridium*, break down proteins in foods like meats, resulting in foul-smelling compounds (Pellissery et al., 2020). While molds convert fats into fatty acids and cause rancidity in dairy products, yeasts ferment carbohydrates into acids and alcohols (Awasti & Anand, 2020). Foods that have been inadequately sealed may also become spoiled by molds, yeasts, and bacteria, which can cause unfavorable changes in their texture, flavor, and appearance (Mazhar et al., 2022). The development of sustainable coating compounds that can successfully preserve the quality and safety of various food items has been the focus of increased scientific and technological advancement in recent years (Ismail et al., 2021). Due to their many advantages over synthetic materials, biopolymers made from proteins like whey, gelatin, and casein, as well as polysaccharides like chitosan, locust bean gum, cellulose, and cassava starch, have become popular options for coating manufacturing (Yaashikaa et al., 2023). These benefits include non-toxicity, edibleness, biodegradability, biocompatibility, renewable nature, and environmental friendliness. Because of its capacity to form films and its antibacterial qualities, chitosan—a polysaccharide derived from the deacetylation (DA) of chitin—found in the shells of crustaceans including crabs, shrimp, and crayfish—is particularly attractive for coating applications (Ismail et al., 2021). Its ability to stop food from spoiling has been studied in great detail because of its strong antibacterial and antifungal qualities, which successfully stop the growth of microorganisms that cause food to deteriorate (Li et al., 2024; Muñoz-Tebar et al., 2023). Chitosan's antibacterial qualities are essential for limiting microbial growth, reducing deterioration, and maintaining the caliber of coated goods. Chitosan-based edible coatings play a major role in preventing food spoiling, improving food safety, and decreasing wastage in the food supply network by creating a regulated environment surrounding the product (Duan et al., 2019). This chapter discusses the prospect for chitosan-based edible coatings as a crucial element of contemporary preserving food approaches, intended to reduce spoilage and enhance food safety, as well as recent developments in chitosan alteration methods to improve its antimicrobial effects and interaction with different types of foods.

6.2 PRODUCTION, ATTRIBUTES, AND ANTIMICROBIAL INFLUENCE OF CHITOSAN EDIBLE COATINGS

The generated discarded products of the maritime food industry are the primary source of chitin and chitosan for commercial usage. Around eight million metric tons of crustaceans were harvested for human consumption worldwide in 2016, with 40% of those tons being leftover exoskeletons with a 15%–40% chitin content. Fungi are the second most prevalent source of chitin, after crustaceans (Gopal et al., 2019). Chitosan, a natural biopolymer obtained from the DA of chitin, is present in the exoskeletons of crustaceans such as shrimp, crabs, and lobsters, as well as in fungal cell wall. A lot of research has been done on fungi and its potential economic value as a result of the recent interest in fungi as a vegan and alternative source of chitin and chitosan (Tan et al., 2020; Islam et al., 2023). Chitin shares a structure with the cell walls of crustaceans and makes up 1%–15% of fungal cell walls. Numerous fungal phyla, such as Ascomycota, Zygomycota, and Basidiomycota, are known to contain chitin. *Aspergillus niger* (*Ascomycetes*), *Absidia* spp. (*Zygomycetes*), *Mucor rouxii* (*Zygomycetes*), *Lentinus edodes* (*Basidiomycetes*), and *Rhizopus oryzae* (*Zygomycetes*) are various species that have been widely studied for their prospect to effectively generate chitosan (Rehman et al., 2023). To produce chitosan for edible coatings, chitin undergoes DA, a process that can be carried out using either chemical or biological methods. This process removes the acetyl groups from chitin, resulting in the deacetylated compound known as chitosan (Ali et al., 2024; Islam et al., 2023). The chemical method for converting chitin to chitosan involves decalcification in which chitin is treated with hydrochloric acid or acetic acid. This treatment removes calcium carbonate and other minerals from the shell that are unnecessary for the final product (Song et al., 2022). Deproteination occurs after decalcification; the decalcified chitin is then treated with sodium hydroxide or potassium hydroxide. The purpose of this treatment is to eliminate proteins and other impurities, ensuring that the chitin is purified and ready for further processing (Gopal et al., 2019). The deproteinated chitin is decolorized by being treated with hydrogen peroxide or sodium hypochlorite to remove any remaining pigments, resulting in a purer and whiter chitin. Chitin is a polysaccharide that can have a slightly off-white or creamy color, even after decolorization. The treatment removes most of the pigments, resulting in a brighter and more uniform color (Rehman et al., 2023). The decolorized chitin is then deacetylated in which the chitin is heated in a strong alkaline solution, such as sodium hydroxide, at high temperatures and pressures. This treatment breaks down the acetyl groups attached to the chitin molecules, converting them into amino groups. This chemical transformation results in the production of chitosan, which has distinct properties and applications compared to chitin (Islam et al., 2023).

Because of changes in the physicochemical characteristics of extracted chitin, green extraction technologies are becoming more and more well-liked for their cleaner and more ecologically friendly methods (Tan et al., 2020; Rehman et al., 2023). The biological process of converting chitin to chitosan can be achieved through various methods such as fermentation, enzymatic treatment, and hybrid processes. The biological process involves the action of enzymes that break down the acetyl groups in chitin, resulting in the formation of chitosan (El-Aidie, 2018). This process typically involves breaking down of chitin by chitinases into smaller fragments, such as chitooligomers. These enzymes are formed by several microorganisms, including bacteria and fungi (Mahajan et al., 2024). Chitosanases are enzymes that can break down chitosan into even smaller fragments, such as monomers. These enzymes are also produced by microorganisms and are involved in the degradation of chitosan (Ali et al., 2024), while chitin deacetylase catalyzed the hydrolysis of *N*-acetamido bonds in chitin, resulting in the formation of chitosan. This enzyme is produced by various microorganisms, which is involved in the biological conversion of chitin to chitosan (Liang et al., 2023).

6.3 COATING FORMULATION OF CHITOSAN EDIBLE COATING

Fungi are the second most common source of chitin, following crustaceans (Gopal et al., 2019). Chitosan is a biopolymer prepared from the DA of chitin, which in turn is derived from *N*-acetyl glucosamine residues in the exoskeleton of crustaceans like shrimp, crab, and lobsters, as well as the cell wall of fungi. As a result of the growing interest in fungus as a vegan and alternative source of chitin and chitosan, much study has been conducted on them and their potential economic worth (Tan et al., 2020; Islam et al., 2023). Chitin, which accounts for 1%–15% of fungal wall cells, has a structure similar to crustacean cell walls. Several fungal phyla, including Ascomycota, Zygomycota, and Basidiomycota, are known to contain chitin. Some of the species that have been widely examined for their ability to successfully manufacture chitosan are *R. oryzae* (*Zygomycetes*) *Absidia* spp. (*Zygomycetes*), *M. rouxii* (*Zygomycetes*), *A. niger* (*Ascomycetes*), and *L. edodes* (*Basidiomycetes*) (Rehman et al., 2023). Chitin is deacetylated to create chitosan for edible coatings, which can be done chemically or biologically. This procedure eliminates chitin's acetyl groups, yielding the deacetylated molecule chitosan (Ali et al., 2024; Islam et al., 2023). Decalcification is a chemical technique for turning chitin to chitosan that uses hydrochloric or acetic acid. This process removes calcium carbonate and other minerals from the shell that are not required for the finished product (Song et al., 2022). Following the decalcification process, the chitin is deproteinized and treated with sodium hydroxide or potassium hydroxide. The goal of this treatment is to remove proteins and other contaminants, ensuring that the chitin is pure and ready for further processing (Gopal et al., 2019). The deproteinated chitin is decolorized by treating it with hydrogen peroxide or sodium hypochlorite to eliminate any remaining hues, yielding a purer and whiter chitin. Chitin is a complex molecule that becomes pale white or creamy after decolorization. The procedure removes the majority of the pigments, yielding a brighter and more uniform color (Rehman et al., 2023). The decolorized chitin is subsequently deacetylated by heating it in a strong alkaline solution, such as sodium hydroxide, at high temperatures and pressure. This procedure degrades the acetyl groups linked to the chitin molecules, turning them to amino groups. This chemical transition produces chitosan, which has different features and applications than chitin (Islam et al., 2023). Based on the alterations to the physicochemical attributes of extracted chitin, green extraction technology is gaining popularity as clean and further ecologically friendly approaches (Tan et al., 2020; Rehman et al., 2023). Fermentation, hybrid procedures, and enzymatic treatment are all possible techniques for converting chitin to chitosan. The biological process involves enzymes breaking down the acetyl groups in chitin, which results in the creation of chitosan (El-Aidie, 2018). This process often involves chitinases breaking down chitin into smaller particles, such as chitooligomers. Microorganisms such as bacteria and fungi manufacture these enzymes (Mahajan et al., 2024). Chitosanases are catalysts that may break breakdown chitosan into smaller particles known as monomers. Microorganisms also manufacture these enzymes, which are involved in the breakdown of chitosan (Ali et al., 2024), whereas chitin deacetylase promotes the hydrolysis of *N*-acetamido linkages in chitin, resulting in the production of chitosan. Various microorganisms produce this enzyme, which is involved in the fermentation of chitin to chitosan (Liang et al., 2023). Chitosan's biocompatibility, biodegradability, film-forming ability, and non-toxicity make it an excellent alternative for edible coatings (Muñoz-Tebar et al., 2023). Its antibacterial and antifungal qualities help to improve the shelf life and preserve the quality of food products. Chitosan's antibacterial activity against microorganisms, especially Gram-positive bacteria, makes it a desirable biological preservative in the processing of food and functional additives (Duan et al., 2019; Ismail et al., 2021; Muñoz-Tebar et al., 2023). Chitosan has been effectively used as a film and packaging material for a variety of fresh and dairy goods. Chitosan prevents microbiological deterioration and contamination by producing a clear layer with good gas barrier qualities (CO_2 and O_2). Other research has demonstrated that chitosan can preserve foods from fungal degradation and alter the atmosphere of fresh fruits (Romanazzi & Moumni, 2022). Chitosan has good mechanical qualities, such as flexibility, durability, high break resistance, strength, and toughness. These properties, together with its chemical ability and natural advantages, make it an adaptable material for the development of edible coatings that improve food quality and shelf stability. The ability of chitosan to form films, as well as its antibacterial characteristics and biodegradability, makes it particularly useful in food preservation. It is non-toxic and biocompatible, ensuring consumer safety. Chitosan, which has been generally recognized as safe (GRAS) by regulatory organizations such as the FDA, is appropriate for usage in food goods and other applications where safety is paramount (Ismail et al., 2021). Chitosan is one of the most widely recognized and commonly used polysaccharides for creating edible films and coatings because of its key qualities that improve food preservation and safety.

6.4 METHODS AND PROPERTIES OF CHITOSAN EDIBLE COATINGS

Chitosan films have a high tensile strength and flexibility, which helps to protect food from mechanical damage during transportation and storage. The films' capacity to withstand stress and stretch without ripping provides a protective barrier for food goods, retaining their freshness (Muñoz-Tebar et al., 2023). Chitosan is recyclable, consumable, and not harmful to consume, making it both environmentally beneficial and ideal for usage in food products. This feature is crucial for minimizing the environmental impact of food packaging and assuring customer safety (Muñoz-Tebar et al., 2023). Chitosan's electrostatic flexibility in polymers complex can be attributed to its intrinsic cation polymer framework, which has positive charges due to the existence of amino groups. This characteristic sets chitosan apart from many polysaccharides, which are neutral in nature or anions. Chitosan's cationic nature permits it to build structures with multiple layers or electrostatic complexes with both artificial and negative-charged organic polymers (Alves et al., 2018). Furthermore, chitosan's film-forming capacity allows for the creation of robust and flexible films that stick well to a variety of food surfaces via a technique known as solvent casting. This method entails dissolving chitosan in an appropriate solvent, such as aqueous acetic acid, and then pouring the mixture to a smooth surface. As the solvent evaporates, a solid film is formed. This feature enables chitosan coatings to efficiently cover food goods, forming a barrier of protection against environmental components such as air, moisture, and microbes. As a result, this protection improves food shelf life by lowering oxidation, avoiding the loss of moisture, and preventing microbial development (Khubiev et al., 2023). Chitosan coatings are good gas and moisture barriers, minimizing oxidation and dryness in food products, which is critical to maintaining their quality and extending their shelf life. Chitosan prevents oxidation by minimizing oxygen exposure, which preserves the flavor, color, and nutritional content of the food. It also inhibits moisture loss, which preserves the food's texture and juiciness. These protective characteristics help to prevent deterioration, allowing food goods to keep fresher for longer (Khubiev et al., 2023). Its intrinsic antibacterial capabilities prevent germs and fungi from growing on coated food. This reduces deterioration and increases the shelf life of food products. Chitosan edible films can be mixed with a variety of green-synthesized materials, including silver nanoparticles (NPs) derived from plant extracts, to improve their antibacterial properties. These extracts help to reduce, stabilize, and cap NPs, increasing their effectiveness in safeguarding food goods from microbial contamination and spoiling (Ali et al., 2022). Chitosan also has prebiotic qualities, which can improve intestinal health. It can boost the number of beneficial microbes in the colon while decreasing detrimental pro-inflammatory bacteria, resulting in a more balanced gut microbiota and less mucosal inflammation. This dual effect promotes digestion and intestinal health (Yan et al., 2021). As a prebiotic, it feeds healthy gut bacteria like Lactobacilli and Bifidobacteria. These bacteria also can break down chitosan and generate short-chain fatty acids (SCFAs) comprising propionate, butyrate, and acetate. SCFAs are absolutely required for the health of the colon because they extend energy to colonocytes, the cells that make up the lining of the colon and its mucosa and submucosa, and facilitate their division and repair (Gao et al., 2020). There are numerous procedures for applying chitosan solution to food goods, each with its own set of benefits, making chitosan a versatile option for increasing food safety, extending shelf life, and providing eco-friendly packaging solutions. Solution: The casting approach is simple and successful for making a uniform film by dissolving chitosan in an acidic solution, often acetic acid, and then spreading it onto a flat surface and allowing it to dry, resulting in a thin and thick film (Wang et al., 2020). Emulsion technique uses lipids or essential oils in order to create a coating (Ismail et al., 2021). While spraying coating involves spraying directly onto the food product, leaving a thin and uniform coating. This approach is effective for covering irregularly shaped products and can be scaled up for industrial use (Muñoz-Tebar et al., 2023). Food goods can also be soaked in chitosan solution and allowed to dry, resulting in a coating about it. This technology ensures complete covering of the food product and makes it appropriate for items with varied shapes (Wang et al., 2020). Electrostatic deposition is a precise coating process that controls the thickness and dispersion of the chitosan film. Chitosan particles are electrically charged and subsequently applied to the food surface through an electric field. This produces a homogenous and adherent covering (Hameed et al., 2022).

6.4.1 Mechanisms of Microbial Inhibition by Chitosan-Based Coatings

Chitosan-based coatings effectively inhibit a broad spectrum of microorganisms, including bacteria, yeast, and molds. The antimicrobial mechanisms involve various interactions and processes that disrupt essential cellular functions, leading to inhibited microbial growth (Ke et al., 2021). Extensive research has shown that chitosan's antimicrobial properties are influenced by its structure, physical and chemical characteristics, environmental conditions, and reactive hydroxyl groups (Dhlamini et al., 2024; Hameed et al., 2022; Ke et al., 2021; Teixeira-Santos et al., 2021). Chitosan can act extracellularly, intracellularly, or both, depending on its target. High-molecular-weight (HMW) chitosan primarily acts outside the cell by chelating essential metals, blocking nutrient uptake, and altering cell permeability. Low-molecular-weight (LMW) chitosan can function both outside and inside the cell, impacting RNA and protein synthesis and mitochondrial function. Thus, the mechanism of chitosan's action, whether extracellular or intracellular, depends on its molecular weight and varies with the type of microorganism targeted (Ke et al., 2021).

6.4.2 Inhibition of Bacterial Growth by Chitosan Edible Coatings

Chitosan-based edible coatings have been proven for inhibiting the growth of several bacteria, including both Gram-positive and Gram-negative species (Dhlamini et al., 2024). Research indicates that surfaces coated with chitosan exhibit stronger inhibitory effects on Gram-negative bacteria when compared to Gram-positive ones (Aktug et al., 2021; Ke et al., 2021). This disparity arises from differences in the cellular structure. Gram-positive bacteria have denser peptidoglycan layers, whereas Gram-negative bacteria have higher levels of lipopolysaccharides (LPS). These structural distinctions influence their susceptibility to chitosan, with Gram-negative bacteria, due to their negatively charged LPS with phosphorylated groups, showing greater affinity for cationic chitosan at pH levels below 6.5. This interaction suggests that Gram-negative bacteria may be more susceptible to chitosan's antimicrobial effects (Ke et al., 2021). In Gram-positive bacteria, teichoic acids, which are negatively charged due to phosphate groups, contribute to chitosan's interaction (Pasquina-Lemonche et al., 2020). However, deleting the teichoic acid biosynthesis pathway in *Staphylococcus aureus* increases resistance to chitosan, signifying that its mechanism is more intricate than simple electrostatic interfaces (Yan et al., 2021). Gram-negative bacteria also feature more hydrophilic surfaces compared to Gram-positive bacteria, enhancing their interaction with chitosan. This increased hydrophilicity leads to significant alterations in cell membrane structure and permeability upon chitosan binding, ultimately resulting in bactericidal effects and bacterial death (Aktug et al., 2021).

The effectiveness of chitosan in inhibiting bacterial growth depends on its molecular weight. HMW chitosan primarily disrupts bacterial functions externally by chelating metal ions, blocking nutrient uptake, and altering cell membrane permeability. Conversely, LMW chitosan can penetrate bacterial cells, disrupting internal processes, such as RNA and protein synthesis, and affecting mitochondrial function (Ke et al., 2021). The thick cell wall of the Gram-positive bacteria can also limit the access that chitosan has toward the bacterial cell membrane, which differs from the Gram-negative bacteria. Nevertheless, small chitosan oligomers (less than 5 kDa) can infiltrate the cell wall and disrupt DNA/RNA or protein synthesis. Some reports reveal that chitosan (less than 50 kDa) is able to penetrate through the cell wall and affect DNA transcription. Therefore, the effectiveness of chitosan is influenced by both its structure and molecular weight, determining whether its action is extracellular, intracellular, or both. Thus, chitosan's mode of action, influenced by its molecular weight, can be classified into extracellular and intracellular actions depending on the specific target microorganism (Ke et al., 2021)

6.4.3 Extracellular Mechanism of Chitosan Inhibition of Bacterial Growth

The extracellular mechanism by which chitosan inhibits bacterial growth involves actions that take place outside the bacterial cell, mainly focusing on the cell membrane and essential nutrient uptake processes. Both HMW and LMW chitosan operate through extracellular mechanisms that can disrupt cell membranes and chelate essential metals, thereby inhibiting nutrient uptake and altering cell permeability (Yilmaz Atay, 2020). Chelation of essential nutrients involves chitosan binding to and sequestering metal ions that are crucial for bacterial growth and metabolism (Figure 6.2). These metal ions, such as magnesium (Mg^{2+}), calcium (Ca^{2+}), and iron (Fe^{2+}), are vital for various bacterial processes. By chelating these ions, chitosan deprives bacteria of essential nutrients, disrupting metal-dependent processes and inhibiting bacterial growth. This nutrient deprivation weakens the bacteria, reducing their viability and growth (Guarnieri et al., 2022).

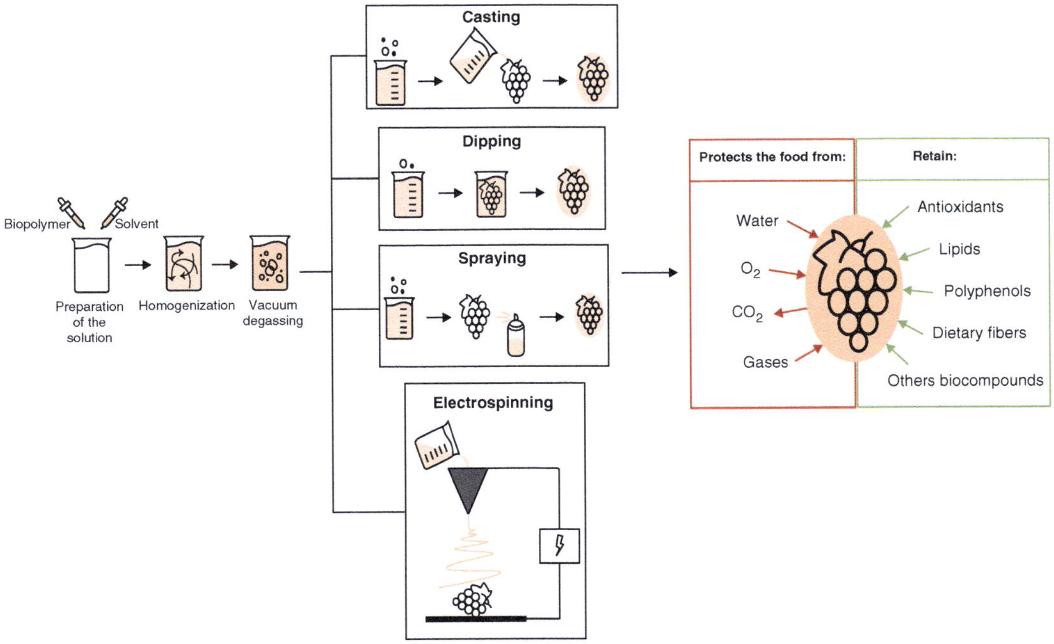

FIGURE 6.2 Application techniques of chitosan edible coatings (Wang et al., 2020).

6.4.4 Intracellular Mechanism of Chitosan Inhibition of Bacterial Growth

Chitosan inhibits bacterial growth through intracellular mechanisms that occur within the bacterial cell, primarily involving LMW chitosan and chitosan oligomers (Guarnieri et al., 2022). These smaller chitosan molecules can penetrate the bacterial cell wall and membrane to access the cytoplasm (Figure 6.3), where they exert their antimicrobial effects in various ways: LMW chitosan can infiltrate bacterial cells and interact with DNA and RNA, resulting in the inhibition of gene expression, transcription, and translation. This disruption hinders the production of essential proteins, ultimately leading to bacterial growth inhibition and cell death (Khubiev et al., 2023). Chitosan can affect mitochondrial activity within bacterial cells, leading to disruptions in energy production. This interference with the mitochondria impairs the cell's ability to generate ATP, the energy currency of the cell, leading to cell death (Ke et al., 2021). By penetrating the cytoplasm, chitosan inhibits various enzymatic activities crucial for bacterial metabolism and survival, including those involved in DNA replication, repair, and other metabolic pathways (Guarnieri et al., 2022). By targeting enzymes like DNA gyrase, topoisomerase, and DNA polymerase, chitosan disrupts the bacterial cell's ability to replicate and repair its genetic material, leading to growth inhibition and cell death. Chitosan also targets enzymes involved in energy production, such as those in the glycolysis and TCA cycle pathways, disrupting these metabolic processes and depriving the bacterial cell of necessary energy and resources for growth and survival. In addition, chitosan interferes with the activity of enzymes responsible for bacterial cell wall synthesis, such as peptidoglycan synthases and hydrolases, weakening the cell wall's structural integrity and making the bacteria more susceptible to osmotic stress and lysis. The effectiveness of chitosan in inhibiting enzymatic activities is influenced by its molecular weight and degree of DA. LMW chitosan and higher degrees of DA enhance the ability of chitosan to penetrate the bacterial cell and interact with intracellular enzymes. It can prompt the formation of reactive oxygen species (ROS) within the bacterial cell

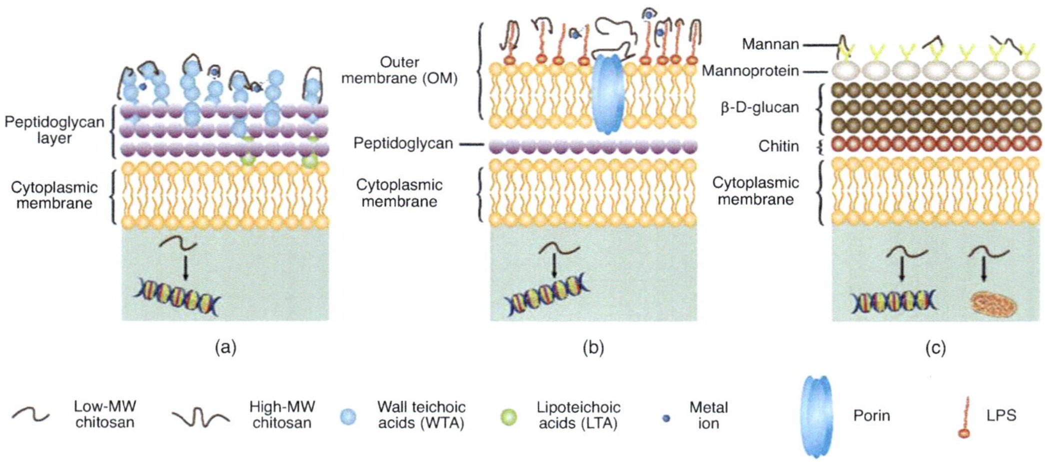

FIGURE 6.3 Schematic synopsis showing the mechanism of chitosan against pathogenic microorganisms. (a) Gram-positive bacteria: The negative charges of the Wall teichoic acids (WTAs) in their cell walls attract positively charged chitosan, leading to leaking of cell contents and cell death. (b) Gram-negative bacteria: Chitosan binds to the negatively charged lipopolysaccharides (LPS) components, chelating metal ions on their surface and causing cell membrane rupture. HMW chitosan blocks nutrient exchange, while low-molecular-weight (LMW) chitosan can penetrate the cells, inhibiting DNA/RNA or protein synthesis and leading to cell death. (c) Fungi: Chitosan interacts with negatively charged components (phosphorylated mannose in fungi), causing leakage of cell contents. LMW chitosan can enter fungal cells, disrupting mitochondrial function and reducing ATP production, ultimately leading to cell death (Yan et al., 2021).

through which it exerts its antimicrobial effects (Khubiev et al., 2023). ROS is descripted as an amount of very reactive molecules that consist of superoxide anion ($O_2^{\bullet-}$), molecular oxygen (O_2), hydroxyl radicals ($^{\bullet}OH$), and hydrogen peroxide (H_2O_2) (Lam et al., 2020).

The processes through which chitosan induces ROS formation include the interface of chitosan with the bacterial cell membrane initiating an electron transfer process that generates ROS. This interaction destabilizes the membrane and disrupts vital cellular functions, weakening the bacterial cell. Chitosan has the ability to chelate metal ions, such as iron (Fe^{2+}), which act as crucial role in ROS formation through the Fenton reaction. In this reaction, Fe^{2+} reacts with hydrogen peroxide produced by the bacteria's metabolic activities to generate highly reactive hydroxyl radicals. Chitosan can also directly induce oxidative stress in bacterial cells by disrupting normal redox homeostasis. This disruption leads to an overproduction of ROS within the cells, damaging essential cellular components and impairing metabolic functions. Furthermore, in eukaryotic cells—and potentially in prokaryotic cells—chitosan can interfere with the respiratory chain. This interference increases electron leakage, leading to the formation of ROS and disrupting ATP production and other energy-dependent processes. The formation of ROS can cause oxidative damage to cellular components, resulting in cell dysfunction and death. The effects of ROS include significant harm to DNA, including strand breaks, cross-linking, and base modifications. These changes can result in mutations and impede DNA replication and repair processes, jeopardizing the genetic integrity and viability of the bacterial cell. Also, oxidation of amino acids in proteins can lead to a loss of enzymatic activity, structural alterations, and protein aggregation. This disruption affects the function of essential cellular proteins, impairing various metabolic and structural processes within the bacterial cell. Cellular membranes or organelle membranes are susceptible to ROS damage by attacking lipids in the bacterial cell membrane. This results in greater permeability and a loss of membrane integrity, which can lead to cell lysis and ultimately cause the death of bacterial cells. Due to the extensive damage, ROS can produce programmed cell death (apoptosis) or can provoke unprogrammed cell death (necrosis). Whereas necrosis is thought of as a passive and unintended cell death caused by environmental disturbances, with uncontrollably released inflammatory cellular contents, apoptosis is an active, regulated process of autonomous cellular deconstruction with no induction of inflammation. Consequently, chitosan induces the production of ROS by binding to the bacterial cell membrane, chelating metal ions, inducing oxidative stress, and inhibiting the respiratory chain. These ROS damages the DNA, proteins and lipids of the bacterial cell leading to cell death (Lam et al., 2020). The dual mechanisms of antibacterial action in chitosan edible coatings occur by harnessing both the extracellular and intracellular antibacterial properties of chitosan. HMW chitosan primarily inhibits bacterial growth from the outside, while LMW chitosan provides a more comprehensive antibacterial effect by targeting both the exterior and interior of bacterial cells. This dual mechanism of action enhances the overall effectiveness of chitosan edible coatings in controlling bacterial contamination and prolonging the shelf life of food products (Ke et al., 2021).

6.4.5 Inhibition of Yeast and Mold Growth by Chitosan

Yeasts and molds are types of fungi that are commonly found in soil, water, plants, and animals. The cell wall of fungi mainly consists of chitin, β-D-glucans, and mannoproteins or mannan (Comitini et al., 2023) (Figure 6.3). Fungi contain a phosphorylated mannosyl side, and chitosan binds to this causing disruption of the plasma membrane and leading to intracellular leakage. It also affects functions such as DNA/RNA replication and synthesis of proteins. The small oligo-chitosan molecules can easily be incorporated into the fungal cell wall and membrane, impairing the mitochondria. The molecular weight of chitosan, the degree of its acetylation, the pH value of an environment, and the specific species of fungi are the factors that determine the antifungal potential of chitosan. Higher degrees of acetylation increase fungicidal activity, while lower molecular weight decreases it. Chitosan-sensitive fungi have higher ratio of polyunsaturated fatty acids (PUFA), which enhances membrane fluidity and negative charge leading to high membrane permeability (Yan et al., 2021). Chitosan-based coatings effectively inhibit the growth of yeasts and molds through extracellular and intracellular mechanisms, similar to their action against bacteria. Chitosan interacts with fungal cell membranes via electrostatic interactions between its positively charged amino groups and the negatively charged components on the fungal cell surface, such as phospholipids and proteins. This interaction disrupts membrane permeability, leading to the leakage of intracellular contents and eventually causing cell death (Yan et al., 2021). In addition, chitosan binds and sequesters vital metal ions, such as magnesium, calcium, and iron, which are crucial for yeast and mold growth and metabolism. By chelating these ions, chitosan disrupts fungal nutrient uptake and metabolic processes, weakening the cells and inhibiting their growth (Yan et al., 2021). Chitosan also inhibits the germination of fungal spores, which are the reproductive units of molds and yeasts. This reduces the likelihood of fungal growth and contamination in food products. Chitosan further prevents the growth of fungal mycelium (Meng et al., 2020). Intracellularly, LMW chitosan and chitosan oligomers can penetrate fungal cell walls and membranes, reaching the cytoplasm. Once inside, they inhibit key enzymes involved in metabolic pathways crucial for fungal survival. This interference disrupts normal cellular functions, inhibits growth, and promotes cell death. Chitosan also induces the formation of ROS within fungal cells. These ROS cause oxidative damage to essential cellular components, including DNA, proteins, and lipids, thereby disrupting cellular processes and contributing to cell dysfunction and death. Moreover, LMW chitosan penetrates fungal cells, disrupting mitochondrial function and impairing energy production. This disruption leads to metabolic impairment, ATP depletion, oxidative stress, and ultimately cell death. These intracellular effects, combined with extracellular actions like membrane disruption and nutrient chelation, enhance the antimicrobial efficacy of LMW chitosan against a broad spectrum of microorganisms (Yan et al., 2021).

6.5 FACTORS INFLUENCING EFFICACY OF CHITOSAN EDIBLE COATINGS

The use of chitosan-based edible coatings is on the rise due to their antimicrobial properties, which offer a natural and effective means to inhibit microbial growth and extend the shelf life of food products. However, the efficacy of these coatings is influenced by several critical factors, including pH and temperature conditions, the concentration and molecular weight of chitosan, and the type and surface properties of the food substrate (Ke et al., 2021; Khubiev et al., 2023). Understanding these factors is crucial for optimizing the application and performance of chitosan-based coatings in food preservation.

The molecular weight of chitosan determines its ability to penetrate microbial cell surfaces for intracellular antimicrobial activity. The complex layers of bacterial and fungal cell walls, mainly composed of polysaccharides and some proteins, serve multiple functions, including pathogenesis, adhesion to surfaces, and defense against the environment (Pasquina-Lemonche et al., 2020). These cell walls regulate the passage of molecules through delicate mechanisms or diffusion, with pore sizes varying widely from 2–4 up to 8 nm (Ke et al., 2021). LMW chitosan can more easily penetrate the microbial cell wall and membrane, allowing it to exert intracellular antimicrobial effects, such as inhibiting DNA/RNA synthesis, protein synthesis, and mitochondrial function (Ke et al., 2021). In contrast, HMW chitosan primarily acts on the cell surface, altering membrane permeability without significant intracellular penetration (Guarnieri et al., 2022). Studies have shown that LMW chitosan and chitosan oligomers are more effective in inhibiting microbial growth compared to HMW chitosan (Guarnieri et al., 2022; Ke et al., 2021; Yang et al., 2023). Research on bacteria such as *Escherichia coli*, *Klebsiella pneumonia*, *S. aureus*, *Salmonella enterica*, *Pseudomonas aeruginosa*, *Bacillus cereus*, *Listeria monocytogenes* and *Bacillus subtilis* indicated that LMW chitosan is more effective in inhibiting microorganism growth. This increased effectiveness is due to the smaller chitosan molecules' ability to interact more easily with microbial cell membranes (Khan et al., 2020).

The degree of DA of chitosan refers to the percentage of acetyl groups ($-COCH_3$) that have been removed from the polymer chain derived from chitin. It measures how extensively chitosan has been deacetylated from its precursor (Laroche et al., 2018). The DA significantly influences chitosan's solubility, charge density, and biological activity. Chitosan with a lower DA tends to be more soluble and exhibits enhanced antimicrobial properties. This is because a lower DA increases the number of positively charged amino groups ($-NH_2$) on the chitosan molecule, which enhances its ability to interact with negatively charged microbial cell membranes (Mumtaz et al., 2023). Through electrostatic interactions, chitosan disrupts the membrane permeability of microorganisms, leading to leakage of intracellular contents and ultimately causing cell death (Ke et al., 2021).

The effectiveness of chitosan edible coatings is significantly influenced by pH and temperature. The antimicrobial effect of chitosan primarily stems from its ability to interact electrostatically with negatively charged cell walls (Yan et al., 2021). Chitosan's amino groups have pKa values around 6.3–6.5, making it insoluble in alkaline solutions and organic solvents, and less soluble in water above pH 6.5 (Duan et al., 2019; Ke et al., 2021). As the pH decreases, chitosan becomes more soluble, increasing the positive charge on its amino groups and enhancing its antimicrobial potency (Kravanja et al., 2019). Numerous studies have highlighted chitosan's robust antimicrobial activity in acidic conditions (Ke et al., 2021; Kravanja et al., 2019; Yan et al., 2021). Temperature can significantly affect the properties and antimicrobial activity of chitosan-based edible coatings. High temperatures can degrade the chitosan polymer, reducing its molecular weight and degree of DA, which diminishes its antimicrobial effectiveness. High temperatures can also decrease chitosan's solubility, making it harder to incorporate into coatings and reducing its ability to disrupt microbial cell membranes (Egorov et al., 2023). Low temperatures, on the other hand, enhance the antimicrobial activity and stability of chitosan-based coatings by slowing microbial growth and metabolism, making them more susceptible to chitosan. They also maintain the structural integrity and physicochemical properties of chitosan, preserving its efficacy over time (Sutharsan & Zhao, 2023). This is crucial for the long-term storage and shelf life of chitosan-coated food products. The combination of low temperatures and chitosan-based coatings can synergistically inhibit microbial growth, offering enhanced protection. Therefore, optimizing temperature conditions is essential for maximizing the effectiveness and stability of chitosan-based edible coatings in food preservation.

The antimicrobial effectiveness of chitosan in edible coatings is dependent on its concentration. Typically, higher concentrations of chitosan result in greater antimicrobial activity, which acts effectively on coats and interacts with microbial cell surfaces, causing significant permeability changes that lead to cell death. The minimum inhibitory concentration (MIC) of chitosan varies depending on the specific microorganism and other influencing factors (Másson, 2024). The MIC is the lowest concentration of an antimicrobial agent, such as chitosan, that prevents visible growth of a microorganism after a specified period. It is a crucial measure in microbiology for determining the effectiveness of antimicrobial substances against specific pathogens. The MIC helps assess the required dosage for inhibiting microbial growth and is used to compare the potency of different antimicrobial agents (Morata-Moreno et al., 2024). The effectiveness of chitosan coatings is influenced by the characteristics of the substrate and its surface. The type of substrate and its surface properties play a crucial role in how well chitosan adheres and spreads as a coating. Variations in surface texture, composition, and wettability among different food products affect the interaction and attachment of chitosan coatings (Kaloper et al., 2024). Smooth surfaces that are hydrophilic generally support better adhesion and even spreading of chitosan coatings, which enhances their antimicrobial and protective qualities (Deng et al., 2020). For instance, research on apples showed that their smooth surfaces facilitated uniform chitosan coating application, resulting in improved antimicrobial effectiveness against pathogens (Aayush et al., 2022). Conversely, rough or hydrophobic surfaces can pose challenges for achieving uniform chitosan coating. Uneven surface textures and poor wettability may lead to inconsistent coating application, diminishing the overall antimicrobial and barrier properties offered by chitosan coatings (Wang et al., 2022) (Figure 6.4).

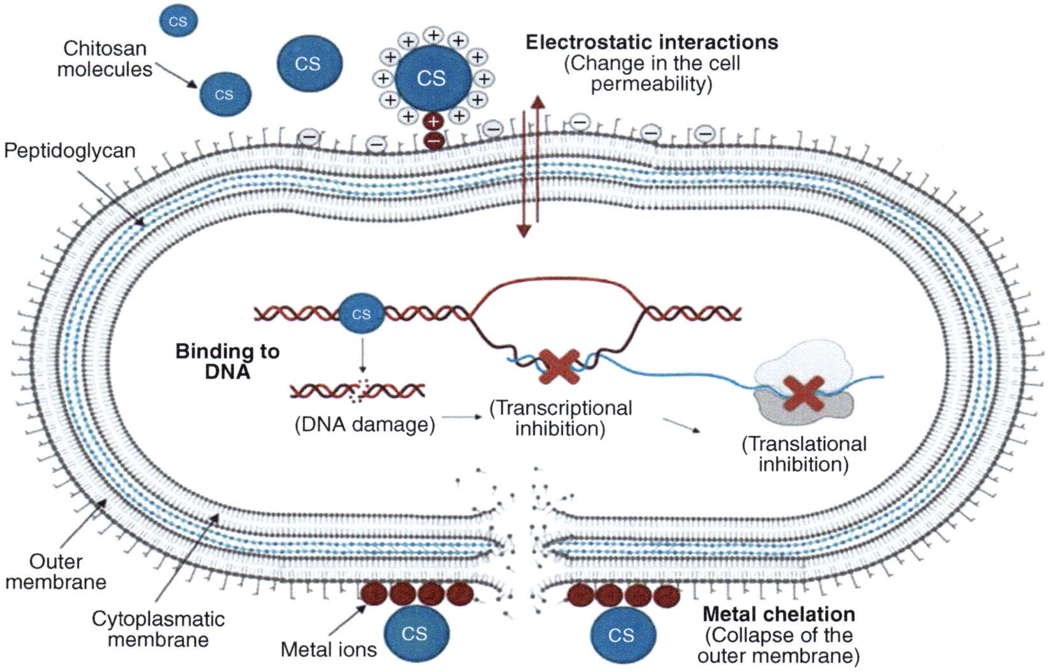

FIGURE 6.4 Diagram presenting the chitosan's antibacterial properties and its allied derivatives (Teixeira-Santos et al., 2021).

6.6 RECENT APPLICATION OF CHITOSAN EDIBLE COATINGS AS VIABLE PRESERVATION ALTERNATIVE

Chitosan-based coatings, in particular, can be applied to various food items, including fruits and vegetables, meat and poultry, dairy, bakery goods, confectionery, and seafood. These coatings provide several advantages, such as decreasing moisture loss, delaying ripening, preventing microbial growth, and improving the overall quality and longevity of food products (Muñoz-Tebar et al., 2023). Numerous studies have explored and confirmed the efficacy of chitosan-based coatings in controlling a wide range of microorganisms that impact food safety and quality. It has been verified that chitosan are active against harmful microbes such as *Shigella dysenteriae*, *E. coli*, *Salmonella typhimurium*, *L. monocytogenes*, and additional microbes that cause food to spoil (mesotrophs and psychrotrophs) (Olaimat et al., 2023). Chitosan has been reported to exhibit broad-spectrum antimicrobial properties against bacteria, yeasts, and molds by disrupting their cell membranes, interfering with metabolic processes and inducing oxidative stress, ultimately leading to the inhibition of microbial growth and cell death (Ke et al., 2021; Yan et al., 2021). The chitosan-based coatings are effective in controlling the growth of *Pseudomonas* spp., a prevalent mesotrophic bacterium known for its spoilage effects on meat products. Đorđević et al. (2022) examined the application of chitosan coatings enriched with *Thymus kotschyanus* essential oil on beef samples. A significant reduction in *Pseudomonas* spp. growth by approximately 2.5 log CFU/g was reported when compared to uncoated samples over 21 days of refrigerated storage (4°C). This represents a substantial antimicrobial effect of chitosan coatings against mesotrophic bacteria. Muñoz-Tebar et al. (2023) also explored chitosan coatings by incorporating nano-encapsulated tarragon essential oil on pork slices. Extension in shelf life by 8–12 days was documented compared to uncoated samples during 16 days of refrigerated storage. The coatings effectively maintained microbial counts below acceptable thresholds and reduced lipid oxidation, as indicated by lower TBAR (Thiobarbituric Acid Reactive Substances) values. TBAR values measure lipid oxidation in food, which contributes to meat degradation. Lower TBAR values indicate reduced lipid oxidation and better food quality preservation, thus showcasing the potential of chitosan-based coatings enriched with bioactive substances to inhibit *Pseudomonas* spp. growth and enhance meat product shelf life. The effectiveness of chitosan-based coatings in controlling the growth of *L. monocytogenes*, a common psychotropic bacterium, on cheese surfaces has been demonstrated by Olaimat et al. (2023) and Brown et al. (2018). For example, the use of chitosan–ZnO nanocomposite coatings on white brined cheese. The study showed that chitosan coatings with 1.0% ZnO NPs significantly reduced *L. monocytogenes* counts by 1.5–3.7 logCFU/g on the cheese surface and by 0.9–1.5 log CFU/g within the cheese during storage at 4°C and 10°C. The ZnO NPs remained on the cheese surface without migrating into the cheese matrix, indicating stability and effective surface-level antimicrobial action. Brown et al. (2018) studied the application of chitosan coatings with various antimicrobials on Queso Fresco. Coatings containing 5% hydrogen peroxide were highly effective, reducing *L. monocytogenes* counts by more than 3 log CFU/g and preventing growth for up to 35 days. The addition of sodium caprylate (SC) to lauric arginate (LAE) or acidified calcium sulfate with lactic acid (ACSL) enhanced the antimicrobial activity, resulting in reductions of over 1 log CFU/g after 24 hours and maintaining listeristatic effects for 28–35 days. By inhibiting *L. monocytogenes* growth, the chitosan coating significantly reduced the risk of foodborne illness associated with cheese consumption. This highlights the potential of chitosan coatings to enhance food safety and protect consumer health, particularly for vulnerable populations (Osek et al., 2022).

Candida spp. are known for causing spoilage and deterioration on fruit surfaces, which results in decreased fruit shelf life and quality. These yeasts can generate extracellular polymeric substances (EPS), aiding in their attachment to fruit surfaces and making their removal challenging (Namangkalakul et al., 2020). Researchers have conducted studies where they applied a coating based on chitosan to fruit surfaces to assess its efficacy in inhibiting the growth of *Candida* spp. Al-Zahrani et al. (2021) demonstrated the effectiveness of chitosan nanoparticles (CSNPs) in inhibiting various fungal species, including *Candida* spp. The researchers assessed the antifungal efficacy of CSNPs prepared and combined with LMW and HMW chitosan against *A. niger*, *Candida albicans*, and *Fusarium solani*. Their findings showed that CSNPs could inhibit the growth of these fungi, highlighting chitosan's potential as a powerful antifungal agent. Specifically, the study found that CSNPs exhibited significant antifungal activity against *C. albicans*. This supports the use of chitosan coatings in reducing *Candida* spp. growth on fruit surfaces, thereby extending their shelf life. Other researches have shown chitosan's effectiveness in preserving fruit by inhibiting spoilage microorganisms like *Candida* spp. (Fatima et al., 2024; Namangkalakul et al. 2020). The antimicrobial properties of chitosan, especially in NP form, significantly enhance its ability to protect fruits and vegetables from fungal contamination and decay.

Aspergillus flavus is a specific species known for its ability to produce aflatoxins. These toxins are highly carcinogenic and pose significant health risks to both humans and animals when ingested. Therefore, controlling the growth of *Aspergillus* spp., especially *A. flavus*, is crucial in food safety to prevent the contamination of nuts with aflatoxins (Awuchi et al., 2020). Jafari and Javadi (2020) applied a chitosan-based coating to nut surfaces against *A. flavus* growth and aflatoxin production with walnut leaf extract in prolonging the shelf life of pistachios (seeds of the pistachio tree). The results indicated significant reductions in pest infestation between coated and uncoated pistachios ($p < 0.05$). Moreover, higher concentrations of chitosan demonstrated enhanced antioxidant and antimicrobial effects, effectively inhibiting fungal activity in pistachios against *Aspergillus parasiticus* and *A. flavus*. Reducing peroxide content with antioxidants such as chitosan and walnut leaf extract coatings, as shown in the study, helps minimize oxidative stress on nuts. This reduction inhibits fungal growth, including aflatoxin-producing molds, thereby lowering the risk of aflatoxin contamination (Rajaram et al., 2023).

Chitosan coatings prevented dehydration and shriveling in apples coated with nanochitosan solutions; the best result using 1.5% (w/v) nanochitosan solution (S3) resulted in minimum weight loss of 1.53% over a 15-day storage period. The apples were stored at (25°C±5°C) and (35%±10%) RH and 10°C and 75% RH (Dasgupta & Mitra, 2024). Wong et al. (2021) found that probiotic coatings effectively reduced *L. monocytogenes* in apples while preserving probiotic levels. Antioxidant in chitosan films from banana peel extract decreased respiration level and mass loss, maintaining apple quality, according to Zhang et al. (2020). Furthermore, Yang et al. (2022) revealed that by preventing fungal growth and preserving antioxidant qualities, chitosan coatings combined with turmeric and green tea extracts prolonged the shelf life of strawberries. While the green tea extract improved its antioxidant content for 4–8 days at 20°C, the turmeric extract prohibited *Botrytis cinerea* from developing for 7 days while it was being kept. Eldib et al. (2020) showed that chitosan-nisin films in silicon dioxide NPs effectively preserved blueberries by controlling shrinkage, decay, and maintaining vitamin C content. The integration of nisin in the glazes aided control shrinkage and deterioration rates while inhibiting microbial populations, including molds, yeast, and mesophilic bacteria. Similarly, Sarengaowa et al. (2022) found that chitosan coatings in cinnamon essential oil enhanced the value and safety of fresh-cut potatoes by preventing browning, weight loss, and reducing microbial counts. In addition, Nagib et al. (2024) demonstrated that a 1% chitosan coating formed a protective layer on broccoli, inhibiting microbial growth and preserving quality. When combined with vacuum packaging and stored under optimal conditions (cold or frozen), this significantly extended the shelf life and quality retention of broccoli heads, resulting in a projected shelf life of 10–14 days under cold storage at 4°C (Nagib et al., 2024).

Roshandel-Hesari et al. (2022) showed that chitosan/casein films with oregano essential oil maintained the quality of cherry tomatoes by reducing shrinkage and weight loss and preserving firmness over extended storage. The study on apples coated with nanochitosan solutions found that the 1.5% (w/v) nanochitosan solution (S3) maintained firmness ranging from 7.3 to 5.409 N and effectively preserved the apples' quality attributes for 15 days (Dasgupta & Mitra, 2024). In another study, treatments namely chitosan, glycine betaine, and chitosan-coated glycine-betaine NPs on strawberries have shown in the maximum tissue firmness, maintained ascorbate peroxidase activity and total anthocyanin content, exhibited the least deterioration percentage and mass loss, and increased antioxidant capacity over a 12-day storage period at 4°C (Bahmani et al., 2024).

In meat and poultry preservation, chitosan-based coatings are increasingly being utilized to extend shelf life and enhance product quality. These coatings offer a natural and safe alternative to synthetic preservatives, aligning with consumer demand for more sustainable and health-conscious food preservation methods (Moller et al., 2023). The utilization of chitosan-based coatings to meat and poultry products serves multiple preservative functions. Chitosan inhibits the growth of spoilage microorganisms and pathogens such as *Listeria* and *Salmonella* (Elsherif et al., 2024). In a study by Elsherif et al. (2024), chicken breasts coated with essential oil nanoemulsions (EONe) and chitosan demonstrated significant antimicrobial efficacy, particularly with thyme oil, which showed the highest inhibitory effect against *E. coli* and *S. aureus*. The EONe-chitosan coating successfully prolonged the shelf life of chicken breast by significantly inhibiting mesophilic bacteria, psychrophilic bacteria, and yeasts, leading to an estimated shelf-life extension of approximately 14–21 days under refrigeration (Elsherif et al., 2024). Similarly, Moller et al. (2023) found that a carrageenan/chitosan coating enhanced with allyl isothiocyanate (AITC) reduced yeast and mold populations and extended aerobic spoilage by 3–12 days over 21 days of storage. Both coatings maintained stable quality indicators, with no significant changes in pH or color parameters. The coating minimizes lipid oxidation, preserving the flavor and nutritional quality of meat products. Moller et al. (2023) evaluated a layered carrageenan/chitosan coating combined with AITC. The coatings did not affect rancidity levels over the 21-day storage period, suggesting that the chitosan coatings help maintain the meat's quality without promoting oxidative rancidity. It also helps to retain moisture, preventing the meat from drying out and improving juiciness. The EONe-chitosan coating from Elsherif et al.

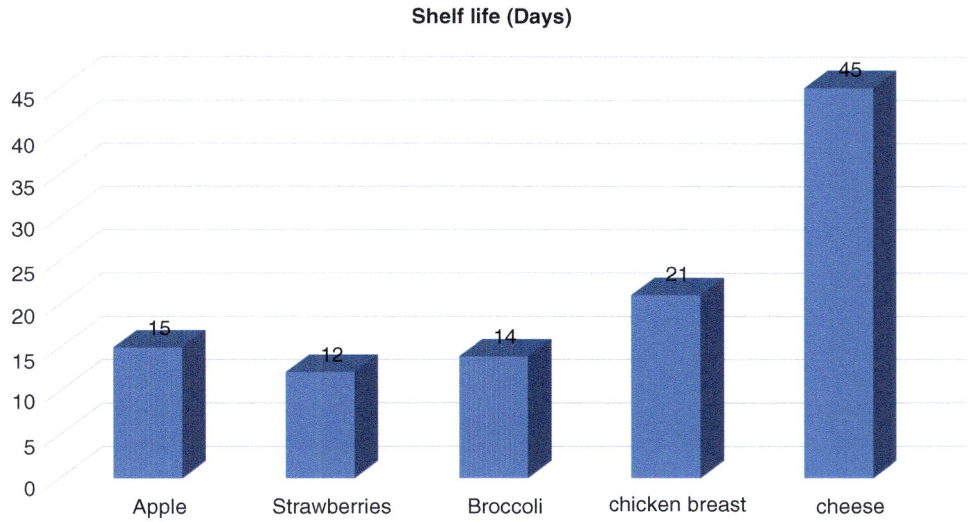

FIGURE 6.5 The chart demonstrates the significant impact of chitosan-based coatings on the shelf-life extension of various food products, with some achieving an extension of up to 45 days compared to uncoated counterparts (Bahmani et al., 2024; Dasgupta & Mitra, 2024; Elsherif et al., 2024; Ghasemian et al., 2024; Nagib et al., 2024).

(2024) showed the highest efficacy in reducing cooking loss, which indicates improved moisture retention and juiciness in the coated chicken breasts compared to untreated samples.

Cheese are an indispensable part of the day-to-day regimen as they contain vital nutrients such as proteins, minerals, lipids, and unsaturated fatty acids. Chitosan-based coatings are becoming more popular for preserving dairy products because of their numerous advantages. These coatings enhance the shelf life of dairy items such as cheese to inhibit loss of quality during the transportation, ripening period, and sales as these products are mostly predisposed to microbial spoilage due to their water activity, pH, and nutritional composition. Acting as a protective layer, chitosan coatings prevent spoilage, retain texture, and protect against harmful bacteria, keeping dairy products fresher and safer for a longer period (Iqbal et al., 2021; Muñoz-Tebar et al., 2023). Chitosan/gelatin glaze with thyme and papaya leave extracts significantly reduced total mesophilic aerobes in the coated samples compared to the untreated ones (from 5.70 to 3.76–4.99 log CFU/g). *Enterobacteriaceae* counts were 4.45 and 6.39 log CFU/g for coatings with thyme extract and papaya leaves, respectively, compared to 8.55 log CFU/g in the control samples (Hassan et al., 2022). The application of chitosan films containing rosemary extract significantly decreased the total count of bacteria, coliforms, lactic acid bacteria, mold, and yeast in treated cheeses. The most effective treatment with combination of 1% rosemary extract and 2% chitosan successfully inhibited microorganism growth, preventing spoilage. The chitosan-rosemary extract coating extended the shelf life of white cheese by 45 days under refrigerated conditions (4°C) compared to untreated white cheese (Ghasemian et al., 2024) (Figure 6.5).

6.7 BENEFITS, LIMITATIONS, AND FUTURE DIRECTIONS

Chitosan edible coatings have received significant attention recently due to their effectiveness in prolonging shelf life and retaining food quality (Muñoz-Tebar et al., 2023). While these methods offer various advantages, including sustainability and enhanced nutritional value, they also come with limitations that need to be considered. Understanding these factors is crucial for optimizing their application in the food industry (Table 6.1). Moreover, investigating future directions and

TABLE 6.1
Comparison of Chitosan Edible Coatings with Other Preservation Methods

Method	Mechanism	Benefits	Drawbacks
Chitosan coatings	Barrier formation	Natural, maintains quality	May vary in effectiveness
Chemical preservatives	Chemical inhibition	Cost-effective	Health concerns, flavor impact
Modified atmosphere	Gas composition alteration	Effective shelf-life extension	Specialized packaging needed
Refrigeration/Freezing	Temperature reduction	Simple and effective	Energy-intensive
High-pressure processing	High-pressure application	Nutrient retention	High costs
Fermentation	Microbial activity	Adds flavor, probiotic benefits	Limited application

potential improvements can drive advancements in food preservation technologies, ultimately benefiting both consumers and producers (Pham et al., 2023). Chitosan, obtained from natural chitin in crustacean shells, is an eco-friendly choice due to its biodegradable properties, which promote decomposition and reduce waste, aligning with sustainability initiatives. Furthermore, regulatory agencies like the FDA recognize chitosan as safe for food applications, awarding it GRAS status. Its non-toxic and biocompatible nature ensures safe direct contact with food, making it a superior option for food-related uses (Ismail et al., 2021). By forming a protective barrier, chitosan coatings also restrict the growth of bacteria and fungi that can lead to spoilage. This dual effect helps maintain the quality, texture, and flavor of the produce, ultimately reducing food waste and enhancing its marketability (Bahmani et al., 2024; Dasgupta & Mitra, 2024). The antimicrobial ability of chitosan to disrupt the cell membranes of microorganisms leads to preventing multiplication and subsequent decay (Muñoz-Tebar et al., 2023). Thus, chitosan coatings help retain the nutritional value of food products by preventing nutrient loss during storage. For instance, when applied to blueberries, chitosan coatings can preserve vitamin C content, which typically degrades over time (Eldib et al., 2020). By forming the barrier that reduces exposure to air and moisture, chitosan coatings protect the essential nutrients in food, ensuring that they remain intact and beneficial for consumption (Khubiev et al., 2023). It can also improve the quality of food by enhancing its appearance, texture, and flavor, thus making it more appealing to consumers. Application of chitosan to fruits helps maintain their shiny look and prevents browning, keeping them visually appealing (Dasgupta & Mitra, 2024), thereby enhancing the overall flavor (Singh et al., 2024). By preserving these quality attributes, chitosan coatings make food products more attractive and acceptable for consumers (Chaudhari et al., 2023).

The acceptance of edible coatings, such as chitosan, can vary widely among consumers, affecting market demand. Many individuals may be unaware of the benefits or may have concerns regarding their safety and effectiveness. Clear labeling and marketing are essential; without proper identification, consumers might opt for traditional products instead. Additionally, cultural factors influence attitudes toward food technologies, leading to differing levels of acceptance in various regions. Consumer awareness and acceptance significantly impact the demand for products that utilize edible coatings (Mauricio et al., 2022). Producing and applying chitosan coatings can be more costly than using traditional preservatives, which can hinder their widespread adoption. For instance, extracting and purifying chitosan from crustacean shells involves a complex and expensive process. Moreover, the application of chitosan coatings requires specialized equipment and technology, further driving up costs (Ikram et al., 2021). Chitosan coatings are not be suitable for all types of food products, particularly those with high moisture content or specific textures. Foods like leafy greens or fresh berries with high water content might experience altered texture or reduced effectiveness of the coating due to excessive moisture. Similarly, certain textured foods such as baked goods or crispy snacks might suffer from changes in texture, making the coating less desirable. As a result, the application of chitosan coatings can be limited, with optimal effectiveness primarily seen in low-moisture foods like apples or potatoes, where the coatings can preserve quality without compromising texture (Iqbal et al., 2021). The effectiveness of chitosan coatings can fluctuate based on various factors, resulting in inconsistent outcomes. The formulation of the coating, such as incorporating natural extracts or modifying the molecular weight, can significantly affect its performance. For example, a chitosan coating mixed with essential oils may offer better antimicrobial properties than a basic formulation (Chaudhari et al., 2023). Environmental factors, especially temperature and humidity, also influence effectiveness. In high-humidity conditions, such as those encountered during the storage of chitosan-coated apples, the coating may absorb moisture, compromising its protective barrier. These various factors together affect how well chitosan coatings perform across different food products and storage environments (Ikram et al., 2021).

6.8 FUTURE DIRECTIONS AND OPPORTUNITIES FOR CHITOSAN EDIBLE COATINGS

Research is focusing on developing biodegradable coatings and exploring the use of waste materials or byproducts as potential sources of chitosan (Ikram et al., 2021). As consumer demand for eco-friendly products increases, prioritizing sustainability becomes crucial (Bhardwaj et al., 2023). Chitosan edible coatings present significant prospects for innovation in food preservation. One primary area of focus is developing enhanced formulations that integrate chitosan with various natural extracts or bioactive compounds, such as probiotics or essential oils, to boost the antimicrobial properties of the coatings (Pham et al., 2023). Raising awareness of the benefits of chitosan coatings is vital for fostering consumer acceptance. Effective labeling and marketing strategies will be essential in this educational initiative (Mauricio et al., 2022). Future research focusing on tailoring chitosan coatings to address the unique requirements of different food products while integrating it with smart packaging technologies can further improve food preservation, thus enabling real-time monitoring of food quality, thereby enhancing preservation outcomes (Liu et al., 2024). The discovery of cost-effective methods for producing chitosan and applying coatings will also make this technology more accessible to food producers, particularly smaller operations (Chaudhari et al., 2023; Ikram et al., 2021) can significantly advance food preservation technologies, ultimately benefiting both consumers and the food industry. Application of chitosan-based edible coatings could profoundly boost food preservation technologies not just decreasing food spoilage, wastage, and dependence on synthetic preservatives but also yield sustainable impacts across the entire food industry.

REFERENCES

Aayush, K., McClements, D. J., Sharma, S., Sharma, R., Singh, G. P., Sharma, K., & Oberoi, K. (2022). Innovations in the development and application of edible coatings for fresh and minimally processed apple. *Food Control*, *141*, 109188.

Aktug, S. L., Durdu, S., Kalkan, S., Cavusoglu, K., & Usta, M. (2021). In vitro biological and antimicrobial properties of chitosan-based bioceramic coatings on zirconium. *Scientific Reports*, *11*(1), 15104.

Ali, G., Sharma, M., Salama, E. S., Ling, Z., & Li, X. (2024). Applications of chitin and chitosan as natural biopolymer: potential sources, pretreatments, and degradation pathways. *Biomass Conversion and Biorefinery*, *14*(4), 4567–4581.

Ali, S. G., Jalal, M., Ahmad, H., Sharma, D., Ahmad, A., Umar, K., & Khan, H. M. (2022). Green synthesis of silver nanoparticles from Camellia sinensis and its antimicrobial and antibiofilm effect against clinical isolates. *Materials*, *15*(19), 6978.

Alves, V. L., Rico, B. P., Cruz, R. M., Vicente, A. A., Khmelinskii, I., & Vieira, M. C. (2018). Preparation and characterization of a chitosan film with grape seed extract-carvacrol microcapsules and its effect on the shelf-life of refrigerated Salmon (Salmo salar). *LWT*, *89*, 525–534.

Al-Zahrani, S. S., Bora, R. S., & Al-Garni, S. M. (2021). Antimicrobial activity of chitosan nanoparticles. *Biotechnology & Biotechnological Equipment*, *35*(1), 1874–1880.

Awasti, N., & Anand, S. (2020). The role of yeast and molds in dairy industry: an update. In: M. Jagrani, S. V. Aparna, & A. Kumari (Eds.), *Dairy Processing: Advanced Research to Applications* (pp. 243–262). Springer, Singapore.

Awuchi, C. G., Awuchi, C. G., Ukpe, A. E., Asoegwu, C. R., Uyo, C. N., & Ngoka, K. E. (2020). Environmental impacts of food and agricultural production: a systematic review. *European Academic Research*, *8*(2), 1120–1135.

Bahmani, R., Razavi, F., Mortazavi, S. N., Gohari, G., & Juárez-Maldonado, A. (2024). Enhancing postharvest quality and shelf life of strawberries through advanced coating technologies: a comprehensive investigation of chitosan and glycine betaine nanoparticle treatments. *Plants*, *13*(8), 1136.

Bhardwaj, M., Mishra, A., & Rivera, K. M. (2023). Consumer perception towards eco-friendly products: a quantitative study. *Journal of Informatics Education and Research*, *3*(2), 119–123.

Brown, S. R., Kozak, S. M., & D'Amico, D. J. (2018). Applications of edible coatings formulated with antimicrobials inhibit Listeria monocytogenes growth on Queso Fresco. *Frontiers in Sustainable Food Systems*, *2*, 1.

Chaudhari, A. K., Das, S., Dwivedi, A., & Dubey, N. K. (2023). Application of chitosan and other biopolymers based edible coatings containing essential oils as green and innovative strategy for preservation of perishable food products: a review. *International Journal of Biological Macromolecules*, *253*, 127688.

Comitini, F., Canonico, L., Agarbati, A., & Ciani, M. (2023). Biocontrol and probiotic function of non-Saccharomyces yeasts: new insights in agri-food industry. *Microorganisms*, *11*(6), 1450.

Dasgupta, P., & Mitra, J. (2024). Feasibility study of nanochitosan solution as a potential coating for shelf-life extension of apples. *Journal of Packaging Technology and Research*, *8*(2), 139–151.

Deng, Z., Wang, T., Chen, X., & Liu, Y. (2020). Applications of chitosan-based biomaterials: a focus on dependent antimicrobial properties. *Marine Life Science & Technology*, *2*(4), 398–413.

Dhlamini, K. S., Selepe, C. T., Ramalapa, B., Tshweu, L., & Ray, S. S. (2024). Reimagining chitosan-based antimicrobial biomaterials to mitigate antibiotic resistance and alleviate antibiotic overuse: a review. *Macromolecular Materials and Engineering*, *309*, 2400018.

Dooshima, U., Idyorough, A. E., Confort, U.-O., & Nwafor, S. C. (2022). Internal displacement of farmers and food losses in Benue State, Nigeria. *Research on Humanities and Social Sciences*, *12*(10), 50–60.

Đorđević, N., Karabegović, I., Cvetković, D., Šojić, B., Savić, D., & Danilović, B. (2022). Assessment of chitosan coating enriched with free and nanoencapsulated Saturejamontana L. essential oil as a novel tool for beef preservation. *Foods*, *11*(18), 2733.

Duan, C., Meng, X., Meng, J., Khan, M. I. H., Dai, L., Khan, A., & Ni, Y. (2019). Chitosan as a preservative for fruits and vegetables: a review on chemistry and antimicrobial properties. *Journal of Bioresources and Bioproducts*, *4*(1), 11–21.

Egorov, A. R., Kirichuk, A. A., Rubanik, V. V., Rubanik Jr, V. V., Tskhovrebov, A. G., & Kritchenkov, A. S. (2023). Chitosan and its derivatives: preparation and antibacterial properties. *Materials*, *16*(18), 6076.

El-Aidie, S. A. A. M. (2018). A review on chitosan: ecofriendly multiple potential applications in the food industry. *International Journal of Advancement in Life Sciences Research*, *1*, 1–14.

Eldib, R., Khojah, E., Elhakem, A., Benajiba, N., & Helal, M. (2020). Chitosan, nisin, silicon dioxide nanoparticles coating films effects on blueberry (Vaccinium myrtillus) quality. *Coatings*, *10*(10), 962.

Elsherif, W. M., Zayed, G. M., & Tolba, A. O. (2024). Antimicrobial activity of chitosan-edible films containing a combination of carvacrol and rosemary nano-emulsion against Salmonella enterica serovar Typhimurium and Listeria monocytogenes for ground meat. *International Journal of Food Microbiology*, *418*, 110713.

Fatima, M., Mir, S., Ali, M., Hassan, S., Khan, Z. U. H., & Waqar, K. (2024). Synthesis and applications of chitosan derivatives in food preservation: a review. *European Polymer Journal*, *216*, 113242.

Gao, J., Azad, M. A., Han, H., Wan, D., & Li, T. (2020). Impact of prebiotics on enteric diseases and oxidative stress. *Current Pharmaceutical Design*, *26*(22), 2630–2641.

Ghasemian, S. O., Ahmadi-Dastgerdi, A., Abdollahi, A., Tirtashi, F. E., Zokaei, M., Fallah, N., & Dolatyari, F. (2024). The effect of active packaging film based on chitosan containing rosemary (Rosmarinus officinalis L.) extract on cheese shelf life. *Journal of Food Biochemistry*, *2024*(1), 2108707.

Gopal, J., Muthu, M., Dhakshanamurthy, T., Kim, K. J., Hasan, N., Kwon, S. J., & Chun, S. (2019). Sustainable ecofriendly phytoextract mediated one pot green recovery of chitosan. *Scientific Reports*, *9*, 1–12.

Guarnieri, A., Triunfo, M., Scieuzo, C., Ianniciello, D., Tafi, E., Hahn, T., & Falabella, P. (2022). Antimicrobial properties of chitosan from different developmental stages of the bioconverter insect Hermetiaillucens. *Scientific Reports*, *12*(1), 8084.

Hameed, A. Z., Raj, S. A., Kandasamy, J., Baghdadi, M. A., & Shahzad, M. A. (2022). Chitosan: a sustainable material for multifarious applications. *Polymers*, *14*(12), 2335.

Hassan, A. H. A., Korany, A. M., Zeinhom, M. M. A., Mohamed, D. S., & Abdel-Atty, N. S. (2022). Effect of chitosan-gelatin coating fortified with papaya leaves and thyme extract on quality and shelf life of chicken breast fillet and Kareish cheese during chilled storage. *International Journal of Food Microbiology*, *371*, 109667. https://doi.org/10.1016/j.ijfoodmicro.2022.109667

Ikram, R., Mohamed Jan, B., Abdul Qadir, M., Sidek, A., Stylianakis, M. M., & Kenanakis, G. (2021). Recent advances in chitin and chitosan/graphene-based bio-nanocomposites for energetic applications. *Polymers*, *13*(19), 3266.

Iqbal, M. W., Riaz, T., Yasmin, I., Leghari, A. A., Amin, S., Bilal, M., & Qi, X. (2021). Chitosan-based materials as edible coating of cheese: a review. *Starch-Stärke*, *73*(11–12), 2100088.

Islam, N., Hoque, M., & Taharat, S. F. (2023). Recent advances in extraction of chitin and chitosan. *World Journal of Microbiology and Biotechnology*, *39*(1), 28.

Ismail, B. B., Mohammed, S. F., Garba, U., & Sanusi, S. N. (2021). Edible films and coatings based on chitosan with essential oils for antimicrobial food packaging application. *RivistaItaliana Delle Sostanze Grasse*, *98*(1), 1–11.

Jafari, F. J., & Javadi, A. (2020). The effect of chitosan coating incorporated with walnut leaf extract on shelf life of pistachio. *Journal of Food Research*, *30*(3), 221–232.

Kaloper, S., Plohl, O., Možina, S. S., Vesel, A., Šimat, V., & Zemljič, L. F. (2024). Exploring chitosan-plant extract bilayer coatings: advancements in active food packaging via polypropylene modification. *International Journal of Biological Macromolecules*, *270*, 132308.

Ke, C. L., Deng, F. S., Chuang, C. Y., & Lin, C. H. (2021). Antimicrobial actions and applications of chitosan. *Polymers*, *13*(6), 904.

Khan, F., Pham, D. T. N., Oloketuyi, S. F., Manivasagan, P., Oh, J., & Kim, Y. M. (2020). Chitosan and their derivatives: antibiofilm drugs against pathogenic bacteria. *Biointerfaces*, 185, 110627.

Khubiev, O. M., Egorov, A. R., Kirichuk, A. A., Khrustalev, V. N., Tskhovrebov, A. G., & Kritchenkov, A. S. (2023). Chitosan-based antibacterial films for biomedical and food applications. *International Journal of Molecular Sciences*, *24*(13), 10738.

Kravanja, G., Primožič, M., Knez, Ž., & Leitgeb, M. (2019). Chitosan-based (nano) materials for novel biomedical applications. *Molecules*, *24*(10), 1960.

Lam, P. L., Wong, R. M., Lam, K. H., Hung, L. K., Wong, M. M., Yung, L. H., Ho, Y.-W., Wong, W.-Y., Hau, D.K.-P., Gambari, R., & Chui, C.-H. (2020). The role of reactive oxygen species in the biological activity of antimicrobial agents: an updated mini review. *Chemico-Biological Interactions*, *320*, 109023.

Laroche, C., Delattre, C., Mati-Baouche, N., Salah, R., Ursu, A. V., Moulti-Mati, F., & Pierre, G. (2018). Bioactivity of chitosan and its derivatives. *Current Organic Chemistry*, *22*(7), 641–667.

Li, L., Xu, Y., Xu, Z., Wu, C., Chen, Q., Xu, K., & Rao, X. (2024). Synthesis, characterization and antifungal properties of maleopimaric anhydride modified chitosan. *International Journal of Biological Macromolecules, 267*, 131373.

Liang, B., Song, W., Xing, R., Liu, S., Yu, H., & Li, P. (2023). The source, activity influencing factors and biological activities for future development of chitin deacetylase. *Carbohydrate Polymers*, *321*, 121335.

Liu, T., Zheng, N., Ma, Y., Zhang, Y., Lei, H., Zhen, X., Wang, Y., Gou, D., & Zhao, J. (2024). Recent advancements in chitosan-based intelligent food freshness indicators: categorization, advantages, and applications. *International Journal of Biological Macromolecules*, *275*,133554.

Mahajan, G., Sharma, V., & Gupta, R. (2024). Chitinase: a potent biocatalyst and its diverse applications. *Biocatalysis and Biotransformation*, *42*(2), 85–109.

Másson, M. (2024). The quantitative molecular weight-antimicrobial activity relationship for chitosan polymers, oligomers, and derivatives. *Carbohydrate Polymers*, *337*, 122159.

Mauricio, R. A., Deliza, R., & Nassu, R. T. (2022). Consumers' attitudes toward the use of an edible coating for lamb meat according to label information. *Foods (Basel, Switzerland)*, *11*(3), 323.

Mazhar, S., Yasmeen, R., Chaudhry, A., Summia, K., Hussain, I., Amjad, S., & Ali, E. (2022). Role of microorganisms in modern food industry. *International Journal of Food Science & Technology*, *4*, 65–77.

McCarthy, U., Uysal, I., Badia-Melis, R., Mercier, S., O'Donnell, C., & Ktenioudaki, A. (2018). Global food security: issues, challenges and technological solutions. *Trends in Food Science & Technology*, *77*, 11–20. https://doi.org/10.1016/j.tifs.2018.05.002

Meng, D., Garba, B., Ren, Y., Yao, M., Xia, X., Li, M., & Wang, Y. (2020). Antifungal activity of chitosan against Aspergillus ochraceus and its possible mechanisms of action. *International Journal of Biological Macromolecules*, *158*, 1063–1070.

Moller, A., Leone, C., Kataria, J., Sidhu, G., Rama, E. N., Kroft, B., Thippareddi, H., & Singh, M. (2023). Effect of a carrageenan/chitosan coating with allyl isothiocyanate on microbial spoilage and quality of chicken breast. *Poultry Science*, *102*(3), 102442.

Morata-Moreno, N., Pérez-Tanoira, R., del Campo-Balguerias, A., Carrillo-Hermosilla, F., Hernando-Gozalo, M., Rescalvo-Casas, C., Ocana, A. V., Segui, P., Alonso-Moreno, C., Pérez-Martínez, F. C., & Molina-Alarcón, M. (2024). A new guanidine-core small-molecule compound as a potential antimicrobial agent against resistant bacterial strains. *Antibiotics*, *13*(7), 609.

Mumtaz, S., Ali, S., Mumtaz, S., Mughal, T. A., Tahir, H. M., & Shakir, H. A. (2023). Chitosan conjugated silver nanoparticles: the versatile antibacterial agents. *Polymer Bulletin*, *80*(5), 4719–4736.

Muñoz-Tebar, N., Pérez-Álvarez, J. A., Fernández-López, J., & Viuda-Martos, M. (2023). Chitosan edible films and coatings with added bioactive compounds: antibacterial and antioxidant properties and their application to food products: a review. *Polymers*, *15*(2), 396.

Nagib, A., Helal, M., Sami, R., Aljumayi, H., Algarni, E., Al-Mushhin, A. A., & Taha, I. M. (2024). Contribution of hurdles technology to preserve the qualities of broccoli and extend their shelf-life. *Journal of Biobased Materials and Bioenergy*, *18*(3), 453–464.

Namangkalakul, W., Benjavongkulchai, S., Pochana, T., Promchai, A., Satitviboon, W., Howattanapanich, S., Phuprasong, R., Ungvijanpunya, N., Supakanjanakanti, D., Chaitrakoonthong, T., Muangsawat, S., Thanyasrisung, P., & Matangkasombut, O. (2020). Activity of chitosan antifungal denture adhesive against common Candida species and Candida albicans adherence on denture base acrylic resin. *The Journal of Prosthetic Dentistry*, *123*(1), 181.e1.

Olaimat, A. N., Al-Holy, M., Al-Nabulsi, A. A., & Osaili, T. M. (2023). Chapter 8- Chitosan: a potential antimicrobial agent to enhance microbial safety and shelf life of salad dips and ethnic foods. In: I. N. Savvaidis (Ed.), *Chitosan: Novel Applications in Food Systems* (pp. 269–278). Academic Press. https://doi.org/10.1016/B978-0-12-821663-7.00003-X

Olaniran, A. F., Afolabi, R. O., Abu, H. E., Owolabi, A., Iranloye, Y. M., & Okolie, C. E. (2020). Lime potentials as biopreservative as alternative to chemical preservatives in pineapple, orange and watermelon juice blend. *Food Research*, *4*(6), 1878–1884.

Osek, J., Lachtara, B., & Wieczorek, K. (2022). Listeria monocytogenes: how this pathogen survives in food-production environments? *Frontiers in Microbiology*, *13*, 866462.

Pasquina-Lemonche, L., Burns, J., Turner, R. D., Kumar, S., Tank, R., Mullin, N., Wilson, J. S., Chakrabarti, B. Bullough, P. A., Foster, S. J., & Hobbs, J. K. (2020). The architecture of the gram-positive bacterial cell wall. *Nature*, *582*(7811), 294–297.

Pellissery, A. J., Vinayamohan, P. G., Amalaradjou, M. A. R., & Venkitanarayanan, K. (2020). Spoilage bacteria and meat quality. In: I. Meat (Ed.) *Meat quality analysis*, chapter 17, (pp. 307–334). Academic Press Cambrigde.

Pham, T. T., Nguyen, L. L. P., Dam, M. S., & Baranyai, L. (2023). Application of edible coating in extension of fruit shelf life. *AgriEngineering*, *5*(1), 520–536.

Porat, R., Lichter, A., Terry, L. A., Harker, R., & Buzby, J. (2018). Postharvest losses of fruit and vegetables during retail and in consumers' homes: quantifications, causes, and means of prevention. *Postharvest Biology and Technology*, *139*, 135–149.

Rajaram, S., Damasceno, N. R. T., Braga, R. A. M., Martinez, R., Kris-Etherton, P., & Sala-Vila, A. (2023). Effect of nuts on markers of inflammation and oxidative stress: a narrative review. *Nutrients*, *15*(5), 1099.

Rehman, K. U., Hollah, C., Wiesotzki, K., Heinz, V., Aganovic, K., Rehman, R. U., Petrusan, J. I., Zheng, L., Zhang, J., Sohail, S., & Cai, M. (2023). Insect-derived chitin and chitosan: a still unexploited resource for the edible insect sector. *Sustainability*, *15*(6), 4864.

Romanazzi, G., & Moumni, M. (2022). Chitosan and other edible coatings to extend shelf life, manage postharvest decay, and reduce loss and waste of fresh fruits and vegetables. *Current Opinion in Biotechnology*, *78*, 102834.

Roshandel-Hesari, N., Mokaber-Esfahani, M., Taleghani, A., & Akbari, R. (2022). Investigation of physicochemical properties, antimicrobial and antioxidant activity of edible films based on chitosan/casein containing Origanum vulgare L. essential oil and its effect on quality maintenance of cherry tomato. *Food Chemistry*, *396*, 133650.

Sarengaowa, W. L., Liu, Y., Yang, C., Feng, K., & Hu, W. (2022). Screening of essential oils and effect of a chitosan-based edible coating containing cinnamon oil on the quality and microbial safety of fresh-cut potatoes. *Coatings*, *12*(10), 1492.

Shaltout, F. A. (2024). Ways of food contamination, its impact and prevention. *Food Science & Nutrition Technology*, *9*, 1–8.

Singh, P., Yadav, V., Sahu, D., Kumar, K., Kim, D., Yang, D., Jayaraman, S., Jarzębski, M., Wieruszewsk, M., & Pal, K. (2024). Exploring chitosan lactate as a multifunctional additive: enhancing quality and extending shelf life of whole wheat bread. *Foods*, *13*(10), 1590.

Song, Y. S., Jo, Y. H., Han, Y. S., & Jung, W. J. (2022). Production of chitin-and chitosan-oligosaccharide using the edible insect, Tenebrio molitor. *Entomological Research*, *52*(4), 207–213.

Sutharsan, J., & Zhao, J. (2023). Physicochemical and biological properties of chitosan based edible films. *Food Reviews International*, *39*(9), 6296–6323.

Tan, Y. N., Lee, P. P., & Chen, W. N. (2020). Microbial extraction of chitin from seafood waste using sugars derived from fruit waste-stream. *AMB Express*, *10*, 17.

Teixeira-Santos, R., Lima, M., Gomes, L. C., & Mergulhão, F. J. (2021). Antimicrobial coatings based on chitosan to prevent implant-associated infections: a systematic review. *Iscience*, *24*(12), 103480.

Wang, P., Li, Y., Zhang, C., Que, F., Weiss, J., & Zhang, H. (2020). Characterization and antioxidant activity of trilayer gelatin/dextran-propyl gallate/gelatin films: electrospinning versus solvent casting. *LWT*, *128*, 109536.

Wang, Y., Zhao, W., Han, L., & Tam, K. C. (2022). Superhydrophobic surfaces from sustainable colloidal systems. *Current Opinion in Colloid & Interface Science*, *57*, 101534.

Wong, C. H., Mak, I. E. K., & Li, D. (2021). Bilayer edible coating with stabilized Lactobacillus plantarum 299v improved the shelf life and safety quality of fresh-cut apple slices. *Food Packaging and Shelf Life*, *30*, 100746.

Yaashikaa, P. R., Kamalesh, R., Kumar, P. S., Saravanan, A., Vijayasri, K., & Rangasamy, G. (2023). Recent advances in edible coatings and their application in food packaging. *Food Research International*, *173*, 113366.

Yan, D., Li, Y., Liu, Y., Li, N., Zhang, X., & Yan, C. (2021). Antimicrobial properties of chitosan and chitosan derivatives in the treatment of enteric infections. *Molecules*, *26*(23), 7136.

Yang, C., Lu, J. H., Xu, M. T., Shi, X. C., Song, Z. W., Chen, T. M., & Wang, S. Y. (2022). Evaluation of chitosan coatings enriched with turmeric and green tea extracts on postharvest preservation of strawberries. *LWT*, *163*, 113551.

Yang, Y., Aghbashlo, M., Gupta, V. K., Amiri, H., Pan, J., Tabatabaei, M., & Rajaei, A. (2023). Chitosan nanocarriers containing essential oils as a green strategy to improve the functional properties of chitosan: a review. *International Journal of Biological Macromolecules*, *236*, 123954.

Yilmaz Atay, H. (2020). Antibacterial activity of chitosan-based systems. In: S. Jana & S. Jana (Eds.), Functional Chitosan*: Drug Delivery and Biomedical Applications* (pp. 457–489). Springer, Singapore.

Zhang, W., Li, X., & Jiang, W. (2020). Development of antioxidant chitosan film with banana peels extract and its application as coating in maintaining the storage quality of apple. *International Journal of Biological Macromolecules*, *154*, 1205–1214.

7 Integration of Chitosan Edible Coating and Modified Atmosphere Packaging

John Tsado Mathew[1], Jonathan Hussaini[1], Abel Inobeme[2], Elijah Yanda Shaba[3], Mary Jummai Jiya[4], Charles Oluwaseun Adetunji[5], Yakubu Azeh[1], Musah Monday[1], Muhammad Aishetu Ibrahim[1], Etsuyankpa Muhammad Bini[6], Musa Safiyanu Tanko[6], Ismaila Aliyu Oga[6], Tanko Ezekiel[3], and Amos Mamman[1]

[1]Department of Chemistry, Ibrahim Badamasi Babangida University Lapai, Niger State, Nigeria

[2]Department of Chemistry, Edo University Uzairue, Auchi, Edo State, Nigeria

[3]Department of Chemistry, Federal University of Technology, Minna, Niger State, Nigeria

[4]Department of Food Science and Technology, Federal University of Technology, Minna, Niger State, Nigeria

[5]Department of Microbiology, Applied Microbiology, Biotechnology and Nanotechnology Laboratory, Edo University Uzairue, Auchi, Edo State, Nigeria

[6]Department of Chemistry, Federal University of Lafia, Nasarawa State, Nigeria

TRODUCTION

st for sustainable and effective food preservation methods, the integration of chito- oatings with modified atmosphere packaging (MAP) has emerged as a compelling h the increasing demand for minimally processed and extended shelf-life foods, this egy offers a synergistic solution to address both preservation and packaging chal- , a natural biopolymer derived from chitin found in crustacean shells, has garnered on for its biocompatibility, biodegradability, and antimicrobial properties. When coating, chitosan forms a protective barrier that reduces moisture loss, inhibits nd extends the shelf life of perishable foods (Díaz-Montes & Castro-Muñoz, l., 2025).

benefits of chitosan coatings, MAP technology allows for precise control nditions surrounding packaged foods. By modifying the composition of oxygen and increasing carbon dioxide levels, MAP creates an environ- proliferation and oxidative reactions, thereby preserving food freshness n of chitosan coatings and MAP presents a multifaceted approach to the inherent properties of chitosan with the tailored atmospheric con- soun et al., 2020).

DOI: 10.1201/9781003355458-7

7.2 OVERVIEW OF CHITOSAN

Chitosan, a naturally occurring biopolymer, is derived from chitin, the second most abundant biopolymer in nature after cellulose. It is primarily extracted from the shells of crustaceans such as shrimp, crab, and lobster, as well as from fungal sources like mushrooms. Chitosan is characterized by its unique chemical structure, consisting of β-(1→4)-linked D-glucosamine and *N*-acetyl-D-glucosamine units. This structure gives chitosan its remarkable properties, including biodegradability, biocompatibility, and antimicrobial activity (Piekarska et al., 2023).

Chitosan's versatility has led to its widespread application in various industries, including pharmaceuticals, agriculture, cosmetics, and food. In the food industry, chitosan is particularly valued for its use as an edible coating, where it forms a protective layer on the surface of food products. This coating helps extend the shelf life of perishable foods by reducing moisture loss, inhibiting microbial growth, and delaying oxidative reactions. Additionally, chitosan has been explored for its potential health benefits, including cholesterol-lowering effects, wound healing properties, and immunomodulatory activities (Thambiliyagodage et al., 2023; Mathew et al., 2025a).

With its numerous advantageous properties and diverse applications, chitosan continues to attract significant attention from researchers, industry professionals, and consumers alike. Ongoing research efforts aim to further explore and optimize the use of chitosan in various fields, including food technology, biomedical engineering, and environmental science, highlighting its potential to address pressing challenges and contribute to sustainable development (El-Araby et al., 2024).

Chitosan's wide-ranging applications stem from its unique properties and biocompatibility. In pharmaceuticals, chitosan serves as a versatile excipient in drug delivery systems due to its mucoadhesive and permeation-enhancing properties. It is used in formulations for wound dressings, oral drug delivery, and tissue engineering scaffolds. Moreover, chitosan's ability to chelate metal ions makes it valuable in water treatment applications, where it is used to remove heavy metals and other pollutants from wastewater (Mura et al., 2022).

In agriculture, chitosan-based products are utilized as biostimulants and biopesticides to enhance plant growth, increase crop yields, and protect against pests and diseases. Chitosan's biocompatibility and biodegradability make it an attractive alternative to synthetic agrochemicals, contributing to sustainable agricultural practices and reducing environmental impact (Stasińska-Jakubas & Hawrylak-Nowak, 2022; Malik et al., 2025).

Furthermore, chitosan's antimicrobial properties have sparked interest in its use as a natural preservative in food packaging materials, where it can help extend the shelf life of packaged foods by inhibiting the growth of spoilage microorganisms. Chitosan films and coatings offer a sustainable alternative to conventional plastic packaging, addressing concerns about plastic pollution and food waste. Chitosan's versatility and biocompatibility make it a valuable resource with diverse applications across multiple industries. Ongoing research efforts continue to explore new uses and optimization strategies for chitosan, highlighting its potential to address pressing challenges in healthcare, agriculture, environmental protection, and beyond. As technology advances and our understanding of chitosan deepens, its role in shaping a more sustainable and resilient future is poised to expand even further (Duda-Chodak et al., 2023).

7.3 CHITOSAN EDIBLE COATING: PROPERTIES AND APPLICATIONS

As shown in Table 7.1, chitosan, a natural biopolymer derived from chitin found in the exoskeletons of crustaceans such as shrimp and crab, has garnered significant interest in the food industry for its versatile properties and various applications. One of its key applications is as an edible coating, where chitosan is used to form a protective layer on the surface of food products. This coating offers several beneficial properties that contribute to the preservation and quality enhancement of perishable foods (Teixeira-Costa & Andrade, 2021; Dawud et al., 2025).

TABLE 7.1
Chitosan Edible Coating: Properties and Applications

Properties	Description	Applications	References
Biodegradability	Chitosan is biodegradable, making it environmentally friendly and suitable for sustainable packaging solutions	Eco-friendly packaging	Zehra et al. (2022)
Film-forming ability	Chitosan has excellent film-forming properties, allowing it to create a protective barrier on the surface of foods	Extending shelf life	Jiang et al. (2023)
Antimicrobial activity	Chitosan exhibits antimicrobial properties, inhibiting the growth of various microorganisms, including bacteria and fungi	Minimizing microbial spoilage	Wu et al. (2023)
Barrier to moisture	Chitosan coatings act as a barrier to moisture, preventing dehydration and maintaining the quality of perishable foods	Preserving freshness	Rajestary et al. (2022)

First and foremost, chitosan is biodegradable, making it an environmentally friendly alternative to synthetic packaging materials. Its biocompatibility and non-toxic nature make it suitable for direct contact with food, addressing consumer concerns about chemical residues in packaging. Moreover, chitosan exhibits excellent film-forming ability, allowing it to create a thin, transparent coating that adheres well to the surface of various food substrates. This film acts as a barrier to moisture, preventing dehydration and maintaining the texture and appearance of the coated food (Jiménez-Gómez & Cecilia, 2020).

In addition to its barrier properties, chitosan possesses inherent antimicrobial activity. Its positively charged amino groups interact with negatively charged microbial cell membranes, disrupting their structure and leading to cell death. As a result, chitosan coatings can effectively inhibit the growth of spoilage microorganisms such as bacteria and fungi, thereby extending the shelf life of perishable foods. This antimicrobial effect is particularly beneficial for fresh produce and seafood, which are prone to microbial contamination during storage and transportation (Khubiev et al., 2023).

Furthermore, chitosan coatings have been shown to exhibit antioxidant properties, scavenging free radicals and inhibiting oxidative reactions that cause food deterioration. By reducing lipid oxidation and enzymatic browning, chitosan helps preserve the sensory attributes and nutritional quality of coated foods, including color, flavor, and vitamin content (Yan et al., 2021; Mathew et al., 2025b).

The properties of chitosan make it a versatile material for edible coatings with various applications in the food industry. From extending the shelf life of fresh fruits and vegetables to preserving the quality of seafood and meat products, chitosan coatings offer a sustainable and effective solution for enhancing food safety and reducing food waste. As research in this field continues to advance, chitosan-based coatings hold great promise for addressing the evolving needs of consumers and the food industry alike (Romanazzi & Moumni, 2022).

7.4 PROPERTIES RELEVANT TO EDIBLE COATING

Chitosan's properties make it highly suitable for edible coating applications, offering a range of benefits that contribute to food preservation, quality enhancement, and sustainability (Matloob et al., 2023).

Film-Forming Ability: Chitosan possesses excellent film-forming properties, allowing it to create a thin, flexible, and transparent coating on the surface of food products. This coating acts as a physical barrier, preventing moisture loss and protecting against external contaminants, including microbial pathogens and environmental pollutants. The uniformity and adherence of chitosan films ensure effective coverage and preservation of the coated food, enhancing its shelf life and maintaining its quality during storage and transportation (Muñoz-Tebar et al., 2023).

Antimicrobial Activity: Chitosan exhibits inherent antimicrobial properties, which arise from its cationic nature and interaction with microbial cell membranes. The positively charged amino groups in chitosan molecules bind to negatively charged sites on microbial cell surfaces, disrupting membrane integrity and leading to cell death. This antimicrobial effect inhibits the growth of spoilage microorganisms and pathogens, reducing the risk of foodborne illnesses and extending the microbiological safety of coated foods. Additionally, chitosan's antimicrobial activity can help minimize the need for synthetic preservatives in food products, aligning with consumer preferences for natural and minimally processed foods (Atay, 2020; Adetunji et al., 2023a).

Barrier to Moisture and Gas: Chitosan coatings act as a barrier to moisture, gases, and other external factors that can accelerate food deterioration. By reducing water vapor transmission and gas exchange between the food and its environment, chitosan helps maintain optimal moisture levels and atmospheric conditions within the packaged food. This barrier function preserves the sensory attributes, texture, and nutritional quality of coated foods, preventing dehydration, oxidation, and flavor loss. Furthermore, chitosan coatings can be tailored to control the permeability of specific gases, such as oxygen and carbon dioxide, to further optimize food preservation and freshness (Felicia et al., 2022).

Biodegradability and Biocompatibility: Chitosan is biodegradable, meaning it can be broken down by natural processes into non-toxic byproducts, making it environmentally friendly and sustainable. Its biocompatibility ensures safety for direct contact with food, without posing risks of chemical leaching or contamination. Chitosan-based edible coatings offer a green and renewable alternative to synthetic packaging materials, reducing reliance on fossil fuels and mitigating the environmental impact of food packaging waste (Hameed et al., 2022; Adetunji et al., 2023b).

Furthermore, the unique properties of chitosan, including its film-forming ability, antimicrobial activity, barrier function, and biodegradability, make it an ideal material for edible coatings in food preservation. By leveraging these properties, chitosan coatings help extend the shelf life, enhance the quality, and ensure the safety of a wide range of perishable food products, while promoting sustainability and consumer confidence in the food supply chain (Muñoz-Tebar et al., 2023).

7.5 APPLICATIONS IN FOOD PRESERVATION

Chitosan's applications in food preservation encompass a range of techniques aimed at extending the shelf life, maintaining quality, and ensuring safety of various food products. First, chitosan coatings are widely employed to extend the shelf life of fresh fruits, vegetables, seafood, and meats by forming a protective barrier against moisture loss, microbial contamination, and oxidative reactions (Table 7.2) (Teshome et al., 2022).

This helps prolong the freshness and visual appeal of the coated foods during storage and transportation. Second, chitosan's inherent antimicrobial properties make it effective in minimizing microbial spoilage and inhibiting the growth of bacteria and fungi that can cause foodborne illnesses and quality deterioration. Additionally, chitosan coatings enhance the freshness and quality of bakery products, dairy items, and ready-to-eat meals by preserving their sensory attributes, texture, and nutritional value. Lastly, chitosan-based coatings offer a natural and sustainable alternative to synthetic preservatives, reducing the reliance on chemical additives in processed meats, dairy products, and sauces, thus meeting consumer demand for clean-label foods. Overall, chitosan's versatile applications in food preservation contribute to enhancing food safety, reducing food waste, and meeting consumer preferences for fresh, high-quality, minimally processed foods (Suvarna et al., 2022; Abimbola et al., 2023).

TABLE 7.2
Applications in Food Preservation

Applications	Description	Examples	References
Extending shelf life	Chitosan coatings are applied to food surfaces to create a protective barrier against moisture loss, microbial contamination, and oxidative reactions, thereby prolonging shelf life	Fresh fruits and vegetables	Perez-Vazquez et al. (2023)
Minimizing microbial spoilage	Chitosan exhibits antimicrobial properties, inhibiting the growth of spoilage microorganisms such as bacteria and fungi, which helps maintain food safety and quality during storage and transportation	Seafood, meat, and poultry products	Vasile and Baican (2021) and Inobeme et al. (2023)
Enhancing freshness and quality	Chitosan coatings preserve the sensory attributes, texture, and nutritional quality of coated foods by preventing dehydration, oxidation, and flavor loss, ensuring freshness and consumer satisfaction	Bakery products, cheese, and ready-to-eat meals	Adiletta et al. (2021) and Muhammad et al. (2025)
Reducing chemical preservatives	Chitosan-based coatings offer a natural and sustainable alternative to synthetic preservatives, minimizing the need for chemical additives in food products, thus meeting consumer demand for clean-label foods	Processed meats, dairy products, and sauces	Kaur et al. (2022)

Furthermore, chitosan's versatility extends to its ability to reduce the need for chemical preservatives, offering a natural and sustainable approach to food preservation. By incorporating chitosan-based coatings in processed meats, dairy products, and sauces, manufacturers can minimize the use of synthetic additives while ensuring product safety and shelf stability. This aligns with consumer preferences for clean-label foods, free from artificial ingredients and chemical residues. Chitosan's efficacy in preserving food quality and safety, coupled with its biodegradability and biocompatibility, positions it as a valuable tool in the pursuit of sustainable food preservation practices. Chitosan's applications in food preservation encompass a broad spectrum of techniques that address key challenges in maintaining the freshness, safety, and quality of food products. From extending shelf life and minimizing microbial spoilage to enhancing sensory attributes and reducing chemical preservatives, chitosan-based coatings offer a multifaceted approach to food preservation that aligns with consumer demands for natural, sustainable, and minimally processed foods. As research and innovation in this field continue to advance, the potential for chitosan to revolutionize food preservation practices and contribute to a more sustainable food supply chain remains promising (Ogundolie et al., 2023; Stefanowska et al., 2023).

7.6 GAS COMPOSITION AND EFFECTS ON FOOD PRESERVATION

The gas composition surrounding packaged foods plays a critical role in food preservation, with various gases exerting different effects on microbial growth, enzymatic reactions, and food quality. MAP involves adjusting the atmospheric conditions within the packaging to create an optimal environment for extending the shelf life of perishable foods (Martin et al., 2023).

Oxygen (O_2): Oxygen is essential for aerobic microbial growth and enzymatic reactions, but it also contributes to oxidative reactions that lead to food spoilage and quality deterioration. In MAP, reducing oxygen levels can slow down oxidative processes, such as lipid oxidation and enzymatic browning, thereby preserving the color, flavor, and nutritional content of foods (Safaei et al., 2022).

Carbon Dioxide (CO_2): Carbon dioxide inhibits the growth of spoilage microorganisms and pathogens by altering the pH of the food matrix and creating an acidic environment. Additionally, CO_2 can suppress the activity of certain enzymes responsible for food deterioration, such as polyphenol oxidase in fruits and vegetables. MAP often involves increasing CO_2 levels to suppress microbial growth and extend the shelf life of packaged foods (Cejudo et al., 2023).

Nitrogen (N_2): Nitrogen is commonly used as an inert gas in MAP to displace oxygen and create a modified atmosphere with reduced oxygen levels. By minimizing oxygen exposure, nitrogen helps prevent oxidative reactions and maintain the freshness and quality of packaged foods. Nitrogen also acts as a filler gas, providing cushioning and protection against physical damage during handling and transportation (Czerwiński et al., 2021; Chimbekujwo et al., 2025).

Ethylene (C_2H_4): Ethylene is a natural plant hormone produced by fruits and vegetables during ripening, which can accelerate the ripening process and promote senescence and decay. Controlling ethylene levels in MAP can help regulate the ripening and shelf life of fresh produce, extending its marketability and reducing food waste (Cocetta & Natalini, 2022).

Furthermore, the manipulation of gas composition in MAP offers a powerful tool for food preservation, allowing for customized packaging solutions tailored to the specific needs of different food products. By optimizing gas composition to minimize microbial growth, oxidative reactions, and enzymatic degradation, MAP helps maintain the safety, quality, and freshness of packaged foods, thereby extending their shelf life and reducing food waste throughout the supply chain. The combination of MAP with chitosan edible coatings can synergistically enhance food preservation by leveraging the benefits of both techniques. Chitosan coatings create a physical barrier against moisture loss and microbial contamination on the surface of food products, while MAP adjusts the gas composition within the packaging to further inhibit microbial growth and oxidative reactions (Versino et al., 2023).

By integrating chitosan coatings with MAP, oxygen levels can be reduced to slow down oxidation processes, while carbon dioxide levels can be increased to suppress microbial proliferation. This combined approach extends the shelf life of perishable foods, preserves their quality, and enhances their safety. Additionally, chitosan coatings can help maintain the integrity of MAP by providing an additional layer of protection against physical damage and punctures during handling and transportation (Kontominas et al., 2021; Mathew et al., 2024a).

The synergy between chitosan coatings and MAP offers a comprehensive solution to the challenges of food preservation, addressing microbial spoilage, oxidative deterioration, and physical damage simultaneously. This integrated approach not only extends the shelf life of packaged foods but also reduces the need for synthetic preservatives and plastic packaging materials, contributing to sustainability and environmental conservation (Gigante et al., 2023).

7.7 CURRENT APPLICATIONS AND LIMITATIONS OF CHITOSAN EDIBLE COATINGS

Current applications of chitosan edible coatings and MAP span various sectors of the food industry, offering solutions for extending shelf life, enhancing food safety, preserving quality, and promoting sustainability. Chitosan coatings combined with MAP are extensively used to prolong the shelf life of perishable foods such as fresh fruits, vegetables, seafood, meats, and dairy products. By forming a protective barrier against moisture loss, microbial contamination, and oxidative reactions, these technologies help maintain food freshness and quality throughout storage and distribution (Perez-Vazquez et al., 2023).

In terms of food safety, the antimicrobial properties of chitosan coatings and MAP contribute to inhibiting the growth of spoilage microorganisms and pathogens, reducing the risk of foodborne illnesses. This is particularly relevant for ready-to-eat meals, deli meats, and minimally processed foods where maintaining microbial safety is crucial (Wrońska et al., 2021).

Preservation of food quality is another key application, with chitosan coatings and MAP helping to preserve the sensory attributes, texture, and nutritional content of foods. These technologies minimize dehydration, oxidation, and flavor loss, ensuring consumer satisfaction and minimizing food waste. Bakery products, cheese, fresh-cut produce, and pre-packaged salads are among the foods benefiting from enhanced quality preservation. Moreover, chitosan-based coatings and MAP offer sustainable packaging solutions by reducing the need for synthetic preservatives and plastic materials. By minimizing environmental impact and promoting eco-friendly packaging alternatives, these technologies contribute to sustainability efforts in the food industry (Gautam et al., 2023).

However, despite their numerous advantages, chitosan edible coatings and MAP also have limitations that may affect their widespread adoption. High production costs and scalability issues pose challenges for small-scale producers and economically constrained regions, limiting access to these technologies. Sensory changes, such as alterations in color, texture, and flavor, may occur in certain food products subjected to chitosan coatings and MAP, impacting consumer acceptance. Compatibility with packaging materials is another consideration, as chitosan coatings may not be suitable for all substrates, necessitating compatibility testing and potential modifications. Additionally, regulatory approval and labeling requirements for chitosan coatings and MAP vary by region, posing challenges for international trade and market access. Despite these limitations, ongoing research and development efforts aim to address these challenges and further optimize the applications of chitosan coatings and MAP in the food industry (Algarni et al., 2022).

Efforts to overcome the limitations of chitosan edible coatings and MAP include advancements in production processes to reduce costs and improve scalability, as well as research into mitigating sensory changes through formulation optimization. Compatibility testing with a wide range of packaging materials is also underway to expand the applicability of chitosan coatings. Additionally, collaboration with regulatory agencies and standardization of labeling requirements can streamline the approval process and facilitate market access for chitosan-based technologies (Lisitsyn et al., 2021; Mathew et al., 2023a).

Furthermore, ongoing research aims to enhance the efficacy and versatility of chitosan coatings and MAP through innovations such as incorporating active ingredients for targeted functionality, developing composite materials for improved performance, and optimizing processing parameters for optimal outcomes. These advancements can address specific challenges in food preservation while capitalizing on the inherent benefits of chitosan and MAP. While chitosan edible coatings and MAP offer promising solutions for food preservation and sustainability, addressing their limitations is essential for broader adoption and integration into the food industry. Through ongoing research, technological innovation, and collaborative efforts across academia, industry, and regulatory bodies, the potential of chitosan coatings and MAP to revolutionize food preservation practices and contribute to a more sustainable food supply chain can be fully realized. As these technologies continue to evolve, they hold the promise of enhancing food safety, reducing food waste, and meeting the changing needs and preferences of consumers in an increasingly complex and dynamic food landscape (Ilyas et al., 2022; Mathew et al., 2024b).

7.8 SYNERGISTIC EFFECTS OF CHITOSAN EDIBLE COATING AND MAP

As shown in Table 7.3, the combination of chitosan edible coating and MAP offers synergistic effects that significantly enhance food preservation, safety, quality, and sustainability. First, this synergistic approach extends the shelf life of packaged foods by providing a dual barrier against

TABLE 7.3
Synergistic Effects of Chitosan Edible Coating and MAP

Synergistic Effects	Description	Examples	References
Extended shelf life	The combination of chitosan coating and MAP creates a synergistic effect by providing a dual barrier against moisture loss, microbial contamination, and oxidative reactions, prolonging the shelf life of packaged foods	Fresh fruits, vegetables, seafood, meats	Berdejo et al. (2023)
Enhanced microbial safety	Chitosan coatings inhibit microbial growth on the surface of food products, while MAP adjusts the atmosphere to further suppress microbial proliferation, enhancing the overall microbial safety and quality of packaged foods	Ready-to-eat meals, deli meats, minimally processed foods	Yu et al. (2021) and Mathew et al. (2023b)
Improved preservation of food quality	The combined action of chitosan coating and MAP helps preserve the sensory attributes, texture, and nutritional content of foods by minimizing dehydration, oxidation, and flavor loss, ensuring higher quality and consumer satisfaction	Bakery products, cheese, fresh-cut produce, pre-packaged salads	Olawuyi et al. (2021) and Mathew et al. (2024c)
Sustainable packaging solutions with added benefits	Chitosan-based coatings in combination with MAP offer sustainable packaging alternatives that not only extend shelf life and enhance food safety but also reduce the need for synthetic preservatives and plastic materials, promoting environmental sustainability	Eco-friendly packaging, reduced plastic waste, minimized food waste	Bose et al. (2023)

moisture loss, microbial contamination, and oxidative reactions. Chitosan coatings create a physical barrier on the food surface, while MAP adjusts the atmosphere within the packaging to minimize microbial growth and oxidative deterioration. This combined action results in prolonged freshness and improved shelf stability, particularly evident in fresh fruits, vegetables, seafood, and meats (Liao et al., 2023).

The antimicrobial properties of chitosan coatings complement the microbial control achieved through MAP, resulting in enhanced food safety. Chitosan inhibits microbial growth on the food surface, while MAP further suppresses microbial proliferation by adjusting gas composition. This combined approach is effective in inhibiting spoilage microorganisms and pathogens, reducing the risk of foodborne illnesses, and ensuring higher microbial safety in ready-to-eat meals, deli meats, and minimally processed foods (Duda-Chodak et al., 2023).

Furthermore, the synergistic effects of chitosan coating and MAP contribute to improved preservation of food quality. By minimizing dehydration, oxidation, and flavor loss, while preserving sensory attributes and nutritional content, this combined approach enhances the overall quality and consumer satisfaction of packaged foods. Bakery products, cheese, fresh-cut produce, and pre-packaged salads are among the food items benefitting from this improved preservation. The integration of chitosan coating with MAP offers sustainable packaging solutions with added benefits. Chitosan-based coatings reduce the need for synthetic preservatives and plastic materials, while

MAP minimizes food waste and extends shelf life. This synergistic approach promotes environmental sustainability by reducing plastic waste, minimizing food waste, and offering eco-friendly packaging alternatives. The synergistic effects of chitosan edible coating and MAP offer comprehensive solutions for food preservation and sustainability. By combining their unique advantages, these technologies address key challenges in the food industry, ensuring extended shelf life, enhanced food safety, improved quality, and reduced environmental impact. As research and innovation in this field continue to advance, the potential for synergistic effects to further optimize food preservation practices and contribute to a more sustainable food supply chain remains promising (Liu et al., 2022). The integration of chitosan edible coatings with MAP facilitates the implementation of tailored packaging solutions to meet specific needs of different food products and market requirements. By adjusting the gas composition within the packaging environment, MAP can optimize conditions for particular food types, such as controlling ethylene levels to regulate the ripening process of fruits and vegetables. This customization ensures optimal preservation outcomes, maximizing the shelf life and quality of diverse food products while minimizing resource use and environmental impact (Moeini et al., 2022).

Moreover, the combined application of chitosan coatings and MAP holds promise for addressing emerging challenges in the food industry, such as reducing reliance on synthetic additives, combating food waste, and meeting evolving consumer preferences for sustainable and minimally processed foods. By harnessing the synergistic effects of these technologies, food manufacturers can offer innovative packaging solutions that enhance food safety, quality, and sustainability throughout the supply chain. Despite the considerable benefits of the synergistic approach, ongoing research is necessary to optimize formulation, application methods, and packaging designs to further enhance effectiveness and scalability. Additionally, efforts to streamline regulatory approval processes and ensure compatibility with existing packaging infrastructure will facilitate widespread adoption of chitosan coatings and MAP in the food industry. Collaboration among stakeholders, including researchers, industry professionals, regulatory agencies, and consumers, is essential to drive innovation, overcome challenges, and realize the full potential of synergistic food preservation technologies (Maliki et al., 2022).

In addition, the synergistic effects of chitosan edible coatings and MAP offer multifaceted solutions for enhancing food preservation, safety, quality, and sustainability. By leveraging the complementary benefits of these technologies, the food industry can address current and future challenges while meeting the evolving needs and preferences of consumers. As research and development efforts continue to advance, the integration of chitosan coatings with MAP holds promise for revolutionizing food packaging and preservation practices, contributing to a more resilient, efficient, and sustainable food supply chain (Nunes et al., 2023).

7.9 MECHANISMS OF ACTION

As shown in Table 7.4, these mechanisms of action work synergistically to preserve the quality, safety, and freshness of packaged foods, offering comprehensive solutions for food preservation in various sectors of the food industry. The mechanisms of action of chitosan edible coatings and MAP contribute synergistically to the preservation of packaged foods, addressing key factors such as moisture control, microbial inhibition, and gas composition regulation (Shonte et al., 2024).

Chitosan edible coatings create a physical barrier on the surface of food products, serving as a protective layer that inhibits moisture loss and prevents the entry of external contaminants. By forming a tight seal, chitosan coatings help maintain optimal moisture levels within the food matrix, reducing the risk of dehydration and maintaining texture and freshness. This physical barrier also impedes gas exchange, preventing the ingress of oxygen and the escape of volatile compounds, which can contribute to flavor loss and oxidative deterioration. Examples of foods benefiting from this mechanism include fresh fruits, vegetables, and seafood, where chitosan coatings help prolong shelf life and preserve quality during storage and transportation (Muñoz-Tebar et al., 2023).

TABLE 7.4
Mechanisms of Action

Mechanisms of Action	Description	Examples	References
Physical barrier	Chitosan coatings create a physical barrier on the surface of food products, preventing moisture loss, gas exchange, and the entry of external contaminants, such as microbes and pollutants	Fresh fruits, vegetables, seafood	Kocira et al. (2021) and Inobeme et al. (2024a)
Antimicrobial activity	Chitosan exhibits inherent antimicrobial properties, disrupting microbial cell membranes through electrostatic interactions and inhibiting microbial growth and proliferation, thereby enhancing food safety and extending shelf life	Meats, dairy products, minimally processed food	Ke et al. (2021)
Gas composition control	Modified atmosphere packaging (MAP) adjusts the gas composition within the packaging environment to optimize conditions for food preservation, including reducing oxygen levels, increasing carbon dioxide levels, and controlling ethylene concentrations	Fresh produce, bakery items, processed meats	Fang and Wakisaka (2021) and Mathew et al. (2023c)

Chitosan exhibits inherent antimicrobial properties, making it effective in inhibiting the growth and proliferation of spoilage microorganisms and pathogens. Chitosan disrupts microbial cell membranes through electrostatic interactions, leading to leakage of cellular contents and eventual cell death. This antimicrobial activity is particularly valuable in extending the shelf life and enhancing the safety of perishable foods, such as meats, dairy products, and minimally processed foods. By reducing microbial contamination, chitosan coatings contribute to food safety assurance and minimize the risk of foodborne illnesses, thus improving consumer confidence and satisfaction (Shineh et al., 2023).

Additionally, MAP complements the action of chitosan coatings by controlling the gas composition within the packaging environment. MAP adjusts oxygen, carbon dioxide, and ethylene levels to create an optimal atmosphere that inhibits microbial growth, slows down oxidative reactions, and regulates the ripening process of fruits and vegetables. By reducing oxygen levels, MAP helps mitigate oxidative deterioration, preserving the color, flavor, and nutritional content of packaged foods. Increasing carbon dioxide levels suppress microbial proliferation, extending shelf life and maintaining product safety. Moreover, controlling ethylene concentrations regulates the ripening process, allowing for better management of fruit and vegetable quality during storage and distribution. The combined mechanisms of action of chitosan edible coatings and MAP provide comprehensive solutions for food preservation, addressing multiple aspects of quality, safety, and shelf life extension. By leveraging physical barriers, antimicrobial activity, and gas composition control, these technologies offer effective strategies for maintaining the freshness, quality, and safety of a wide range of packaged foods, contributing to improved food security, reduced food waste, and enhanced consumer satisfaction (Iturralde-García et al., 2022).

7.10 CHALLENGES AND OPPORTUNITIES

Chitosan edible coatings and MAP offer promising solutions for food preservation, but they also face several challenges and present opportunities for further development and innovation in the food industry (Onyeaka et al., 2022).

One of the primary challenges is the cost-effectiveness and scalability of chitosan-based technologies. Chitosan, derived from shellfish exoskeletons or fungal sources, can be expensive to produce, limiting its widespread adoption, particularly among small-scale producers and in economically constrained regions. However, opportunities exist to optimize production processes, explore alternative sources of chitosan, and develop cost-effective formulations to overcome these barriers. Additionally, advancements in manufacturing techniques and increased demand for sustainable packaging solutions may drive down costs and improve accessibility (Popa et al., 2023).

Another challenge is the compatibility of chitosan coatings with existing packaging materials and manufacturing processes. Chitosan coatings may not adhere well to certain substrates or may require specialized equipment for application, posing logistical challenges for food manufacturers. Addressing these compatibility issues requires collaboration between researchers, packaging manufacturers, and food industry stakeholders to develop innovative packaging solutions that integrate seamlessly with chitosan coatings while maintaining efficiency and cost-effectiveness (Saberi Riseh et al., 2023).

Regulatory approval and labeling requirements also present challenges for the adoption of chitosan-based technologies. Variations in regulatory frameworks across different regions and uncertainty surrounding the safety and efficacy of chitosan coatings may hinder market acceptance and international trade. However, opportunities exist to work closely with regulatory agencies to establish clear guidelines and standards for chitosan-based packaging materials, ensuring compliance with food safety regulations and facilitating market access (Kiss & Borbély, 2023).

Despite these challenges, chitosan edible coatings and MAP offer significant opportunities for innovation and advancement in the food industry. One key opportunity lies in enhancing the functionality and performance of chitosan-based coatings through the incorporation of active ingredients, such as antimicrobial agents, antioxidants, and flavor enhancers. By leveraging synergistic effects, researchers can develop multifunctional coatings that provide enhanced preservation, safety, and sensory attributes, meeting the evolving needs and preferences of consumers (Singh et al., 2022; Inobeme et al., 2024b).

Furthermore, there is potential to explore novel applications of chitosan coatings and MAP in niche markets and specialized food products. For example, chitosan coatings may be utilized in the pharmaceutical industry for controlled-release drug delivery systems or in the cosmetics industry for encapsulating active ingredients in skincare products. These emerging applications represent untapped opportunities for expanding the use of chitosan-based technologies beyond traditional food packaging (Kulka & Sionkowska, 2023).

Another opportunity lies in promoting sustainability and environmental stewardship through the development of eco-friendly packaging solutions. Chitosan coatings and MAP offer biodegradable alternatives to conventional plastic packaging materials, reducing plastic waste and environmental pollution. By embracing sustainable packaging practices, food manufacturers can enhance their brand reputation, attract environmentally conscious consumers, and contribute to a more sustainable food supply chain (Ibrahim et al., 2022).

In addition, while chitosan edible coatings and MAP face challenges related to cost, compatibility, and regulatory approval, they also present significant opportunities for innovation, market expansion, and sustainability in the food industry. By addressing these challenges and capitalizing on emerging opportunities, researchers, manufacturers, and policymakers can unlock the full potential of chitosan-based technologies and drive positive change in the way food is preserved, packaged, and consumed (Bose et al., 2023).

7.11 CONCLUSION

The integration of chitosan edible coating with MAP presents a promising approach to extend the shelf life and maintain the quality of perishable food products. Chitosan, a biopolymer derived from chitin, offers excellent barrier properties against moisture, oxygen, and microbial growth, making it an effective material for edible coatings. When combined with MAP, which involves modifying the gaseous environment surrounding the food product, synergistic benefits can be achieved. Chitosan coatings create a protective barrier that helps to reduce moisture loss and inhibit microbial growth on the food surface, thus delaying spoilage. This is particularly beneficial when used in conjunction with MAP, as the controlled atmosphere further slows down microbial proliferation and enzymatic reactions, preserving the freshness of the product for a longer duration. The incorporation of chitosan in MAP enhances the mechanical strength and stability of the packaging, thereby reducing the risk of punctures or tears during handling and transportation. This ensures the integrity of the packaging system, preventing external contaminants from compromising the quality and safety of the enclosed food. The combination of chitosan edible coating and MAP offers a multifaceted approach to prolonging the shelf life of perishable foods by providing a protective barrier against environmental factors, inhibiting microbial growth, and enhancing packaging integrity. This integrated system not only extends the product's freshness but also contributes to reducing food waste and enhancing food safety throughout the supply chain. Therefore, it represents a promising solution for the preservation and quality maintenance of various food products.

REFERENCES

Abimbola, O. F., Adetunji, C. O., Dare, B. J., Mathew, J. T., Inobeme, A., Ghazanfar, S., Olaniyan, O. T, Ajiboye, M. D., & Dauda, W. (2023). Packaging of Animal-Based Food Products: The Need for Intelligent Freshness Sensors. In: *Sensing and Artificial Intelligence Solutions for Food Manufacturing* (pp. 89–102). CRC Press. https://doi.org/10.1201/9781003207955-6

Adetunji, C. O., Mathew, J. T., Inobeme, A., Taiwo, O., Olaniyan, O., Shakira, G., & Wadazani, D. (2023a). Application of Biosensors for Detection and Monitoring of Water Quality. In: *Organic and Inorganic Materials Based Sensors* (pp. 1065–1077). John Wiley & Sons, Inc. https://doi.org/10.1002/9783527834266.ch46

Adetunji, C. O., Olaniyan, O. T., Shariati, M. A., Olugbenga, S. M., Osemwegie, O. O., Rebezov, M., Ajayi, O. O., Makhmudov, F., Bodunrinde, R. E., Adetunji, J. B., Adeniyi, M. J., Hefft, D. I., Inobeme, A., & Mathew, J. T. (2023b). Medical Attributes of *Breynia disticha* J.R. Forst. & G. Forst. In: T. Pullaiah (Ed.), *Phytochemical Composition and Pharmacy of Medicinal Plants* (Vol. 2, pp. 531–538). Apple Academic Press, Inc. Co-published with CRC Press (Taylor & Francis), Palm Bay, FL.

Adiletta, G., Di Matteo, M., & Petriccione, M. (2021). Multifunctional Role of Chitosan Edible Coatings on Antioxidant Systems in Fruit Crops: A Review. *International Journal of Molecular Sciences*, 22(5), 2633. https://doi.org/10.3390/ijms22052633

Algarni, E. H. A., Elnaggar, I. A., Abd El-Wahed, A. E.´N., Taha, I. M., Al-Jumayi, H. A., Elhamamsy, S. M., Mahmoud, S. F., & Fahmy, A. (2022). Effect of Chitosan Nanoparticles as Edible Coating on the Storability and Quality of Apricot Fruits. *Polymers*, 14(11), 2227. https://doi.org/10.3390/polym14112227

Berdejo, D., García-Gonzalo, D., Oulahal, N., Denkova-Kostova, R., Shopska, V., Kostov, G., Degraeve, P., & Pagan, R. (2023). Minimal Processing Technologies for Production and Preservation of Tailor-Made Foods. *Food Technology and Biotechnology*, 61(3), 357–377. https://doi.org/10.17113/ftb.61.03.23.8013

Bodunrinde, R. E., Chimbekujwo, K. I., Oyewole, O. A., Adetunji, C. O., Inobeme, A., Eniola, K. I. T., Yerima, M. B., & Mathew, J. T. (2025). The Use of Nanochitosan for Enhancement in the Quality and Yield of Fruit Crops. In: *Nanochitosan Applications for Enhanced Crop Production and Food Security* (pp. 111–122). John Wiley & Sons. https://doi.org/10.1002/9781394214945.ch5

Bose, I., Roy, S., Pandey, V. K., & Singh, R. (2023). A Comprehensive Review on Significance and Advancements of Antimicrobial Agents in Biodegradable Food Packaging. *Antibiotics* (Basel, Switzerland), 12(6), 968. https://doi.org/10.3390/antibiotics12060968

Cejudo, C., Díaz, A. B., Casas, L., Martínez de la Ossa, E., & Mantell, C. (2023). Supercritical CO_2 Processing of White Grape Must as a Strategy to Reduce the Addition of SO_2. *Foods*, 12, 3085. https://doi.org/10.3390/foods12163085

Chimbekujwo, K. I., Yakubu, J. G., Omoregie, I. P., Oyewole, O. A., Muhammad, W. M., Adetunji, C. O., Eniola, K. I. T., Yerima, M. B., & Mathew, J. T. (2025). Synthesis of Nano-Chitosan Using Agricultural Waste. In: *Nanochitosan Applications for Enhanced Crop Production and Food Security* (pp. 39–68). John Wiley & Sons, Inc. https://doi.org/10.1002/9781394214945.ch2

Cocetta, G., & Natalini, A. (2022). Ethylene: Management and Breeding for Postharvest Quality in Vegetable Crops: A Review. *Frontiers in Plant Science*, 13, 968315. https://doi.org/10.3389/fpls.2022.968315

Czerwiński, K., Rydzkowski, T., Wróblewska-Krepsztul, J., & Thakur, V. K. (2021). Towards Impact of Modified Atmosphere Packaging (MAP) on Shelf-Life of Polymer-Film-Packed Food Products: Challenges and Sustainable Developments. *Coatings*, 11, 1504. https://doi.org/10.3390/coatings11121504

Dawud, I., Bodunrinde, R. E., Oyewole, O. A., Adetunji, C. O., Inobeme, A., Eniola, K. I. T., Yerima, M. B., & Mathew, J. T. (2025). Application of Nanochitosan for Effective Fruit Production. In: *Nanochitosan Applications for Enhanced Crop Production and Food Security* (pp. 123–135). John Wiley & Sons. https://doi.org/10.1002/9781394214945.ch6

Díaz-Montes, E., & Castro-Muñoz, R. (2021). Edible Films and Coatings as Food-Quality Preservers: An Overview. *Foods (*Basel, Switzerland*)*, 10(2), 249. https://doi.org/10.3390/foods10020249

Duda-Chodak, A., Tarko, T., & Petka-Poniatowska, K. (2023). Antimicrobial Compounds in Food Packaging. *International Journal of Molecular Sciences*, 24(3), 2457. https://doi.org/10.3390/ijms24032457

El-Araby, A., Janati, W., Ullah, R., Ercisli, S., & Errachidi, F. (2024). Chitosan, Chitosan Derivatives, and Chitosan-Based Nanocomposites: Eco-Friendly Materials for Advanced Applications (A Review). *Frontiers in Chemistry*, 11, 1327426. https://doi.org/10.3389/fchem.2023.1327426

Fang, Y., & Wakisaka, M. A. (2021). Review on the Modified Atmosphere Preservation of Fruits and Vegetables with Cutting-Edge Technologies. *Agriculture*, 11, 992. https://doi.org/10.3390/agriculture11100992

Felicia, W. X. L., Rovina, K., Nur'Aqilah, M. N., Vonnie, J. M., Erna, K. H., Misson, M., & Halid, N. F. A. (2022). Recent Advancements of Polysaccharides to Enhance Quality and Delay Ripening of Fresh Produce: A Review. *Polymers*, 14(7), 1341. https://doi.org/10.3390/polym14071341

Gautam, S., Lapčík, L., Lapčíková, B., & Gál, R. (2023). Emulsion-Based Coatings for Preservation of Meat and Related Products. *Foods*, 12, 832. https://doi.org/10.3390/foods12040832

Gigante, V., Aliotta, L., Ascrizzi, R., Pistelli, L., Zinnai, A., Batoni, G., Coltelli, M. B., & Lazzeri, A. (2023). Innovative Biobased and Sustainable Polymer Packaging Solutions for Extending Bread Shelf Life: A Review. *Polymers*, 15(24), 4700. https://doi.org/10.3390/polym15244700

Hameed, A. Z., Raj, S. A., Kandasamy, J., Baghdadi, M. A., & Shahzad, M. A. (2022). Chitosan: A Sustainable Material for Multifarious Applications. *Polymers*, 14(12), 2335. https://doi.org/10.3390/polym14122335

Hassoun, A., Carpena, M., Prieto, M. A., Simal-Gandara, J., Özogul, F., Özogul, Y., Çoban, Ö. E., Guðjónsdóttir, M., Barba, F. J., Marti-Quijal, F. J., Jambrak, A. R., Maltar-Strmečki, N., Kljusurić, J. G., & Regenstein, J. M. (2020). Use of Spectroscopic Techniques to Monitor Changes in Food Quality during Application of Natural Preservatives: A Review. *Antioxidants (*Basel, Switzerland*)*, 9(9), 882. https://doi.org/10.3390/antiox9090882

Ibrahim, I. D., Hamam, Y., Sadiku, E. R., Ndambuki, J. M., Kupolati, W. K., Jamiru, T., Eze, A. A., & Snyman, J. (2022). Need for Sustainable Packaging: An Overview. *Polymers*, 14(20), 4430. https://doi.org/10.3390/polym14204430

Ilyas, R. A., Aisyah, H. A., Nordin, A. H., Ngadi, N., Zuhri, M. Y. M., Asyraf, M. R. M., Sapuan, S. M., Zainudin, E. S., Sharma, S., Abral, H., Asrofi, M., Syafri, E., Sari, N. H., Rafidah, M., Zakaria, S. Z. S., Razman, M. R., Majid, N. A., Ramli, Z., Azmi, A., Bangar, S. P., & Ibrahim, R. (2022). Natural-Fiber-Reinforced Chitosan, Chitosan Blends and Their Nanocomposites for Various Advanced Applications. *Polymers*, 14(5), 874. https://doi.org/10.3390/polym14050874

Inobeme, A., Adetunji, C. O., Mathew, J. T., Ajai, A. I., Inobeme, J., Bamigboye, M. O., Onyeaku, S., Maliki, M., Eziukwu, C., & Tawa, K. (2023). Nanotechnology for Bioremediation of Heavy Metals. In: *Microbial Technologies in Industrial Wastewater Treatment* (pp. 19–30). Springer Nature Singapore. https://doi.org/10.1007/978-981-99-2435-6_2

Inobeme, A., Mathew, J. T., Ajai, I. A., Adetunji, C. O., Inobeme, J., Maliki, M., Adekoya, M. A., Shaba, E. Y., Eric, O., Akhor, S., & Eziukwu, C. (2024a). Common Techniques in Food Processing Technologies. In: *Nanobiotechnology for Food Processing and Packaging* (pp. 223–234). Elsevier Inc. https://doi.org/10.1016/B978-0-323-91749-0.00002-2

Inobeme, A., Shahnawaz, M., Adetunji, C. O., Mathew, J. T., Adetuyi, B. O., Popoola, O. A., Olaitan, F. Y., Akinbo, O., Kolawole, O. M., Oyewole, O. A., Eniola, K. I. T., & Yerima, B. M. (2024b). Microplastic as a Pathogenic Vector to Cause Diseases in Marine Biota. In: *Microplastic Pollution* (Vol. 1, pp. 321–349). Springer, Singapore. https://doi.org/10.1007/978-981-99-8357-5_19

Iturralde-García, R. D., Cinco-Moroyoqui, F. J., Martínez-Cruz, O., Ruiz-Cruz, S., Wong-Corral, F. J., Borboa-Flores, J., Cornejo-Ramírez, Y. I., Bernal-Mercado, A. T., & Del-Toro-Sánchez, C. L. (2022). Emerging Technologies for Prolonging Fresh-Cut Fruits' Quality and Safety during Storage. *Horticulturae*, 8, 731. https://doi.org/10.3390/horticulturae8080731

Jiang, A., Patel, R., Padhan, B., Palimkar, S., Galgali, P., Adhikari, A., Varga, I., & Patel, M. (2023). Chitosan Based Biodegradable Composite for Antibacterial Food Packaging Application. *Polymers*, 15(10), 2235. https://doi.org/10.3390/polym15102235

Jiménez-Gómez, C. P., & Cecilia, J. A. (2020). Chitosan: A Natural Biopolymer with a Wide and Varied Range of Applications. *Molecules (*Basel, Switzerland*)*, 25(17), 3981. https://doi.org/10.3390/molecules25173981

Kaur, R., Kaur, L., Gupta, T. B., Singh, J., & Bronlund, J. (2022). Multitarget Preservation Technologies for Chemical-Free Sustainable Meat Processing. *Journal of Food Science*, 87(10), 4312–4328. https://doi.org/10.1111/1750-3841.16329

Ke, C.-L., Deng, F.-S., Chuang, C.-Y., & Lin, C.-H. (2021). Antimicrobial Actions and Applications of Chitosan. *Polymers*, 13, 904. https://doi.org/10.3390/polym13060904

Khubiev, O. M., Egorov, A. R., Kirichuk, A. A., Khrustalev, V. N., Tskhovrebov, A. G., & Kritchenkov, A. S. (2023). Chitosan-Based Antibacterial Films for Biomedical and Food Applications. *International Journal of Molecular Sciences*, 24(13), 10738. https://doi.org/10.3390/ijms241310738

Kiss, B., & Borbély, J. (2023). Business Risk Mitigation in the Development Process of New Monoclonal Antibody Drug Conjugates for Cancer Treatment. *Pharmaceutics*, 15(6), 1761. https://doi.org/10.3390/pharmaceutics15061761

Kocira, A., Kozłowicz, K., Panasiewicz, K., Staniak, M., Szpunar-Krok, E., & Hortyńska, P. (2021). Polysaccharides as Edible Films and Coatings: Characteristics and Influence on Fruit and Vegetable Quality: A Review. *Agronomy*, 11, 813. https://doi.org/10.3390/agronomy11050813

Kontominas, M. G., Badeka, A. V., Kosma, I. S., & Nathanailides, C. I. (2021). Recent Developments in Seafood Packaging Technologies. *Foods (*Basel, Switzerland*)*, 10(5), 940. https://doi.org/10.3390/foods10050940

Kulka, K., & Sionkowska, A. (2023). Chitosan Based Materials in Cosmetic Applications: A Review. *Molecules (*Basel, Switzerland*)*, 28(4), 1817. https://doi.org/10.3390/molecules28041817

Liao, X., Xing, Y., Fan, X., Qiu, Y., Xu, Q., & Liu, X. (2023). Effect of Composite Edible Coatings Combined with Modified Atmosphere Packaging on the Storage Quality and Microbiological Properties of Fresh-Cut Pineapple. *Foods*, 12, 1344. https://doi.org/10.3390/foods12061344

Lisitsyn, A., Semenova, A., Nasonova, V., Polishchuk, E., Revutskaya, N., Kozyrev, I., & Kotenkova, E. (2021). Approaches in Animal Proteins and Natural Polysaccharides Application for Food Packaging: Edible Film Production and Quality Estimation. *Polymers*, 13, 1592. https://doi.org/10.3390/polym13101592

Liu, T., Li, J., Tang, Q., Qiu, P., Gou, D., & Zhao, J. (2022). Chitosan-Based Materials: An Overview of Potential Applications in Food Packaging. *Foods*, 11, 1490. https://doi.org/10.3390/foods11101490

Malik, S. I., Chimbekujwo, K. I., Maku, A. M., Oyewole, O. A., Omoregie, I. P., Adetunji, C. O., Eniola, K. I. T., Yerima, M. B., & Mathew, J. T. (2025). The Application of Nanochitosan Biopesticides as a Replacement to Synthetic Pesticides. In: *Nanochitosan Applications for Enhanced Crop Production and Food Security* (pp. 89–110). John Wiley & Sons. https://doi.org/10.1002/9781394214945.ch4

Maliki, S., Sharma, G., Kumar, A., Moral-Zamorano, M., Moradi, O., Baselga, J., Stadler, F. J., & García-Peñas, A. (2022). Chitosan as a Tool for Sustainable Development: A Mini Review. *Polymers*, 14(7), 1475. https://doi.org/10.3390/polym14071475

Martin, D., Joly, C., Dupas-Farrugia, C., Adt, I., Oulahal, N., & Degraeve, P. (2023). Volatilome Analysis and Evolution in the Headspace of Packed Refrigerated Fish. *Foods*, 12, 2657. https://doi.org/10.3390/foods12142657

Mathew, J. T., Adetunji, C. O., Inobeme, A., Musah, M., Azeh, Y., Etsuyankpa, M. B., Otori, A. A., Chado, I. M., Shaba, E. Y., Akoji, J. N., Mamman, A., & Oyewole, O. A. (2025b). Application of Nanochitosan for the Biodegradation of Agricultural Wastes. In: *Nanochitosan Applications for Enhanced Crop Production and Food Security* (pp. 361–382). John Wiley & Sons. https://doi.org/10.1002/9781394214945.ch16

Mathew, J. T., Adetunji, C. O., Inobeme, A., Musah, M., Azeh, Y., Otori, A. A., Shaba, E. Y., Mamman, A., & Tanko, E. (2023b). Removal of Heavy Metals Using Bio-remedial Techniques. In: M. P. Shah (Ed.), *Modern Approaches in Waste Bioremediation* (pp. 117–130). Springer Nature, Switzerland AG. https://doi.org/10.1007/978-3-031-24086-7_6

Mathew, J. T., Adetunji, C. O., Inobeme, A., Otori, A. A., Musah, M., Azeh, Y., Shaba, E. Y., Muhammad, I. A., Mamman, A., & Oyewole, O. A. (2025a). Relevance of Nanochitosan in Food Sector and Food Packaging Sectors: Current Trends. In: *Nanochitosan Applications for Enhanced Crop Production and Food Security* (pp. 345–360). John Wiley & Sons. https://doi.org/10.1002/9781394214945.ch15

Mathew, J. T., Inobeme, A., Etsuyankpa, M. B., Adetunji, C. O., Shaba, E. Y., Musah, M., Azeh, Y., Jiya, M. J., Tanko, E., Muhammad, A. I., Mamman, A., & Maurice, J. (2024b). Applications of Diatom for Treatment of Industrial Effluents: Waste to Wealth Approach. In: *Marine Biomass: Biorefinery, Bioproducts and Environmental Bioremediation* (pp. 1–20). Walter de Gruyter GmbH & Co KG. https://doi.org/10.1515/9783111353951-001

Mathew, J. T., Inobeme, A., Hussaini, J., Shaba, E. Y., Azeh, Y., Musah, M., Muhammad, E. B., Tanko, M. S., Ismail, H., Adetunji, C. O., Kanwa, A. M., & Mamman, A. (2024a). Properties of Nanomaterials for Corrosion Control. In A. Thakur & A. Kumar (Eds.), *Innovations in Nanomaterials-Based Corrosion Inhibitors* (pp. 74–99). IGI Global. https://doi.org/10.4018/979-8-3693-3088-3.ch004

Mathew, J. T., Inobeme, A., Musah, M., Azeh, Y., Etsuyankpa, M. B., Adetunji, C. O., Shaba, E. Y., Jiya, M. J., Josiah, J. G., Tanko, E., Muhammad, A. I., Mamman, A., & Jemkur, M. (2024c). Utilization of Marine Green Algae for Development of Bioplastics. In: *Marine Biomass: Biorefinery, Bioproducts and Environmental Bioremediation* (pp. 39–55). Walter de Gruyter GmbH & Co KG. https://doi.org/10.1515/9783111353951-003

Mathew, J. T., Musah, M., Azeh, Y., & Muhammed, M. (2023a). Adsorptive Removal of Selected Toxic Metals from Pharmaceutical Wastewater using Fe_3O_4/ZnO Nanocomposite. *Dutse Journal of Pure and Applied Sciences*, 9(4a), 236–248. https://doi.org/10.4314/dujopas.v9i4a.22

Mathew, J. T., Musah, M., Azeh, Y., & Muhammed, M. (2023c). Kinetic Study of Heavy Metals Removal from Pharmaceutical Wastewater Using Geopolymer/Fe_3O_4 Nanocomposite. *Bima Journal of Science and Technology*, 7(4), 152–163. https://doi.org/10.56892/bima.v7i4.512

Matloob, A., Ayub, H., Mohsin, M., Ambreen, S., Khan, F. A., Oranab, S., Rahim, M. A., Khalid, W., Nayik, G. A., Ramniwas, S., & Ercisli, S. (2023). A Review on Edible Coatings and Films: Advances, Composition, Production Methods, and Safety Concerns. *ACS Omega*, 8(32), 28932–28944. https://doi.org/10.1021/acsomega.3c03459

Moeini, A., Pedram, P., Fattahi, E., Cerruti, P., & Santagata, G. (2022). Edible Polymers and Secondary Bioactive Compounds for Food Packaging Applications: Antimicrobial, Mechanical, and Gas Barrier Properties. *Polymers*, 14(12), 2395. https://doi.org/10.3390/polym14122395

Muñoz-Tebar, N., Pérez-Álvarez, J. A., Fernández-López, J., & Viuda-Martos, M. (2023). Chitosan Edible Films and Coatings with Added Bioactive Compounds: Antibacterial and Antioxidant Properties and Their Application to Food Products: A Review. *Polymers*, 15(2), 396. https://doi.org/10.3390/polym15020396

Mura, P., Maestrelli, F., Cirri, M., & Mennini, N. (2022). Multiple Roles of Chitosan in Mucosal Drug Delivery: An Updated Review. *Marine Drugs*, 20(5), 335. https://doi.org/10.3390/md20050335

Nunes, C., Silva, M., Farinha, D., Sales, H., Pontes, R., & Nunes, J. (2023). Edible Coatings and Future Trends in Active Food Packaging-Fruits' and Traditional Sausages' Shelf Life Increasing. Foods (Basel, Switzerland*)*, 12(17), 3308. https://doi.org/10.3390/foods12173308

Ogundolie, F. A., Adetunji, C. O., Mathew, J. T., Inobeme, A., Ghazanfar, S., Olaniyan, O. T., Ajiboye, M. D., Dauda, W. P., Adetunji, J. B., Akinbo, O., & Adetuyi, B. O. (2023). Intelligent Freshness Indicators in Packaging of Plant-Based Food Products. In: *Evaporative Coolers for the Postharvest Management of Fruits and Vegetables* (pp. 323–331). Academic Press. https://doi.org/10.1016/B978-0-323-89864-5.00008-4

Olawuyi, I. F., Park, J. J., Lee, J. J., & Lee, W. Y. (2021). Combined Effect of Chitosan Coating and Modified Atmosphere Packaging on Fresh-Cut Cucumber. *Food Science & Nutrition*, 7(3), 1043–1052. https://doi.org/10.1002/fsn3.937

Onyeaka, H., Obileke, K., Makaka, G., & Nwokolo, N. (2022). Current Research and Applications of Starch-Based Biodegradable Films for Food Packaging. *Polymers*, 14(6), 1126. https://doi.org/10.3390/polym14061126

Perez-Vazquez, A., Barciela, P., Carpena, M., & Prieto, M. A. (2023). Edible Coatings as a Natural Packaging System to Improve Fruit and Vegetable Shelf Life and Quality. Foods (Basel, Switzerland*)*, 12(19), 3570. https://doi.org/10.3390/foods12193570

Piekarska, K., Sikora, M., Owczarek, M., & Jóźwik-Pruska, J. (2023). Wiśniewska-Wrona, M. Chitin and Chitosan as Polymers of the Future-Obtaining, Modification, Life Cycle Assessment and Main Directions of Application. *Polymers*, 15, 793. https://doi.org/10.3390/polym15040793

Popa, L., Ghica, M. V., & Dinu-Pîrvu, C. E. (2023). Chitosan Biomaterials: Advances and Challenges. *International Journal of Molecular Sciences*, 24(22), 16150. https://doi.org/10.3390/ijms242216150

Rajestary, R., Xylia, P., Chrysargyris, A., Romanazzi, G., & Tzortzakis, N. (2022). Preharvest Application of Commercial Products Based on Chitosan, Phosphoric Acid Plus Micronutrients, and Orange Essential Oil on Postharvest Quality and Gray Mold Infections of Strawberry. *International Journal of Molecular Sciences*, 23(24), 15472. https://doi.org/10.3390/ijms232415472

Romanazzi, G., & Moumni, M. (2022). Chitosan and Other Edible Coatings to Extend Shelf Life, Manage Postharvest Decay, and Reduce Loss and Waste of Fresh Fruits and Vegetables. *Current Opinion in Biotechnology*, 78, 102834. https://doi.org/10.1016/j.copbio.2022.102834

Saberi Riseh, R., Vatankhah, M., Hassanisaadi, M., & Kennedy, J. F. (2023). Chitosan-Based Nanocomposites as Coatings and Packaging Materials for the Postharvest Improvement of Agricultural Product: A Review. *Carbohydrate Polymers*, 309, 120666. https://doi.org/10.1016/j.carbpol.2023.120666

Safaei, F. S., Sedaghat, N., & Sharif, A. (2022). Effect of Packaging Material and Oxygen Concentration on Quality of Ghavoot during the Storage. *Journal of Food Science and Technology*, 59(3), 879–889. https://doi.org/10.1007/s13197-021-05083-8

Shineh, G., Mobaraki, M., Perves Bappy, M. J., & Mills, D. K. (2023). Biofilm Formation, and Related Impacts on Healthcare, Food Processing and Packaging, Industrial Manufacturing, Marine Industries, and Sanitation: A Review. *Applied Microbiology*, 3, 629–665. https://doi.org/10.3390/applmicrobiol3030044

Shonte, T. T., Mulla, M. F., Foley, L., & Pathania, S. (2024). Mechanisms of Action and Preservation Effects of Packaging Systems for Mushrooms: Novel Approaches to Preserve Irish Edible Mushrooms. *Coatings*, 14, 172. https://doi.org/10.3390/coatings14020172

Singh, A. K., Kim, J. Y., & Lee, Y. S. (2022). Phenolic Compounds in Active Packaging and Edible Films/Coatings: Natural Bioactive Molecules and Novel Packaging Ingredients. *Molecules*, 27, 7513. https://doi.org/10.3390/molecules27217513

Stasińska-Jakubas, M., & Hawrylak-Nowak, B. (2022). Protective, Biostimulating, and Eliciting Effects of Chitosan and Its Derivatives on Crop Plants. *Molecules (*Basel, Switzerland*)*, 27(9), 2801. https://doi.org/10.3390/molecules27092801

Stefanowska, K., Woźniak, M., Dobrucka, R., & Ratajczak, I. (2023). Chitosan with Natural Additives as a Potential Food Packaging. Materials (Basel, Switzerland*)*, 16(4), 1579. https://doi.org/10.3390/ma16041579

Suvarna, V., Nair, A., Mallya, R., Khan, T., & Omri, A. (2022). Antimicrobial Nanomaterials for Food Packaging. *Antibiotics (*Basel, Switzerland*)*, 11(6), 729. https://doi.org/10.3390/antibiotics11060729

Teixeira-Costa, B. E., & Andrade, C. T. (2021). Chitosan as a Valuable Biomolecule from Seafood Industry Waste in the Design of Green Food Packaging. *Biomolecules*, 11(11), 1599. https://doi.org/10.3390/biom11111599

Teshome, E., Forsido, S. F., Rupasinghe, H. P. V., & Olika Keyata, E. (2022). Potentials of Natural Preservatives to Enhance Food Safety and Shelf Life: A Review. *The Scientific World Journal*, 2022, 9901018. https://doi.org/10.1155/2022/9901018

Thambiliyagodage, C., Jayanetti, M., Mendis, A., Ekanayake, G., Liyanaarachchi, H., & Vigneswaran, S. (2023). Recent Advances in Chitosan-Based Applications: A Review. Materials (Basel, Switzerland*)*, 16(5), 2073. https://doi.org/10.3390/ma16052073

Vasile, C., & Baican, M. (2021). Progresses in Food Packaging, Food Quality, and Safety-Controlled-Release Antioxidant and/or Antimicrobial Packaging. *Molecules (*Basel, Switzerland*)*, 26(5), 1263. https://doi.org/10.3390/molecules26051263

Versino, F., Ortega, F., Monroy, Y., Rivero, S., López, O. V., & García, M. A. (2023). Sustainable and Bio-Based Food Packaging: A Review on Past and Current Design Innovations. *Foods*, 12, 1057. https://doi.org/10.3390/foods12051057

Wrońska, N., Katir, N., Miłowska, K., Hammi, N., Nowak, M., Kędzierska, M., Anouar, A., Zawadzka, K., Bryszewska, M., El Kadib, A., & Lisowska, K. (2021). Antimicrobial Effect of Chitosan Films on Food Spoilage Bacteria. *International Journal of Molecular Sciences*, 22(11), 5839. https://doi.org/10.3390/ijms22115839

Wu, W., Li, Y., Zhu, X., Wang, L., Wang, J., Qin, Y., Zhang, M., Yu, C., Gou, C., & Yan, X. (2023). Antimicrobial Activity Enabled by Chitosan-ε-Polylysine-Natamycin and Its Effect on Microbial Diversity of Tomato Scrambled Egg Paste. *Food Chemistry:* X, 19, 100872. https://doi.org/10.1016/j.fochx.2023.100872

Yan, D., Li, Y., Liu, Y., Li, N., Zhang, X., & Yan, C. (2021). Antimicrobial Properties of Chitosan and Chitosan Derivatives in the Treatment of Enteric Infections. *Molecules (*Basel, Switzerland*)*, 26(23), 7136. https://doi.org/10.3390/molecules26237136

Yilmaz Atay, H. (2020). Antibacterial Activity of Chitosan-Based Systems. In: *Functional Chitosan: Drug Delivery and Biomedical Applications* (pp. 457–489). https://doi.org/10.1007/978-981-15-0263-7_15

Yu, H. H., Chin, Y. W., & Paik, H. D. (2021). Application of Natural Preservatives for Meat and Meat Products against Food-Borne Pathogens and Spoilage Bacteria: A Review. Foods (Basel, Switzerland*)*, 10(10), 2418. https://doi.org/10.3390/foods10102418

Zehra, A., Wani, S. M., Jan, N., Bhat, T. A., Rather, S. A., Malik, A. R., & Hussain, S. Z. (2022). Development of Chitosan-Based Biodegradable Films Enriched with Thyme Essential Oil and Additives for Potential Applications in Packaging of Fresh Collard Greens. *Scientific Reports*, 12(1), 16923. https://doi.org/10.1038/s41598-022-20751-1

8 Application of Chitosan for the Extension of the Shelf Life of Poultry, Meat, and Fishery Products

John Tsado Mathew[1], *Abel Inobeme*[2], *Yakubu Azeh*[1], *Musah Monday*[1], *Elijah Yanda Shaba*[3], *Mary Jumai Jiya*[4], *Etsuyankpa Muhammad Bini*[5], *Ismaila Aliyu Oga*[5], *Musa Safiyanu Tanko*[5], *Tanko Ezekiel*[3], *Jonathan Hussaini*[1], *Mohammed Jibrin Ndejiko*[6], *Adamu Isah Abubakar*[8], *Charles Oluwaseun Adetunji*[7], *and Amos Mamman*[1]

[1]Department of Chemistry, Ibrahim Badamasi Babangida University Lapai, Niger State, Nigeria

[2]Department of Chemistry, Edo University Uzairue, Auchi, Edo State, Nigeria

[3]Department of Chemistry, Federal University of Technology, Minna, Niger State, Nigeria

[4]Department of Food Science and Technology, Federal University of Technology, Minna, Niger State, Nigeria

[5]Department of Chemistry, Federal University of Lafia, Nasarawa State, Nigeria

[6]Department of Microbiology, Ibrahim Badamasi Babangida University Lapai, Niger State, Nigeria

[7]Department of Microbiology, Applied Microbiology, Biotechnology and Nanotechnology Laboratory, Edo University Uzairue, Auchi, Edo State, Nigeria

[8]Department of Computer Sciences, Ibrahim Badamasi, Babangida University Lapai, Niger State, Nigeria

8.1 INTRODUCTION

The preservation of perishable food products, such as poultry, meat, and fishery items, is a critical concern in the food industry to meet the demands of global supply chains and ensure consumer access to safe and high-quality products. Among various preservation methods, the application of chitosan, a natural polymer derived from crustacean shells, has emerged as a promising solution to extend the shelf life of these products. This introduction explores the background, rationale, and significance of employing chitosan-based coatings in addressing the challenges associated with the shelf life of poultry, meat, and fishery products (Teshome et al., 2022; Malik et al., 2025).

Perishable food items, including poultry, meat, and fishery products, are susceptible to microbial contamination, oxidative processes, and moisture loss, all of which contribute to their rapid

DOI: 10.1201/9781003355458-8

deterioration. The demand for fresh and high-quality products has heightened with increasing consumer awareness of food safety and nutrition. Traditional preservation methods, such as refrigeration and freezing, are effective but may not sufficiently address the complex interactions leading to product degradation during storage and transportation. In recent years, attention has turned to innovative approaches, with a particular focus on natural compounds. Chitosan, a derivative of chitin, exhibits unique properties that make it an attractive candidate for food preservation. Its antimicrobial, antioxidant, and film-forming characteristics position chitosan as a versatile material for coating perishable food items, offering a protective layer against external factors that compromise product quality (Ogundolie et al., 2023).

The shelf life of poultry, meat, and fishery products is a critical factor influencing both producers and consumers. Shortened shelf life not only contributes to food waste but also poses challenges in meeting market demands, especially in the context of global trade. Addressing the factors that lead to product deterioration is essential for extending shelf life, ensuring product safety, and meeting consumer expectations for freshness and quality. Microbial contamination, including bacteria and fungi, is a primary contributor to the spoilage of these products. Oxidative processes, such as lipid oxidation, lead to changes in color, flavor, and nutritional content. Additionally, moisture loss can result in undesirable textural changes and compromises the overall quality of the product. Chitosan, with its multifaceted attributes, presents an opportunity to mitigate these challenges and extend the shelf life of poultry, meat, and fishery items (Li et al., 2022; Muhammad et al., 2025).

Chitosan's significance in food preservation stems from its inherent properties, which directly address the factors contributing to product degradation. As an antimicrobial agent, chitosan exhibits inhibitory effects against a broad spectrum of microorganisms, including bacteria and fungi. This property is particularly advantageous in preventing spoilage and enhancing the safety of perishable food products. Moreover, chitosan's antioxidant properties contribute to the prevention of oxidative reactions. Lipid oxidation, in particular, is a common concern in meat and fishery products, leading to rancidity and off-flavors. The ability of chitosan to create a protective barrier against oxygen helps maintain the freshness of these products and preserves their sensory attributes (Khubiev et al., 2023).

The film-forming characteristics of chitosan enable its application as an edible coating on the surface of food items. This coating acts as a physical barrier, reducing moisture loss and preventing the ingress of external contaminants. By forming a protective layer, chitosan-based coatings contribute to the overall preservation of product quality. In addition to its functional attributes, the biodegradability and eco-friendly nature of chitosan align with the increasing consumer preference for sustainable and natural food preservation methods. As the food industry strives to balance efficiency with environmental responsibility, chitosan emerges as a promising solution that addresses both objectives (Matloob et al., 2023).

More so, the introduction sets the stage for an in-depth exploration of the application of chitosan in extending the shelf life of poultry, meat, and fishery products. The background highlights the challenges associated with product deterioration, emphasizing the need for innovative preservation methods. The rationale establishes the significance of addressing these challenges to meet market demands and reduce food waste. Finally, the significance of chitosan in food preservation is underscored, emphasizing its multifaceted properties that make it a valuable tool in enhancing the safety, quality, and sustainability of perishable food products (Zhang et al., 2023).

8.2 PROPERTIES AND CHARACTERISTICS OF CHITOSAN: A VERSATILE BIOPOLYMER FOR FOOD PRESERVATION

Chitosan, derived from chitin found in the exoskeletons of crustaceans like shrimp and crab shells, has emerged as a versatile biopolymer with unique properties that make it an attractive material for various applications, particularly in the realm of food preservation. Its properties and characteristics contribute to its effectiveness in extending the shelf life of perishable food products, including poultry, meat, and fishery items (Santos et al., 2020; Mathew et al., 2025a).

Origin and Source: Chitosan's journey begins with its sustainable and renewable source—the shells of crustaceans. The extraction process involves deacetylating chitin through the removal of acetyl groups. This results in the formation of chitosan, a biopolymer with diverse applications. The origin of chitosan from seafood byproducts aligns with sustainable practices, utilizing waste materials to create value-added products. In food preservation, the sourcing of chitosan from crustacean shells is particularly advantageous due to its availability as a byproduct of the seafood industry. This ensures a sustainable supply chain and reduces reliance on synthetic or non-renewable materials (Egorov et al., 2023).

Antimicrobial Properties: One of the defining characteristics of chitosan is its inherent antimicrobial activity. Chitosan exhibits effectiveness against a broad spectrum of microorganisms, including bacteria and fungi. This property arises from its cationic nature, allowing interactions with negatively charged microbial cell walls. In the context of food preservation, the antimicrobial properties of chitosan play a pivotal role in inhibiting the growth of spoilage microorganisms and pathogens. By incorporating chitosan-based coatings on food surfaces, the biopolymer acts as a protective shield, preventing microbial contamination and extending the product's shelf life. This is particularly crucial in the preservation of poultry, meat, and fishery products, where microbial spoilage is a common concern (Atay, 2020; Chimbekujwo et al., 2025).

Biodegradability: Chitosan's biodegradability is a noteworthy feature that aligns with the increasing emphasis on environmentally friendly materials. Being a natural polymer, chitosan undergoes degradation into non-toxic components, making it an eco-friendly option for various applications, including food packaging. In food preservation, the biodegradability of chitosan contributes to sustainable practices within the food industry. As consumers and producers alike prioritize environmentally conscious choices, chitosan-based solutions offer an attractive alternative to synthetic and non-biodegradable materials. This characteristic makes chitosan an ideal candidate for the development of biodegradable films and coatings used in the packaging and preservation of poultry, meat, and fish products (Jiang et al., 2023).

Film-forming Characteristics: Chitosan possesses remarkable film-forming characteristics, allowing it to create a thin, protective layer on the surface of food products. This property is particularly valuable in the development of edible coatings for various perishable items. In food preservation, the film-forming nature of chitosan serves as a physical barrier against external factors that contribute to product deterioration. By forming a protective film, chitosan helps reduce moisture loss, prevent oxidation, and inhibit microbial contamination. This is of significant importance in extending the shelf life of poultry, meat, and fishery products, where maintaining the quality and freshness of the products is paramount (Perez-Vazquez et al., 2023).

Application in Poultry, Meat, and Fishery Products: The properties and characteristics of chitosan find direct application in enhancing the preservation of poultry, meat, and fishery products. The antimicrobial activity of chitosan inhibits the growth of spoilage microorganisms, preventing bacterial contamination and extending the product's shelf life. In meat preservation, chitosan's film-forming properties create a protective coating that minimizes moisture loss and oxidation, preserving the color, texture, and nutritional content of the meat. Similarly, in fishery products, where maintaining freshness is critical, chitosan coatings act as a barrier against microbial spoilage and oxidative processes, contributing to prolonged shelf life (Muñoz-Tebar et al., 2023).

In addition, the properties and characteristics of chitosan, including its sustainable origin, antimicrobial activity, biodegradability, and film-forming nature, collectively position it as a versatile biopolymer for food preservation. The unique attributes of chitosan address key challenges in extending the shelf life of perishable products, making it an invaluable resource in the development of innovative solutions for the food industry, particularly in the preservation of poultry, meat, and fishery items. As research and applications continue to evolve, chitosan remains at the forefront of sustainable and effective strategies for enhancing food safety and quality (Jiang et al., 2023).

8.3 CHALLENGES IN POULTRY, MEAT, AND FISHERY PRODUCT SHELF LIFE

Poultry, meat, and fishery products are highly perishable, and their shelf life is influenced by various factors, presenting challenges for producers, distributors, and retailers. These challenges impact the safety, quality, and overall acceptability of these products, necessitating the development and implementation of effective preservation strategies (Dawud et al., 2025) (Table 8.1).

Microbial Spoilage: One of the primary challenges in extending the shelf life of poultry, meat, and fishery products is microbial spoilage. Bacteria, yeasts, and molds can proliferate on the surface and within the products, leading to changes in texture, color, flavor, and overall quality. Controlling microbial growth is crucial to preventing spoilage and ensuring product safety (Karanth et al., 2023).

Oxidative Processes: Lipid oxidation and enzymatic reactions contribute to the deterioration of poultry, meat, and fishery products. Exposure to oxygen and light accelerates these processes, leading to off-flavors, discoloration, and the development of undesirable odors. Managing oxidative reactions is essential for maintaining the sensory attributes and nutritional value of these products (Shahidi and Hossain, 2022).

Temperature Control: Maintaining appropriate temperatures throughout the supply chain is critical for preventing bacterial growth and maintaining product quality. Temperature fluctuations

TABLE 8.1
Challenges in Poultry, Meat, and Fishery Product Shelf Life

Product	Challenges	Impact on Shelf Life	Mitigation Strategies	References
Poultry	• Microbial contamination (bacteria, molds) • Lipid oxidation leading to off-flavors • Temperature fluctuations during storage and transportation	• Reduced shelf life due to spoilage • Altered taste and decreased product quality • Accelerated bacterial growth and spoilage	• Strict adherence to hygiene and sanitation practices • Use of antioxidants in packaging or processing • Implementing a cold chain system for temperature control	Li et al. (2022), Inobeme et al. (2024b)
Meat (beef, pork)	• Presence of pathogens like *Salmonella* and *Escherichia coli* • Oxidation of myoglobin causing color changes (brown discoloration) • Moisture loss due to improper packaging	• Risk of foodborne illnesses and shorter shelf life • Reduced visual appeal and consumer rejection • Texture changes and increased likelihood of freezer burn	• Implementation of HACCP protocols for safe processing • Modified atmosphere packaging to control oxygen exposure • Vacuum packaging to minimize moisture loss	Shaik et al. (2022)
Fishery products	• Rapid bacterial and enzymatic activity in fish • Formation of histamine in certain fish species • Oxidation of omega-3 fatty acids in oily fish	• Quick deterioration, unpleasant odor • Risk of histamine poisoning and shortened shelf life • Rancidity, off-flavors, and reduced nutritional value	• Immediate chilling and storage at low temperatures • Proper handling and storage to prevent histamine formation • Use of antioxidants and storage in dark, airtight containers	Madejska et al. (2022), Mathew et al. (2024b)

during storage, transportation, or retail display can compromise the safety and shelf life of poultry, meat, and fishery products. Cold chain management is essential to mitigate this challenge (Rani et al., 2023).

Packaging Issues: Packaging plays a crucial role in preserving the quality of these products. Inadequate packaging may result in moisture loss, freezer burn, or increased susceptibility to contamination. On the other hand, excessive packaging may lead to anaerobic conditions, contributing to spoilage. Striking the right balance in packaging design and materials is essential (Abimbola et al., 2023).

Contamination Risks: Poultry, meat, and fishery products are susceptible to contamination from various sources, including pathogens, chemicals, and cross-contamination during processing and handling. Stringent hygiene practices and adherence to food safety standards are necessary to minimize contamination risks and ensure consumer safety (Cudjoe et al., 2022).

Inadequate Preservation Methods: Traditional preservation methods like salting, smoking, and drying, while effective in the past, may not align with the demands of modern supply chains and evolving consumer preferences. These methods often fall short in maintaining the extended shelf life required for global distribution, leading to potential spoilage and compromised product quality. Additionally, traditional techniques may not always meet contemporary safety standards, posing a risk of contamination and foodborne illnesses. The reliance on heavy salting or smoking can also impact the taste and nutritional value of products, which may not appeal to health-conscious consumers. As the global food market evolves, there is a growing need for more advanced preservation methods that ensure safety, quality, and freshness while meeting the logistical requirements of longer supply chains and diverse consumer expectations (Speranza et al., 2021; Bodunrinde et al., 2025).

Consumer Expectations: Consumer preferences are increasingly shifting toward minimally processed, natural, and preservative-free products, posing significant challenges for the food industry. Meeting these expectations while ensuring the safety and extending the shelf life of poultry, meat, and fishery products requires innovative solutions and a careful re-evaluation of traditional processing and preservation methods. Consumers seek products that are free from artificial additives and preservatives, pushing manufacturers to explore natural alternatives that do not compromise quality or safety. This demand drives the development of advanced preservation techniques, such as the use of natural antimicrobial agents, modified atmosphere packaging (MAP), and edible coatings like chitosan. These methods can help maintain freshness, prevent microbial growth, and extend shelf life without relying on synthetic chemicals, thus aligning with consumer preferences for clean-label products that are both safe and natural (Li et al., 2022).

Supply Chain Complexity: The complexity of modern supply chains presents significant challenges in maintaining optimal conditions for product storage and transportation. From farm to fork, products often pass through multiple stages and are exposed to varying temperature and humidity conditions, which can negatively impact their shelf life and quality. Fluctuations in temperature and inadequate humidity control can accelerate spoilage, leading to food waste and potential safety issues. To overcome these challenges, efficient logistics and seamless communication across the supply chain are essential. Implementing real-time monitoring systems, cold chain management practices, and robust packaging solutions are crucial for maintaining product integrity. Coordination between suppliers, transporters, and retailers ensures that products remain in optimal conditions throughout their journey, ultimately delivering safe and high-quality food to consumers while minimizing waste and spoilage (Mathew et al., 2025b).

Globalization and Export Challenges: Products intended for international markets face additional challenges due to diverse regulations, varying climatic conditions, and extended transportation durations. Each country may have different quality and safety standards, which necessitate careful consideration and compliance to ensure market access. Navigating these regulatory frameworks can pose logistical challenges, requiring strategic planning and coordination to meet the requirements of multiple regions. Furthermore, varying climatic conditions during transport can affect the quality and stability of products, demanding packaging solutions that offer robust protection against temperature fluctuations and humidity. Prolonged transportation durations also require materials that can maintain product integrity over time. To successfully navigate these challenges, companies must adopt comprehensive strategies that prioritize regulatory compliance, quality assurance, and effective packaging solutions to meet the diverse needs of global markets (Raj et al., 2022).

Emerging Pathogens and Contaminants: The emergence of new pathogens and contaminants, along with evolving food safety regulations, poses a continuous challenge for the food industry. As food safety concerns become more complex, proactive monitoring and testing are crucial to identifying and managing potential risks. Regular surveillance for pathogens, contaminants, and chemical residues helps ensure that products meet safety standards. Adherence to best practices, such as implementing Hazard Analysis and Critical Control Points (HACCP) systems and maintaining stringent hygiene protocols, is essential for minimizing contamination. Additionally, staying updated with changes in food safety regulations across different regions is vital to remain compliant and avoid disruptions. This proactive approach not only safeguards public health but also helps build consumer trust and confidence in the safety and quality of food products (Bodunrinde et al., 2025).

In addressing these challenges, the poultry, meat, and fishery industries are increasingly turning to innovative preservation technologies, such as high-pressure processing, MAP, and natural antimicrobial agents like chitosan. Research and development efforts continue to focus on finding sustainable, effective, and consumer-friendly solutions to enhance the shelf life of these perishable products. Additionally, education and awareness campaigns on proper storage, handling, and cooking practices contribute to improving overall food safety and quality for consumers worldwide (Bose et al., 2023).

8.4 CHITOSAN-BASED COATINGS: MECHANISMS OF ACTION

Chitosan-based coatings have gained prominence in the food industry for their remarkable ability to extend the shelf life of perishable products, including fruits, vegetables, poultry, meat, and fishery items. The effectiveness of these coatings lies in their multifaceted mechanisms of action, addressing key factors that contribute to product deterioration. This article explores the diverse ways in which chitosan-based coatings exert their influence, from antimicrobial effects to oxidation prevention, and their specific applications in preserving various food products (Riseh et al., 2023).

Antimicrobial Effects: One of the primary mechanisms of action for chitosan-based coatings is their potent antimicrobial activity. Chitosan, with its cationic nature, interacts with negatively charged microbial cell walls, leading to membrane disruption and cell death. This antimicrobial property is particularly advantageous in preventing the growth of spoilage microorganisms and pathogens that contribute to the deterioration of food products. In the case of fruits and vegetables, chitosan-based coatings create a protective barrier that inhibits the growth of fungi and bacteria, extending the freshness and shelf life of these perishable items. Similarly, in poultry, meat, and fishery products, where microbial contamination is a significant concern, chitosan coatings act as a formidable defense against bacteria and fungi, preserving product safety and quality (Yan et al., 2021; Mathew et al., 2025b).

Oxidation Prevention: Chitosan-based coatings play a crucial role in preventing oxidative processes that lead to the degradation of lipids and other sensitive components in food products. Oxidation is a common challenge in the preservation of meat, poultry, and fishery items, resulting in rancidity, off-flavors, and a decline in nutritional quality. The antioxidant properties of chitosan help combat oxidation by scavenging free radicals and inhibiting lipid peroxidation. By forming a protective barrier on the surface of food products, chitosan coatings reduce exposure to oxygen, preventing the oxidation of fats and oils. This is especially relevant in the context of meat preservation, where the prevention of lipid oxidation contributes to the maintenance of flavor, color, and overall product quality (Muñoz-Tebar et al., 2023).

Moisture Retention: Chitosan-based coatings contribute to moisture retention, addressing another critical aspect of food preservation. Moisture loss can lead to undesirable textural changes, affecting the quality and appeal of various food products. Chitosan's film-forming characteristics create a thin layer that acts as a barrier, reducing the evaporation of water from the product. In fruits and vegetables, chitosan coatings help maintain crispness and firmness, preserving the sensory attributes that consumers value. In meat and fishery products, moisture retention is crucial for preventing dehydration, which can result in undesirable changes in texture and juiciness. The ability of chitosan to create a protective film contributes to the overall quality and succulence of coated products (Popescu et al., 2022).

Nutritional Preservation: Beyond its role in safety and quality, chitosan-based coatings contribute to the nutritional preservation of coated food products. By preventing microbial contamination and oxidation, chitosan helps maintain the levels of essential nutrients, including vitamins and minerals, in fruits, vegetables, and meat products. In the context of meat preservation, where the nutritional content is a key consideration for consumers, chitosan coatings assist in retaining the levels of protein, vitamins, and minerals. This is particularly relevant in addressing consumer preferences for minimally processed foods that retain their nutritional value (Adiletta et al., 2021).

Applications in Poultry, Meat, and Fishery Products: The mechanisms of action of chitosan-based coatings find specific applications in the preservation of poultry, meat, and fishery products. In poultry and meat, chitosan coatings inhibit bacterial growth, prevent oxidation of fats, and contribute to moisture retention, collectively extending the shelf life and preserving product quality. In fishery products, where maintaining freshness is paramount, chitosan acts as a barrier against microbial spoilage and oxidative processes, contributing to prolonged shelf life and enhanced nutritional value (Zhang et al., 2023).

In addition, chitosan-based coatings showcase a combination of antimicrobial effects, oxidation prevention, moisture retention, and nutritional preservation. These mechanisms collectively contribute to their efficacy in extending the shelf life of a diverse range of perishable food products. As the food industry continues to explore sustainable and effective preservation methods, chitosan-based coatings stand out as a versatile solution with wide-ranging applications, aligning with the evolving preferences of consumers for safe, high-quality, and nutritionally rich foods (Hoa et al., 2024).

8.5 APPLICATION OF CHITOSAN IN THE POULTRY INDUSTRY: A PARADIGM SHIFT IN SAFETY AND PRESERVATION

The poultry industry faces inherent challenges related to microbial contamination, oxidation, moisture loss, and the preservation of nutritional quality. In addressing these challenges, chitosan, a biopolymer derived from crustacean shells, has emerged as a versatile solution. Its unique properties make it a valuable tool in enhancing the safety and shelf life of poultry products, transforming the landscape of poultry preservation (El-Araby et al., 2024).

Antimicrobial Activity: Chitosan's remarkable antimicrobial properties are pivotal in addressing the microbial challenges prevalent in the poultry industry. With its cationic nature, chitosan interacts with the negatively charged cell walls of bacteria and fungi, disrupting membrane integrity and inhibiting their growth. This antimicrobial effect is particularly relevant in preventing spoilage and enhancing the safety of poultry products. In practice, chitosan-based coatings are applied to the surfaces of raw chicken cuts. These coatings act as a protective barrier, reducing bacterial contamination and extending the shelf life of poultry. Studies have shown a significant reduction in bacterial counts on chitosan-coated poultry, underscoring its effectiveness in ensuring microbiological safety (Atay, 2020).

Oxidation Prevention: Oxidation is a common challenge in poultry preservation, leading to the deterioration of flavor, color, and nutritional quality. Chitosan's antioxidant properties play a crucial role in preventing lipid oxidation, a process that results in off-flavors and rancidity in meat.

By scavenging free radicals and inhibiting lipid peroxidation, chitosan helps maintain the freshness and overall quality of poultry products. Chitosan-based films are applied to poultry surfaces, creating a protective layer that minimizes exposure to oxygen. This application has proven successful in inhibiting lipid oxidation and preserving the sensory attributes of poultry. The prevention of oxidative reactions is particularly vital in processed poultry products, where lipid oxidation can compromise the quality of value-added items (Huang and Ahn, 2019; Dawud et al., 2025).

Moisture Retention: Chitosan's film-forming characteristics contribute to moisture retention, addressing the challenge of dehydration in poultry products. The application of chitosan coatings creates a thin protective layer that reduces water loss from poultry cuts. This is especially crucial for maintaining the juiciness, tenderness, and overall texture of poultry, factors highly valued by consumers. In poultry processing, where various cuts and products undergo freezing and refrigeration, chitosan-coated surfaces act as a barrier against moisture loss. This helps prevent undesirable textural changes and ensures that the poultry products retain their appealing attributes throughout storage and distribution (Yaghoubi et al., 2021).

Nutritional Preservation: Beyond safety and quality, chitosan contributes to the nutritional preservation of poultry products. By inhibiting microbial contamination and oxidation, chitosan helps retain the levels of essential nutrients, including proteins, vitamins, and minerals. This aspect is particularly crucial in meeting consumer expectations for nutritious and wholesome poultry offerings. Chitosan coatings are applied to processed poultry products, such as marinated or seasoned cuts. These coatings not only enhance safety and quality but also contribute to the preservation of nutritional content. This is aligned with consumer preferences for minimally processed foods that retain their natural nutritional value (Li et al., 2022).

Challenges and Future Perspectives: While the application of chitosan in the poultry industry holds immense promise, certain challenges and considerations must be acknowledged. Adhesion to poultry surfaces, ensuring uniform coating, and optimizing application methods are areas of ongoing research. Additionally, addressing consumer perceptions and regulatory aspects related to chitosan-coated products is essential for widespread acceptance. Future perspectives include exploring innovative formulations, such as chitosan blends with other natural polymers or bioactive compounds, to enhance the overall efficacy of coatings. Advances in nanotechnology may also offer opportunities to create nanoscale delivery systems for chitosan, optimizing its interaction with poultry surfaces and improving its antimicrobial and antioxidant effects (Matloob et al., 2023).

The application of chitosan in the poultry industry represents a paradigm shift in the quest for safer, higher quality, and more sustainable poultry products. Its multifaceted mechanisms of action, including antimicrobial effects, oxidation prevention, moisture retention, and nutritional preservation, address key challenges faced by the industry. As research and development in this field continue, chitosan stands out as a beacon of innovation, offering a natural and effective solution for enhancing the safety and shelf life of poultry products, ultimately meeting the evolving demands of consumers for high-quality, nutritious, and sustainable food options (Inobeme et al., 2023a).

8.6 CHITOSAN IN MEAT PRESERVATION

Meat preservation is a critical aspect of the food industry, ensuring the safety and quality of meat products for consumers. Various methods and technologies are employed to extend the shelf life of meat and prevent spoilage. One emerging and promising ingredient in meat preservation is chitosan, a natural polysaccharide derived from chitin, a compound found in the exoskeletons of crustaceans such as shrimp and crab (Zang et al., 2023).

Chitosan possesses unique properties that make it suitable for meat preservation. One of its key attributes is its antimicrobial activity. Chitosan exhibits antibacterial properties against a broad spectrum of microorganisms, including bacteria, yeast, and molds. This antimicrobial action is attributed to its ability to disrupt the cell membranes of these microorganisms, leading to their inactivation.

In the context of meat preservation, this property is particularly valuable in inhibiting the growth of spoilage bacteria and pathogens, thereby extending the meat's shelf life. The application of chitosan in meat preservation involves incorporating it into various forms, such as coatings, films, and edible films, which are directly applied to the surface of meat products. These chitosan-based materials create a protective barrier that helps reduce oxygen permeability and moisture loss, both of which are crucial factors influencing the deterioration of meat. By forming a barrier, chitosan coatings can slow down the oxidation and microbial growth processes, contributing to the preservation of meat quality (Yan et al., 2021).

Moreover, chitosan has been demonstrated to possess antioxidant properties. Oxidation is a common cause of quality deterioration in meat, leading to flavor changes, color loss, and the development of off-odors. The antioxidant activity of chitosan helps counteract oxidative reactions, preserving the sensory attributes of meat products. This dual functionality—antimicrobial and antioxidant—makes chitosan a versatile ingredient for enhancing the safety and quality of preserved meats. In addition to its preservation capabilities, chitosan is environmentally friendly and biodegradable. This aligns with the growing consumer demand for sustainable and eco-friendly food packaging solutions. Chitosan-based films and coatings can serve as alternatives to traditional packaging materials, not only offering preservation benefits but also addressing concerns related to plastic waste and environmental impact (Pateiro et al., 2021; Inobeme et al., 2024a).

Research studies have explored the effectiveness of chitosan in preserving different types of meat, including beef, poultry, and fish. The results indicate that chitosan applications can lead to a reduction in microbial contamination, delay spoilage, and extend the overall shelf life of meat products. Furthermore, the sensory attributes of the preserved meats, such as taste, texture, and color, remain satisfactory, demonstrating the potential of chitosan as a viable option in meat preservation. Despite its promising characteristics, challenges exist in the widespread adoption of chitosan in the meat industry. Issues such as cost, availability, and the potential impact on sensory characteristics need to be addressed. Researchers and industry professionals are actively working to optimize chitosan applications, exploring combinations with other natural antimicrobial agents and refining processing techniques to enhance its efficacy while minimizing any negative effects on meat quality (Elgadir, and Mariod, 2022).

However, chitosan holds significant promise as an effective and sustainable ingredient in meat preservation. Its antimicrobial and antioxidant properties, coupled with its biodegradability, make it a valuable tool for extending the shelf life of meat products. As research in this field continues, chitosan may play a pivotal role in shaping the future of meat preservation practices, aligning with the industry's evolving focus on safety, quality, and environmental responsibility (El-Araby et al., 2024).

8.7 CHITOSAN'S ROLE IN EXTENDING FISHERY PRODUCT SHELF LIFE

The preservation of fishery products is a critical concern for the seafood industry, aiming to ensure the freshness, safety, and quality of products until they reach consumers. Among various preservation methods, chitosan, derived from the exoskeletons of crustaceans, has emerged as a promising natural ingredient with remarkable properties beneficial for extending the shelf life of fishery products.

Chitosan, a polysaccharide derived from chitin, exhibits unique characteristics that make it suitable for applications in seafood preservation. Its primary attributes include antimicrobial properties, biodegradability, and versatility in different application forms such as coatings, films, and injections. These properties collectively contribute to its effectiveness in combating spoilage and prolonging the shelf life of fish products. One of the common applications of chitosan in fishery product preservation is through coatings. Salmon fillets, for instance, have been subjected to chitosan coatings, resulting in a noticeable extension of shelf life by approximately 3 days. The antimicrobial action of chitosan helps inhibit the growth of spoilage microorganisms on the surface of the fish, delaying the deterioration process and preserving its quality during storage and transportation. In the case of shrimp, chitosan has been employed in the form of edible films. The application of chitosan edible films on shrimp has demonstrated a significant extension of shelf life, with an increase of approximately 5 days. These films create a protective barrier, reducing the permeability of oxygen and moisture, which are key factors influencing the freshness of seafood. By slowing down oxidation and moisture loss, chitosan contributes to maintaining the texture, flavor, and overall quality of shrimp (Teixeira-Costa and Andrade, 2021).

Cod fish, another popular seafood item, has also benefited from chitosan applications. Chitosan coatings on cod fish have been shown to increase shelf life by around 4 days. The antimicrobial activity of chitosan helps control the growth of bacteria and spoilage microorganisms, preventing the onset of off-flavors and textures commonly associated with fish deterioration. This extension of shelf life is particularly crucial in ensuring that the cod fish reaches consumers in optimal condition. In addition to coatings, chitosan injections have been explored as a preservation method, particularly in tuna steaks. Chitosan injections into tuna steaks have extended their shelf life by approximately 2 days. This method allows for the direct introduction of chitosan into the fish tissue, enhancing its antimicrobial and antioxidant effects internally. The result is a delayed onset of spoilage and improved preservation of the tuna's sensory attributes. The success of chitosan in extending the shelf life of fishery products can be attributed to its multifaceted functionality. Chitosan's antimicrobial properties are effective against a broad spectrum of microorganisms, including bacteria, yeast, and molds (Adetunji et al., 2023a). Its ability to disrupt cell membranes hinders the growth and proliferation of these microorganisms, thereby retarding spoilage processes (Table 8.2).

Furthermore, chitosan exhibits antioxidant properties, countering oxidative reactions that contribute to flavor changes, color loss, and the development of off-odors in fish. By mitigating these oxidative processes, chitosan helps maintain the natural characteristics of seafood, ensuring that consumers receive products with optimal flavor and visual appeal. The biodegradability of chitosan is an additional advantage, aligning with the growing demand for sustainable and environmentally friendly packaging solutions. Chitosan-based films and coatings not only contribute to shelf-life extension but also address concerns related to plastic waste, making them an attractive option for environmentally conscious consumers. While chitosan shows immense promise in fishery product preservation, challenges such as cost, scalability, and potential impacts on sensory characteristics must be addressed for widespread adoption in the industry. Ongoing research aims to optimize chitosan applications, exploring combinations with other natural antimicrobial agents and refining processing techniques to enhance efficacy while minimizing any adverse effects on product quality (Teshome et al., 2022).

In addition, chitosan stands out as a valuable and natural ingredient in the preservation of fishery products. Its antimicrobial, antioxidant, and biodegradable properties make it a versatile solution for extending shelf life, ensuring that consumers can enjoy seafood with confidence in its freshness, safety, and overall quality. As research and development in this field continue, chitosan is poised to play a crucial role in shaping the future of sustainable and effective seafood preservation practices (Teixeira-Costa and Andrade, 2021).

TABLE 8.2
Chitosan's Role in Extending Fishery Product Shelf Life

Type of Fishery Product	Chitosan Application	Shelf-Life Extension	References
Salmon fillets	Chitosan coating	Extended by 3 days	Liu et al. (2023)
Shrimp	Chitosan edible film	Prolonged by 5 days	Zhang et al. (2019)
Cod fish	Chitosan coating	Increased by 4 days	Nuzaiba et al. (2023)
Tuna steaks	Chitosan injection	Extended by 2 days	Grande-Tovar et al. (2018)

8.8 ENHANCING SHELF LIFE IN MULTI-COMPONENT PRODUCTS OF FISH AND MEAT

Multi-component products, combining fish and meat, not only offer unique culinary experiences but also present challenges in terms of shelf life due to the diverse microbial and chemical reactions that can occur. The preservation of these complex products is crucial to maintain their safety, quality, and appeal to consumers. Various strategies, including the use of antimicrobial agents, MAP, and innovative processing techniques, play a vital role in extending the shelf life of multi-component fish and meat products. One effective approach to enhance the shelf life of these products is the utilization of natural antimicrobial agents, such as chitosan. Chitosan, derived from chitin found in crustacean exoskeletons, exhibits antimicrobial properties against a broad spectrum of microorganisms. When applied as coatings or incorporated into packaging materials, chitosan forms a protective barrier that inhibits the growth of spoilage bacteria and pathogens, thus extending the shelf life of multi-component products (Karbowiak et al., 2023).

In the case of multi-component fish and meat products, combining chitosan coatings with other natural antimicrobial agents, such as plant extracts or essential oils, can provide a synergistic effect. This combination enhances the overall antimicrobial activity, offering a broader spectrum of protection against various microorganisms that may affect different components of the product. MAP is another effective strategy to extend the shelf life of multi-component fish and meat products. This technique involves adjusting the composition of gases within the packaging to create an environment that slows down the deterioration processes. For example, by reducing oxygen levels and increasing carbon dioxide, MAP inhibits the growth of aerobic bacteria and delays oxidation reactions, preserving the freshness of both the fish and meat components (Casalini and Baschetti, 2023).

Furthermore, the use of vacuum packaging can contribute to shelf-life extension. Vacuum packaging removes air from the packaging, minimizing the risk of oxidation and reducing the potential for microbial growth. This method is particularly beneficial for multi-component products, as it helps maintain the distinct qualities of both fish and meat components, preventing cross-contamination and preserving their individual flavors and textures. Innovative processing techniques also play a pivotal role in enhancing the shelf life of multi-component fish and meat products. High-pressure processing (HPP), for instance, is a non-thermal method that effectively reduces microbial load while preserving the nutritional and sensory attributes of the products. HPP can be tailored to the specific requirements of multi-component products, ensuring that both fish and meat components undergo sufficient processing to guarantee safety and quality (Inobeme et al., 2023b).

Furthermore, incorporating natural antioxidants into the formulation of multi-component products can mitigate oxidative reactions and extend shelf life. Antioxidants derived from sources such as herbs, spices, or plant extracts can help prevent lipid oxidation and maintain the color and flavor stability of both fish and meat components. This approach aligns with consumer preferences for clean-label products, as natural antioxidants contribute to the overall safety and quality without the need for synthetic additives. The challenge in enhancing shelf life in multi-component fish and meat products lies in maintaining a balance between preserving the individual characteristics of each component and ensuring overall product safety. Consumer expectations regarding freshness, taste, and nutritional value add complexity to the preservation process. Therefore, a holistic approach that considers the synergies between antimicrobial agents, packaging techniques, and processing methods is essential (López-Pedrouso et al., 2022).

In addition, the preservation of multi-component fish and meat products requires a combination of effective strategies to address the diverse challenges associated with each component. Natural antimicrobial agents like chitosan, MAP, vacuum packaging, and innovative processing techniques contribute collectively to extend shelf life while preserving the distinctive qualities of both fish and meat. As technology advances and consumer preferences evolve, ongoing research and development will continue to refine these preservation methods, ensuring that multi-component products meet the highest standards of safety, quality, and sensory satisfaction (Gumienna and Górna, 2021; Mathew et al., 2024a).

8.9 REGULATORY CONSIDERATIONS AND SAFETY OF CHITOSAN IN FOOD PRESERVATION

Chitosan, a natural polysaccharide derived from chitin, has gained prominence in the food industry for its diverse applications, especially in food preservation. As a food additive, chitosan's safety and regulatory compliance are critical aspects that must be carefully evaluated to ensure consumer protection and meet industry standards. From a regulatory perspective, the use of chitosan in food preservation is subject to oversight by food safety agencies worldwide. In the United States, the Food and Drug Administration (FDA) regulates food additives, including those derived from chitosan. Chitosan is generally recognized as safe (GRAS) when used within specified limits and for approved purposes. Manufacturers are required to adhere to the FDA's guidelines, providing scientific evidence to support the safety of chitosan in intended applications. Similarly, the European Food Safety Authority (EFSA) assesses the safety of food additives within the European Union. Chitosan has been evaluated by the EFSA, and its safety is affirmed within established usage levels and applications. Compliance with these regulatory standards ensures that chitosan meets the necessary criteria for use in food preservation without posing undue risks to consumers (Stefanowska et al., 2023).

The safety of chitosan is rooted in its natural origin, as it is derived from the shells of crustaceans such as shrimp and crab. However, considerations must be given to potential allergenicity for individuals with shellfish allergies. Manufacturers are required to clearly label products containing chitosan to alert consumers with shellfish allergies, allowing them to make informed choices and avoid potential adverse reactions. Chitosan's safety profile is further supported by its biodegradability and low toxicity. Unlike some synthetic preservatives, chitosan is considered environmentally friendly, aligning with growing consumer preferences for sustainable and eco-friendly food products. This characteristic has contributed to the adoption of chitosan in food packaging materials, addressing concerns related to plastic waste.

In the context of food preservation, chitosan's antimicrobial properties are pivotal for inhibiting the growth of bacteria, yeast, and molds. Its mechanism of action involves disrupting the cell membranes of microorganisms, rendering them inactive. While this antimicrobial activity is advantageous for extending the shelf life of foods, it also underscores the importance of using chitosan within established limits to prevent any unintended effects. Chitosan's applications in food preservation include coatings, films, and edible films applied directly to the surface of foods. The development of chitosan-based materials for food packaging is an area of active research. These materials serve as barriers, reducing oxygen permeability and moisture loss, thus preserving the quality and safety of food products. The migration of chitosan from packaging materials into foods is carefully controlled to ensure that it remains within acceptable levels (Muthu et al., 2021).

Research studies and toxicological assessments continue to contribute to the understanding of chitosan's safety in food applications. While chitosan is generally regarded as safe, ongoing studies explore potential interactions with other food components and evaluate its safety under various conditions. To enhance consumer confidence, clear labeling practices and transparent communication about the use of chitosan in food products are essential. This includes providing information on the source of chitosan and any potential allergenic risks for susceptible individuals. Additionally, manufacturers must adhere to Good Manufacturing Practices (GMP) to ensure the quality and safety of products containing chitosan (Thambiliyagodage et al., 2023).

In addition, the regulatory considerations and safety of chitosan in food preservation are fundamental to its widespread use in the food industry. Regulatory bodies such as the FDA and EFSA play a crucial role in assessing the safety of chitosan, ensuring that it meets established standards and poses no undue risks to consumers. As a natural and versatile ingredient, chitosan continues to contribute to the development of innovative and sustainable food preservation practices while maintaining a commitment to safety and regulatory compliance (Biswas et al., 2022).

8.10 EMERGING TECHNOLOGIES IN FOOD PRESERVATION

Food preservation is a critical aspect of the food industry, ensuring that food products remain safe, nutritious, and flavorful for extended periods. As technology continues to advance, several emerging technologies are reshaping the landscape of food preservation, offering innovative solutions to address challenges such as spoilage, contamination, and quality degradation. These technologies not only enhance the shelf life of food products but also contribute to sustainability efforts and meet consumer demands for minimally processed, safe, and high-quality foods (Adetunji et al., 2023b).

High-Pressure Processing (HPP): HPP involves subjecting food products to elevated pressures, typically between 100 and 800 MPa. This non-thermal preservation method effectively inactivates microorganisms, enzymes, and spoilage agents while maintaining the nutritional and sensory qualities of the food. HPP has found applications in a variety of products, including juices, meats, and ready-to-eat meals. Its ability to extend shelf life without the use of heat makes it particularly suitable for preserving fresh and minimally processed foods (Koutsoumanis et al., 2022).

Pulsed Electric Fields (PEF): Pulsed Electric Fields (PEF) technology applies short bursts of electrical energy to food products, leading to microbial inactivation and enzymatic inactivation. PEF is known for its ability to preserve the nutritional content and sensory attributes of foods while extending shelf life. This technology is applicable to liquids, such as juices and dairy products, and offers advantages in terms of energy efficiency compared to traditional thermal processing (Nowosad et al., 2021; Mathew et al., 2024c).

Cold Plasma Technology: Cold plasma, a partially ionized gas, is being explored for its antimicrobial properties in food preservation. This technology generates reactive oxygen and nitrogen species, effectively eliminating pathogens and spoilage microorganisms on the surface of food products. Cold plasma technology has been applied to fruits, vegetables, and meats, offering a non-thermal alternative for enhancing food safety and extending shelf life (Rao et al., 2023).

Nanotechnology in Packaging: Nanotechnology has made significant strides in food packaging, contributing to improved preservation. Nanoparticles, such as silver or titanium dioxide, are incorporated into packaging materials to impart antimicrobial properties and inhibit the growth of microorganisms on the food surface. Additionally, nanocomposites enhance barrier properties, reducing oxygen and moisture permeability and thereby extending the shelf life of packaged foods (Inobeme et al., 2023c).

Modified Atmosphere Packaging (MAP): While not a new concept, MAP continues to evolve with innovations in gas mixtures and packaging materials. This technology involves adjusting the composition of gases surrounding the food product to slow down deterioration processes. MAP is widely used for fruits, vegetables, meats, and seafood, helping to maintain product quality and freshness during storage and transportation.

Biopreservation: Biopreservation involves the use of beneficial microorganisms or their metabolic products to control the growth of spoilage and pathogenic microorganisms in food. Probiotics and lactic acid bacteria are examples of microorganisms utilized for their antimicrobial properties. Biopreservation not only enhances food safety but also contributes to the development of functional foods with potential health benefits (Mathew et al., 2023).

Edible Coatings and Films: Edible coatings and films, often made from natural polymers such as chitosan, alginate, or starch, are gaining popularity as an eco-friendly and sustainable preservation solution. These coatings create a protective barrier on the surface of food products, reducing oxygen permeability and moisture loss. Edible films can extend the shelf life of fruits, vegetables, and meats while minimizing the need for traditional plastic packaging (Matloob et al., 2023).

Smart Packaging: Advancements in sensor technologies have led to the development of smart packaging systems that monitor and communicate information about the condition of the food. Time-temperature indicators, gas sensors, and freshness indicators integrated into packaging help consumers and retailers assess the quality and safety of the product. These innovations contribute to reducing food waste by providing accurate information on the freshness of the packaged food (Müller and Schmid, 2019).

Ozone Treatment: Ozone, a powerful oxidizing agent, is being explored for its antimicrobial properties in food preservation. Ozone treatment can inactivate bacteria, molds, and viruses on the surface of fruits, vegetables, and meats. Ozone has the advantage of leaving no chemical residues, making it an environmentally friendly option for enhancing food safety.

Microbial Fermentation: Controlled fermentation with specific strains of lactic acid bacteria or yeast can be employed to preserve foods. Fermentation not only extends shelf life by lowering the pH and inhibiting spoilage microorganisms but also imparts unique flavors and textures to the products. Fermented foods, such as sauerkraut, kimchi, and yogurt, exemplify the successful application of this traditional preservation method (Mathew et al., 2018).

In addition, the landscape of food preservation is evolving rapidly with the advent of innovative technologies. These emerging technologies offer sustainable, efficient, and often non-thermal alternatives to traditional preservation methods. As consumer preferences shift toward minimally processed and clean-label foods, the integration of these advanced preservation technologies becomes crucial for meeting the demands of a dynamic and ever-changing food industry (Mathew et al., 2021).

8.11 CONCLUSION

In conclusion, the application of chitosan emerges as a promising and versatile solution for extending the shelf life of poultry, meat, and fishery products. Its antimicrobial and antioxidant properties play a pivotal role in inhibiting the growth of spoilage microorganisms and delaying oxidative processes, thereby preserving the freshness and quality of these perishable products. Chitosan's effectiveness is evident in various applications, including coatings, films, and edible films, where it forms protective barriers against oxygen and moisture. As a natural polysaccharide derived from chitin, chitosan aligns with consumer preferences for sustainable and eco-friendly food preservation methods. The biodegradability of chitosan further contributes to its appeal as an environmentally conscious choice. While challenges such as cost and potential impact on sensory characteristics exist, ongoing research and development aim to optimize chitosan applications and address these concerns. In the poultry, meat, and fishery industries, where maintaining the safety and quality of products is paramount, chitosan stands out as a valuable tool. Its multifaceted benefits make it not only a preservative but also a key contributor to the industry's pursuit of safe, sustainable, and high-quality food products for consumers worldwide.

REFERENCES

Abimbola, O. F., Adetunji, C. O., Dare, B. J., Mathew, J. T., Inobeme, A., Ghazanfar, S., Olaniyan, O. T, Ajiboye, M. D., & Dauda, W. (2023). Packaging of animal-based food products: The need for intelligent freshness sensors. In: *Sensing and Artificial Intelligence Solutions for Food Manufacturing*. CRC Press, 89–102. https://doi.org/10.1201/9781003207955-6

Adetunji, C. O., Mathew, J. T., Inobeme, A., Taiwo, O., Olaniyan, O., Shakira, G., & Wadazani, D. (2023b). Application of biosensors for detection and monitoring of water quality. In: *Organic and Inorganic Materials Based Sensors*. John Wiley & Sons, Inc., 1065–1077. https://doi.org/10.1002/9783527834266.ch46

Adetunji, C. O., Ogundolie, F. A., Mathew, J. T., Inobeme, A., Olotu, T., Adetunji, J. B., Dauda, W. P., Aborode, A. T., Ajayi, O. O., Ghazanfar, S., Aliyu, A., Imoisi, S. E., Moses-Oke, R. O., Egbuna, C., Uzualu, E. J., Ajenifujah-Solebo, S. O., Akinbo, O., Adetuyi, B. O., & Edetalehn, O. I. (2023a). Patenting protocols, toxicity, risk assessments, and policy issues of nanomaterials with diverse applications in food, biomedical, and other relevant sectors. In: *Evaporative Coolers for the Postharvest Management of Fruits and Vegetables*. Academic Press, 421–446. https://doi.org/10.1016/B978-0-323-89864-5.00005-9

Adiletta, G., Di Matteo, M., & Petriccione, M. (2021). Multifunctional role of chitosan edible coatings on antioxidant systems in fruit crops: A review. *International Journal of Molecular Sciences*, *22*(5), 2633. https://doi.org/10.3390/ijms22052633

Atay, Y. H. (2020). Antibacterial activity of chitosan-based systems. In: *Functional Chitosan: Drug Delivery and Biomedical Applications*. Springer, 457–489. https://doi.org/10.1007/978-981-15-0263-7_15

Biswas, R., Alam, M., Sarkar, A., Haque, M. I., Hasan, M. M., & Hoque, M. (2022). Application of nanotechnology in food: processing, preservation, packaging and safety assessment. *Heliyon*, *8*(11), e11795. https://doi.org/10.1016/j.heliyon.2022.e11795

Bodunrinde, R. E., Chimbekujwo, K. I., Oyewole, O. A., Adetunji, C. O., Inobeme, A., Eniola, K. I. T., Yerima, M. B., & Mathew, J. T. (2025). The use of nanochitosan for enhancement in the quality and yield of fruit crops. In: *Nanochitosan Applications for Enhanced Crop Production and Food Security*. John Wiley & Sons, 111–122. https://doi.org/10.1002/9781394214945.ch5

Bose, I., Roy, S., Pandey, V. K., & Singh, R. (2023). A comprehensive review on significance and advancements of antimicrobial agents in biodegradable food packaging. *Antibiotics (*Basel, Switzerland), *12*(6), 968. https://doi.org/10.3390/antibiotics12060968

Casalini, S., & Baschetti, G. M. (2023). The use of essential oils in chitosan or cellulose-based materials for the production of active food packaging solutions: A review. *Journal of the Science of Food and Agriculture*, *103*(3), 1021–1041. https://doi.org/10.1002/jsfa.11918

Chimbekujwo, K. I., Yakubu, J. G., Omoregie, I. P., Oyewole, O. A., Muhammad, W. M., Adetunji, C. O., Eniola, K. I. T., Yerima, M. B., & Mathew, J. T. (2025). Synthesis of nano-chitosan using agricultural waste. In: *Nanochitosan Applications for Enhanced Crop Production and Food Security*. John Wiley & Sons, Inc., 39–68. https://doi.org/10.1002/9781394214945.ch2

Cudjoe, C. D., Balali, G. I., Titus, O. O., Osafo, R., & Taufiq, M. (2022). Food safety in Sub-Sahara Africa: An insight into Ghana and Nigeria. *Environmental Health Insights*, *16*, 11786302221142484. https://doi.org/10.1177/11786302221142484

Dawud, I., Bodunrinde, R. E., Oyewole, O. A., Adetunji, C. O., Inobeme, A., Eniola, K. I. T., Yerima, M. B., & Mathew, J. T. (2025). Application of nanochitosan for effective fruit production. In: *Nanochitosan Applications for Enhanced Crop Production and Food Security*. John Wiley & Sons, 123–135. https://doi.org/10.1002/9781394214945.ch6

Egorov, A. R., Anatoly, A. K., Vasili, V. R., Vasili, V. R., Alexander, G. T., & Andreii, S. K. (2023). Chitosan and its derivatives: Preparation and antibacterial properties. *Materials,* 16(18), 6076. https://doi.org/10.3390/ma16186076

El-Araby, A., Janati, W., Ullah, R., Ercisli, S., & Errachidi, F. (2024). Chitosan, chitosan derivatives, and chitosan-based nanocomposites: Eco-friendly materials for advanced applications (a review). *Frontiers in Chemistry*, *11*, 1327426. https://doi.org/10.3389/fchem.2023.1327426

Elgadir, M. A., & Mariod, A. A. (2022). Gelatin and chitosan as meat by-products and their recent applications. *Foods (*Basel, Switzerland), *12*(1), 60. https://doi.org/10.3390/foods12010060

Grande-Tovar, C. D., Serio, A., Delgado-Ospina, J., Paparella, A., Rossi, C., & Chaves-López, C. (2018). Chitosan films incorporated with *Thymus capitatus* essential oil: Mechanical properties and antimicrobial activity against degradative bacterial species isolated from tuna (*Thunnus* sp.) and swordfish (*Xiphias gladius*). *Journal of Food Science and Technology*, *55*(10), 4256–4265. https://doi.org/10.1007/s13197-018-3364-y

Gumienna, M., & Górna, B. (2021). Antimicrobial food packaging with biodegradable polymers and bacteriocins. *Molecules (*Basel, Switzerland), *26*(12), 3735. https://doi.org/10.3390/molecules26123735

Hoa, V. B., Song, D. H., Seol, K. H., Kim, Y. S., Kim, H. W., Bae, I. S., & Cho, S. H. (2024). Effect of coating with combined chitosan and gallic acid on shelf-life stability of Jeju black cattle beef. *Animal Bioscience*, *37*(1), 123–130. https://doi.org/10.5713/ab.23.0180

Huang, X., & Ahn, D. U. (2019). Lipid oxidation and its implications to meat quality and human health. *Food Science and Biotechnology*, *28*(5), 1275–1285. https://doi.org/10.1007/s10068-019-00631-7

Inobeme, A., Adetunji, C. O., Ajai, A. I., Inobeme, J., Mathew, J. T., Obar, A., Maliki, M., Nwakife, N., & Eziukwu, C. (2023b). Chemical nanosensors for monitoring environmental pollution. In: *Advanced Application of Nanotechnology to Industrial Wastewater*. Springer Nature Singapore, 93–103. https://doi.org/10.1007/978-981-99-3292-4_6

Inobeme, A., Adetunji, C. O., Mathew, J. T., Ajai, A. I., Inobeme, J., Bamigboye, M. O., Onyeaku, S., Maliki, M., Eziukwu, C., & Tawa, K. (2023c). Nanotechnology for bioremediation of heavy metals. In: *Microbial Technologies in Industrial Wastewater Treatment*. Springer Nature Singapore, 19–30. https://doi.org/10.1007/978-981-99-2435-6_2

Inobeme, A., Ajai, I. A., Inobeme, J., Adetunji, C. O., Obar, A., Mathew, J. T., Jacob, J. O., & Nwakife, N. C. (2023a). Superabsorbent polymers for the development of nanofiltration. In: *Properties and Applications of Superabsorbent Polymers*. Springer, Singapore, 157–170. https://doi.org/10.1007/978-981-99-1102-8_7

Inobeme, A., Mathew, J. T., Ajai, I. A., Adetunji, C. O., Inobeme, J., Maliki, M., Adekoya, M. A., Shaba, E. Y., Eric, O., Akhor, S., & Eziukwu, C. (2024a). Common techniques in food processing technologies. In: *Nanobiotechnology for Food Processing and Packaging*. Elsevier Inc., 223–234. https://doi.org/10.1016/B978-0-323-91749-0.00002-2

Inobeme, A., Shahnawaz, M., Adetunji, C. O., Mathew, J. T., Adetuyi, B. O., Popoola, O. A., Olaitan, F. Y., Akinbo, O., Kolawole, O. M., Oyewole, O. A., Eniola, K. I. T., & Yerima, B. M. (2024b). Microplastic as a pathogenic vector to cause diseases in Marine Biota. In: *Microplastic Pollution*. Springer, Singapore, 1, 321–349. https://doi.org/10.1007/978-981-99-8357-5_19

Jiang, A., Patel, R., Padhan, B., Palimkar, S., Galgali, P., Adhikari, A., Varga, I., & Patel, M. (2023). Chitosan based biodegradable composite for antibacterial food packaging application. *Polymers*, *15*(10), 2235. https://doi.org/10.3390/polym15102235

Karanth, S., Feng, S., Patra, D., & Pradhan, A. K. (2023). Linking microbial contamination to food spoilage and food waste: The role of smart packaging, spoilage risk assessments, and date labeling. *Frontiers in Microbiology*, *14*, 1198124. https://doi.org/10.3389/fmicb.2023.1198124

Karbowiak, M., Szymański, P., & Zielińska, D. (2023). Synergistic effect of combination of various microbial hurdles in the biopreservation of meat and meat products-systematic review. *Foods* (Basel, Switzerland), *12*(7), 1430. https://doi.org/10.3390/foods12071430

Khubiev, O. M., Egorov, A. R., Kirichuk, A. A., Khrustalev, V. N., Tskhovrebov, A. G., & Kritchenkov, A. S. (2023). Chitosan-based antibacterial films for biomedical and food applications. *International Journal of Molecular Sciences*, *24*(13), 10738. https://doi.org/10.3390/ijms241310738

Koutsoumanis, K., Alvarez-Ordóñez, A., Bolton, D., Bover-Cid, S., Chemaly, M., Davies, R., De Cesare, A., Herman, L., Hilbert, F., Lindqvist, R., Nauta, M., Peixe, L., Ru, G., Simmons, M., Skandamis, P., Suffredini, E., Castle, L., Crotta, M., Grob, K., … Allende, A. (2022). The efficacy and safety of high-pressure processing of food. *EFSA Journal. European Food Safety Authority*, *20*(3), e07128. https://doi.org/10.2903/j.efsa.2022.7128

Li, X., Zhang, R., Hassan, M. M., Cheng, Z., Mills, J., Hou, C., Realini, C. E., Chen, L., Day, L., Zheng, X., Zhang, D., & Hicks, T. M. (2022). Active packaging for the extended shelf-life of meat: Perspectives from consumption habits, market requirements and packaging practices in China and New Zealand. *Foods* (Basel, Switzerland), *11*(18), 2903. https://doi.org/10.3390/foods11182903

Liu, Y., Chen, J., Li, H., & Wang, Y. (2023). Nanocomplexes film composed of gallic acid loaded ovalbumin/chitosan nanoparticles and pectin with excellent antibacterial activity: Preparation, characterization and application in coating preservation of salmon fillets. *International Journal of Biological Macromolecules*, *259*(Pt 1), 128934. Advance online publication. https://doi.org/10.1016/j.ijbiomac.2023.128934

López-Pedrouso, M., Lorenzo, J. M., & Franco, D. (2022). Advances in natural antioxidants for food improvement. *Antioxidants* (Basel, Switzerland), *11*(9), 1825. https://doi.org/10.3390/antiox11091825

Madejska, A., Pawul-Gruba, M., & Osek, J. (2022). Histamine content in selected production stages of fish products. *Journal of Veterinary Research*, *66*(4), 599–604. https://doi.org/10.2478/jvetres-2022-0063

Malik, S. I., Chimbekujwo, K. I., Maku, A. M., Oyewole, Oluwafemi, A., Omoregie, I. P., Adetunji, C. O., Eniola, K. I. T., Yerima, M. B., & Mathew, J. T. (2025). The application of nanochitosan biopesticides as a replacement to synthetic pesticides. In: *Nanochitosan Applications for Enhanced Crop Production and Food Security*. John Wiley & Sons, 89–110. https://doi.org/10.1002/9781394214945.ch4

Mathew, J. T., Adetunji, C. O., Inobeme, A., Musah, M., Azeh, Y., Etsuyankpa, M. B., Otori, A. A., Chado, I. M., Shaba, E. Y., Akoji, J. N., Mamman, A., & Oyewole, O. A. (2025b). Application of nanochitosan for the biodegradation of agricultural wastes. In: *Nanochitosan Applications for Enhanced Crop Production and Food Security*. John Wiley & Sons, 361–382. https://doi.org/10.1002/9781394214945.ch16

Mathew, J. T., Adetunji, C. O., Inobeme, A., Otori, A. A., Musah, M., Azeh, Y., Shaba, E. Y., Muhammad, I. A., Mamman, A., & Oyewole, O. A. (2025a). Relevance of nanochitosan in food sector and food packaging sectors: Current trends. In: *Nanochitosan Applications for Enhanced Crop Production and Food Security*. John Wiley & Sons, 345–360. https://doi.org/10.1002/9781394214945.ch15

Mathew, J. T., Dauda, B. E. N., Mann, A., Ndamitso, M. M., Etsuyankpa, M. B., & Shaba, E. Y. (2018). Assessment of the nutritive and anti-nutritive compositions of fermented and unfermented African Custard Apple (*Annona senegalensis*) seeds from Niger State, Nigeria. *FUW Trends in Science & Technology Journal*, *3*(2A), 471–477.

Mathew, J. T., Inobeme, A., Etsuyankpa, M. B., Adetunji, C. O., Shaba, E. Y., Musah, M., Azeh, Y., Jiya, M. J., Tanko, E., Muhammad, A. I., Mamman, A., & Maurice, J. (2024b). Applications of diatom for treatment of industrial effluents: Waste to wealth approach. In: *Marine Biomass: Biorefinery, Bioproducts and Environmental Bioremediation*. Walter de Gruyter GmbH & Co KG, 1–20. https://doi.org/10.1515/9783111353951-001

Mathew, J. T., Inobeme, A., Hussaini, J., Shaba, E. Y., Azeh, Y., Musah, M., Muhammad, E. B., Tanko, M. S., Ismail, H., Adetunji, C. O., Kanwa, A. M., & Mamman, A. (2024a). Properties of nanomaterials for corrosion control. In: A. Thakur & A. Kumar (Eds.), *Innovations in Nanomaterials-Based Corrosion Inhibitors*. IGI Global, 74–99. https://doi.org/10.4018/979-8-3693-3088-3.ch004

Mathew, J. T., Inobeme, A., Musah, M., Azeh, Y., Etsuyankpa, M. B., Adetunji, C. O., Shaba, E. Y., Jiya, M. J., Josiah, J. G., Tanko, E., Muhammad, A. I., Mamman, A., & Jemkur, M. (2024c). Utilization of marine green algae for development of bioplastics. In: *Marine Biomass: Biorefinery, Bioproducts and Environmental Bioremediation*. Walter de Gruyter GmbH & Co KG, 39–55. https://doi.org/10.1515/9783111353951-003

Mathew, J. T., Musah, M., Azeh, Y., & Muhammed, M. (2023). Kinetic study of heavy metals removal from pharmaceutical wastewater using geopolymer/Fe_3O_4 nanocomposite. *Bima Journal of Science and Technology*, *7*(4), 152–163.

Mathew, J. T., Musah, M., Shaba, E. Y., Mohammed, M., Umar, M. T., Otori, A. A., Tsado, A. N., & Muhammad, A. I. (2021). Effect of fermentation on chemical compositions of African Custard Apple seeds (Annona senegalensis) obtained from Niger State, Nigeria. *Lapai Journal of Science and Technology*, 7(1), 14–31.

Matloob, A., Ayub, H., Mohsin, M., Ambreen, S., Khan, F. A., Oranab, S., Rahim, M. A., Khalid, W., Nayik, G. A., Ramniwas, S., & Ercisli, S. (2023). A review on edible coatings and films: Advances, composition, production methods, and safety concerns. *ACS Omega*, *8*(32), 28932–28944. https://doi.org/10.1021/acsomega.3c03459

Muhammad, W. M., Ibrahim, A. B., Samuel, J. O., Khalifa, A. S., Oyewole, O. A., Adetunji, C. O., Eniola, K. I. T., Yerima, M. B., & Mathew, J. T. (2025). The role of nanomaterials in agriculture as nanofertilizers. In: *Nanochitosan Applications for Enhanced Crop Production and Food Security*. John Wiley & Sons, Inc., 1–38. https://doi.org/10.1002/9781394214945.ch1

Müller, P., & Schmid, M. (2019). Intelligent packaging in the food sector: A brief overview. *Foods (*Basel, Switzerland), *8*(1), 16. https://doi.org/10.3390/foods8010016

Muñoz-Tebar, N., Pérez-Álvarez, J. A., Fernández-López, J., & Viuda-Martos, M. (2023). Chitosan edible films and coatings with added bioactive compounds: Antibacterial and antioxidant properties and their application to food products: A review. *Polymers*, *15*(2), 396. https://doi.org/10.3390/polym15020396

Muthu, M., Gopal, J., Chun, S., Devadoss, A. J. P., Hasan, N., & Sivanesan, I. (2021). Crustacean waste-derived chitosan: Antioxidant properties and future perspective. Antioxidants (Basel, Switzerland), *10*(2), 228. https://doi.org/10.3390/antiox10020228

Nowosad, K., Sujka, M., Pankiewicz, U., & Kowalski, R. (2021). The application of PEF technology in food processing and human nutrition. *Journal of Food Science and Technology*, *58*(2), 397–411. https://doi.org/10.1007/s13197-020-04512-4

Nuzaiba, P.M., Gupta, S., & Gupta, S. (2023). Synthesis of L-methionine-loaded chitosan nanoparticles for controlled release and their in vitro and in vivo evaluation. *Scientific Reports*, *13*, 7606. https://doi.org/10.1038/s41598-023-34448-6

Ogundolie, F. A., Adetunji, C. O., Mathew, J. T., Inobeme, A., Ghazanfar, S., Olaniyan, O. T., Ajiboye, M. D., Dauda, W. P., Adetunji, J. B., Akinbo, O., & Adetuyi, B. O. (2023). Intelligent freshness indicators in packaging of plant-based food products. In: *Evaporative Coolers for the Postharvest Management of Fruits and Vegetables*. Academic Press, 323–331. https://doi.org/10.1016/B978-0-323-89864-5.00008-4

Pateiro, M., Gómez-Salazar, J. A., Jaime-Patlán, M., Sosa Morales, M. E., & Lorenzo, J. M. (2021). Plant extracts obtained with green solvents as natural antioxidants in fresh meat products. *Antioxidants (*Basel, Switzerland), *10*(2), 181. https://doi.org/10.3390/antiox10020181

Perez-Vazquez, A., Barciela, P., Carpena, M., & Prieto, M. A. (2023). Edible coatings as a natural packaging system to improve fruit and vegetable shelf life and quality. *Foods (*Basel, Switzerland), *12*(19), 3570. https://doi.org/10.3390/foods12193570

Popescu, P. A., Palade, L. M., Nicolae, I. C., Popa, E. E., Miteluţ, A. C., Drăghici, M. C., Matei, F., & Popa, M. E. (2022). Chitosan-based edible coatings containing essential oils to preserve the shelf life and postharvest quality parameters of organic strawberries and apples during cold storage. Foods (Basel, Switzerland), *11*(21), 3317. https://doi.org/10.3390/foods11213317

Raj, A., Mukherjee, A. A., de Sousa Jabbour, A. B. L., & Srivastava, S. K. (2022). Supply chain management during and post-COVID-19 pandemic: Mitigation strategies and practical lessons learned. *Journal of Business Research*, *142*, 1125–1139. https://doi.org/10.1016/j.jbusres.2022.01.037

Rani, Z. T., Mhlongo, L. C., & Hugo, A. (2023). Microbial profiles of meat at different stages of the distribution chain from the abattoir to retail outlets. *International Journal of Environmental Research and Public Health*, *20*(3), 1986. https://doi.org/10.3390/ijerph20031986

Rao, W., Li, Y., Dhaliwal, H., Feng, M., Xiang, Q., Roopesh, M. S., Pan, D., & Du, L. (2023). The application of cold plasma technology in low-moisture foods. *Food Engineering Reviews*, *15*(1), 86–112. https://doi.org/10.1007/s12393-022-09329-9

Riseh, S. R., Vatankhah, M., Hassanisaadi, M., & Kennedy, J. F. (2023). Chitosan-based nanocomposites as coatings and packaging materials for the postharvest improvement of agricultural product: A review. *Carbohydrate Polymers*, *309*, 120666. https://doi.org/10.1016/j.carbpol.2023.120666

Santos, V. P., Marques, N. S. S., Maia, P. C. S. V., Lima, M. A. B., Franco, L. O., & Campos-Takaki, G. M. (2020). Seafood waste as attractive source of chitin and chitosan production and their applications. *International Journal of Molecular Sciences*, *21*(12), 4290. https://doi.org/10.3390/ijms21124290

Shahidi, F., & Hossain, A. (2022). Role of lipids in food flavor generation. *Molecules (*Basel, Switzerland), *27*(15), 5014. https://doi.org/10.3390/molecules27155014

Shaik, M. I., Azhari, M. F., & Sarbon, N. M. (2022). Gelatin-based film as a color indicator in food-spoilage observation: A review. *Foods*, 11(23), 3797. https://doi.org/10.3390/foods11233797

Speranza, B., Racioppo, A., Bevilacqua, A., Buzzo, V., Marigliano, P., Mocerino, E., Scognamiglio, R., Corbo, M. R., Scognamiglio, G., & Sinigaglia, M. (2021). Innovative preservation methods improving the quality and safety of fish products: Beneficial effects and limits. *Foods (*Basel, Switzerland), *10*(11), 2854. https://doi.org/10.3390/foods10112854

Stefanowska, K., Woźniak, M., Dobrucka, R., & Ratajczak, I. (2023). Chitosan with natural additives as a potential food packaging. *Materials (*Basel, Switzerland), *16*(4), 1579. https://doi.org/10.3390/ma16041579

Teixeira-Costa, B. E., & Andrade, C. T. (2021). Chitosan as a valuable biomolecule from seafood industry waste in the design of green food packaging. *Biomolecules*, *11*(11), 1599. https://doi.org/10.3390/biom11111599

Teshome, E., Forsido, S. F., Rupasinghe, H. P. V., & Olika Keyata, E. (2022). Potentials of natural preservatives to enhance food safety and shelf life: A review. *The Scientific World Journal*, *2022*, 9901018. https://doi.org/10.1155/2022/9901018

Thambiliyagodage, C., Jayanetti, M., Mendis, A., Ekanayake, G., Liyanaarachchi, H., & Vigneswaran, S. (2023). Recent advances in chitosan-based applications: A review. *Materials (*Basel, Switzerland), *16*(5), 2073. https://doi.org/10.3390/ma16052073

Yaghoubi, M., Ayaseh, A., Alirezalu, K., Nemati, Z., Pateiro, M., & Lorenzo, J. M. (2021). Effect of chitosan coating incorporated with *Artemisia fragrans* essential oil on fresh chicken meat during refrigerated storage. *Polymers*, *13*(5), 716. https://doi.org/10.3390/polym13050716

Yan, D., Li, Y., Liu, Y., Li, N., Zhang, X., & Yan, C. (2021). Antimicrobial properties of chitosan and chitosan derivatives in the treatment of enteric infections. *Molecules (*Basel, Switzerland), *26*(23), 7136. https://doi.org/10.3390/molecules26237136

Zhang, J., Wei, Z., Lu, T., Qi, X., Xie, L., Vincenzetti, S., Polidori, P., Li, L., & Liu, G. (2023). The research field of meat preservation: A scientometric and visualization analysis based on the web of science. Foods (Basel, Switzerland), *12*(23), 4239. https://doi.org/10.3390/foods12234239

Zhang, Z., Xia, G., Yang, Q., Fan, X., & Lyu, S. (2019). Effects of chitosan-based coatings on storage quality of Chinese shrimp. *Food Science & Nutrition*, *7*(12), 4085–4094. https://doi.org/10.1002/fsn3.1275

9 Biodegradability Studies of Chitosan-Based Packaging

John Tsado Mathew[1], *Jonathan Hussaini*[1], *Abel Inobeme*[2], *Elijah Yanda Shaba*[3], *Charles Oluwaseun Adetunji*[4], *and Amos Mamman*[1]
[1]Department of Chemistry, Ibrahim Badamasi Babangida University Lapai, Niger State, Nigeria
[2]Department of Chemistry, Edo University Uzairue, Auchi, Edo State, Nigeria
[3]Department of Chemistry, Federal University of Technology, Minna, Niger State, Nigeria
[4]Department of Microbiology, Applied Microbiology, Biotechnology and Nanotechnology Laboratory, Edo University Uzairue, Auchi, Edo State, Nigeria

9.1 INTRODUCTION

The escalating concerns over environmental degradation and the accumulation of non-biodegradable waste have spurred a global quest for sustainable alternatives in various industries, particularly in packaging. Traditional packaging materials, primarily derived from fossil fuels, pose significant challenges due to their non-renewable nature and resistance to degradation. As a result, the exploration of eco-friendly alternatives has gained substantial traction in recent years, with biopolymers emerging as promising candidates for sustainable packaging solutions (Bhatia et al., 2023).

Among these biopolymers, chitosan, a derivative of chitin found abundantly in the exoskeletons of crustaceans, insects, and fungi, has garnered considerable attention. Chitosan exhibits excellent biocompatibility, biodegradability, and antimicrobial properties, making it an attractive material for various applications, including food packaging, biomedical devices, and wastewater treatment. Its abundance, low cost, and non-toxic nature further enhance its appeal as a sustainable alternative to conventional packaging materials (Rehman et al., 2023).

The utilization of chitosan-based packaging holds immense potential to mitigate the environmental impact associated with traditional packaging materials. By harnessing the inherent biodegradability of chitosan, researchers aim to develop packaging solutions that not only offer effective protection and preservation of products but also facilitate their eco-friendly disposal at the end of their lifecycle (Wrońska et al., 2023).

This introduction explores biodegradability studies conducted on chitosan-based packaging materials, highlighting their significance in advancing sustainable packaging practices and addressing environmental concerns. Through a comprehensive review of existing literature and research findings, this paper aims to elucidate the key factors influencing the biodegradation of chitosan-based packaging and assess its environmental implications (Bose et al., 2023; Inobeme et al., 2023a).

The biodegradability of chitosan-based packaging is influenced by various factors, including the molecular weight of chitosan, degree of deacetylation, processing techniques, and environmental conditions. Higher-molecular-weight chitosan polymers tend to exhibit slower degradation rates due to their reduced susceptibility to enzymatic hydrolysis. Conversely, chitosan with a higher degree of

DOI: 10.1201/9781003355458-9

deacetylation, resulting in increased amino group density, is more readily degraded by microorganisms (Azeh et al., 2023; Desai et al., 2023).

The processing techniques employed during the fabrication of chitosan-based packaging also play a crucial role in determining its biodegradability. Methods such as blending, casting, and extrusion can affect the structural integrity and porosity of the packaging material, influencing its susceptibility to microbial degradation. Additionally, the incorporation of additives, such as plasticizers and reinforcing agents, may alter the biodegradation kinetics of chitosan-based packaging (Souza et al., 2024).

Environmental factors, including temperature, humidity, pH, and the presence of microbial populations, significantly impact the biodegradation of chitosan-based packaging materials. Studies have shown that favorable conditions, such as higher temperatures and moisture levels, accelerate the degradation process by promoting microbial activity and enzymatic hydrolysis. However, the extent and rate of biodegradation may vary depending on the specific environmental conditions and microbial communities present in the disposal environment (Pires et al., 2022; Mathew et al., 2023a).

Biodegradation studies of chitosan-based packaging involve the assessment of various parameters, including weight loss, structural changes, microbial colonization, and degradation products. Techniques such as gravimetric analysis, scanning electron microscopy (SEM), Fourier-transform infrared spectroscopy (FTIR), and microbial culture assays are commonly employed to evaluate the biodegradation kinetics and mechanisms of chitosan-based packaging materials (Liu et al., 2024).

The environmental implications of chitosan-based packaging biodegradation extend beyond waste management to encompass broader sustainability considerations. By promoting the adoption of renewable and biodegradable materials, chitosan-based packaging contributes to reducing the reliance on finite fossil resources and mitigating carbon emissions associated with traditional packaging production and disposal processes. Moreover, the utilization of chitosan-based packaging in various applications can foster circular economy principles by facilitating the transition toward a more sustainable and resource-efficient model of production and consumption (Cristofoli et al., 2023).

Biodegradability studies of chitosan-based packaging represent a crucial area of research aimed at advancing sustainable packaging solutions and addressing environmental challenges. By elucidating the factors influencing the biodegradation of chitosan-based materials and assessing their environmental implications, researchers can contribute to the development of innovative packaging technologies that promote resource conservation, waste reduction, and environmental stewardship. Through interdisciplinary collaboration and continuous innovation, chitosan-based packaging holds the promise of ushering in a more sustainable future for the packaging industry and beyond (Teixeira-Costa and Andrade, 2023).

Moreover, the biodegradability studies of chitosan-based packaging also shed light on the importance of lifecycle assessments (LCAs) in evaluating the overall environmental performance of packaging materials. LCAs enable researchers and industry stakeholders to quantify the environmental impacts associated with the entire lifecycle of packaging, including raw material extraction, production, use, and end-of-life scenarios. By considering factors such as energy consumption, greenhouse gas emissions, and resource depletion, LCAs provide valuable insights into the environmental sustainability of different packaging options and inform decision-making processes aimed at minimizing environmental footprints (Asim et al., 2022).

Furthermore, the biodegradation of chitosan-based packaging presents opportunities for synergistic approaches that integrate waste management and resource recovery strategies. As chitosan-based materials degrade into non-toxic byproducts, they can potentially serve as substrates for biogas production through anaerobic digestion or as feedstocks for composting processes. By valorizing chitosan-based packaging waste, these integrated approaches contribute to closing the loop in material flows, promoting resource efficiency, and reducing the environmental burden associated with conventional waste disposal methods (Ali and Ahmed, 2021).

The development and commercialization of chitosan-based packaging also face several challenges and opportunities on the path toward widespread adoption. Challenges such as cost competitiveness, scalability of production, and regulatory compliance need to be addressed to facilitate the market penetration of chitosan-based packaging solutions. Additionally, efforts to enhance the mechanical properties, barrier properties, and shelf-life stability of chitosan-based materials through innovative formulations and processing techniques are essential to meet the performance requirements of diverse packaging applications (Alvarado et al., 2023).

9.2 MOTIVATION FOR BIODEGRADABLE PACKAGING

Biodegradable packaging is gaining momentum in the modern era due to increasing environmental concerns and the need for sustainable solutions in various industries. This packaging material holds promise in mitigating the adverse effects of traditional non-biodegradable packaging on ecosystems, human health, and overall sustainability. Here, we delve into motivations driving the adoption of biodegradable packaging (Inobeme et al., 2023b; Shaikh et al., 2024).

Environmental Impact: One of the primary motivations for biodegradable packaging is its reduced environmental impact compared to conventional plastics. Traditional plastics, derived from non-renewable resources like petroleum, pose significant challenges in terms of pollution, resource depletion, and waste management. Biodegradable packaging, on the other hand, offers a viable alternative by breaking down naturally into organic compounds through microbial action. This minimizes the accumulation of plastic waste in landfills, oceans, and other ecosystems, thereby reducing environmental pollution and preserving natural resources (Atiwesh et al., 2021; Mathew et al., 2023b).

Waste Management: Biodegradable packaging plays a crucial role in waste management by facilitating composting and organic recycling processes. Unlike traditional plastics that persist in the environment for hundreds of years, biodegradable materials decompose relatively quickly, thereby reducing the burden on waste management infrastructure. By diverting organic waste from landfills and incinerators, biodegradable packaging contributes to the creation of nutrient-rich compost, which can be used to enrich soil fertility and support agricultural sustainability (Ncube et al., 2021, Mathew et al., 2024).

Consumer Awareness: Growing consumer awareness and demand for eco-friendly products have fueled the adoption of biodegradable packaging by businesses across various sectors. As consumers become more environmentally conscious, they seek products packaged in materials that minimize environmental harm and promote sustainability. Biodegradable packaging aligns with these preferences, offering businesses a competitive edge and enhancing brand reputation by demonstrating a commitment to environmental stewardship. Consequently, companies are increasingly incorporating biodegradable packaging into their product offerings to meet consumer expectations and drive market growth (Bojanowska and Sulimierska, 2023).

Regulatory Pressures: Stringent regulations and policies aimed at reducing plastic pollution and promoting sustainable practices have accelerated the adoption of biodegradable packaging globally. Governments and regulatory bodies are implementing measures such as bans on single-use plastics, extended producer responsibility (EPR) schemes, and incentives for eco-friendly alternatives to address the environmental challenges posed by conventional packaging materials. Compliance with these regulations and adherence to sustainable packaging standards have become imperative for businesses seeking to operate in a socially responsible manner and avoid fines or penalties associated with non-compliance (Cruz et al., 2022; Muhammad et al., 2025).

Innovations in Material Science: Advancements in material science and technology have led to the development of innovative biodegradable materials with enhanced properties and functionalities. Researchers and industry experts are continually exploring novel biomaterials, bio-based polymers, and biocomposites derived from renewable sources such as plant fibers, starches, cellulose, and algae. These bio-based materials offer comparable performance to conventional plastics in terms of strength, durability, and barrier properties while being biodegradable and compostable.

Moreover, biodegradable packaging solutions are becoming more cost-effective and scalable as manufacturing processes mature and economies of scale are realized (Alaswad et al., 2022).

Climate Change Mitigation: Biodegradable packaging contributes to climate change mitigation efforts by reducing greenhouse gas emissions associated with the production, transportation, and disposal of traditional plastics. The production of biodegradable materials from renewable resources requires less energy and generates fewer carbon emissions compared to petroleum-based plastics. Additionally, the decomposition of biodegradable packaging in composting facilities or natural environments results in the release of carbon dioxide, water, and organic matter, which can be reabsorbed by plants and soil, thus closing the carbon cycle and mitigating the overall environmental impact (Yadav et al., 2024; Malik et al., 2025).

Circular Economy Principles: The adoption of biodegradable packaging aligns with the principles of a circular economy, wherein resources are utilized efficiently, waste is minimized, and products are designed for reuse, recycling, or regeneration. Biodegradable materials enable the transition toward a circular economy by providing a sustainable alternative to linear consumption patterns and disposable packaging practices. Through initiatives such as closed-loop recycling systems, bio-based packaging can be collected, processed, and reintegrated into the production cycle, reducing reliance on virgin resources and promoting a more sustainable and resilient economic model (Kassab et al., 2023).

In addition, the motivation for biodegradable packaging stems from its potential to address pressing environmental challenges, meet consumer preferences, comply with regulations, drive innovation, mitigate climate change, and promote the transition toward a circular economy. By embracing biodegradable packaging solutions, businesses can minimize their environmental footprint, enhance brand value, and contribute to the creation (Jain et al., 2023).

9.3 CHITOSAN AS A PROMISING BIOPOLYMER FOR PACKAGING

Chitosan, a biopolymer derived from chitin, is emerging as a promising alternative for packaging materials due to its unique properties and various advantages. From its abundance in nature to its biodegradability and antimicrobial properties, chitosan offers a sustainable solution to the challenges posed by conventional packaging materials. Here, we explore the potential of chitosan in packaging applications and the reasons behind its growing popularity (Perera et al., 2023).

Abundance and Renewability: Chitosan is derived from chitin, a polysaccharide found abundantly in the exoskeletons of crustaceans such as shrimp, crab, and lobster, as well as in the cell walls of fungi. This natural abundance makes chitosan a renewable resource, unlike petroleum-based plastics that rely on finite fossil fuel reserves. The extraction of chitosan from byproducts of the seafood industry or agricultural waste streams contributes to resource efficiency and waste reduction, aligning with sustainable practices and circular economy principles (Muthu et al., 2024).

Biodegradability and Environmental Compatibility: One of the most significant advantages of chitosan in packaging is its biodegradability. Chitosan is inherently biodegradable, meaning it can be broken down into harmless substances by microorganisms in the environment, such as bacteria and fungi. Unlike traditional plastics that persist in landfills and oceans for centuries, chitosan-based packaging decomposes relatively quickly, reducing the burden on waste management infrastructure and minimizing environmental pollution. This environmental compatibility makes chitosan an attractive option for eco-conscious consumers and businesses seeking to reduce their environmental footprint (Jiang et al., 2023a).

Antimicrobial Properties: Chitosan exhibits inherent antimicrobial properties, making it suitable for food packaging applications where microbial growth and spoilage are concerns. The antimicrobial activity of chitosan is attributed to its positively charged amino groups, which interact with negatively charged microbial cell membranes, disrupting their structure and inhibiting growth. By incorporating chitosan into packaging materials, food products can be preserved for longer periods, reducing food waste and extending shelf life. Additionally, chitosan-based films can help maintain food safety by inhibiting the growth of pathogenic bacteria and fungi, thus enhancing consumer confidence in the quality and safety of packaged products (Gomes et al., 2024; Adetunji et al., 2023a).

Barrier Properties: Chitosan possesses excellent barrier properties against gases such as oxygen and carbon dioxide, as well as moisture vapor transmission. This barrier functionality helps protect packaged products from external factors such as moisture, oxygen, and light, which can degrade product quality and shelf life. Chitosan-based films can be tailored to specific packaging requirements by adjusting factors such as polymer concentration, molecular weight, and processing techniques. As a result, chitosan offers versatility in packaging applications, ranging from fresh produce and meats to pharmaceuticals and medical devices (Petraru and Amariei, 2023).

Biocompatibility and Non-toxicity: Chitosan is biocompatible and non-toxic, making it suitable for direct contact with food and pharmaceutical products. Unlike some synthetic polymers that may leach harmful chemicals into packaged goods, chitosan is derived from natural sources and is generally recognized as safe (GRAS) by regulatory authorities such as the U.S. Food and Drug Administration (FDA). This biocompatibility makes chitosan an attractive option for sensitive applications where consumer safety is paramount, such as food packaging, medical devices, and drug delivery systems (Reddy et al., 2023).

Potential for Value-added Applications: Beyond its primary role as a packaging material, chitosan offers potential for value-added applications in various industries. For example, chitosan-based coatings can enhance the quality and shelf life of fruits and vegetables by reducing moisture loss and microbial growth. Chitosan nanoparticles have shown promise in drug delivery systems, where they can improve the solubility, stability, and bioavailability of pharmaceutical compounds. Additionally, chitosan-based membranes and scaffolds have applications in wound healing, tissue engineering, and environmental remediation, demonstrating the versatility and multifunctionality of this biopolymer (Pham et al., 2023).

Innovation and Research: Ongoing research and innovation in the field of chitosan-based materials are driving advancements in packaging technology and expanding the range of potential applications. Researchers are exploring novel formulations, processing techniques, and functional additives to enhance the performance and sustainability of chitosan-based packaging materials. Collaborations between academia, industry, and government agencies are facilitating the development of scalable production methods, quality standards, and regulatory frameworks to support the commercialization of chitosan-based products (Thambiliyagodage et al., 2023).

Furthermore, chitosan holds immense promise as a sustainable and versatile biopolymer for packaging applications. Its abundance in nature, biodegradability, antimicrobial properties, barrier functionality, biocompatibility, and potential for value-added applications make it an attractive alternative to conventional packaging materials. With continued research, innovation, and collaboration, chitosan-based packaging has the potential to drive positive environmental and economic impacts while meeting the growing demand for sustainable packaging solutions (Jiménez-Gómez and Cecilia, 2024).

9.4 OVERVIEW OF BIODEGRADABLE PACKAGING MATERIALS

Biodegradable packaging materials represent a crucial innovation in the realm of sustainable packaging, offering alternatives to traditional plastics that address environmental concerns while maintaining functionality and performance. These materials are designed to break down naturally into harmless compounds, reducing their environmental impact and supporting circular economy principles. In this overview, we'll explore various types of biodegradable packaging materials, including their characteristics, applications, and potential benefits (Shaikh et al., 2024; Ogundolie et al., 2023).

Polylactic Acid (PLA): Polylactic acid, or PLA, is a biodegradable polymer derived from renewable resources such as corn starch, sugarcane, or other plant-based sources. It offers properties similar to conventional plastics while being compostable and biodegradable under certain conditions, such as industrial composting facilities. PLA is commonly used in food packaging, disposable tableware, and compostable bags due to its transparency, stiffness, and heat resistance. However, its limited heat resistance and moisture barrier properties may restrict its suitability for certain applications (Rezvani Ghomi et al., 2023).

Polyhydroxyalkanoates (PHA): Polyhydroxyalkanoates, or PHA, are a family of biodegradable polymers produced by microorganisms through fermentation of renewable feedstocks such as plant sugars or agricultural waste. PHA polymers exhibit a wide range of properties, including flexibility, biodegradability, and thermal stability, making them suitable for various packaging applications. PHA-based materials can be tailored to specific requirements, offering opportunities for customization and optimization based on end-use applications (Vicente et al., 2023).

Starch-Based Polymers: Starch-based polymers are derived from renewable resources such as corn, potatoes, or tapioca and can be processed into biodegradable packaging materials. These materials typically consist of a blend of starch and biodegradable polymers such as PLA or polybutylene adipate terephthalate (PBAT). Starch-based packaging offers good barrier properties against moisture and oxygen, making it suitable for food packaging applications. Additionally, starch-based materials are compostable and biodegradable, contributing to waste reduction and environmental sustainability (Jayarathna et al., 2022; Mathew et al., 2025a).

Cellulose-Based Materials: Cellulose-based materials, derived from plant sources such as wood pulp, cotton, or hemp, represent another category of biodegradable packaging materials. These materials include cellulose films, fibers, and coatings that offer biodegradability, barrier properties, and strength. Cellulose-based packaging can be composted or recycled, offering a sustainable alternative to traditional plastics in various applications, including food packaging, paper products, and textiles (Xu et al., 2024).

Polyhydroxyurethanes (PHUs): Polyhydroxyurethanes (PHUs) are a class of biodegradable polymers synthesized from renewable resources such as vegetable oils, amino alcohols, or carbon dioxide. These polymers exhibit properties similar to conventional polyurethanes but are biodegradable and environmentally friendly. PHUs can be processed into films, foams, coatings, and adhesives for packaging applications, offering versatility and performance while reducing environmental impact (Vieira et al., 2024).

Aliphatic Polyesters: Aliphatic polyesters such as PBAT and polybutylene succinate (PBS) are biodegradable polymers derived from renewable resources. These polymers exhibit good mechanical properties, flexibility, and biodegradability, making them suitable for various packaging applications. PBAT and PBS are often used as additives or blends with other biodegradable polymers to enhance properties such as flexibility, toughness, and biodegradation rate (Aliotta et al., 2022).

Alginate: Alginate is a biopolymer derived from seaweed or algae and is used in packaging applications due to its biodegradability, water solubility, and film-forming properties. Alginate films can be used as coatings, wrappers, or encapsulation materials for food products, pharmaceuticals, and other applications. Alginate-based packaging offers biodegradability and environmental compatibility, making it a sustainable alternative to conventional plastics (Abourehab et al., 2023).

Chitosan: Chitosan is a biopolymer derived from chitin, a natural polymer found in the exoskeletons of crustaceans. Chitosan exhibits antimicrobial properties, biodegradability, and biocompatibility, making it suitable for packaging applications where food safety and environmental concerns are paramount. Chitosan-based packaging materials offer barrier properties, flexibility, and compostability, contributing to waste reduction and environmental sustainability (Kumar et al., 2023).

Furthermore, biodegradable packaging materials offer a sustainable alternative to traditional plastics, addressing environmental concerns while maintaining functionality and performance. These materials, including PLA, PHA, starch-based polymers, cellulose-based materials, PHUs, aliphatic polyesters, alginate, and chitosan, offer a diverse range of properties and applications, from food packaging to pharmaceuticals to consumer goods. With ongoing research, innovation, and collaboration, biodegradable packaging materials have the potential to drive positive environmental

TABLE 9.1
Properties and Applications of Chitosan-Based Packaging

Properties	Applications	References
Biodegradable	Food packaging	Panou and Karabagias (2023)
Antimicrobial	Pharmaceutical packaging	Kamarudin et al. (2024)
Biocompatible	Medical device packaging	Seok (2022)
Barrier properties	Cosmetics packaging	Marano et al. (2022)
Renewable	Agricultural product packaging	Hofmann et al. (2023)
Non-toxic	Biodegradable plastic alternatives	Malhi et al. (2021)
Compostable	Sustainable packaging solutions	Ibrahim et al. (2022)
Versatile	Packaging for sensitive products	Pascall et al. (2022)
Flexible	Single-use packaging	Piano et al. (2023)
Cost-effective	Eco-friendly packaging solutions	Boz et al. (2023)

and economic impacts while meeting the growing demand for sustainable packaging solutions (Dirpan et al., 2023).

9.5 PROPERTIES AND APPLICATIONS OF CHITOSAN-BASED PACKAGING

As shown in Table 9.1, Chitosan-based packaging materials possess a range of desirable properties, including biodegradability, antimicrobial activity, biocompatibility, barrier properties, renewability, non-toxicity, compostability, versatility, flexibility, and cost-effectiveness. These properties make chitosan-based packaging suitable for various applications across industries, including food packaging, pharmaceutical packaging, medical device packaging, cosmetics packaging, agricultural product packaging, and as alternatives to conventional plastics (Jiang et al., 2023).

Chitosan-based packaging offers a range of properties that make it ideal for diverse applications. Its biodegradability makes it environmentally friendly and suitable for single-use packaging and agricultural product packaging. Additionally, chitosan's antimicrobial properties ensure food safety, making it ideal for food packaging. Its biocompatibility and non-toxic nature allow for pharmaceutical and medical device packaging, while its barrier properties make it suitable for cosmetics packaging. Chitosan is renewable and compostable, contributing to sustainable packaging solutions. Its versatility enables packaging for sensitive products, and its flexibility allows for various packaging formats. Moreover, chitosan-based packaging is cost-effective, providing eco-friendly alternatives to traditional plastics. Overall, chitosan-based packaging aligns with the growing demand for sustainable packaging across industries, offering solutions that prioritize environmental responsibility without compromising functionality or performance (Babaei-Ghazvini et al., 2023).

9.6 PREVIOUS STUDIES ON BIODEGRADABILITY OF CHITOSAN-BASED PACKAGING

Several studies have investigated the biodegradability of chitosan-based packaging materials, assessing their environmental impact and potential for waste reduction. These studies have examined various aspects of chitosan-based packaging, including degradation rates, decomposition products, microbial activity, and compostability (Mann & Sooch, 2023).

Degradation Rates: Studies have evaluated the degradation rates of chitosan-based packaging under different environmental conditions, such as soil burial, composting, and marine environments. Research found that chitosan-based films degraded rapidly in soil, with significant weight loss observed within weeks, indicating their suitability for soil burial applications. Similarly, the degradation of chitosan-based films in composting conditions were investigated and complete degradation were reported within a few months, highlighting their compostability (Ncube et al., 2024).

Decomposition Products: Researchers have analyzed the decomposition products generated during the biodegradation of chitosan-based packaging materials. A study by Ferreira et al. (2026) examined the breakdown products of chitosan-based films during composting and identified carbon dioxide, water, and organic matter as the primary decomposition products, indicating complete mineralization. This study demonstrated that chitosan-based materials undergo microbial degradation, leading to the formation of harmless compounds (Versino et al., 2023).

Studies have examined the environmental impact of chitosan-based packaging compared to conventional plastics. A study by Versino et al. (2023) conducted an LCA of chitosan-based films and polyethylene films used for food packaging and found that chitosan-based films had lower environmental impacts in terms of energy consumption, greenhouse gas emissions, and resource depletion. The study suggested that chitosan-based packaging could contribute to reducing the environmental burden associated with packaging materials (Jain et al., 2023).

9.7 PREPARATION OF CHITOSAN-BASED PACKAGING MATERIALS

The preparation of chitosan-based packaging materials involves several steps, including raw material sourcing, extraction of chitosan, formulation of packaging materials, and processing techniques (Pellis et al., 2022; Chimbekujwo et al., 2025).

Raw Material Sourcing: Chitosan, the primary component of chitosan-based packaging materials, is derived from chitin, a natural polymer found in the exoskeletons of crustaceans such as shrimp, crab, and lobster, as well as in the cell walls of fungi. The raw materials for chitosan extraction typically include shrimp or crab shells, which are byproducts of the seafood industry, or fungal biomass from fermentation processes. These raw materials must undergo cleaning and deproteinization to remove impurities and proteins before chitosan extraction (Román-Doval et al., 2023).

Extraction of Chitosan: Chitosan extraction from chitin-containing raw materials involves a multi-step process to isolate pure chitosan. The first step, demineralization, uses acidic solutions such as hydrochloric acid (HCl) or acetic acid to remove minerals and inorganic impurities like calcium carbonate from the raw material, such as crustacean shells. This is followed by deproteinization, where the demineralized material is treated with alkaline solutions like sodium hydroxide (NaOH) to eliminate proteins and other organic compounds. The final step is deacetylation, which involves exposing the purified chitin to a concentrated alkaline solution, often at elevated temperatures, to remove acetyl groups. This conversion transforms chitin into chitosan, a biopolymer with improved solubility and bioactivity, making it suitable for various industrial and biomedical applications (Younes and Rinaudo, 2024).

Demineralization: The raw material, such as crustacean shells, undergoes treatment with acidic solutions like hydrochloric acid (HCl) or acetic acid (CH_3COOH) to remove minerals and inorganic impurities. This process, known as demineralization, dissolves calcium carbonate and other mineral components present in the shells. Typically, the shells are soaked in an acid solution, causing a reaction that releases carbon dioxide gas and forms soluble salts, which are then washed away. The concentration of the acid, temperature, and duration of treatment are carefully controlled to maximize the removal of inorganic matter while preserving the organic components, such as chitin. This demineralized material, free from most of its inorganic content, is then ready for subsequent processing to extract purified chitin (El-Araby et al., 2022).

Deproteinization: The demineralized material, obtained by removing minerals from sources like crustacean shells, undergoes further processing to isolate chitin. This involves treatment with alkaline solutions, commonly sodium hydroxide (NaOH), to eliminate proteins and other organic compounds that are tightly bound to the chitin structure. The deproteinization process typically involves soaking the material in a hot NaOH solution, which breaks down and solubilizes the proteins and other organics, leaving behind a more purified form of chitin. The temperature, concentration of NaOH, and duration of treatment are carefully controlled to ensure the effective removal of unwanted compounds while maintaining the integrity of the chitin. This purified chitin serves as a precursor for producing chitosan through subsequent deacetylation processes (Triunfo et al., 2022).

Deacetylation: Chitin, a natural biopolymer found in the exoskeletons of crustaceans and insects, undergoes alkaline hydrolysis to convert it into chitosan. This process involves treating chitin with a concentrated alkaline solution, such as sodium hydroxide, at elevated temperatures. The alkaline conditions break down the acetyl groups attached to the chitin chains, leading to deacetylation and resulting in the formation of chitosan, which is more soluble in acidic solutions. The degree of deacetylation can be controlled by adjusting the concentration of the alkali, temperature, and reaction time, directly influencing the properties of the resulting chitosan. Chitosan, with its increased solubility and bioactivity, finds applications in biomedicine, agriculture, water treatment, and packaging due to its biodegradability, biocompatibility, and antimicrobial properties (Mathew et al., 2023a).

Formulation of Packaging Materials: Once chitosan is extracted, it can be formulated into various types of packaging materials, including films, coatings, fibers, and composites. The formulation process involves blending chitosan with other ingredients such as plasticizers, cross-linking agents, antioxidants, antimicrobial agents, and reinforcing fillers to achieve desired properties such as flexibility, strength, barrier properties, and antimicrobial activity. The formulation may also involve adjusting the molecular weight, degree of deacetylation, and concentration of chitosan to optimize the performance of the packaging material for specific applications (Pal et al., 2021).

Processing Techniques: After formulation, chitosan-based packaging materials are subjected to various processing techniques to shape them into the desired forms. Common techniques include extrusion, where chitosan mixtures are heated and extruded through a die to create continuous films, sheets, or fibers. Another method is casting, which involves dissolving chitosan in a solvent, such as acetic acid, pouring the solution onto a flat surface, and drying it to form thin films. Compression molding is also widely used, where chitosan powders or pellets are heated and compressed in molds under high pressure to produce solid shapes like trays or containers. These processing techniques enhance the functionality of chitosan-based packaging, making them suitable for applications in food preservation and other sustainable packaging solutions (Mohammed et al., 2023).

Solution casting: In this technique, chitosan is first dissolved in a solvent like acetic acid or formic acid to create a homogeneous solution. This solution is then cast onto a flat surface, such as a glass plate, and allowed to dry, forming thin, transparent films. These chitosan-based films exhibit antimicrobial properties, making them suitable for food packaging applications. Alternatively, compression molding is used for creating solid shapes like trays or containers. In this process, chitosan-based powders or pellets are placed into molds and subjected to high heat and pressure. The heat softens the material, while the pressure forces it into the desired shape. Once cooled, the molded chitosan maintains its solid form, offering biodegradable and antimicrobial packaging solutions.

Extrusion: Chitosan-based formulations can be processed using extrusion, a technique where the material is heated and forced through a die to create continuous profiles such as films, sheets, or fibers. During extrusion, chitosan is typically blended with plasticizers and other additives to improve its processability and mechanical properties. The formulation is then heated to a specific temperature to achieve a molten state before being pushed through a die, which shapes the material into thin films, wide sheets, or fine fibers. Once extruded, the chitosan-based profiles are cooled and solidified, resulting in flexible, durable, and biodegradable materials. These chitosan-based films and fibers are used for various applications, including food packaging, biomedical devices, and environmentally friendly products, offering antimicrobial properties and sustainability (Mathew et al., 2025b).

Coating: Chitosan-based coatings are applied to substrate surfaces like paper, cardboard, or plastic films using various techniques, including spraying, dipping, or brushing. In the spraying method, a chitosan solution is evenly sprayed over the substrate to create a uniform coating layer. Dipping involves immersing the substrate into a chitosan solution, ensuring that the entire surface is coated before it is left to dry. Brushing, on the other hand, requires manually applying the chitosan solution onto the surface using a brush, offering greater control over the thickness and uniformity of the coating. These techniques enhance the antimicrobial and barrier properties of the substrate, making them more suitable for applications in food packaging and extending the shelf life of perishable products (Lieder et al., 2023).

Characterization and Testing: Chitosan-based packaging materials undergo characterization and testing to assess their physical, mechanical, thermal, barrier, and antimicrobial properties. Common tests include:

Mechanical properties: Tensile strength, elongation at break, and puncture resistance are measured to evaluate the mechanical performance of chitosan-based films and coatings.

Barrier properties: Oxygen permeability, moisture vapor transmission rate (MVTR), and water vapor permeability are measured to assess the barrier properties of chitosan-based packaging materials against gases and moisture.

Thermal properties: Differential scanning calorimetry (DSC) and thermogravimetric analysis (TGA) are used to evaluate the thermal stability and decomposition behavior of chitosan-based materials (Salavati et al., 2022).

Antimicrobial activity: Microbiological tests like the agar diffusion assay and direct contact tests are commonly used to evaluate the antimicrobial efficacy of chitosan-based packaging materials. In the agar diffusion assay, small disks or wells are filled with the chitosan-based material and placed on agar plates inoculated with pathogenic bacteria and fungi. The presence of inhibition zones around the material indicates antimicrobial activity. Direct contact tests involve placing the packaging material in direct contact with the microorganisms on a solid or liquid medium. The microbial load is then measured to assess the reduction in viability due to the antimicrobial properties of the chitosan. These tests provide insights into the potential use of chitosan-based packaging for food safety and preservation (Adetunji et al., 2023b).

Optimization and Scale-up: Once the properties of chitosan-based packaging materials are characterized and tested, optimization and scale-up processes may be employed to improve production efficiency, reduce costs, and ensure consistency in quality and performance. Optimization may involve fine-tuning formulation parameters, processing conditions, and quality control measures to meet industry standards and regulatory requirements (Suryanegara et al., 2021; Dawud et al., 2025).

In addition, the preparation of chitosan-based packaging materials involves sourcing raw materials, extracting chitosan, formulating packaging materials, processing techniques, characterization and testing, and optimization and scale-up processes. These steps are essential for producing chitosan-based packaging materials with desired properties and performance for various applications in industries such as food packaging, pharmaceuticals, cosmetics, and agriculture (Pellis et al., 2022).

9.8 FACTORS INFLUENCING BIODEGRADABILITY

Table 9.2 shows how these factors interact in complex ways to influence the biodegradability of materials. Understanding these factors is essential for designing and developing biodegradable materials with tailored properties and performance for specific applications and environmental conditions (Bher et al., 2022).

Biodegradability is influenced by several factors including material composition, molecular weight, degree of crystallinity, presence of additives, environmental conditions, microbial community, processing techniques, surface modification, and environmental exposure. Materials derived from renewable resources like starch or chitin are inherently more biodegradable than synthetic polymers. Lower molecular weight and amorphous structure enhance biodegradability, while additives and surface modifications can either facilitate or inhibit microbial degradation.

TABLE 9.2
Factors Influencing Biodegradability

Factors	Description	Examples	References
Material composition	The chemical composition and structure of the material significantly affect its biodegradability. Biopolymers derived from renewable resources such as starch, cellulose, chitin, or PHA are inherently more biodegradable than synthetic polymers derived from petroleum	Starch-based plastics, cellulose-based materials, chitosan-based packaging	Baranwal et al. (2022)
Molecular weight	The molecular weight of the polymer chains influences the rate of biodegradation. Generally, lower-molecular-weight polymers are more susceptible to microbial degradation due to increased surface area and accessibility to enzymes	Low-molecular-weight polyesters, oligomeric compounds	Mohanan et al. (2020) and Bodunrinde et al. (2025)
Degree of crystallinity	The degree of crystallinity affects the accessibility of polymer chains to microbial enzymes. Amorphous regions are more readily biodegradable than crystalline regions, which may hinder microbial degradation	Amorphous polymers, semi-crystalline polymers	Inobeme et al. (2023c) and Trofimchuk et al. (2023)
Presence of additives	Additives such as plasticizers, fillers, antioxidants, and antimicrobial agents may affect the biodegradability of the material. Some additives may enhance or inhibit microbial activity, depending on their chemical composition and interactions with microorganisms	PLA with plasticizers, chitosan films with antimicrobial agents	Hernández-García et al. (2021)
Environmental conditions	Environmental factors such as temperature, humidity, pH, oxygen availability, and microbial activity influence the rate and extent of biodegradation. Optimal conditions for microbial growth and activity accelerate biodegradation, while extreme conditions may inhibit degradation	Composting facilities, soil environments, and marine environments	Bala et al. (2022)
Microbial community	The composition and activity of microbial communities play a crucial role in biodegradation processes. Different microorganisms possess specific enzymes capable of degrading certain polymers, and their abundance and diversity vary depending on environmental conditions	Bacteria, fungi, algae, and archaea	Zhai et al. (2023)

(Continued)

TABLE 9.2 (*Continued*)
Factors Influencing Biodegradability

Factors	Description	Examples	References
Processing techniques	Processing methods used to manufacture the material may affect its biodegradability. Factors such as temperature, pressure, shear forces, and exposure to chemicals during processing can alter the structure and properties of the material, influencing its susceptibility to microbial degradation	Extrusion, injection molding, and compression molding	Zarski et al. (2023)
Surface modification	Surface treatments or modifications can affect the biodegradability of materials by altering surface properties such as hydrophobicity, roughness, and charge. Surface modifications may enhance or inhibit microbial adhesion and enzyme accessibility, thereby impacting biodegradation	Plasma treatment, chemical grafting, and surface coating	Hu et al. (2023)
Environmental exposure	The duration and intensity of exposure to environmental conditions affect the rate of biodegradation. Materials exposed to harsh or extreme environments may degrade more slowly than those exposed to optimal conditions for microbial activity	Long-term outdoor exposure, accelerated aging tests	Bher et al. (2022)

Environmental factors such as temperature, moisture, and microbial activity impact biodegradation rates. The composition and diversity of microbial communities determine the enzymatic degradation of materials. Processing techniques and environmental exposure also affect biodegradability. Understanding these factors is crucial for developing biodegradable materials with tailored properties for specific applications and environmental conditions, supporting sustainable waste management and environmental conservation efforts (Siracusa, 2023).

9.9 BIODEGRADABILITY IN AQUATIC ENVIRONMENTS

Biodegradability in aquatic environments is a critical aspect of environmental sustainability, as it influences the fate and impact of materials released into water bodies. Understanding how materials degrade in aquatic ecosystems is essential for assessing their environmental impact and developing strategies for waste management and pollution prevention (Bashir et al., 2023).

Factors Influencing Biodegradability in Aquatic Environments:

Material Composition: The chemical composition of materials significantly affects their biodegradability in aquatic environments. Biopolymers derived from natural sources such as starch, cellulose, chitin, or algae are more readily biodegradable than synthetic polymers derived from petroleum. Natural polymers contain chemical bonds that are more easily broken down by aquatic microorganisms, facilitating degradation. Environmental Conditions: Aquatic environments exhibit varying conditions such as temperature, pH, oxygen levels, and nutrient availability, which influence microbial activity and degradation rates. Warm temperatures and high levels of oxygen and nutrients generally support microbial growth and enhance biodegradation. However, extreme conditions such as low oxygen levels or acidic pH may inhibit degradation (Silva et al., 2023).

Microbial Community: Aquatic ecosystems are home to diverse microbial communities with the capability to degrade a wide range of organic compounds, including biodegradable polymers. Bacteria, fungi, algae, and protozoa are integral to the breakdown of organic matter, each contributing to different stages and types of decomposition. Bacteria are often involved in the initial stages of polymer degradation, while fungi can further break down complex organic structures. Algae and protozoa also play roles in nutrient cycling and organic matter processing. The composition and activity of these microbial communities are influenced by factors such as water quality, nutrient availability, and environmental disturbances. Variations in these factors can affect the efficiency of organic matter degradation and the overall health of aquatic ecosystems (Adetunji et al., 2023c).

Particle Size and Surface Area: The size and surface area of materials play a crucial role in their biodegradability by influencing microbial colonization and enzymatic degradation. Smaller particles and materials with higher surface area-to-volume ratios provide more accessible sites for microbial attachment and enzyme activity, thereby accelerating the biodegradation process. This increased surface area allows microbes to more effectively break down the material into simpler, more manageable components. In aquatic environments, filter feeders such as certain fish and bivalves may ingest microplastics. Within their digestive tracts, these microplastics can be exposed to microorganisms that facilitate further microbial degradation. Thus, the physical characteristics of materials, combined with biological interactions, significantly impact the rate and efficiency of their breakdown in aquatic systems (Yun et al., 2022).

Presence of Additives and Contaminants: Additives and contaminants in materials can significantly impact their biodegradability and toxicity in aquatic environments. Certain additives, such as plasticizers, stabilizers, or antimicrobial agents, may inhibit microbial activity or interfere with natural biodegradation processes, reducing the material's breakdown rate. These additives can create barriers that slow down microbial colonization and enzymatic degradation, leading to prolonged persistence in the environment. Additionally, contaminants such as heavy metals, including lead or cadmium, and persistent organic pollutants (POPs) can leach from materials into aquatic systems. These contaminants can accumulate in aquatic organisms through bioaccumulation and biomagnification, posing serious health risks to wildlife and potentially disrupting ecosystem functioning by affecting reproductive, neurological, and immune systems. Monitoring and regulating these additives and contaminants is essential to ensure environmental safety and sustainability (Olaniran et al., 2023).

Biodegradability of Common Materials in Aquatic Environments: Biodegradable polymers, including polylactic acid (PLA), polyhydroxyalkanoates (PHA), and chitosan, demonstrate varying degrees of biodegradability in aquatic environments. These biopolymers undergo microbial degradation, breaking down into benign end products such as carbon dioxide and water. PLA and PHA are commonly used due to their ability to degrade in industrial composting conditions, though their degradation rates can differ in natural aquatic environments. Chitosan, derived from chitin, also decomposes relatively quickly due to its natural origin and high susceptibility to microbial action. However, the rate and extent of degradation for these polymers depend on factors such as polymer composition, environmental conditions (temperature, pH), and microbial activity. Understanding these variables is crucial for assessing the environmental impact and effectiveness of biodegradable materials (Naser et al., 2021).

Conventional Plastics: Conventional plastics such as polyethylene (PE), polypropylene (PP), and polystyrene (PS) are typically non-biodegradable in aquatic environments due to their chemical stability and resistance to microbial degradation. Their inert nature prevents them from breaking down easily, leading to their persistence in water bodies for extended periods. This longevity poses significant risks to marine life, as these plastics can be ingested by aquatic organisms, causing physical harm or toxicological effects. Additionally, they can entangle marine animals and contribute to habitat degradation by disrupting ecosystems. The accumulation of these plastics in aquatic environments highlights the need for effective waste management strategies and the development of biodegradable alternatives to mitigate their environmental impact (Yang et al., 2023).

Natural Fibers: Natural fibers like cotton, jute, and hemp are biodegradable in aquatic environments due to their composition of cellulose and other organic compounds, which are amenable to microbial breakdown. Microorganisms in aquatic systems can effectively degrade these fibers, leading to their eventual conversion into harmless substances like carbon dioxide and water. However, the biodegradability of natural fibers can be compromised by synthetic dyes or chemical treatments applied during processing. These additives may hinder microbial degradation and introduce pollutants into aquatic ecosystems. For instance, synthetic dyes can release harmful chemicals into the water, potentially affecting water quality and aquatic life. Therefore, ensuring that natural fibers are free from harmful treatments is essential for maintaining their environmental benefits (Brunšek et al., 2023).

9.10 POTENTIAL APPLICATIONS OF CHITOSAN-BASED PACKAGING

Table 9.3 shows chitosan-based packaging materials that offer a wide range of applications across industries, from food and pharmaceuticals to cosmetics and agriculture. Their unique combination

TABLE 9.3
Potential Applications of Chitosan-Based Packaging

Application	Description	Examples	References
Food packaging	Chitosan-based packaging materials offer excellent barrier properties against moisture, oxygen, and microbial contamination, making them ideal for extending the shelf life of food products	Fresh produce packaging, meat and poultry packaging, and cheese and dairy packaging	Abimbola et al. (2023) and Duda-Chodak et al. (2023)
Pharmaceutical packaging	Chitosan's biocompatibility, antimicrobial properties, and ability to form films and coatings make it suitable for pharmaceutical packaging applications. Chitosan-based packaging can protect sensitive drugs from moisture, oxygen, and light degradation while ensuring product safety and stability	Capsules and tablets blister packs, drug delivery systems, and medical device packaging	Mathew et al. (2023a) and Baharlouei and Rahman (2022)
Cosmetic packaging	Chitosan-based packaging materials provide a sustainable and biodegradable alternative to conventional plastics in the cosmetics industry. Chitosan's antimicrobial properties help preserve the quality and efficacy of cosmetic products, while its versatility allows for customized packaging solutions	Cream jars and bottles, shampoo and conditioner bottles, and soap packaging	Kulka and Sionkowska (2023)

(Continued)

TABLE 9.3 (*Continued*)
Potential Applications of Chitosan-Based Packaging

Application	Description	Examples	References
Agricultural product packaging	Chitosan-based packaging materials can help reduce postharvest losses and extend the shelf life of agricultural products such as fruits, vegetables, and grains. Chitosan's antimicrobial properties inhibit mold growth and decay, while its biodegradability ensures minimal environmental impact	Fruit and vegetable trays, grain bags, and seed coatings	Riseh et al. (2023)
Biomedical packaging	Chitosan's biocompatibility, non-toxicity, and ability to form films and scaffolds make it suitable for biomedical packaging applications. Chitosan-based materials can be used to package and protect medical devices, implants, and tissue-engineered constructs, ensuring sterility and biocompatibility	Surgical implants packaging, tissue scaffolds packaging, and wound dressings packaging	Thambiliyagodage et al. (2023)
Single-use packaging	Chitosan-based materials offer a sustainable and compostable alternative to single-use plastics in various packaging applications. Chitosan's biodegradability ensures minimal environmental impact and supports circular economy principles by promoting waste reduction and resource efficiency	Takeaway food containers, disposable cutlery, and coffee cups	Wrońska et al. (2021)
Eco-friendly packaging	Chitosan-based packaging materials align with the growing demand for eco-friendly and sustainable packaging solutions across industries. Chitosan's renewable nature, biodegradability, and low environmental impact make it a preferred choice for environmentally conscious consumers and businesses seeking to reduce their carbon footprint	Eco-friendly pouches, compostable bags, and biodegradable films	Ali and Ahmed (2021)

of properties, including biodegradability, barrier properties, antimicrobial activity, and biocompatibility, makes them versatile and environmentally friendly packaging solutions (Thambiliyagodage et al., 2023).

Chitosan-based packaging materials hold promise for various applications due to their unique properties. In food packaging, they offer excellent barrier properties against moisture and oxygen, extending shelf life and preserving freshness. Additionally, their antimicrobial properties help inhibit microbial growth, enhancing food safety. Pharmaceutical packaging benefits from chitosan's biocompatibility and ability to form films, protecting sensitive drugs and medical devices while ensuring stability. Cosmetic packaging utilizes chitosan's sustainability and antimicrobial benefits to maintain product quality and efficacy. In agriculture, chitosan-based packaging reduces postharvest losses by inhibiting mold growth and extending the shelf life of fruits, vegetables, and grains. Biomedical packaging relies on chitosan's biodegradability and non-toxicity to safeguard surgical implants, tissue scaffolds, and wound dressings. Moreover, chitosan-based materials offer eco-friendly solutions for single-use packaging, supporting sustainability initiatives and reducing environmental impact. Overall, chitosan-based packaging demonstrates versatility and effectiveness across various industries, addressing the demand for sustainable and functional packaging solutions (Jiang et al., 2023a).

9.11 CONCLUSION

Chitosan-based packaging represents a promising solution to address the growing demand for sustainable and environmentally friendly packaging materials. With its unique combination of properties, including biodegradability, barrier functionality, antimicrobial activity, and biocompatibility, chitosan offers versatile applications across various industries. The versatility of chitosan-based packaging is evident in its potential applications. In the food industry, chitosan-based materials offer superior barrier properties against moisture and oxygen, extending the shelf life of perishable products and reducing food waste. Its antimicrobial properties further enhance food safety by inhibiting microbial growth, ensuring product quality and consumer satisfaction. In pharmaceutical packaging, chitosan's biocompatibility and ability to form films provide an ideal solution for protecting sensitive drugs and medical devices. Its biodegradability ensures minimal environmental impact, aligning with regulatory requirements and sustainability goals in the healthcare sector.

Chitosan-based packaging also finds applications in cosmetics, where its sustainability and antimicrobial benefits contribute to preserving product quality and efficacy. Additionally, chitosan's renewable nature and low environmental footprint make it an attractive choice for eco-conscious consumers and businesses seeking sustainable packaging alternatives. Agricultural product packaging benefits from chitosan's ability to inhibit mold growth and extend the shelf life of fruits, vegetables, and grains, reducing postharvest losses and supporting food security initiatives. Its biodegradability ensures minimal environmental impact, promoting circular economy principles and reducing plastic pollution in the environment.

Furthermore, chitosan-based materials offer eco-friendly solutions for single-use packaging, supporting sustainability initiatives and reducing environmental impact. By replacing conventional plastics with biodegradable alternatives, chitosan-based packaging contributes to waste reduction and resource conservation, addressing the global challenge of plastic pollution. Chitosan-based packaging materials demonstrate versatility, effectiveness, and environmental responsibility across various industries. With ongoing research, innovation, and collaboration, chitosan-based packaging has the potential to drive positive environmental and economic impacts while meeting the growing demand for sustainable packaging solutions. As businesses and consumers increasingly prioritize sustainability, chitosan-based packaging stands poised to play a significant role in shaping the future of packaging materials and contributing to a more sustainable world.

9.12 FUTURE DIRECTIONS

Looking ahead, several future directions hold promise for advancing the field of chitosan-based packaging and further enhancing its sustainability, functionality, and widespread adoption.

Innovation in Material Development: Future research efforts will likely focus on developing novel chitosan-based materials with enhanced properties, such as improved barrier functionality, mechanical strength, and flexibility. Advances in material science and nanotechnology may lead to the development of chitosan nanocomposites and bio-based additives, further enhancing the performance and versatility of chitosan-based packaging materials (Souza et al., 2020).

Sustainable Production Methods: Efforts to develop more sustainable and resource-efficient production methods for chitosan extraction and processing are underway. Utilizing alternative sources of chitosan, such as fungal biomass or waste streams from aquaculture and food processing industries, can reduce reliance on crustacean shells and minimize environmental impact. Additionally, exploring green chemistry approaches and renewable energy sources in chitosan production can further enhance sustainability (Pellis et al., 2022).

Functional Additives and Coatings: Research into incorporating functional additives and coatings into chitosan-based packaging materials can expand their applications and performance. Antimicrobial agents, antioxidants, and barrier enhancers can be incorporated to improve shelf life, safety, and quality of packaged products. Moreover, developing smart packaging technologies, such as indicators for freshness and spoilage, can enhance consumer confidence and satisfaction (Dawud et al., 2025).

Bioprocessing and Biotechnological Advances: Advancements in bioprocessing and biotechnological techniques hold promise for optimizing chitosan production and processing. Biorefinery approaches, enzyme engineering, and fermentation technologies can improve yield, purity, and cost-effectiveness of chitosan production. Additionally, genetic engineering of microorganisms for tailored chitosan production and modification may lead to the development of designer chitosans with specific properties for targeted applications (Aguilar et al., 2019).

Circular Economy and Waste Management: Promoting circular economy principles and implementing effective waste management strategies are crucial for maximizing the sustainability of chitosan-based packaging. Developing closed-loop systems for chitosan recovery, recycling, and reuse can minimize waste generation and resource depletion. Furthermore, exploring innovative end-of-life options such as composting, anaerobic digestion, and biodegradation can facilitate the complete biodegradability and environmental compatibility of chitosan-based materials (Bodunrinde et al., 2025).

Regulatory Framework and Market Adoption: Establishing clear regulatory frameworks and standards for chitosan-based packaging materials is essential to ensure safety, quality, and market acceptance. Collaboration between industry stakeholders, policymakers, and regulatory agencies can streamline approval processes and promote market adoption of chitosan-based packaging. Moreover, raising awareness among consumers and businesses about the environmental benefits and applications of chitosan-based packaging can drive demand and market growth.

REFERENCES

Abimbola, O. F., Adetunji, C. O., Dare, B. J., Mathew, J. T., Inobeme, A., Ghazanfar, S., Olaniyan, O. T., Ajiboye, M. D., & Dauda, W. (2023). Packaging of Animal-Based Food Products: The Need for Intelligent Freshness Sensors. In: *Sensing and Artificial Intelligence Solutions for Food Manufacturing*. CRC Press, 89–102. https://doi.org/10.1201/9781003207955-6

Abourehab, M. A. S., Rajendran, R. R., Singh, A., Pramanik, S., Shrivastav, P., Ansari, M. J., Manne, R., Amaral, L. S., & Deepak, A. (2023). Alginate as a Promising Biopolymer in Drug Delivery and Wound Healing: A Review of the State-of-the-Art. *International Journal of Molecular Sciences*, *23*(16), 9035. https://doi.org/10.3390/ijms23169035

Adetunji, C. O., Mathew, J. T., Inobeme, A., Taiwo, O., Olaniyan, O., Shakira, G., & Wadazani, D. (2023a). Application of Biosensors for Detection and Monitoring of Water Quality. In: *Organic and Inorganic Materials Based Sensors*. John Wiley & Sons, Inc., 1065–1077. https://doi.org/10.1002/9783527834266.ch46

Adetunji, C. O., Olaniyan, O. T., Shariati, M. A., Bodunrinde, R. E., Ajayi, O. O., Adetunji, J. B., Hefft, D. I., Olugbenga, S. M., Rebezov, M., Khayrullin, M., Adeniyi, M. J., Inobeme, A., & Mathew, J. T. (2023c). Pharmacology, Phytochemistry, and Medicinal Usefulness of *Basella alba* L. In: T. Pullaiah (Ed.), *Phytochemical Composition and Pharmacy of Medicinal Plants* (Vol. 2, pp. 499–516). Apple Academic Press, Inc. Co-published with CRC Press (Taylor & Francis), Palm Bay, FL.

Adetunji, C. O., Olaniyan, O. T., Shariati, M. A., Olugbenga, S. M., Osemwegie, O. O., Rebezov, M., Ajayi, O. O., Makhmudov, F., Bodunrinde, R. E., Adetunji, J. B., Adeniyi, M. J., Hefft, D. I., Inobeme, A., & Mathew, J. T. (2023b). Medical Attributes of *Breynia disticha* J.R. Forst. & G. Forst. In: T. Pullaiah (Ed.), *Phytochemical Composition and Pharmacy of Medicinal Plants* (Vol. 2, pp. 531–538). Apple Academic Press, Inc. Co-published with CRC Press (Taylor & Francis), Palm Bay, FL.

Aguilar, C. N., Ruiz, H. A., Rubio Rios, A., Chávez-González, M., Sepúlveda, L., Rodríguez-Jasso, R. M., Loredo-Treviño, A., Flores-Gallegos, A. C., Govea-Salas, M., & Ascacio-Valdes, J. A. (2019). Emerging Strategies for the Development of Food iindustries. *Bioengineered, 10*(1), 522–537. https://doi.org/10.1080/21655979.2019.1682109

Alaswad, S. O., Mahmoud, A. S., & Arunachalam, P. (2022). Recent Advances in Biodegradable Polymers and Their Biological Applications: A Brief Review. *Polymers, 14*(22), 4924. https://doi.org/10.3390/polym14224924

Ali, A. A., & Ahmed, S. (2021). Eco-friendly natural extract loaded antioxidative chitosan/polyvinyl alcohol based active films for food packaging. *Heliyon, 7*(3), e06550. https://doi.org/10.1016/j.heliyon.2021.e06550

Ali, S., Isha, & Chang, Y.-C. (2024). Ecotoxicological Impact of Bioplastics Biodegradation: A Comprehensive Review. *Processes, 11*, 3445. https://doi.org/10.3390/pr11123445

Aliotta, L., Seggiani, M., Lazzeri, A., Gigante, V., & Cinelli, P. (2022). A Brief Review of Poly (Butylene Succinate) (PBS) and Its Main Copolymers: Synthesis, Blends, Composites, Biodegradability, and Applications. *Polymers, 14*(4), 844. https://doi.org/10.3390/polym14040844

Alvarado, N., Abarca, R. L., & Linares-Flores, C. (2023). Use of Chitosan-Based Polyelectrolyte Complexes for Its Potential Application in Active Food Packaging: A Review of Recent Literature. *International Journal of Molecular Sciences, 24*, 11535. https://doi.org/10.3390/ijms241411535

Asim, Z., Shamsi, I. R. A., Wahaj, M., Raza, A., Abul Hasan, S., Siddiqui, S. A., Aladresi, A., Sorooshian, S., & Seng Teck, T. (2022). Significance of Sustainable Packaging: A Case-Study from a Supply Chain Perspective. *Applied System Innovation, 5*, 117. https://doi.org/10.3390/asi5060117

Atiwesh, G., Mikhael, A., Parrish, C. C., Banoub, J., & Le, T. T. (2021). Environmental Impact of Bioplastic Use: A Review. *Heliyon, 7*(9), e07918. https://doi.org/10.1016/j.heliyon.2021.e07918

Azeh, Y., Musah, M., Mathew, J. T., Badeggi, U. M., Estuyankpa, M. B., Muhammad, Z. D., Oluwaseyi, D. S., Umar, M. T., Abdulkarim, M. A., Muhammad, A. I., Ibrahim, A. A., Ismail, H., Muhammad, K. T., Umar, G., & Ahmed, T. (2023). In Situ Synthesis of Polyaniline Nanohybrid and Formulation of Polyaniline/Carboxymethyl Cellulose/Ethylene Glycol Nanocomposite: Study of Its Conducting and Antibacterial Properties. *Journal of Materials and Polymer Science, 3*(4), 1–9. https://doi.org/10.47485/2832-9384.1041

Babaei-Ghazvini, A., Acharya, B., & Korber, D. R. (2023). Antimicrobial Biodegradable Food Packaging Based on Chitosan and Metal/Metal-Oxide Bio-Nanocomposites: A Review. *Polymers, 13*(16), 2790. https://doi.org/10.3390/polym13162790

Baharlouei, P., & Rahman, A. (2022). Chitin and Chitosan: Prospective Biomedical Applications in Drug Delivery, Cancer Treatment, and Wound Healing. *Marine Drugs, 20*(7), 460. https://doi.org/10.3390/md20070460

Bala, S., Garg, D., Thirumalesh, B. V., Sharma, M., Sridhar, K., Inbaraj, B. S., & Tripathi, M. (2022). Recent Strategies for Bioremediation of Emerging Pollutants: A Review for a Green and Sustainable Environment. *Toxics*, 10(8), 484. https://doi.org/10.3390/toxics10080484

Baranwal, J., Barse, B., Fais, A., Delogu, G. L., & Kumar, A. (2022). Biopolymer: A Sustainable Material for Food and Medical Applications. *Polymers*, 14, 983. https://doi.org/10.3390/polym14050983

Bashir, I., Lone, F. A., Bhat, R. A., Mir, S. A., Dar, Z. A., & Dar, S. A. (2023). Concerns and Threats of Contamination on Aquatic Ecosystems. In: *Bioremediation and Biotechnology: Sustainable Approaches to Pollution Degradation*, 1–26. https://doi.org/10.1007/978-3-030-35691-0_1

Bhatia, L., Jha, H., Sarkar, T., & Sarangi, P.K. (2023). Food Waste Utilization for Reducing Carbon Footprints towards Sustainable and Cleaner Environment: A Review. *International Journal of Environmental Research and Public Health, 20*, 2318. https://doi.org/10.3390/ijerph20032318

Bher, A., Mayekar, P. C., Auras, R. A., & Schvezov, C. E. (2022). Biodegradation of Biodegradable Polymers in Mesophilic Aerobic Environments. *International Journal of Molecular Sciences, 23*(20), 12165. https://doi.org/10.3390/ijms232012165

Bodunrinde, R. E., Chimbekujwo, K. I., Oyewole, O. A., Adetunji, C. O., Inobeme, A., Eniola, K. I. T., Yerima, M. B., & Mathew, J. T. (2025). The Use of Nanochitosan for Enhancement in the Quality and Yield of Fruit Crops. In: *Nanochitosan Applications for Enhanced Crop Production and Food Security*. John Wiley & Sons, 111–122. https://doi.org/10.1002/9781394214945.ch5

Bojanowska, A., & Sulimierska, A. (2023). Consumer Awareness of Biodegradability of Food Products Packaging. *Sustainability*, *15*, 13980. https://doi.org/10.3390/su151813980

Bose, I., Roy, S., Pandey, V. K., & Singh, R. (2023). A Comprehensive Review on Significance and Advancements of Antimicrobial Agents in Biodegradable Food Packaging. *Antibiotics (*Basel, Switzerland*)*, *12*(6), 968. https://doi.org/10.3390/antibiotics12060968

Boz, Z., Korhonen, V., & Koelsch Sand, C. (2023). Consumer Considerations for the Implementation of Sustainable Packaging: A Review. *Sustainability*, *12*, 2192. https://doi.org/10.3390/su12062192

Brunšek, R., Kopitar, D., Schwarz, I., & Marasović, P. (2023). Biodegradation Properties of Cellulose Fibers and PLA Biopolymer. *Polymers*, *15*(17), 3532. https://doi.org/10.3390/polym15173532

Chimbekujwo, K. I., Yakubu, J. G., Omoregie, I. P., Oyewole, O. A., Muhammad, W. M., Adetunji, C. O., Eniola, K. I. T., Yerima, M. B., & Mathew, J. T. (2025). Synthesis of Nano-Chitosan Using Agricultural Waste. In: *Nanochitosan Applications for Enhanced Crop Production and Food Security*. John Wiley & Sons, Inc., 39–68. https://doi.org/10.1002/9781394214945.ch2

Cristofoli, N. L., Lima, A. R., Tchonkouang, R. D. N., Quintino, A. C., & Vieira, M. C. (2023). Advances in the Food Packaging Production from Agri-Food Waste and By-Products: Market Trends for a Sustainable Development. *Sustainability*, *15*, 6153. https://doi.org/10.3390/su15076153

Cruz, R. M. S., Krauter, V., Krauter, S., Agriopoulou, S., Weinrich, R., Herbes, C., Scholten, P. B. V., Uysal-Unalan, I., Sogut, E., Kopacic, S., Lahti, J., Rutkaite, R., & Varzakas, T. (2022). Bioplastics for Food Packaging: Environmental Impact, Trends and Regulatory Aspects. Foods (Basel, Switzerland*)*, *11*(19), 3087. https://doi.org/10.3390/foods11193087

Dawud, I., Bodunrinde, R. E., Oyewole, O. A., Adetunji, C. O., Inobeme, A., Eniola, K. I. T., Yerima, M. B., & Mathew, J. T. (2025). Application of Nanochitosan for Effective Fruit Production. In: *Nanochitosan Applications for Enhanced Crop Production and Food Security*. John Wiley & Sons, , 123–135. https://doi.org/10.1002/9781394214945.ch6

Desai, N., Rana, D., Salave, S., Gupta, R., Patel, P., Karunakaran, B., Sharma, A., Giri, J., Benival, D., & Kommineni, N. (2023). Chitosan: A Potential Biopolymer in Drug Delivery and Biomedical Applications. *Pharmaceutics*, *15*, 1313. https://doi.org/10.3390/pharmaceutics15041313

Dirpan, A., Ainani, A. F., & Djalal, M. A. (2023). Review on Biopolymer-Based Biodegradable Film for Food Packaging: Trends over the Last Decade and Future Research. *Polymers*, *15*, 2781. https://doi.org/10.3390/polym15132781

Duda-Chodak, A., Tarko, T., & Petka-Poniatowska, K. (2023). Antimicrobial Compounds in Food Packaging. *International Journal of Molecular Sciences*, *24*(3), 2457. https://doi.org/10.3390/ijms24032457

El-Araby, A., El Ghadraoui, L., & Errachidi, F. (2022). Physicochemical Properties and Functional Characteristics of Ecologically Extracted Shrimp Chitosans with Different Organic Acids during Demineralization Step. *Molecules*, *27*, 8285. https://doi.org/10.3390/molecules27238285

Ferreira, D. C. M., Carneiro, C. R., Miranda Júnior, J. R., Coimbra, J. S. D. R., Pelissari, F. M., & de Oliveira, E. B. (2026). Environmental Fate, Biodegradation Pathways, and Phytotoxicological Risks of Biopolymer-Based Packaging: A Critical Review. *Environmental Science and Pollution Research International*, *33*(11), 4657–4676. https://doi.org/10.1007/s11356-026-37565-7

Gomes, L. C., Faria, S. I., Valcarcel, J., Vázquez, J. A., Cerqueira, M. A., Pastrana, L., Bourbon, A. I., & Mergulhão, F. J. (2024). The Effect of Molecular Weight on the Antimicrobial Activity of Chitosan from Loligo opalescens for Food Packaging Applications. *Marine Drugs*, *19*, 384. https://doi.org/10.3390/md19070384

Hernández-García, E., Vargas, M., González-Martínez, C., & Chiralt, A. (2021). Biodegradable Antimicrobial Films for Food Packaging: Effect of Antimicrobials on Degradation. *Foods (*Basel, Switzerland), 10(6), 1256. https://doi.org/10.3390/foods10061256

Hofmann, T., Ghoshal, S., Tufenkji, N., Adamowskim, J. F., Bayen, S., Chen, Q., Demokritou, P., Flury, M., Hüffer, T., Ivleva, N. P., Ji, R., Leask, R. L., Maric, M., Mitrano, D. M., Sander, M., Pahl, S., Rillig, M. C., Walker, T. R., White, J. C., & Wilkinson, K. J. (2023). Plastics Can Be Used More Sustainably in Agriculture. *Communications Earth & Environment*, *4*, 332. https://doi.org/10.1038/s43247-023-00982-4

Hu, X., Wang, T., Li, F., & Mao, X. (2023). Surface Modifications of Biomaterials in Different Applied Fields. *RSC Advances*, *13*(30), 20495–20511. https://doi.org/10.1039/d3ra02248j

Ibrahim, I. D., Hamam, Y., Sadiku, E. R., Ndambuki, J. M., Kupolati, W. K., Jamiru, T., Eze, A. A., & Snyman, J. (2022). Need for Sustainable Packaging: An Overview. *Polymers*, *14*(20), 4430. https://doi.org/10.3390/polym14204430

Inobeme, A., Adetunji, C. O., Ajai, A. I., Inobeme, J., Mathew, J. T., Obar, A., Maliki, M., Nwakife, N., & Eziukwu, C. (2023a). Chemical Nanosensors for Monitoring Environmental Pollution. In: *Advanced Application of Nanotechnology to Industrial Wastewater*. Springer Nature Singapore, 93–103. https://doi.org/10.1007/978-981-99-3292-4_6

Inobeme, A., Mathew, J. T., Adetunji, C. O., Ajai, A. I., Inobeme, J., Maliki, M., Okonkwo, S., Adekoya, M. A., Bamigboye, M. O., Jacob, J. O., & Eziukwu, C. A. (2023b). Recent Advances in Nanotechnology for Remediation of Heavy Metals. *Environmental Monitoring Assessment*, *195*(111), 1–24. https://doi.org/10.1007/s10661-022-10614-7

Inobeme, A., Mathew, J. T., Jatto, E., Inobeme, J., Adetunji, C. O., Muniratu, M., Onyeachu, B. I., Adekoya, M. A., Ajai, A. I., Mann, A., Olori, E., Akhor, S. O., Eziukwu, C. A., Kelani, T., & Omali, P. I. (2023c). Recent Advances in Instrumental Techniques for Heavy Metal Quantification. *Environmental Monitoring and Assessment*, *195*(4), 452. https://doi.org/10.1007/s10661-023-11058-3

Jain, P., Chou, M. C., Fan, F., & Santoso, M. P. (2023). Embedding Sustainability in the Consumer Goods Innovation Cycle and Enabling Tools to Measure Progress and Capabilities. *Sustainability*, *13*, 6662. https://doi.org/10.3390/su13126662

Jayarathna, S., Andersson, M., & Andersson, R. (2022). Recent Advances in Starch-Based Blends and Composites for Bioplastics Applications. *Polymers*, *14*, 4557. https://doi.org/10.3390/polym14214557

Jiang, A., Patel, R., Padhan, B., Palimkar, S., Galgali, P., Adhikari, A., Varga, I., & Patel, M. (2023a). Chitosan Based Biodegradable Composite for Antibacterial Food Packaging Application. *Polymers*, *15*(10), 2235. https://doi.org/10.3390/polym15102235

Jiang, Y., Zhang, Y., & Deng, Y. (2023b). Latest Advances in Active Materials for Food Packaging and Their Application. *Foods*, *12*, 4055. https://doi.org/10.3390/foods12224055

Jiménez-Gómez, C. P., & Cecilia, J. A. (2024). Chitosan: A Natural Biopolymer with a Wide and Varied Range of Applications. *Molecules (*Basel, Switzerland*)*, 25(17), 3981. https://doi.org/10.3390/molecules25173981

Kamarudin, S. H., Rayung, M., Abu, F., Ahmad, S., Fadil, F., Karim, A. A., Norizan, M. N., Sarifuddin, N., Mat Desa, M. S. Z., Mohd Basri, M. S., Samsudin, H., & Abdullah, L. C. (2024). A Review on Antimicrobial Packaging from Biodegradable Polymer Composites. *Polymers*, *14*(1), 174. https://doi.org/10.3390/polym14010174

Kassab, A., Al Nabhani, D., Mohanty, P., Pannier, C., & Ayoub, G.Y. (2023). Advancing Plastic Recycling: Challenges and Opportunities in the Integration of 3D Printing and Distributed Recycling for a Circular Economy. *Polymers*, 15, 3881. https://doi.org/10.3390/polym15193881

Kulka, K., & Sionkowska, A. (2023). Chitosan Based Materials in Cosmetic Applications: A Review. *Molecules*, *28*, 1817. https://doi.org/10.3390/molecules28041817

Kumar, A., Yadav, S., Pramanik, J., Sivamaruthi, B. S., Jayeoye, T. J., Prajapati, B. G., & Chaiyasut, C. (2023). Chitosan-Based Composites: Development and Perspective in Food Preservation and Biomedical Applications. *Polymers*, *15*, 3150. https://doi.org/10.3390/polym15153150

Lieder, R., Darai, M., Orlygsson, G., & Sigurjonsson, O. E. (2023). Solution Casting of Chitosan Membranes for In Vitro Evaluation of Bioactivity. *Biological Procedures Online*, *15*(1), 11. https://doi.org/10.1186/1480-9222-15-11

Liu, T., Li, J., Tang, Q., Qiu, P., Gou, D., & Zhao, J. (2024). Chitosan-Based Materials: An Overview of Potential Applications in Food Packaging. *Foods (*Basel, Switzerland*)*, 11(10), 1490. https://doi.org/10.3390/foods11101490

Malik, S. I., Chimbekujwo, K. I., Maku, A. M., Oyewole, Oluwafemi, A., Omoregie, I. P., Adetunji, C. O., Eniola, K. I. T., Yerima, M. B., & Mathew, J. T. (2025). The Application of Nanochitosan Biopesticides as a Replacement to Synthetic Pesticides. In: *Nanochitosan Applications for Enhanced Crop Production and Food Security*. John Wiley & Sons, 89–110. https://doi.org/10.1002/9781394214945.ch4

Mann, M. K., & Sooch, B. S. (2023). Biodegradable Nano-reinforced Packaging with Improved Functionality to Extend the Freshness and Longevity of Plums Oemleria Cerasiformis. *Scientific Reports*, *13*, 14583. https://doi.org/10.1038/s41598-023-41640-1

Marano, S., Laudadio, E., Minnelli, C., & Stipa, P. (2022). Tailoring the Barrier Properties of PLA: A State-of-the-Art Review for Food Packaging Applications. *Polymers*, *14*, 1626. https://doi.org/10.3390/polym14081626

Mathew, J. T., Adetunji, C. O., Inobeme, A., Musah, M., Azeh, Y., Etsuyankpa, M. B., Otori, A. A., Chado, I. M., Shaba, E. Y., Akoji, J. N., Mamman, A., & Oyewole, O. A. (2025a). Application of Nanochitosan for the Biodegradation of Agricultural Wastes. In: *Nanochitosan Applications for Enhanced Crop Production and Food Security*. John Wiley & Sons, 361–382. https://doi.org/10.1002/9781394214945.ch16

Mathew, J. T., Adetunji, C. O., Inobeme, A., Otori, A. A., Musah, M., Azeh, Y., Shaba, E. Y., Muhammad, I. A., Mamman, A., & Oyewole, O. A. (2025b). Relevance of Nanochitosan in Food Sector and Food Packaging Sectors: Current Trends. In: *Nanochitosan Applications for Enhanced Crop Production and Food Security*. John Wiley & Sons, 345–360. https://doi.org/10.1002/9781394214945.ch15

Mathew, J. T., Inobeme, A., Musah, M., Azeh, Y., Abdullahi, A., Shaba E. Y., Salihu, A. M., Muhammad, E. B., Josiah, J. G., Jibrin, N. A., Ismail, H., Muhammad, A. I., Maurice, J., Mamman, A., & Ndamitso, M. M. (2024). A Critical Review of Green Approach on Wastewater Treatment Strategies. *Journal of Applied Science and Environmental Management, 28*(2), 363–391. https://doi.org/10.4314/jasem.v28i2.9

Mathew, J. T., Musah, M., Azeh, Y., & Muhammed, M. (2023a). Kinetic Study of Heavy Metals Removal from Pharmaceutical Wastewater Using Geopolymer/Fe_3O_4 Nanocomposite. *Bima Journal of Science and Technology*, *7*(4), 152–163.

Mathew, J. T., Musah, M., Azeh, Y., & Muhammed, M. (2023b). Adsorptive Removal of Selected Toxic Metals from Pharmaceutical Wastewater using Fe_3O_4/ZnO Nanocomposite. *Dutse Journal of Pure and Applied Sciences*, *9*(4a), 236–248. https://doi.org/10.4314/dujopas.v9i4a.22

Mohammed, M. A., Syeda, J. T. M., Wasan, K. M., & Wasan, E. K. (2023). An Overview of Chitosan Nanoparticles and Its Application in Non-Parenteral Drug Delivery. *Pharmaceutics*, *9*(4), 53. https://doi.org/10.3390/pharmaceutics9040053

Mohanan, N., Montazer, Z., Sharma, P. K., & Levin, D. B. (2020). Microbial and Enzymatic Degradation of Synthetic Plastics. *Frontiers in Microbiology*, *11*, 580709. https://doi.org/10.3389/fmicb.2020.580709

Muhammad, W. M., Ibrahim, A. B., Samuel, J. O., Khalifa, A. S., Oyewole, O. A., Adetunji, C. O., Eniola, K. I. T., Yerima, M. B., & Mathew, J. T. (2025). The Role of Nanomaterials in Agriculture as Nanofertilizers. In: *Nanochitosan Applications for Enhanced Crop Production and Food Security*. John Wiley & Sons, Inc., 1–38. https://doi.org/10.1002/9781394214945.ch1

Muthu, M., Gopal, J., Chun, S., Devadoss, A. J. P., Hasan, N., & Sivanesan, I. (2024). Crustacean Waste-Derived Chitosan: Antioxidant Properties and Future Perspective. Antioxidants (Basel, Switzerland), *10*(2), 228. https://doi.org/10.3390/antiox10020228

Naser, A. Z., Deiab, I., Defersha, F., & Yang, S. (2021). Expanding Poly(lactic acid) (PLA) and Polyhydroxyalkanoates (PHAs) Applications: A Review on Modifications and Effects. *Polymers*, *13*(23), 4271. https://doi.org/10.3390/polym13234271

Ncube, L. K., Ude, A. U., Ogunmuyiwa, E. N., Zulkifli, R., & Beas, I. N. (2021). An Overview of Plastic Waste Generation and Management in Food Packaging Industries. *Recycling*, *6*, 12. https://doi.org/10.3390/recycling6010012

Ncube, L. K., Ude, A. U., Ogunmuyiwa, E. N., Zulkifli, R., & Beas, I. N. (2024). Environmental Impact of Food Packaging Materials: A Review of Contemporary Development from Conventional Plastics to Polylactic Acid Based Materials. Materials (Basel, Switzerland), *13*(21), 4994. https://doi.org/10.3390/ma13214994

Ogundolie, F. A., Adetunji, C. O., Mathew, J. T., Inobeme, A., Ghazanfar, S., Olaniyan, O. T., Ajiboye, M. D., Dauda, W. P., Adetunji, J. B., Akinbo, O., & Adetuyi, B. O. (2023). Intelligent Freshness Indicators in Packaging of Plant-Based Food Products. In: *Evaporative Coolers for the Postharvest Management of Fruits and Vegetables*. Academic Press, 323–331. https://doi.org/10.1016/B978-0-323-89864-5.00008-4

Olaniran, A. O., Balgobind, A., & Pillay, B. (2023). Bioavailability of Heavy Metals in Soil: Impact on Microbial Biodegradation of Organic Compounds and Possible Improvement Strategies. *International Journal of Molecular Sciences*, *14*(5), 10197–10228. https://doi.org/10.3390/ijms140510197

Pal, K., Bharti, D., Sarkar, P., Anis, A., Kim, D., Chałas, R., Maksymiuk, P., Stachurski, P., & Jarzębski, M. (2021). Selected Applications of Chitosan Composites. *International Journal of Molecular Sciences*, *22*(20), 10968. https://doi.org/10.3390/ijms222010968

Panou, A., & Karabagias, I. K. (2023). Biodegradable Packaging Materials for Foods Preservation: Sources, Advantages, Limitations, and Future Perspectives. *Coatings*, *13*, 1176. https://doi.org/10.3390/coatings13071176

Pascall, M. A., DeAngelo, K., Richards, J., & Arensberg, M. B. (2022). Role and Importance of Functional Food Packaging in Specialized Products for Vulnerable Populations: Implications for Innovation and Policy Development for Sustainability. *Foods*, *11*, 3043. https://doi.org/10.3390/foods11193043

Pellis, A., Guebitz, G. M., & Nyanhongo, G. S. (2022). Chitosan: Sources, Processing and Modification Techniques. *Gels* (Basel, Switzerland), 8(7), 393. https://doi.org/10.3390/gels8070393

Perera, K. Y., Jaiswal, A. K., & Jaiswal, S. (2023). Biopolymer-Based Sustainable Food Packaging Materials: Challenges, Solutions, and Applications. *Foods*, *12*, 2422. https://doi.org/10.3390/foods12122422

Petraru, A., & Amariei, S. A. (2023). Novel Approach about Edible Packaging Materials Based on Oilcakes: A Review. *Polymers*, *15*, 3431. https://doi.org/10.3390/polym15163431

Pham, T. T., Nguyen, L. L. P., Dam, M. S., & Baranyai, L. (2023). Application of Edible Coating in Extension of Fruit Shelf Life: Review. *AgriEngineering*, *5*, 520–536. https://doi.org/10.3390/agriengineering5010034

Pires, J. R. A., Souza, V. G. L., Fuciños, P., Pastrana, L., & Fernando, A. L. (2022). Methodologies to Assess the Biodegradability of Bio-Based Polymers-Current Knowledge and Existing Gaps. *Polymers*, *14*(7), 1359. https://doi.org/10.3390/polym14071359

Reddy, M. S. B., Ponnamma, D., Choudhary, R., & Sadasivuni, K. K. (2023). A Comparative Review of Natural and Synthetic Biopolymer Composite Scaffolds. *Polymers*, *13*, 1105. https://doi.org/10.3390/polym13071105

Rehman, K. U., Hollah, C., Wiesotzki, K., Heinz, V., Aganovic, K., Rehman, R. U., Petrusan, J.-I., Zheng, L., Zhang, J., Sohail, S., Mansoor, M. K., Rumbos, C. I., Athanassiou, C., & Cai, M. (2023). Insect-Derived Chitin and Chitosan: A Still Unexploited Resource for the Edible Insect Sector. *Sustainability*, *15*, 4864. https://doi.org/10.3390/su15064864

Rezvani Ghomi, E., Khosravi, F., Saedi Ardahaei, A., Dai, Y., Neisiany, R. E., Foroughi, F., Wu, M., Das, O., & Ramakrishna, S. (2023). The Life Cycle Assessment for Polylactic Acid (PLA) to Make It a Low-Carbon Material. *Polymers*, *13*(11), 1854. https://doi.org/10.3390/polym13111854

Riseh, S. R., Vatankhah, M., Hassanisaadi, M., & Kennedy, J. F. (2023). Chitosan-Based Nanocomposites as Coatings and Packaging Materials for the Postharvest Improvement of Agricultural Product: A Review. *Carbohydrate Polymers*, *309*, 120666. https://doi.org/10.1016/j.carbpol.2023.120666

Román-Doval, R., Torres-Arellanes, S. P., Tenorio-Barajas, A. Y., Gómez-Sánchez, A., & Valencia-Lazcano, A. A. (2023). Chitosan: Properties and Its Application in Agriculture in Context of Molecular Weight. *Polymers*, *15*, 2867. https://doi.org/10.3390/polym15132867

Salavati Hamedani, M., Rezaeigolestani, M., & Mohsenzadeh, M. (2022). Optimization of Antibacterial, Physical and Mechanical Properties of Novel Chitosan/Olibanum Gum Film for Food Packaging Application. *Polymers*, *14*(19), 3960. https://doi.org/10.3390/polym14193960

Seok, S. (2022). Polymer-Based Biocompatible Packaging for Implantable Devices: Packaging Method, Materials, and Reliability Simulation. *Micromachines*, *12*, 1020. https://doi.org/10.3390/mi12091020

Shaikh, S., Yaqoob, M., & Aggarwal, P. (2024). An Overview of Biodegradable Packaging in Food Industry. *Current Research in Food Science*, *4*, 503–520. https://doi.org/10.1016/j.crfs.2021.07.005

Silva, R. R. A., Marques, C. S., Arruda, T. R., Teixeira, S. C., & de Oliveira, T. V. (2023). Biodegradation of Polymers: Stages, Measurement, Standards and Prospects. *Macromol*, *3*, 371–399. https://doi.org/10.3390/macromol3020023

Siracusa V. (2023). Microbial Degradation of Synthetic Biopolymers Waste. *Polymers*, *11*(6), 1066. https://doi.org/10.3390/polym11061066

Souza, V. G. L., Pires, J. R. A., Rodrigues, C., Coelhoso, I. M., & Fernando, A. L. (2020). Chitosan Composites in Packaging Industry-Current Trends and Future Challenges. *Polymers*, *12*(2), 417. https://doi.org/10.3390/polym12020417

Souza, V. G. L., Pires, J. R. A., Rodrigues, C., Coelhoso, I. M., & Fernando, A. L. (2024). Chitosan Composites in Packaging Industry-Current Trends and Future Challenges. *Polymers*, *12*(2), 417. https://doi.org/10.3390/polym12020417

Suryanegara, L., Fatriasari, W., Zulfiana, D., Anita, S. H., Masruchin, N., Gutari, S., & Kemala, T. (2021). Novel Antimicrobial Bioplastic Based on PLA-Chitosan by Addition of TiO_2 and ZnO. *Journal of Environmental Health Science & Engineering*, *19*(1), 415–425. https://doi.org/10.1007/s40201-021-00614-z

Teixeira-Costa, B. E., & Andrade, C. T. (2023). Chitosan as a Valuable Biomolecule from Seafood Industry Waste in the Design of Green Food Packaging. *Biomolecules*, *11*(11), 1599. https://doi.org/10.3390/biom11111599

Thambiliyagodage, C., Jayanetti, M., Mendis, A., Ekanayake, G., Liyanaarachchi, H., & Vigneswaran, S. (2023). Recent Advances in Chitosan-Based Applications: A Review. *Materials (*Basel, Switzerland), 16(5), 2073. https://doi.org/10.3390/ma16052073

Triunfo, M., Tafi, E., Guarnieri, A. et al. (2022). Characterization of Chitin and Chitosan Derived from Hermetia illucens, a Further Step in a Circular Economy Process. *Scientific Reports*, *12*, 6613. https://doi.org/10.1038/s41598-022-10423-5

Trofimchuk, E., Ostrikova, V., Ivanova, O., Moskvina, M., Plutalova, A., Grokhovskaya, T., Shchelushkina, A., Efimov, A., Chernikova, E., Zhang, S., & Mironov, V. (2023). Degradation of Structurally Modified Polylactide under the Controlled Composting of Food Waste. *Polymers*, *15*(19), 4017. https://doi.org/10.3390/polym15194017

Versino, F., Ortega, F., Monroy, Y., Rivero, S., López, O. V., & García, M. A. (2023). Sustainable and Bio-Based Food Packaging: A Review on Past and Current Design Innovations. *Foods (Basel, Switzerland)*, *12*(5), 1057. https://doi.org/10.3390/foods12051057

Vicente, D., Proença, D. N., & Morais, P. V. (2023). The Role of Bacterial Polyhydroalkanoate (PHA) in a Sustainable Future: A Review on the Biological Diversity. *International Journal of Environmental Research and Public Health*, *20*(4), 2959. https://doi.org/10.3390/ijerph20042959

Vieira, F. R., Magina, S., Evtuguin, D. V., & Barros-Timmons, A. (2024). Lignin as a Renewable Building Block for Sustainable Polyurethanes. *Materials (*Basel, Switzerland*)*, *15*(17), 6182. https://doi.org/10.3390/ma15176182

Wrońska, N., Katir, N., Nowak-Lange, M., El Kadib, A., & Lisowska, K. (2023). Biodegradable Chitosan-Based Films as an Alternative to Plastic Packaging. *Foods (*Basel, Switzerland), 12(18), 3519. https://doi.org/10.3390/foods12183519

Xu, Y., Wu, Z., Li, A., Chen, N., Rao, J., & Zeng, Q. (2024). Nanocellulose Composite Films in Food Packaging Materials: A Review. *Polymers*, *16*, 423. https://doi.org/10.3390/polym16030423

Yadav, K., & Nikalje, G. C. (2024). Comprehensive Analysis of Bioplastics: Life Cycle Assessment, Waste Management, Biodiversity Impact, and Sustainable Mitigation Strategies. *PeerJ*, *12*, e18013. https://doi.org/10.7717/peerj.18013

Yang, X. G., Wen, P. P., Yang, Y. F., Jia, P. P., Li, W. G., & Pei, D. S. (2023). Plastic Biodegradation by in Vitro Environmental Microorganisms and In Vivo Gut Microorganisms of Insects. *Frontiers in Microbiology*, *13*, 1001750. https://doi.org/10.3389/fmicb.2022.1001750

Younes, I., & Rinaudo, M. (2024). Chitin and Chitosan Preparation from Marine Sources. Structure, Properties and Applications. *Marine Drugs*, *13*(3), 1133–1174. https://doi.org/10.3390/md13031133

Yun, H. S., Kim, D. H., Kim, J. G., Kim, Y. S., & Yoon, H. S. (2022). The Microbial Communities (Bacteria, Algae, Zooplankton, and Fungi) Improved Biofloc Technology Including the Nitrogen-Related Material Cycle in Litopenaeus Vannamei Farms. *Frontiers in Bioengineering and Biotechnology*, *10*, 883522. https://doi.org/10.3389/fbioe.2022.883522

Zarski, A., Bajer, K., & Kapuśniak, J. (2023). Review of the Most Important Methods of Improving the Processing Properties of Starch toward Non-Food Applications. *Polymers*, *13*(5), 832. https://doi.org/10.3390/polym13050832

Zhai, X., Zhang, X. H., & Yu, M. (2023). Microbial Colonization and Degradation of Marine Microplastics in the Plastisphere: A Review. *Frontiers in Microbiology*, *14*, 1127308. https://doi.org/10.3389/fmicb.2023.1127308

10 Application of Chitosan-Based Composite Coatings When Combined with Antimicrobial Agents with Nanocompounds

John Tsado Mathew[1], Jonathan Hussaini[1], Yakubu Azeh[1], Musah Monday[1], Abel Inobeme[2], Elijah Yanda Shaba[3], Mary Jumai Jiya[4], Muhammad Aishetu Ibrahim[1], Charles Oluwaseun Adetunji, Etsuyankpa Muhammad Bini[5], Ismaila Aliyu Oga[5], Musa Safiyanu Tanko[5], Tanko Ezekiel[3], Muhammed Jibrin Ndejiko[6], and Amos Mamman[1]

[1]Department of Chemistry, Ibrahim Badamasi Babangida University Lapai, Niger State, Nigeria

[2]Department of Chemistry, Edo University Uzairue, Auchi, Edo State, Nigeria

[3]Department of Chemistry, Federal University of Technology, Minna, Niger State, Nigeria

[4]Department of Food Science and Technology, Federal University of Technology, Minna, Niger State, Nigeria

[5]Department of Chemistry, Federal University of Lafia, Nasarawa State, Nigeria

[6]Department of Microbiology, Ibrahim Badamasi Babangida University Lapai, Niger State, Nigeria

10.1 INTRODUCTION

In recent years, the development of innovative technologies in food packaging has gained significant attention due to the increasing demand for safer and more sustainable packaging solutions. One notable advancement in this domain is the utilization of chitosan-based composite coatings combined with antimicrobial agents and nanocompounds. Chitosan, a derivative of chitin, has emerged as a promising biopolymer for various applications, particularly in the food industry, owing to its biodegradability, non-toxicity, and antimicrobial properties (Chaudhary et al., 2022).

Chitosan, derived from the deacetylation of chitin, is a linear polysaccharide composed of β-(1–4)-linked D-glucosamine and *N*-acetyl-D-glucosamine units. This biopolymer is sourced from crustacean exoskeletons, fungal cell walls, and other natural materials. Chitosan's unique properties, such as film-forming ability, biocompatibility, and bioadhesiveness, make it an excellent candidate for diverse applications, particularly in the development of composite coatings. Food safety remains a paramount concern, and the incorporation of antimicrobial agents in food packaging materials is crucial to extend the shelf life of perishable goods. Antimicrobial agents inhibit the growth of microorganisms, including bacteria, fungi, and molds, thereby preventing food spoilage

DOI: 10.1201/9781003355458-10

and reducing the risk of foodborne illnesses. Various synthetic and natural antimicrobial agents have been explored for their effectiveness in food packaging applications (Pellis et al., 2022; Malik et al., 2025).

Nanotechnology has revolutionized the field of materials science, offering unprecedented opportunities to enhance the properties of various materials. In the context of food packaging, nanocompounds, such as nanoparticles and nanoclays, have been integrated into composite coatings to impart unique functionalities. These nanocompounds enhance mechanical strength, barrier properties, and, in some cases, provide additional antimicrobial effects. Chitosan-based composite coatings have gained attention as a sustainable and biodegradable alternative to conventional packaging materials. These coatings exhibit excellent film-forming properties, flexibility, and biocompatibility. By combining chitosan with other materials, such as proteins, lipids, or polysaccharides, researchers have developed composite coatings with tailored properties to meet specific packaging requirements (Singh et al., 2023).

The incorporation of antimicrobial agents into chitosan-based composites enhances their efficacy in preventing microbial growth on food surfaces. Commonly used antimicrobial agents include essential oils, organic acids, and antimicrobial peptides. These agents not only inhibit microbial growth but also contribute to the overall safety and quality of packaged food products. The integration of nanocompounds into chitosan coatings offers a multifaceted approach to enhancing packaging performance. Nanoparticles, such as silver nanoparticles, zinc oxide nanoparticles, and nanoclays, have been shown to provide antimicrobial effects, improve mechanical strength, and enhance barrier properties against gases and moisture. This synergistic combination of chitosan, antimicrobial agents, and nanocompounds creates a robust protective barrier for food products (Duda-Chodak et al., 2023; Mathew et al., 2025a).

Furthermore, the integration of chitosan-based composite coatings with antimicrobial agents and nanocompounds represents a cutting-edge approach in the development of advanced food packaging materials. This combination leverages the inherent properties of chitosan, the antimicrobial efficacy of selected agents, and the enhanced functionalities provided by nanocompounds. The resulting composite coatings offer improved shelf life, safety, and sustainability for packaged food products. As research in this field progresses, it is expected that these innovative materials will find widespread application, contributing to the evolution of a safer and more environmentally friendly food packaging industry (Riseh et al., 2023).

10.2 FUNDAMENTALS OF CHITOSAN-BASED COMPOSITE COATINGS

Chitosan, a versatile biopolymer derived from chitin, has garnered significant attention in the development of composite coatings due to its unique properties and environmental benefits. Chitosan-based composite coatings combine the advantages of chitosan with other materials, such as polymers, proteins, and nanoparticles, to create advanced coatings with diverse functionalities. Understanding the fundamentals of chitosan-based composite coatings is essential for exploring their wide-ranging applications in various industries, particularly in food packaging and biomedical fields (Ibrahim et al., 2023).

Chitosan as a Base Material: Chitosan is derived from chitin, a natural polymer found abundantly in the exoskeletons of crustaceans and insects, through a process called deacetylation. This conversion produces a biocompatible and biodegradable polymer with several distinctive properties, including the ability to form films, antimicrobial activity, and biocompatibility. Chitosan's versatility is largely due to its cationic nature, which allows it to interact effectively with negatively charged surfaces. This characteristic makes chitosan an excellent candidate for use in composite coatings, as it can enhance the performance and functionality of the coatings by improving adhesion, providing antimicrobial protection, and offering a sustainable, eco-friendly option for various applications. Its unique properties make chitosan valuable in industries ranging from food packaging to medical devices (Thambiliyagodage et al., 2023; Chimbekujwo et al., 2025).

Film-Forming Properties of Chitosan: One of the key attributes that make chitosan a preferred material for coatings is its exceptional film-forming ability. Chitosan can be processed into thin, flexible films that are ideal for coating a wide range of surfaces, including food products, pharmaceuticals, and medical devices. These chitosan films exhibit excellent adhesion, forming a robust protective barrier that enhances the safety and shelf life of the underlying substrates. The ability of chitosan films to create an effective barrier against contaminants and environmental factors ensures the preservation and integrity of the coated products. This makes chitosan an invaluable material in applications where protection, safety, and extended freshness are crucial, providing both functional and protective benefits across various industries (El-Araby et al., 2024).

Enhancement through Composite Formation: Chitosan's inherent properties can be significantly enhanced by creating composite coatings through the incorporation of other materials. By blending chitosan with proteins like gelatin or soy protein, polysaccharides such as starch or alginate, and synthetic polymers, the resulting composites gain additional functionalities. This synergistic approach allows for the customization of coating properties to meet specific application needs. For instance, combining chitosan with gelatin can improve film flexibility and moisture retention, while blending with synthetic polymers can enhance mechanical strength and durability. These tailored composites enable the development of coatings with optimized performance characteristics, such as improved barrier properties, enhanced adhesion, and targeted release mechanisms. This versatility makes chitosan-based composites suitable for a wide range of applications, including food packaging, medical devices, and protective coatings (Mikušová and Mikuš, 2021; Muhammad et al., 2025).

Antimicrobial Properties of Chitosan Composites: Chitosan inherently possesses antimicrobial properties that can be further enhanced when used in composite coatings. By incorporating additional antimicrobial agents, such as essential oils, organic acids, or antimicrobial peptides, into chitosan-based composites, the coatings' ability to inhibit the growth of bacteria, fungi, and other microorganisms is significantly improved. This enhancement is particularly beneficial for applications like food packaging, where extending the shelf life of perishable products is crucial. The synergistic effect of chitosan and these antimicrobial agents provides a powerful defense against microbial contamination, helping to maintain the safety and quality of the packaged goods. This makes chitosan-based composite coatings an effective solution for reducing spoilage and enhancing the overall shelf life of a wide range of perishable products (Ke et al., 2021; Inobeme et al., 2024a).

Nanocompounds for Functional Enhancement: The performance of chitosan-based composite coatings is significantly enhanced by integrating nanocompounds. Incorporating nanoparticles such as silver nanoparticles, zinc oxide nanoparticles, and nanoclays into chitosan composites can improve various properties. Silver nanoparticles and zinc oxide nanoparticles enhance antimicrobial effects, providing stronger protection against bacteria and fungi. Nanoclays contribute to improved mechanical strength and barrier properties, making the coatings more resistant to damage and environmental factors. The nanoscale features of these compounds play a crucial role in elevating the overall efficiency of the composite coatings. By enhancing both functional and protective attributes, these nanocompounds make chitosan-based coatings more robust, versatile, and suitable for a wide range of applications, including advanced food packaging, medical devices, and protective coatings (Zanca et al., 2022).

Applications Across Industries: Chitosan-based composite coatings are versatile and find applications across various industries. In the food packaging sector, these coatings are used to preserve freshness and quality by preventing microbial contamination and enhancing barriers against external factors like moisture and oxygen. This helps in extending the shelf life of perishable products. In the biomedical field, chitosan composites are valued for their biocompatibility and controlled release capabilities, making them suitable for wound dressings, drug delivery systems, and implant coatings.

The natural antimicrobial properties of chitosan, combined with its ability to support controlled drug release, make it an ideal material for applications requiring both safety and efficacy. This versatility underscores the potential of chitosan-based composites in improving product performance and functionality across different domains (Khubiev et al., 2023).

Sustainability and Environmental Considerations: Chitosan, derived from natural sources like crustacean shells, introduces a sustainable dimension to composite coatings. Its biodegradability and minimal environmental impact position it as an eco-friendly alternative to conventional synthetic materials. Unlike many synthetic polymers, chitosan breaks down naturally without leaving harmful residues, contributing to reduced environmental pollution. This makes it an attractive option for industries seeking to meet the growing demand for sustainable and environmentally responsible solutions. The incorporation of chitosan-based coatings aligns with global sustainability goals by promoting the use of renewable resources and minimizing the ecological footprint of products. As industries increasingly focus on green technologies, chitosan's natural origin and low environmental impact make it a compelling choice for advancing eco-friendly innovations across various applications (Maliki et al., 2022; Mathew et al., 2025b).

In addition, the fundamentals of chitosan-based composite coatings lie in harnessing the unique properties of chitosan and combining them with other materials to create versatile coatings with enhanced functionalities. These coatings have broad applications, ranging from food packaging to biomedical devices, and their sustainable nature further underscores their importance in the evolving landscape of materials science and technology. As research continues to delve into optimizing formulations and exploring novel combinations, the potential for chitosan-based composite coatings to revolutionize diverse industries remains promising (Riseh et al., 2023).

10.3 ANTIMICROBIAL AGENTS IN COMPOSITE COATINGS

Antimicrobial agents incorporated into composite coatings play a pivotal role in enhancing the functionality and durability of various materials, finding applications in diverse fields such as healthcare, food packaging, textiles, and construction. These agents are specifically designed to inhibit the growth and proliferation of microorganisms, including bacteria, fungi, and viruses, thus contributing to improved hygiene and safety. Silver nanoparticles, a prominent metallic antimicrobial agent, are often integrated into polymer-based composite coatings, such as polyethylene, to create antimicrobial films for food packaging. The antimicrobial efficacy of silver nanoparticles arises from their ability to disrupt microbial cell membranes and interfere with cellular processes (Adetunji et al., 2023a,b).

Metal oxides, such as zinc oxide, constitute another class of antimicrobial agents. These are commonly utilized in composite coatings with materials such as polystyrene, particularly in healthcare settings. Zinc oxide exerts antimicrobial effects by generating reactive oxygen species, which are toxic to microorganisms. Organic antimicrobial agents, such as quaternary ammonium compounds and triclosan, are frequently employed in composite coatings for textiles and building materials. These compounds provide a versatile solution, offering antimicrobial protection while being compatible with a range of coating materials (Rosli et al., 2021).

In addition, the integration of antimicrobial agents into composite coatings represents a proactive approach toward creating materials with enhanced resistance to microbial colonization, thereby addressing crucial aspects of public health and product longevity across various industries (Table 10.1).

10.4 MECHANISMS OF ANTIMICROBIAL ACTION IN COATINGS

Table 10.2 provides a brief overview of different antimicrobial agents, their mechanisms of action, the target microorganisms, the coating material with which they are commonly used, and their specific applications

TABLE 10.1
Antimicrobial Agents in Composite Coatings

Antimicrobial Agent	Type of Agent	Composite Coating Material	Application	References
Silver nanoparticles	Metallic	Polyethylene	Food packaging	Zorraquín-Peña et al. (2020)
Zinc oxide	Metal oxide	Polystyrene	Healthcare surfaces	Mathew et al. (2023a)
Quaternary ammonium compounds	Organic	Polypropylene	Textiles	Schneider et al. (2023)
Triclosan	Organic	Epoxy resin	Building materials	Seidi et al. (2020)

TABLE 10.2
Mechanisms of Antimicrobial Action in Coatings

Antimicrobial Agent	Mechanism of Action	Target Microorganisms	Coating Material	Application	References
Silver nanoparticles	Disruption of cell membranes	Bacteria, fungi, viruses	Polyethylene	Food packaging	Bruna et al. (2021)
Zinc oxide	Generation of reactive oxygen species	Bacteria, fungi	Polystyrene	Healthcare surfaces	Mendes et al. (2022)
Quaternary ammonium compounds	Disruption of cell membranes and interference with cellular processes	Bacteria, viruses	Polypropylene	Textiles	Mohapatra et al. (2023)
Triclosan	Inhibition of bacterial fatty acid synthesis	Bacteria	Epoxy resin	Building materials	Adetunji et al. (2022a)
Copper-based compounds	Induction of oxidative stress	Bacteria, fungi	Stainless steel	HVAC systems	Gao and Zhang (2023)

10.5 INTEGRATION STRATEGIES WITH CHITOSAN-BASED MATRICES

Integration strategies with chitosan-based matrices are pivotal in harnessing the unique properties of chitosan for various applications. Chitosan, derived from chitin, is a biopolymer known for its biocompatibility, antimicrobial activity, and versatility. By employing integration strategies, researchers and industries can tailor chitosan-based matrices to meet specific needs in areas such as drug delivery, wound healing, tissue engineering, and food packaging (Kumar et al., 2023).

Drug Delivery: One prominent application of chitosan-based matrices lies in drug delivery systems. Chitosan's ability to form nanoparticles, microparticles, and hydrogels makes it an ideal candidate for controlled drug release. Integration strategies often involve the incorporation of therapeutic agents into chitosan matrices, taking advantage of its mucoadhesive properties for improved bioavailability. Additionally, surface modification or crosslinking techniques can be employed to tailor drug release kinetics, enhancing the efficiency of drug delivery systems (Kim et al., 2023).

Wound Healing: Chitosan's natural hemostatic properties and biocompatibility make it valuable in wound healing applications. Integration with growth factors, antimicrobial agents, or stem cells can enhance the wound healing properties of chitosan-based matrices. Composite materials, such as chitosan-based hydrogels with nanoparticles, offer a supportive environment for cell proliferation and tissue regeneration. These integration strategies accelerate wound closure and minimize the risk of infections, showcasing the potential of chitosan in advanced wound care (Kim et al., 2023).

Tissue Engineering: In tissue engineering, chitosan-based matrices serve as scaffolds for cell attachment, proliferation, and differentiation. Integration with bioactive molecules, such as growth factors or peptides, enhances the biological functionality of these matrices. By mimicking the extracellular matrix, chitosan-based scaffolds support tissue regeneration. Incorporating cells, such as mesenchymal stem cells, into chitosan matrices promotes tissue-specific differentiation, making them invaluable for engineering bone, cartilage, and other tissues (Islam et al., 2020; Mathew et al., 2024a).

Food Packaging: Chitosan's antimicrobial properties make it an excellent candidate for food packaging materials. Integration strategies involve incorporating chitosan into films or coatings to inhibit the growth of spoilage microorganisms and pathogens. The positively charged amino groups in chitosan interact with negatively charged microbial cell membranes, disrupting their structure. Additionally, the integration of other natural antimicrobial agents or nanomaterials further enhances the effectiveness of chitosan-based packaging materials in preserving food quality and safety (Abimbola et al., 2023; Ogundolie et al., 2023).

Biomedical Implants: Chitosan's biocompatibility and potential to promote tissue integration make it suitable for biomedical implants. Integration strategies focus on enhancing the mechanical properties and bioactivity of chitosan-based materials. Incorporating reinforcing agents, such as nanoparticles or fibers, strengthens the implants, while surface modifications with bioactive molecules facilitate better integration with surrounding tissues. These strategies aim to improve the long-term performance and biocompatibility of chitosan-based implants in orthopedic and dental applications (Adetunji et al., 2023c).

Environmental Remediation: Chitosan's adsorption capacity for heavy metals and dyes makes it valuable in environmental remediation. Integration strategies involve modifying chitosan matrices with functional groups that enhance adsorption capabilities. Chitosan-based composites, incorporating materials like activated carbon or nanoparticles, exhibit improved sorption properties. These integrated matrices can be utilized for the removal of pollutants from water and soil, showcasing chitosan's potential in sustainable environmental applications (Wang et al., 2023).

Furthermore, integration strategies with chitosan-based matrices encompass a diverse range of applications, capitalizing on chitosan's unique properties. Whether it's tailoring drug delivery systems, accelerating wound healing, supporting tissue engineering, preserving food, enhancing biomedical implants, or contributing to environmental remediation, these strategies showcase the adaptability and versatility of chitosan in various fields. As research continues to unravel the full potential of chitosan-based matrices, the development of innovative integration strategies will further broaden their impact across different industries (Kim et al., 2023).

10.6 NANOCOMPOUNDS AND THEIR ROLE IN COATINGS

Nanocompounds have emerged as groundbreaking additives in coatings, revolutionizing the field with their unique properties and diverse applications. These materials, typically nanoparticles or nanocomposites, are integrated into coating formulations to impart enhanced functionalities, ranging from improved mechanical properties to advanced functionalities like antimicrobial activity and UV resistance. Understanding the role of nanocompounds in coatings is crucial for harnessing their full potential in various industries (Alshangiti et al., 2023).

Enhanced Mechanical Properties: One of the primary roles of nanocompounds in coatings is to enhance mechanical properties. Nanoparticles, such as silica or clay nanotubes, when dispersed within a coating matrix, reinforce the material by providing increased tensile strength, hardness, and scratch resistance. The high surface area and aspect ratio of nanoparticles contribute to efficient load distribution, resulting in coatings with improved durability and mechanical performance (Mathew et al., 2023b).

Barrier Properties: Nanocompounds are crucial in improving the barrier properties of coatings by incorporating nanoparticles that form a dense and uniform network within the coating matrix.

This structure effectively restricts the penetration of gases, moisture, and other environmental factors, significantly enhancing the protective qualities of the coating. Such improved barrier properties are particularly beneficial in applications like corrosion protection for metal surfaces, where preventing moisture and oxygen ingress is critical to avoid rust and deterioration. Additionally, nanocompound-based coatings can prevent the ingress of harmful substances, making them ideal for protective applications in various industries, including automotive, aerospace, and construction. By enhancing the durability and resilience of coated materials, nanocompounds help extend the lifespan of products and structures, reducing maintenance costs and contributing to sustainability by minimizing the need for frequent replacements (Inobeme et al., 2023a).

UV Resistance and Photostability: Incorporating nanocompounds, particularly metal oxide nanoparticles such as titanium dioxide (TiO_2) and zinc oxide (ZnO), significantly enhances the UV resistance and photostability of coatings. These nanoparticles are highly effective in scattering and absorbing UV radiation, which helps prevent the degradation of both the coating itself and the underlying substrate. By mitigating the harmful effects of UV exposure, such as color fading, cracking, and structural weakening, these nanoparticles prolong the lifespan of coated materials. UV-resistant coatings are widely used in applications where exposure to sunlight is a concern, including outdoor structures, automotive finishes, and building facades. The improved UV protection offered by nanocompound-enhanced coatings not only maintains aesthetic appeal but also contributes to the long-term durability and performance of materials exposed to harsh environmental conditions (Mathew et al., 2023a).

Antimicrobial Properties: Nanocompounds, especially silver nanoparticles and other antimicrobial agents, are increasingly utilized to imbue coatings with strong antimicrobial properties. This capability is particularly crucial in environments where preventing microbial growth is essential, such as healthcare settings, food packaging, and public facilities. Silver nanoparticles are known for their broad-spectrum antimicrobial activity, effectively inhibiting the growth of bacteria, fungi, and viruses. The small size and high surface area of these nanoparticles enhance their contact with microorganisms, boosting their antimicrobial efficacy. By integrating nanocompounds into coatings, a proactive approach to maintaining hygiene and preventing the spread of pathogens is achieved. This innovation not only improves surface hygiene but also contributes to overall safety and health by reducing the risk of contamination and infection, making it a valuable asset in various critical applications (Suvarna et al., 2022).

Self-Healing Capabilities: Innovative nanocompounds can contribute to the development of self-healing coatings. Microcapsules containing healing agents or nanoparticles with intrinsic healing properties can be incorporated into coatings. When damage occurs, these capsules rupture or nanoparticles release healing agents, repairing the coating and restoring its protective function. Self-healing coatings find applications in sectors where maintenance and longevity are critical, such as the aerospace and automotive industries (Shchukina and Shchukin, 2019; Bodunrinde et al., 2025).

Thermal Stability and Fire Resistance: Nanocompounds, such as flame retardant nanoparticles, significantly enhance the thermal stability and fire resistance of coatings. By incorporating nanofillers like nanoclays or metal oxides, the flammability of coatings is reduced, and the ignition process is delayed. These nanoparticles work by forming a protective char layer when exposed to heat, which acts as a barrier, slowing down the transfer of heat and oxygen to the underlying material. This effect is particularly crucial in applications where fire safety is a top priority, such as in construction materials, textiles, electronics, and transportation. By improving the fire-resistant properties of coatings, nanocompounds help to enhance the overall safety of buildings, vehicles, and fabrics, providing additional protection against fire hazards and contributing to safer living and working environments (Rabajczyk et al., 2021).

Improved Adhesion and Surface Modification: Nanocompounds play a vital role in improving the adhesion between coatings and substrates. By using nanoparticles for surface modification, the compatibility between different materials is enhanced, leading to stronger adhesion and cohesion within the coating system. Nanoparticles create a rougher surface texture at the microscopic level, increasing the contact area between the coating and the substrate, which results in better mechanical interlocking. This improved adhesion is crucial for ensuring the long-term performance of coatings, particularly in preventing common issues such as delamination, peeling, and cracking. Enhanced adhesion properties are essential in applications involving diverse substrates, such as metals, plastics, and ceramics, where maintaining a durable bond is critical for the coating's effectiveness. As a result, nanocompound-enhanced coatings offer greater reliability and durability across various industrial and commercial applications (Sanità et al., 2020; Dawud et al., 2025).

Nanocompounds have revolutionized the field of coatings by offering a multitude of benefits, ranging from mechanical reinforcement to advanced functionalities like antimicrobial properties and self-healing capabilities. As research and development in nanotechnology continue to advance, the potential for nanocompounds in coatings is likely to expand, leading to even more innovative applications across various industries. Understanding and harnessing the unique properties of nanocompounds will continue to be a key driver in the evolution of high-performance coatings (Malik et al., 2022).

10.7 CONTRIBUTION OF NANOCOMPOUNDS TO COATING PERFORMANCE

Table 10.3 provides a concise overview of various types of nanocompounds, their specific contributions to coating performance, common applications, and examples of nanocompounds within each category. It highlights the versatility of nanocompounds in enhancing different aspects of coating functionality, making them invaluable in various industries.

TABLE 10.3
Contribution of Nanocompounds to Coating Performance

Nanocompound Type	Contribution to Coating Performance	Applications	Examples of Nanocompounds	References
Reinforcement nanoparticles	Enhanced mechanical properties, increased strength, hardness, and scratch resistance	Automotive coatings and protective coatings for metal surfaces	Silica nanoparticles and clay nanotubes	Farooq et al. (2022)
Barrier nanoparticles	Improved barrier properties, resistance to gases, moisture, and environmental factors	Corrosion protection and protective coatings for electronics	Nanoclays and graphene nanoparticles	Huang et al. (2022)
UV-resistant nanoparticles	Efficient UV absorption and scattering, preventing degradation of coatings and substrates	Exterior coatings for buildings and automotive finishes	Titanium dioxide and zinc oxide nanoparticles	Andrady et al. (2023)
Antimicrobial nanoparticles	Imparting antimicrobial properties and preventing microbial growth on coated surfaces	Healthcare settings, food packaging, and surfaces requiring hygiene control	Silver nanoparticles and copper nanoparticles	Inobeme et al. (2023b)
Self-healing nanocompounds	Capability to repair damage and restore coating integrity, extending service life	Aerospace coatings, automotive finishes, and coatings for critical infrastructure	Microcapsules with healing agents and nanospheres with intrinsic healing properties	Paladugu et al. (2022)

10.8 APPLICATIONS OF CHITOSAN-BASED IN THE FOOD PACKAGING INDUSTRY

Chitosan, derived from chitin found in the exoskeletons of crustaceans and insects, has gained significant attention in the food packaging industry due to its unique properties and eco-friendly nature. Its applications in food packaging stem from its biodegradability, biocompatibility, antimicrobial activity, and film-forming capabilities. As the industry seeks sustainable and innovative solutions, chitosan-based materials offer a promising avenue for enhancing food safety, extending shelf life, and reducing environmental impact (Baharlouei and Rahman, 2022).

Antimicrobial Packaging: One of the key applications of chitosan-based materials in the food packaging industry is their use as antimicrobial packaging. Chitosan exhibits inherent antimicrobial properties, particularly against bacteria and fungi. Incorporating chitosan into packaging films creates a protective barrier that inhibits the growth of spoilage microorganisms and pathogens, contributing to the preservation of food quality and safety. This is especially crucial in extending the shelf life of perishable products (Khubiev et al., 2023).

Edible Coatings: Chitosan-based edible coatings have gained popularity as a sustainable solution for enhancing the quality and safety of fresh produce. These coatings can be applied directly onto fruits and vegetables to create a protective layer that reduces water loss, retards microbial growth, and minimizes oxidation. Edible coatings contribute to maintaining the freshness of produce, reducing postharvest losses, and providing an alternative to traditional non-biodegradable packaging materials (Martins et al., 2024).

Biodegradable Films: Chitosan's biodegradable nature makes it an attractive material for producing eco-friendly packaging films. Chitosan-based films have been developed as an alternative to conventional plastic films, addressing concerns related to plastic pollution. These films exhibit good mechanical properties and oxygen barrier capabilities and can be tailored for specific applications, offering a sustainable option for single-use packaging (Wrońska et al., 2023).

Gas Barrier Properties: Chitosan-based films demonstrate excellent gas barrier properties, particularly against oxygen and carbon dioxide. This makes them suitable for packaging applications where maintaining the freshness and quality of food products is critical. Chitosan films act as effective barriers, preventing the ingress of oxygen and the escape of volatile compounds, thereby extending the shelf life of packaged goods (Iversen et al., 2022).

Active Packaging with Chitosan Derivatives: Chitosan derivatives, such as chitosan nanoparticles or chitosan coatings incorporating active agents, enable the development of active packaging systems. These systems can release antimicrobial compounds, antioxidants, or other functional ingredients to actively interact with the packaged food. This active approach contributes to food preservation, quality enhancement, and even offers the potential for smart packaging that responds to specific environmental conditions (El-Araby et al., 2024).

Biocompatible Sachets and Pouches: Chitosan's biocompatibility allows for the development of biocompatible sachets or pouches that can be used for the controlled release of preservatives, antioxidants, or other functional additives. These sachets can be placed inside food packages to enhance the overall quality and safety of the packaged products. This approach minimizes the need for direct contact between additives and the food, addressing concerns related to migration and safety (Stefanowska et al., 2023).

Seafood and Meat Preservation: Chitosan-based films and coatings find specific applications in the preservation of seafood and meat products. Given that chitosan is derived from seafood waste, its use in packaging aligns with the principles of circular economy and sustainability. Chitosan's antimicrobial properties contribute to inhibiting the growth of spoilage microorganisms and pathogens, reducing the risk of contamination, and extending the shelf life of these perishable products (Teixeira-Costa and Andrade, 2021).

In addition, chitosan-based materials have emerged as versatile and sustainable solutions in the food packaging industry. From antimicrobial packaging to edible coatings and biodegradable films, the applications of chitosan contribute to addressing key challenges such as food safety, shelf life extension, and environmental impact. As the industry continues to prioritize sustainable practices, chitosan-based innovations are likely to play a crucial role in shaping the future of food packaging (Matloob et al., 2023).

10.9 APPLICATIONS OF CHITOSAN-BASED IN MEDICAL AND HEALTHCARE DEVICES

Chitosan, a natural biopolymer derived from chitin, has found widespread applications in the medical and healthcare industry due to its unique properties such as biocompatibility, antimicrobial activity, and versatility. Its ability to form films, gels, and scaffolds makes it suitable for various medical devices and applications (Jiménez-Gómez and Cecilia, 2020). Here are some main applications of chitosan-based materials in medical and healthcare devices:

Wound Dressings: Chitosan's hemostatic properties make it particularly suitable for wound dressings. Chitosan-based dressings help promote blood clotting and accelerate the wound healing process. The biocompatible nature of chitosan reduces the risk of adverse reactions, making it an excellent choice for various types of wounds, including burns and chronic wounds. Additionally, chitosan's antimicrobial properties contribute to preventing infections in the wound site (Aranaz et al., 2021).

Sutures and Suture Coatings: Chitosan has been explored for use in sutures and coatings for sutures. Its biodegradability and biocompatibility make it suitable for internal applications, where the material gradually degrades as the tissue heals. Chitosan coatings on sutures may provide additional benefits, such as antimicrobial activity, reducing the risk of post-surgical infections (Thambiliyagodage et al., 2023).

Tissue Engineering Scaffolds: Chitosan's ability to form porous structures and its biocompatibility make it a valuable material for tissue engineering scaffolds. These scaffolds provide a three-dimensional framework for cell growth and tissue regeneration. Chitosan-based scaffolds can be tailored to mimic the extracellular matrix, facilitating cell attachment, proliferation, and differentiation. This application is particularly significant in regenerative medicine for the repair or replacement of damaged tissues and organs (Signorini et al., 2023).

Drug Delivery Systems: Chitosan-based drug delivery systems have gained attention for their ability to provide controlled and targeted release of therapeutic agents. Chitosan's mucoadhesive properties make it suitable for oral drug delivery, allowing sustained release and improved bioavailability. Chitosan nanoparticles or microparticles can be used to encapsulate drugs, protecting them from degradation and enabling precise delivery to specific tissues or cells (Adetunji et al., 2022b).

Dental Applications: Chitosan has applications in the field of dentistry, including the development of antimicrobial coatings for dental implants and oral care products. Chitosan's ability to inhibit the growth of bacteria makes it valuable in preventing infections associated with dental procedures. Chitosan-based materials may also be used in oral drug delivery systems for localized treatment (Anggani et al., 2021).

Hemodialysis Membranes: Chitosan has been explored for the development of hemodialysis membranes due to its biocompatibility and selective permeability. Chitosan-based membranes can effectively remove waste products from the blood during hemodialysis while minimizing the risk of blood clotting. This application contributes to improving the efficiency and safety of hemodialysis procedures (Thambiliyagodage et al., 2023).

Antimicrobial Coatings for Medical Devices: Chitosan's inherent antimicrobial properties make it an ideal candidate for coatings on medical devices to prevent infections. Medical devices such as catheters, implants, and surgical instruments can be coated with chitosan or chitosan derivatives to create a protective layer that inhibits the growth of bacteria and reduces the risk of device-associated infections (Yan et al., 2021).

Furthermore, chitosan-based materials play a crucial role in various medical and healthcare devices, contributing to wound healing, drug delivery, tissue engineering, and infection prevention. The versatility of chitosan, along with its biocompatibility and antimicrobial properties, positions it as a valuable biomaterial in the development of innovative solutions for medical applications. As research and technology continue to advance, chitosan-based materials are likely to play an even more significant role in shaping the future of medical devices and therapies (Baharlouei and Rahman, 2022).

10.10 AGRICULTURAL AND CROP PROTECTION APPLICATIONS OF CHITOSAN-BASED

Chitosan, a natural biopolymer derived from chitin, has gained attention for its diverse applications in agriculture and crop protection. This eco-friendly and biodegradable material offers several benefits, including its antimicrobial properties, biocompatibility, and the ability to stimulate plant growth (Maliki et al., 2022). Here are various agricultural and crop protection applications of chitosan-based materials:

Plant Disease Management: Chitosan is well-regarded for its antifungal and antibacterial properties, making it an effective tool for managing plant diseases. Chitosan-based formulations can be utilized as biopesticides or fungicides to control a variety of plant pathogens. These formulations work by inhibiting the growth and development of harmful microbes, offering a sustainable alternative to conventional synthetic chemical pesticides. Chitosan's antifungal activity has been demonstrated against several plant pathogens, including those causing powdery mildew, gray mold, and damping-off. By providing a natural and environmentally friendly approach to disease management, chitosan supports integrated pest management strategies and contributes to more sustainable agricultural practices. Its use in plant disease control highlights its potential to enhance crop health while minimizing the ecological impact of traditional chemical treatments (Lemke et al., 2022).

Seed Treatment: Chitosan-based seed coatings or treatments offer significant benefits for seed germination and early plant development. The film-forming ability of chitosan creates a protective layer around the seeds, which enhances water retention and provides a favorable microenvironment for germination. This protective layer helps maintain optimal moisture levels, promoting uniform and robust seed sprouting. Additionally, chitosan's antimicrobial properties safeguard seeds from soil-borne pathogens, reducing the risk of seed rot and damping-off diseases. By offering both physical protection and microbial defense, chitosan-based coatings improve seed health and vigor, contributing to more successful plant establishment and growth. This application of chitosan supports enhanced seed performance and helps ensure better crop yields, aligning with sustainable agricultural practices aimed at improving plant productivity and resilience (Lyalina et al., 2023).

Biofertilizer Enhancer: Chitosan is being investigated as a biofertilizer enhancer due to its ability to promote nutrient absorption and utilization by plants. Chitosan-based formulations improve the efficiency of fertilizers by enhancing their adhesion to plant roots and soil particles. This increased adhesion ensures better nutrient availability and uptake by plants, leading to improved crop yields. Additionally, chitosan can positively impact soil properties, fostering changes that enhance microbial activity and nutrient cycling. By supporting more efficient use of fertilizers and enriching soil health, chitosan contributes to sustainable agricultural practices and helps optimize plant growth and productivity. Its role as a biofertilizer enhancer underscores its potential to support more effective and environmentally friendly farming techniques (Michalik and Wandzik, 2020; Mathew et al., 2024b).

Crop Residue Management: Chitosan-based materials offer promising benefits for crop residue management. When applied to crop residues, chitosan can enhance their decomposition, accelerating the breakdown process and promoting the release of valuable nutrients back into the soil. This contributes to improved soil health and fertility, supporting sustainable agricultural practices by replenishing soil nutrients and enhancing soil structure. Additionally, chitosan's biodegradability ensures that any residues from its application do not persist in the environment, minimizing potential environmental impact. By facilitating the natural recycling of organic matter, chitosan-based materials help maintain a balanced and productive soil ecosystem, further aligning with the principles of sustainable farming and reducing the need for synthetic soil amendments (Liu et al., 2022).

Biostimulant for Plant Growth: Chitosan acts as a biostimulant, promoting plant growth and development when applied to crops. Its benefits include enhanced root development, increased photosynthesis, and improved nutrient uptake, all of which contribute to better crop yield and quality. By stimulating these growth processes, chitosan supports healthier plants and more robust agricultural outputs. This role is particularly valuable in sustainable agriculture practices, as it offers an alternative to chemical fertilizers and growth-promoting substances. Chitosan's natural origin and effectiveness in enhancing plant performance align with eco-friendly farming goals, helping to reduce environmental impact and improve soil health. Its use as a biostimulant underscores its potential to support more sustainable and productive agricultural systems (Stasińska-Jakubas et al., 2022).

Postharvest Preservation: Chitosan-based coatings are effective in extending the shelf life of harvested fruits and vegetables. These coatings form a protective barrier on the produce, significantly reducing water loss and preventing microbial growth. The antimicrobial properties of chitosan are particularly advantageous, as they inhibit the growth of spoilage microorganisms that contribute to decay. By slowing down the onset of microbial contamination and preserving moisture, chitosan coatings help maintain the freshness and quality of fruits and vegetables during storage and transportation. This application is crucial for minimizing food waste and ensuring that produce remains in optimal condition until it reaches consumers. The use of chitosan-based coatings thus supports longer shelf life and contributes to more efficient food supply chains (Adiletta et al., 2021).

Soil Amendment for Water Retention: Chitosan-based hydrogels serve as effective soil amendments, especially in arid or sandy soils where water retention is a challenge. The hydrophilic nature of chitosan allows the hydrogel to absorb and hold significant amounts of water. Once incorporated into the soil, the hydrogel gradually releases water, providing sustained moisture availability to plant roots. This slow release helps to maintain optimal soil moisture levels, enhancing plant growth and reducing the need for frequent irrigation. In regions facing water scarcity, this application contributes to more efficient water use in agriculture, promoting better crop yields and supporting sustainable farming practices. By improving soil water retention, chitosan-based hydrogels play a crucial role in addressing water management issues and enhancing agricultural productivity in drought-prone areas (Michalik and Wandzik, 2020).

The agricultural and crop protection applications of chitosan-based materials underscore their potential to revolutionize sustainable farming practices. From disease management and seed treatment to biostimulation and postharvest preservation, chitosan offers a range of solutions that align with the growing demand for environmentally friendly and effective agricultural practices. As research continues, chitosan-based innovations are likely to play a significant role in addressing challenges related to plant health, food security, and sustainable agriculture (Stasińska-Jakubas and Hawrylak-Nowak, 2022).

10.11 EXISTING CHALLENGES OF CHITOSAN-BASED IN THE DEVELOPMENT AND APPLICATION

While chitosan-based materials hold great promise in various industries, including agriculture, medicine, and food packaging, their development and application face several challenges. Overcoming these challenges is crucial for unlocking the full potential of chitosan-based innovations and ensuring their successful integration into diverse fields. Here are some of the existing challenges associated with chitosan-based materials:

Extraction and Purity: The extraction of chitosan from chitin, commonly obtained from crustacean shells, presents several challenges. Achieving high purity levels of chitosan is complex, as impurities in the extraction process can affect its properties and performance. Variability in chitosan quality can arise from differences in source materials, extraction methods, and processing conditions. This variability impacts the reproducibility and consistency of the final product. Ensuring uniformity in chitosan quality requires meticulous control over extraction parameters and thorough purification processes. Addressing these challenges is crucial for optimizing chitosan's performance across various applications and ensuring reliable outcomes in industrial uses. Improved extraction techniques and standardized processing conditions can help mitigate these issues and enhance the overall quality and consistency of chitosan products (Bandara et al., 2020).

Cost Considerations: The cost of chitosan production is a significant challenge, primarily due to the reliance on traditional sources such as crustacean shells. These sources can result in limited availability and higher production costs. While alternative sources, such as fungal chitosan, are being explored to potentially lower costs, achieving economic feasibility remains a key consideration. The scale of production needs to be optimized to make chitosan more accessible and affordable for widespread applications. Addressing these economic challenges involves improving production efficiency, exploring more cost-effective raw materials, and developing innovative processing techniques. By overcoming these cost barriers, chitosan can become a more viable option for various industrial and commercial uses (Abidin et al., 2020; Inobeme et al., 2024b).

Molecular Weight and Polydispersity: The molecular weight of chitosan plays a critical role in determining its functional properties, including solubility, film-forming ability, and biological activity. Achieving and maintaining a consistent molecular weight during extraction and processing is challenging, as variations can significantly impact the performance of chitosan-based materials. Polydispersity, or the distribution of molecular weights in a sample, can lead to inconsistent properties and performance. Managing this variability requires careful control over extraction conditions, processing methods, and quality control measures. Ensuring a narrow-molecular-weight distribution helps optimize the desired characteristics of chitosan for specific applications, enhancing its effectiveness and reliability in various industrial and commercial uses (Aranaz et al., 2021).

Solubility and Processing Challenges: Chitosan's poor solubility in neutral and alkaline conditions limits its use in various formulations. This solubility issue presents a significant challenge, as achieving sufficient solubility without compromising its beneficial properties is crucial. Researchers are actively exploring modifications and derivatives of chitosan to enhance its solubility, such as by altering its chemical structure or using solubilizing agents. However, balancing enhanced solubility with the preservation of chitosan's desired functionalities, such as its antimicrobial properties or film-forming ability, remains an ongoing challenge. Continued research is essential to develop chitosan formulations that are both effective and versatile, expanding its potential applications across different industries (Wang et al., 2020).

Standardization of Testing Methods: Standardizing testing methods for chitosan-based materials is crucial to ensure accurate characterization and comparability across different studies and applications. Currently, there are no universally accepted standards for assessing key properties such as molecular weight, degree of deacetylation, and antimicrobial activity. This lack of standardization can lead to discrepancies in results and hinder the ability to compare findings or assess product performance reliably. Establishing standardized testing protocols is essential for advancing research, improving the consistency of results, and facilitating the commercialization of chitosan-based products. Such standards would enhance the credibility of research findings and support the development of more effective and reliable chitosan applications in various industries (Thambiliyagodage et al., 2023).

Interaction with Target Environment: Understanding how chitosan-based materials interact with their target environment—whether in the human body, soil, or food packaging systems—is crucial for optimizing their performance and stability. Environmental factors such as pH, temperature, and humidity can significantly affect the behavior of chitosan. For instance, in the human body, pH and temperature can influence chitosan's drug release rates, while in soil, these factors affect its biodegradation and interaction with plant roots. Similarly, in food packaging, environmental conditions impact the material's barrier properties and shelf life. Developing a comprehensive understanding of these interactions is essential for predicting how chitosan-based materials will behave in real-world applications, ensuring their effectiveness and reliability across various uses (Atay, 2020; Mathew et al., 2024c).

Regulatory Approval and Consumer Acceptance: Navigating regulatory frameworks for chitosan-based products presents significant challenges. The approval process often requires comprehensive evidence of safety, efficacy, and compliance with industry-specific regulatory standards. For applications involving direct contact with food or medical devices, additional scrutiny is necessary to ensure that chitosan-based materials meet rigorous safety requirements. Gaining consumer acceptance is also crucial, as public perception and awareness of chitosan's safety and benefits influence its adoption. Effective communication about chitosan's advantages and safety profile, coupled with meeting regulatory demands, is essential for successful integration into various industries. Addressing these challenges involves thorough testing, clear regulatory pathways, and educational efforts to build trust and facilitate the widespread use of chitosan-based products (Marques et al., 2020).

Addressing these challenges requires collaborative efforts from researchers, industry stakeholders, and regulatory bodies. Ongoing research on alternative chitosan sources, the development of scalable and cost-effective extraction methods, and advancements in processing technologies are essential. Standardization of testing methods and a clearer regulatory pathway will facilitate the development and commercialization of chitosan-based innovations. As these challenges are overcome, chitosan-based materials have the potential to contribute significantly to sustainable, eco-friendly, and effective solutions across diverse applications (Muteeb et al., 2023).

10.12 CONCLUSION

The synergistic combination of chitosan-based composite coatings with antimicrobial agents and nanocompounds presents a formidable approach with multifaceted benefits across various applications. The antimicrobial properties inherent to chitosan are amplified when integrated with nanocompounds, enhancing the overall efficacy of the coating system. This synergy not only contributes to inhibiting microbial growth but also augments mechanical strength, barrier properties, and other functionalities critical for diverse applications. In food packaging, the chitosan-based composite coatings, fortified with antimicrobial agents and nanocompounds, offer an innovative solution for extending shelf life and ensuring food safety. The enhanced barrier properties inhibit the ingress of gases and moisture, preserving the freshness of packaged goods. In healthcare settings, these coatings provide antimicrobial surfaces, contributing to infection control on medical devices and surfaces.

Moreover, the combined benefits of chitosan, antimicrobial agents, and nanocompounds extend to industries such as agriculture, where seed coatings and protective films enhance plant health and disease resistance. The broad spectrum of applications showcases the versatility and potential impact of this composite approach on sustainable practices, environmental protection, and advancements in material science. As research and development continue, this integrated strategy holds promise for addressing contemporary challenges and shaping the future of functional coatings across diverse sectors.

REFERENCES

Abidin, Z. N. A., Kormin, F., Zainol Abidin, N. A., Mohamed Anuar, N. A. F., & Abu Bakar, M. F. (2020). The Potential of Insects as Alternative Sources of Chitin: An Overview on the Chemical Method of Extraction from Various Sources. *International Journal of Molecular Sciences*, *21*(14), 4978. https://doi.org/10.3390/ijms21144978

Abimbola, O. F., Adetunji, C. O., Dare, B. J., Mathew, J. T., Inobeme, A., Ghazanfar, S., Olaniyan, O. T., Ajiboye, M. D., & Dauda, W. (2023). Packaging of Animal-Based Food Products: The Need for Intelligent Freshness Sensors. In: *Sensing and Artificial Intelligence Solutions for Food Manufacturing*. CRC Press, 89–102. https://doi.org/10.1201/9781003207955-6

Adetunji, C. O., Egbuna, C., Ajiboye, A. A. A., Ajayi, O. O., Dauda, W. P., Ghazanfar, S., Ogundolie, F. A., Mathew, J. T., Adetunji, J. B., Inobeme, A., Olotu, T., Olaniyan, O. T., Ijabadeniyi, O. A., Aborode, A. T., Ficai, A., Akram, M., Bodunrinde, R. E., Laila, U., Olaniyan, O. P., Micheal, O. S., Aliyu, A., Akinbo, O., & Adetuyi, B. O. (2023c). Biogenic Nanomaterials with Diverse Biological Activities in the Food and Biomedical Industries. In: *Evaporative Coolers for the Postharvest Management of Fruits and Vegetables*. Academic Press, 395–420. https://doi.org/10.1016/B978-0-323-89864-5.00001-1

Adetunji, C. O., Mathew, J. T., Inobeme, A., Olaniyan, O., Singh, K. R. B., Ogundolie, F. A., Vanya, N., Singh, J., & Singh, R. P. (2022b). Microbial and Plant Cell Biosensors for Environmental Monitoring. In: *Nanobiosensors for Environmental Monitoring*. Springer, Cham, 175–190. https://doi.org/10.1007/978-3-031-16106-3_9

Adetunji, C. O., Mathew, J. T., Singh, K. R. B., Inobeme, A., Olaniyan, O., and Singh, J. (2022a). Antibiofilm Potential of Zero-Dimensional Carbon Nanomaterials. In: *Zero-dimensional Carbon Nanomaterials: Fundamentals and Applications*. IOP Publishing. https://doi.org/10.1088/978-0-7503-4048-9ch12

Adetunji, C. O., Olaniyan, O. T., Olugbenga, S. M., Shariati, M. A., Osarenkhoe, O. O., Odoh, U. E., Oyedara, O. O., Rebezov, M., Ajayi, O. O., Babaeva, M., Inobeme, A., Mathew, J. T., Adetunji, J. B., Ogechukwu, H. U., Bodunrinde, R. E., & Adeniyi, M. J. (2023b). Pharmacology, Photochemical, and Healing Effectiveness of *Chromolaena odorata* (L.) R. M. King and H. Robinson. In: *Phytochemical Composition and Pharmacy of Medicinal Plants*: T. Pullaiah (Ed.). Apple Academic Press, Inc. Co-published with CRC Press (Taylor & Francis), Vol. 2, 461–482.

Adetunji, C. O., Shariati, M. A., Olugbenga, S. M., Osarenkhoe, O. O., Odoh, U. E., Olaniyan, O. T., Rebezov, M., Ajayi, O. O., Baibalinova, G., Bodunrinde, R. E., Adetunji, J. B., Adeniyi, M. J., Ugwu, P. N., Inobeme, A., Mathew, J. T., Elufisan, T. O., & Omotayo, O. O. (2023a). Pineapple [*Ananas comosus* (L.) Merr.]: A Biological and Pharmacological Active Medicinal Plant. In: *Phytochemical Composition and Pharmacy of Medicinal Plants*: T. Pullaiah (Ed.). Apple Academic Press, Inc. Co-published with CRC Press (Taylor & Francis), Vol. 1, 537–567.

Adiletta, G., Di Matteo, M., & Petriccione, M. (2021). Multifunctional Role of Chitosan Edible Coatings on Antioxidant Systems in Fruit Crops: A Review. *International Journal of Molecular Sciences*, *22*(5), 2633. https://doi.org/10.3390/ijms22052633

Alshangiti, D. M., El-Damhougy, T. K., Zaher, A., Madani, M., & Mohamady Ghobashy, M. (2023). Revolutionizing Biomedicine: Advancements, Applications, and Prospects of Nanocomposite Macromolecular Carbohydrate-Based Hydrogel Biomaterials: A Review. *RSC Advances*, *13*(50), 35251–35291. https://doi.org/10.1039/d3ra07391b

Andrady, A. L., Heikkilä, A. M., Pandey, K. K., Bruckman, L. S., White, C. C., Zhu, M., & Zhu, L. (2023). Effects of UV radiation on natural and synthetic materials. *Photochemical & Photobiological Sciences: Official Journal of the European Photochemistry Association and the European Society for Photobiology*, *22*(5), 1177–1202. https://doi.org/10.1007/s43630-023-00377-6

Anggani, H. S., Perdana, R. G., Siregar, E., & Bachtiar, E. W. (2021). The effect of coating chitosan on Porphyromonas gingivalis biofilm formation in the surface of orthodontic mini-implant. *Journal of Advanced Pharmaceutical Technology & Research*, *12*(1), 84–88. https://doi.org/10.4103/japtr.JAPTR_95_20

Aranaz, I., Alcántara, A. R., Civera, M. C., Arias, C., Elorza, B., Heras Caballero, A., & Acosta, N. (2021). Chitosan: An Overview of Its Properties and Applications. *Polymers*, *13*(19), 3256. https://doi.org/10.3390/polym13193256

Atay, Y. H. (2020). Antibacterial Activity of Chitosan-Based Systems. In: *Functional Chitosan: Drug Delivery and Biomedical Applications*, 457–489. https://doi.org/10.1007/978-981-15-0263-7_15

Baharlouei, P., & Rahman, A. (2022). Chitin and Chitosan: Prospective Biomedical Applications in Drug Delivery, Cancer Treatment, and Wound Healing. *Marine Drugs*, *20*(7), 460. https://doi.org/10.3390/md20070460

Bandara, S., Du, H., Carson, L., Bradford, D., & Kommalapati, R. (2020). Agricultural and Biomedical Applications of Chitosan-Based Nanomaterials. *Nanomaterials (*Basel, Switzerland), *10*(10), 1903. https://doi.org/10.3390/nano10101903

Bodunrinde, R. E., Chimbekujwo, K. I., Oyewole, O. A., Adetunji, C. O., Inobeme, A., Eniola, K. I. T., Yerima, M. B., & Mathew, J. T. (2025). The Use of Nanochitosan for Enhancement in the Quality and Yield of Fruit Crops. In: *Nanochitosan Applications for Enhanced Crop Production and Food Security*. John Wiley & Sons, 111–122. https://doi.org/10.1002/9781394214945.ch5

Bruna, T., Maldonado-Bravo, F., Jara, P., & Caro, N. (2021). Silver Nanoparticles and Their Antibacterial Applications. *International Journal of Molecular Sciences*, *22*(13), 7202. https://doi.org/10.3390/ijms22137202

Chaudhary, V., Punia Bangar, S., Thakur, N., & Trif, M. (2022). Recent Advancements in Smart Biogenic Packaging: Reshaping the Future of the Food Packaging Industry. *Polymers*, *14*(4), 829. https://doi.org/10.3390/polym14040829

Chimbekujwo, K. I., Yakubu, J. G., Omoregie, I. P., Oyewole, O. A., Muhammad, W. M., Adetunji, C. O., Eniola, K. I. T., Yerima, M. B., & Mathew, J. T. (2025). Synthesis of Nano-Chitosan Using Agricultural Waste. In: *Nanochitosan Applications for Enhanced Crop Production and Food Security*. John Wiley & Sons, Inc., 39–68. https://doi.org/10.1002/9781394214945.ch2

Dawud, I., Bodunrinde, R. E., Oyewole, O. A., Adetunji, C. O., Inobeme, A., Eniola, K. I. T., Yerima, M. B., & Mathew, J. T. (2025). Application of Nanochitosan for Effective Fruit Production. In: Nanochitosan Applications for Enhanced Crop Production and Food Security. John Wiley & Sons, 123–135. https://doi.org/10.1002/9781394214945.ch6

Duda-Chodak, A., Tarko, T., & Petka-Poniatowska, K. (2023). Antimicrobial Compounds in Food Packaging. *International Journal of Molecular Sciences*, *24*(3), 2457. https://doi.org/10.3390/ijms24032457

El-Araby, A., Janati, W., Ullah, R., Ercisli, S., & Errachidi, F. (2024). Chitosan, Chitosan Derivatives, and Chitosan-Based Nanocomposites: Eco-Friendly Materials for Advanced Applications (A Review). *Frontiers in Chemistry*, *11*, 1327426. https://doi.org/10.3389/fchem.2023.1327426

Farooq, S. A., Raina, A., Mohan, S., Arvind Singh, R., Jayalakshmi, S., & Irfan Ul Haq, M. (2022). Nanostructured Coatings: Review on Processing Techniques, Corrosion Behaviour and Tribological Performance. Nanomaterials (Basel, Switzerland), *12*(8), 1323. https://doi.org/10.3390/nano12081323

Gao, L., & Zhang, A. (2023). Copper-Instigated Modulatory Cell Mortality Mechanisms and Progress in Oncological Treatment Investigations. *Frontiers in Immunology*, *14*, 1236063. https://doi.org/10.3389/fimmu.2023.1236063

Huang, J. Q., Liu, K., Song, X., Zheng, G., Chen, Q., Sun, J., Jin, H., Jiang, L., Jiang, Y., Zhang, Y., Jiang, P., & Wu, W. (2022). Incorporation of Al_2O_3, GO, and Al_2O_3@GO Nanoparticles into Water-Borne Epoxy Coatings: Abrasion and Corrosion Resistance. *RSC Advances*, *12*(38), 24804–24820. https://doi.org/10.1039/d2ra04223a

Ibrahim, M. A., Alhalafi, M. H., Emam, E. M., Ibrahim, H., & Mosaad, R. M. (2023). A Review of Chitosan and Chitosan Nanofiber: Preparation, Characterization, and Its Potential Applications. *Polymers*, *15*(13), 2820. https://doi.org/10.3390/polym15132820

Inobeme, A., Adetunji, C. O., Ajai, A. I., Inobeme, J., Mathew, J. T., Obar, A., Maliki, M., Nwakife, N., & Eziukwu, C. (2023b). Chemical Nanosensors for Monitoring Environmental Pollution. In: *Advanced Application of Nanotechnology to Industrial Wastewater*. Springer Nature Singapore, 93–103. https://doi.org/10.1007/978-981-99-3292-4_6

Inobeme, A., Adetunji, C. O., Mathew, J. T., Ajai, A. I., Inobeme, J., Bamigboye, M. O., Onyeaku, S., Maliki, M., Eziukwu, C., & Tawa, K. (2023a). Nanotechnology for Bioremediation of Heavy Metals. In: *Microbial Technologies in Industrial Wastewater Treatment*. Springer Nature Singapore, 19–30. https://doi.org/10.1007/978-981-99-2435-6_2

Inobeme, A., Mathew, J. T., Ajai, I. A., Adetunji, C. O., Inobeme, J., Maliki, M., Adekoya, M. A., Shaba, E. Y., Eric, O., Akhor, S., & Eziukwu, C. (2024a). Common Techniques in Food Processing Technologies. In: *Nanobiotechnology for Food Processing and Packaging*. Elsevier Inc., 223–234. https://doi.org/10.1016/B978-0-323-91749-0.00002-2

Inobeme, A., Shahnawaz, M., Adetunji, C. O., Mathew, J. T., Adetuyi, B. O., Popoola, O. A., Olaitan, F. Y., Akinbo, O., Kolawole, O. M., Oyewole, O. A., Eniola, K. I. T., & Yerima, B. M. (2024b). Microplastic as a Pathogenic Vector to Cause Diseases in Marine Biota. In: *Microplastic Pollution*. Springer, Singapore, 1, 321–349. https://doi.org/10.1007/978-981-99-8357-5_19

Islam, M. M., Shahruzzaman, M., Biswas, S., Nurus Sakib, M., & Rashid, T. U. (2020). Chitosan Based Bioactive Materials in Tissue Engineering Applications: A Review. *Bioactive Materials*, *5*(1), 164–183. https://doi.org/10.1016/j.bioactmat.2020.01.012

Iversen, L. J. L., Rovina, K., Vonnie, J. M., Matanjun, P., Erna, K. H., 'Aqilah, N. M. N., Felicia, W. X. L., & Funk, A. A. (2022). The Emergence of Edible and Food-Application Coatings for Food Packaging: A Review. *Molecules (*Basel, Switzerland), *27*(17), 5604. https://doi.org/10.3390/molecules27175604

Jiménez-Gómez, C. P., & Cecilia, J. A. (2020). Chitosan: A Natural Biopolymer with a Wide and Varied Range of Applications. *Molecules (*Basel, Switzerland), *25*(17), 3981. https://doi.org/10.3390/molecules25173981

Ke, C. L., Deng, F. S., Chuang, C. Y., & Lin, C. H. (2021). Antimicrobial Actions and Applications of Chitosan. *Polymers*, *13*(6), 904. https://doi.org/10.3390/polym13060904

Khubiev, O. M., Egorov, A. R., Kirichuk, A. A., Khrustalev, V. N., Tskhovrebov, A. G., & Kritchenkov, A. S. (2023). Chitosan-Based Antibacterial Films for Biomedical and Food Applications. *International Journal of Molecular Sciences*, *24*(13), 10738. https://doi.org/10.3390/ijms241310738

Kim, Y., Zharkinbekov, Z., Raziyeva, K., Tabyldiyeva, L., Berikova, K., Zhumagul, D., Temirkhanova, K., & Saparov, A. (2023). Chitosan-Based Biomaterials for Tissue Regeneration. *Pharmaceutics*, *15*(3), 807. https://doi.org/10.3390/pharmaceutics15030807

Kumar, A., Yadav, S., Pramanik, J., Sivamaruthi, B. S., Jayeoye, T. J., Prajapati, B. G., & Chaiyasut, C. (2023). Chitosan-Based Composites: Development and Perspective in Food Preservation and Biomedical Applications. *Polymers*, *15*(15), 3150. https://doi.org/10.3390/polym15153150

Lemke, P., Jünemann, L., & Moerschbacher, B. M. (2022). Synergistic Antimicrobial Activities of Chitosan Mixtures and Chitosan-Copper Combinations. *International Journal of Molecular Sciences*, *23*(6), 3345. https://doi.org/10.3390/ijms23063345

Liu, T., Li, J., Tang, Q., Qiu, P., Gou, D., & Zhao, J. (2022). Chitosan-Based Materials: An Overview of Potential Applications in Food Packaging. *Foods (*Basel, Switzerland), *11*(10), 1490. https://doi.org/10.3390/foods11101490

Lyalina, T., Shagdarova, B., Zhuikova, Y., Il'ina, A., Lunkov, A., & Varlamov, V. (2023). Effect of Seed Priming with Chitosan Hydrolysate on Lettuce (*Lactuca sativa*) Growth Parameters. *Molecules (*Basel, Switzerland), *28*(4), 1915. https://doi.org/10.3390/molecules28041915

Malik, S. I., Chimbekujwo, K. I., Maku, A. M., Oyewole, Oluwafemi, A., Omoregie, I. P., Adetunji, C. O., Eniola, K. I. T., Yerima, M. B., & Mathew, J. T. (2025). The Application of Nanochitosan Biopesticides as a Replacement to Synthetic Pesticides. In: *Nanochitosan Applications for Enhanced Crop Production and Food Security*. John Wiley & Sons, 89–110. https://doi.org/10.1002/9781394214945.ch4

Maliki, S., Sharma, G., Kumar, A., Moral-Zamorano, M., Moradi, O., Baselga, J., Stadler, F. J., & García-Peñas, A. (2022). Chitosan as a Tool for Sustainable Development: A Mini Review. *Polymers*, *14*(7), 1475. https://doi.org/10.3390/polym14071475

Marques, C., Som, C., Schmutz, M., Borges, O., & Borchard, G. (2020). How the Lack of Chitosan Characterization Precludes Implementation of the Safe-by-Design Concept. *Frontiers in Bioengineering and Biotechnology*, *8*, 165. https://doi.org/10.3389/fbioe.2020.00165

Martins, V. F. R., Pintado, M. E., Morais, R. M. S. C., & Morais, A. M. M. B. (2024). Recent Highlights in Sustainable Bio-Based Edible Films and Coatings for Fruit and Vegetable Applications. *Foods (*Basel, Switzerland), *13*(2), 318. https://doi.org/10.3390/foods13020318

Mathew, J. T., Adetunji, C. O., Inobeme, A., Musah, M., Azeh, Y., Etsuyankpa, M. B., Otori, A. A., Chado, I. M., Shaba, E. Y., Akoji, J. N., Mamman, A., & Oyewole, O. A. (2025b). Application of Nanochitosan for the Biodegradation of Agricultural Wastes. In: *Nanochitosan Applications for Enhanced Crop Production and Food Security*. John Wiley & Sons, 361–382. https://doi.org/10.1002/9781394214945.ch16

Mathew, J. T., Adetunji, C. O., Inobeme, A., Otori, A. A., Musah, M., Azeh, Y., Shaba, E. Y., Muhammad, I. A., Mamman, A., & Oyewole, O. A. (2025a). Relevance of Nanochitosan in Food Sector and Food Packaging Sectors: Current Trends. In: *Nanochitosan Applications for Enhanced Crop Production and Food Security*. John Wiley & Sons, 345–360. https://doi.org/10.1002/9781394214945.ch15

Mathew, J. T., Inobeme, A., Etsuyankpa, M. B., Adetunji, C. O., Shaba, E. Y., Musah, M., Azeh, Y., Jiya, M. J., Tanko, E., Muhammad, A. I., Mamman, A., & Maurice, J. (2024b). Applications of Diatom for Treatment of Industrial Effluents: Waste to Wealth Approach. In: *Marine Biomass: Biorefinery, Bioproducts and Environmental Bioremediation*. Walter de Gruyter GmbH & Co KG, 1–20. https://doi.org/10.1515/9783111353951-001

Mathew, J. T., Inobeme, A., Hussaini, J., Shaba, E. Y., Azeh, Y., Musah, M., Muhammad, E. B., Tanko, M. S., Ismail, H., Adetunji, C. O., Kanwa, A. M., & Mamman, A. (2024a). Properties of Nanomaterials for Corrosion Control. In *Innovations in Nanomaterials-Based Corrosion Inhibitors*: A. Thakur & A. Kumar (Eds.). IGI Global, 74–99. https://doi.org/10.4018/979-8-3693-3088-3.ch004

Mathew, J. T., Inobeme, A., Musah, M., Azeh, Y., Etsuyankpa, M. B., Adetunji, C. O., Shaba, E. Y., Jiya, M. J., Josiah, J. G., Tanko, E., Muhammad, A. I., Mamman, A., & Jemkur, M. (2024c). Utilization of Marine Green Algae for Development of Bioplastics. In: *Marine Biomass: Biorefinery, Bioproducts and Environmental Bioremediation*. Walter de Gruyter GmbH & Co KG, 39–55. https://doi.org/10.1515/9783111353951-003

Mathew, J. T., Musah, M., Azeh, Y., & Muhammed, M. (2023a). Adsorptive Removal of Selected Toxic Metals from Pharmaceutical Wastewater using Fe_3O_4/ZnO Nanocomposite. *Dutse Journal of Pure and Applied Sciences*, 9(4a), 236–248. https://doi.org/10.4314/dujopas.v9i4a.22

Mathew, J. T., Musah, M., Azeh, Y., & Muhammed, M. (2023b). Kinetic Study of Heavy Metals Removal from Pharmaceutical Wastewater Using Geopolymer/Fe_3O_4 Nanocomposite. *Bima Journal of Science and Technology*, *7*(4), 152–163. https://doi.org/10.56892/bima.v7i4.539

Matloob, A., Ayub, H., Mohsin, M., Ambreen, S., Khan, F. A., Oranab, S., Rahim, M. A., Khalid, W., Nayik, G. A., Ramniwas, S., & Ercisli, S. (2023). A Review on Edible Coatings and Films: Advances, Composition, Production Methods, and Safety Concerns. *ACS Omega*, *8*(32), 28932–28944. https://doi.org/10.1021/acsomega.3c03459

Mendes, C. R., Dilarri, G., Forsan, C. F., Sapata, V. M. R., Lopes, P. R. M., de Moraes, P. B., Montagnolli, R. N., Ferreira, H., & Bidoia, E. D. (2022). Antibacterial Action and Target Mechanisms of Zinc Oxide Nanoparticles against Bacterial Pathogens. *Scientific Reports*, *12*(1), 2658. https://doi.org/10.1038/s41598-022-06657-y

Michalik, R., & Wandzik, I. (2020). A Mini-Review on Chitosan-Based Hydrogels with Potential for Sustainable Agricultural Applications. *Polymers*, *12*(10), 2425. https://doi.org/10.3390/polym12102425

Mikušová, V., & Mikuš, P. (2021). Advances in Chitosan-Based Nanoparticles for Drug Delivery. *International Journal of Molecular Sciences*, *22*(17), 9652. https://doi.org/10.3390/ijms22179652

Mohapatra, S., Yutao, L., Goh, S. G., Ng, C., Luhua, Y., Tran, N. H., & Gin, K. Y. (2023). Quaternary Ammonium Compounds of Emerging Concern: Classification, Occurrence, Fate, Toxicity and Antimicrobial Resistance. *Journal of Hazardous Materials*, *445*, 130393. https://doi.org/10.1016/j.jhazmat.2022.130393

Muhammad, W. M., Ibrahim, A. B., Samuel, J. O., Khalifa, A. S., Oyewole, O. A., Adetunji, C. O., Eniola, K. I. T., Yerima, M. B., & Mathew, J. T. (2025). The Role of Nanomaterials in Agriculture as Nanofertilizers. In: *Nanochitosan Applications for Enhanced Crop Production and Food Security*. John Wiley & Sons, Inc., 1–38. https://doi.org/10.1002/9781394214945.ch1

Muteeb, G., Rehman, M. T., Shahwan, M., & Aatif, M. (2023). Origin of Antibiotics and Antibiotic Resistance, and Their Impacts on Drug Development: A Narrative Review. *Pharmaceuticals (*Basel, Switzerland), *16*(11), 1615. https://doi.org/10.3390/ph16111615

Ogundolie, F. A., Adetunji, C. O., Mathew, J. T., Inobeme, A., Ghazanfar, S., Olaniyan, O. T., Ajiboye, M. D., Dauda, W. P., Adetunji, J. B., Akinbo, O., & Adetuyi, B. O. (2023). Intelligent Freshness Indicators in Packaging of Plant-Based Food Products. In: *Evaporative Coolers for the Postharvest Management of Fruits and Vegetables*. Academic Press, 323–331. https://doi.org/10.1016/B978-0-323-89864-5.00008-4

Paladugu, S. R. M., Sreekanth, P. S. R., Sahu, S. K., Naresh, K., Karthick, S. A., Venkateshwaran, N., Ramoni, M., Mensah, R. A., Das, O., & Shanmugam, R. (2022). A Comprehensive Review of Self-Healing Polymer, Metal, and Ceramic Matrix Composites and Their Modeling Aspects for Aerospace Applications. *Materials (*Basel, Switzerland), *15*(23), 8521. https://doi.org/10.3390/ma15238521

Pellis, A., Guebitz, G. M., & Nyanhongo, G. S. (2022). Chitosan: Sources, Processing and Modification Techniques. *Gels (*Basel, Switzerland), *8*(7), 393. https://doi.org/10.3390/gels8070393

Rabajczyk, A., Zielecka, M., Popielarczyk, T., & Sowa, T. (2021). Nanotechnology in Fire Protection-Application and Requirements. *Materials (*Basel, Switzerland), *14*(24), 7849. https://doi.org/10.3390/ma14247849

Riseh, S. R., Vatankhah, M., Hassanisaadi, M., & Kennedy, J. F. (2023). Chitosan-Based Nanocomposites as Coatings and Packaging Materials for the Postharvest Improvement of Agricultural Product: A Review. *Carbohydrate Polymers*, *309*, 120666. https://doi.org/10.1016/j.carbpol.2023.120666

Rosli, N. A., Teow, Y. H., & Mahmoudi, E. (2021). Current Approaches for the Exploration of Antimicrobial Activities of Nanoparticles. *Science and Technology of Advanced Materials*, *22*(1), 885–907. https://doi.org/10.1080/14686996.2021.1978801

Sanità, G., Carrese, B., & Lamberti, A. (2020). Nanoparticle Surface Functionalization: How to Improve Biocompatibility and Cellular Internalization. *Frontiers in Molecular Biosciences*, *7*, 587012. https://doi.org/10.3389/fmolb.2020.587012

Schneider, G., Vieira, L. G., de Carvalho, H. E. F., de Sousa, Á. F. L., Watanabe, E., de Andrade, D., & Silveira, R. C. C. P. (2023). Textiles Impregnated with Antimicrobial Substances in Healthcare Services: Systematic Review. *Frontiers in Public Health*, *11*, 1130829. https://doi.org/10.3389/fpubh.2023.1130829

Seidi, F., Jouyandeh, M., Taghizadeh, M., Taghizadeh, A., Vahabi, H., Habibzadeh, S., Formela, K., & Saeb, M. R. (2020). Metal-Organic Framework (MOF)/Epoxy Coatings: A Review. *Materials* (Basel, Switzerland), *13*(12), 2881. https://doi.org/10.3390/ma13122881

Shchukina, E., & Shchukin, D. G. (2019). Nanocontainer-Based Active Systems: From Self-Healing Coatings to Thermal Energy Storage. *Langmuir: The ACS Journal of Surfaces and Colloids*, *35*(26), 8603–8611. https://doi.org/10.1021/acs.langmuir.9b00151

Signorini, L., Marenzi, G., Facente, A., Marrelli, B., Marano, R. M., Valletta, A., Pacifici, L., Gasparro, R., Sammartino, G., & Severino, M. (2023). Critical Overview on Pure Chitosan-Based Scaffolds for Bone Tissue Engineering: Clinical insights in Dentistry. *International Journal of Medical Sciences*, *20*(12), 1527–1534. https://doi.org/10.7150/ijms.87978

Singh, R., Dutt, S., Sharma, P., Sundramoorthy, A. K., Dubey, A., Singh, A., & Arya, S. (2023). Future of Nanotechnology in Food Industry: Challenges in Processing, Packaging, and Food Safety. *Global Challenges* (Hoboken, NJ), *7*(4), 2200209. https://doi.org/10.1002/gch2.202200209

Stasińska-Jakubas, M., & Hawrylak-Nowak, B. (2022). Protective, Biostimulating, and Eliciting Effects of Chitosan and Its Derivatives on Crop Plants. *Molecules* (Basel, Switzerland), *27*(9), 2801. https://doi.org/10.3390/molecules27092801

Stefanowska, K., Woźniak, M., Dobrucka, R., & Ratajczak, I. (2023). Chitosan with Natural Additives as a Potential Food Packaging. *Materials* (Basel, Switzerland), *16*(4), 1579. https://doi.org/10.3390/ma16041579

Suvarna, V., Nair, A., Mallya, R., Khan, T., & Omri, A. (2022). Antimicrobial Nanomaterials for Food Packaging. *Antibiotics* (Basel, Switzerland), *11*(6), 729. https://doi.org/10.3390/antibiotics11060729

Teixeira-Costa, B. E., & Andrade, C. T. (2021). Chitosan as a Valuable Biomolecule from Seafood Industry Waste in the Design of Green Food Packaging. *Biomolecules*, *11*(11), 1599. https://doi.org/10.3390/biom11111599

Thambiliyagodage, C., Jayanetti, M., Mendis, A., Ekanayake, G., Liyanaarachchi, H., & Vigneswaran, S. (2023). Recent Advances in Chitosan-Based Applications: A Review. *Materials* (Basel, Switzerland), *16*(5), 2073. https://doi.org/10.3390/ma16052073

Wang, K., Zhang, F., Xu, K., Che, Y., Qi, M., & Song, C. (2023). Modified Magnetic Chitosan Materials for Heavy Metal Adsorption: A Review. *RSC Advances*, *13*(10), 6713–6736. https://doi.org/10.1039/d2ra07112f

Wang, W., Meng, Q., Li, Q., Liu, J., Zhou, M., Jin, Z., & Zhao, K. (2020). Chitosan Derivatives and Their Application in Biomedicine. *International Journal of Molecular Sciences*, *21*(2), 487. https://doi.org/10.3390/ijms21020487

Wrońska, N., Katir, N., Nowak-Lange, M., El Kadib, A., & Lisowska, K. (2023). Biodegradable Chitosan-Based Films as an Alternative to Plastic Packaging. *Foods* (Basel, Switzerland), *12*(18), 3519. https://doi.org/10.3390/foods12183519

Yan, D., Li, Y., Liu, Y., Li, N., Zhang, X., & Yan, C. (2021). Antimicrobial Properties of Chitosan and Chitosan Derivatives in the Treatment of Enteric Infections. *Molecules* (Basel, Switzerland), *26*(23), 7136. https://doi.org/10.3390/molecules26237136

Zanca, C., Carbone, S., Patella, B., Lopresti, F., Aiello, G., Brucato, V., Carfì Pavia, F., La Carrubba, V., & Inguanta, R. (2022). Composite Coatings of Chitosan and Silver Nanoparticles Obtained by Galvanic Deposition for Orthopedic Implants. *Polymers*, *14*(18), 3915. https://doi.org/10.3390/polym14183915

Zorraquín-Peña, I., Cueva, C., Bartolomé, B., & Moreno-Arribas, M. V. (2020). Silver Nanoparticles against Foodborne Bacteria. Effects at Intestinal Level and Health Limitations. *Microorganisms*, *8*(1), 132. https://doi.org/10.3390/microorganisms8010132

11 Chitosan and Its Derivatives as Potential Biomaterials for Medical and Food Technologies

Osarenkhoe Omorefosa Osemwegie[1], Damilare Rotimi[2], Charles Obiora Nwonuma[2], Marvellous Amarachi Acho[2], Adeyemi Adetunde Adeyanju[1], Bamidele Sunday Fakinle[3], and Charles Oluwaseun Adetunji[4]

[1]Department of Food Science and Microbiology, Landmark University, Omu Aran, Kwara State

[2]Department of Biochemistry, Landmark University, Omu Aran, Kwara State, Nigeria

[3]Department of Chemical Engineering, Landmark University, Omu Aran, Kwara State, Nigeria

[4]Department of Microbiology, Edo University, Uzaure, Edo State, Nigeria

11.1 INTRODUCTION

The increasing bioeconomic applications of chitosan polymer coupled with the need to improve its end-use biofunctions or modify it to resolve a specialized need, and concomitant quests for the improvement of its properties for adaptable value-added applications have resulted in the generation of its derivatives. Thermostability, elasticity, polydispersity, crystallinity, permeability to gas, solubility, biodegradability, and mechanical strength are some of the chitosan properties crucial to their biotechnology applications. Consequently, studies on chitosan are rapidly focusing on benign techniques (copolymerization, depolymerization, ultrasonication, chemical coupling, and oligomerization) to achieve numerous derivatives to overcome the limitation of chitosan in the food and medical sectors. Even though chitosan derivatives have attracted a higher percentage of patents in chitosan science, their processing may be hinged on the manipulation/modification of its functional group arrangements without any significant change to its skeleton. This is achieved by using different characteristic ligands ($-NH_2$, $-OH$) to generate numerous chemical reactions like quaternization, ketonization, aldehydization, metal chelation, oxidation of the C-C backbone, alkylation, carboxymethylation, sulfonation, grafting acetylation, thiolation, phosphorylation, and hydroxyethylation (Brasselet et al. 2019). In addition, chitosan derivatives may be formed by intrinsic ionic or hydrophobic interactions of chemical groups along the chitosan structural backbone resulting in cross-linking or molecular entanglement. Although a change influences these mechanisms in the local environment, they invariably help to create both medium (190–310 kDa) and lower (50–190 kDa) molecular weight (Mw) derivatives with more amenable properties. Therefore, chitosan in its structural variations (polymer, oligomer, native, or chemically modified) has been made more amenable for use as a bioadhesive, biosorbent, biopreservative, texturing agent, delivery agent, immobilizer, tissue scaffold, and chelation through numerous

DOI: 10.1201/9781003355458-11

processing. Chitosan derivatives are well reported in the body of literature, and with a paucity of information on their various classification, they have been arbitrarily classified based on their chemistry (trimethyl, glycol, carboxymethyl, alkyl, phosphorylated, mannosylated, sulfonated, deoxycholic acid-modified, thiolate, ethylenediaminetetraacetic acid–chitosan, acrylated, etc.) as well as utility (nanocarrier, preservation, tissue scaffold, and film) (Figure 11.1). Also, the underlying biological mechanism that defines the functional property of each derivative of chitosan, the relative impact of the numerous processes involved in the synthesis of the derivatives, and the influence of Mw, as well as the degree of deacetylation (DD) on their functionality, are yet to be fully understood (Kou et al. 2022).

Chitosan and its derivatives have gained more versatile applications in the food and medical industries. This follows the various structural-based techniques emerging in the last few decades for generating chitosan derivatives for a limited range of biotechnological applications. Because of its malleable properties and non-toxic derivatives, chitosan became a promising biotechnology material that extends the shelf life of fresh and minimally processed foods worldwide. The acquiescent polymer ranks among other bioresources used in food processing as clarifying agents in beverages (fruit juice, wines, beer, water, and tea), preservative agents in the form of antibiodeteriorant (antimicrobials), antioxidant, and packaging/biofilm material (Abd El-Hack et al. 2020; Elzamly et al. 2021, Mondal et al. 2022; Patil et al. 2022), and adsorbent, a replacement for activated carbon, in the removal of impurities from water, foods, and beverages (Omer et al. 2022). Meanwhile, Rocha et al. (2017) observed variations in the efficiency of chitosan derivatives as a biopreservative due to their interactive propensity with compounds in the food matrix and Mw factor. For example, they observed that chitosan showed higher preservative efficiency in apple fruit juice than in dairy milk due to the limiting electrostatic effect caused by its chemical reaction with the protein. Ma et al. (2021) affirmed chitosan potential and its grafted-derived products as an emulsifying agent and animal feed additive (Ayman et al. 2022). While the biological mechanism of chitosan or its derivatives as a feed additive is still enigmatic, it could be attributed to its capacity to function as a prebiotic or substrate for the proliferation of microorganisms beneficial to controlled nutrient delivery, bioavailability, and digestion (Moine et al. 2021; Chen et al. 2022). Chitosan derivatives are enzymatically digested by several alimentary proteases like papain, lysozyme, pepsin, and chitinase to produce crucial physiological intermediates (glycosaminoglycan and glucosamine) that feedstock the animals' metabolic cycles (Aranaz et al. 2009; Jain et al. 2013). This, along with their ability to influence, antimicrobial, salivary, and gut properties, may have accounted for their use, even in the grafted form, as an animal feed additive and gave attraction to their more recent exploration as a functional food or nutraceutical (Gallo et al. 2016; Alkabli 2022; Crini, 2022; Manocha et al. 2022).

Biomedically, chitosan and its derivatives are valued as an adjuvant, mucoadhesive, analgesic, and permeability enhancer, as well as for their use in the control of blood loss (hemostatic effect) or healing of open wounds (especially burns), tissue-engineered medical products coupled with tissue regeneration, and nutraceuticals (Ways et al. 2018). Even though dose-dependent in vivo toxicity of chitosan derivatives has been sporadically reported in the body of scientific literature (Zubareva et al. 2017; Kazemi et al. 2019; Zou et al. 2022) and linked to their degree of deacetylation (DD) coupled with their polymer chain length or Mw, they have become relevant as antitumour, antimicrobial, anticholesterolemic, antioxidant, and antidiabetic in medicine due to their broad spectrum of biological properties (Li et al. 2020; Crini 2022; Priyanka et al. 2022; Xia et al. 2022; Zhang et al. 2022c). By extension, chitosan derivatives have found attraction in controlled drug delivery (intranasal, transdermal, and site-specific) and therapeutic (anti-SARS-CoV-2) applications, as well as in the creation of super-dispersive or disintegrating pharmaceutical matrices and the pharmacological treatment of glaucoma-related neurodegeneration (Al-Remawi 2015; Luo et al. 2020; Kedir et al. 2022; Tan et al. 2022).

11.2 CHITOSAN AND CHITOSAN DERIVATIVES

A foremost attraction to chitosan and its derivative for a wide range of uses in industries is its adaptable property. Chemical functionalization of chitosan is popular relative to biological and physical means of adapting the primary chitosan for new biotechnological applications. These chemical procedures enable the production of various chitosan derivatives with changed but specialized features targeting specific end-use in the biotechnological, biomedical, and pharmaceutical industries. The chemical, mechanical, and physical characteristics influencing the antibacterial, antioxidant, and solubility of chitosan are altered by graft copolymerization, depolymerization, esterification, metal chelation, deacetylation, and enzymatic reactions for the desired end-use application. The functionalization process also entails structural modification by the addition of small functional groups such as acryl, alkyl, carboxymethyl (Alves and Mano 2008), and carboxyl or phenolic compounds (caffeic acid, catechin, ferulic acid, or gallic acid). This facilitates the improvement of chitosan properties like solubility at a wider pH range without affecting its cationic character, antimicrobial, antipesticidal, and antioxidant activities (Curcio et al. 2009; Aytekin et al. 2011; Aljawish et al. 2015; Zhang et al. 2018). As a result, different techniques of chemical functionalization targeting different functional groups will provide enhanced derivatives with diverse physicochemical, biochemical, and biological activities that are value-added in bioeconomic industries (Tungtong et al. 2012; Zhang and Jin 2012). Information on how chitosan end-use functions as an antioxidant (Chen et al. 2015; Schreiber et al. 2013; Woranuch et al. 2015), antimicrobial (Çelebi et al. 2013; Lee and Je 2013), adhesive (Kim et al. 2013; Yin et al. 2013) and adsorbent (Auta and Hameed 2013; Yang et al. 2013) relates to its chemical skeleton remains scanty. However, chemical derivatives may also be modified by acylation, alkylation, hydroxy alkylation, carboxy quaternization, phosphorylation, sulfation, and thiolation (Figure 11.1). Similarly, they may be chemically fabricated or copolymerized using compounds such as methyl methacrylate (MMA) and potassium persulfate (Brasselet

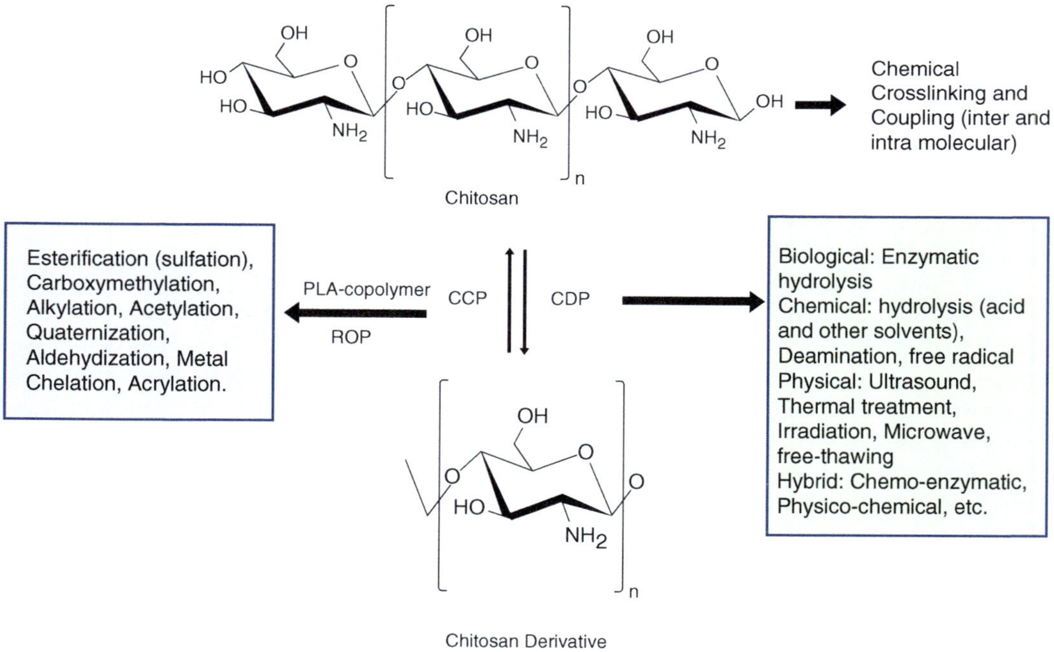

FIGURE 11.1 Schematic illustration of chitosan modification methods utilized to produce chitosan derivatives. CDP, chitosan depolymerization; CCP, chitosan copolymerization; ROP, ring opening polymerization; PL-copolymer, polylactide copolymerization.

et al. 2019), ascorbic acid (Vc) and hydrogen peroxide (H_2O_2) redox pair under an inert atmosphere (Li et al. 2020; Panda et al. 2021), and carbodiimide-mediated coupling method (Moreno-Vásquez et al. 2017). Radiation methods such as electron beam or microwave irradiations (Branca et al. 2013), thermo-treatment, and ultrasound technology are some physical techniques of producing chitosan derivatives (Yoshikawa et al. 1998).

11.3 CHEMICAL NATURE AND END-USE FUNCTION OF CHITOSAN DERIVATES

Many processes are studied to create chitosan derivatives that can be adapted for specific or broad-spectrum use. However, these processes are fraught with difficulties such as functional group interference, which may result in unwanted products. This may be prevented if protective substances are used to mask the functional groups that interfere with the reactions (Carey and Sundberg 2007). Consequently, an excellent protective group reacts selectively with specific functional groups to develop a protected and stable substrate for the next stage of the reaction. It may also be effectively removed to offer a product with a high yield (Wuts and Greene 2006). Chitosan has three nucleophilic functional groups: an –OH group at the C-6 position, a secondary –OH group at the C-3 position, and a $-NH_2$ group at the C-2 position. The highly reactive nature of the hydroxyl and amino moieties is implicated in chitosan's organic chemical reactions to create new pendular moieties in the amino group. For example, the triphenyl methyl group can be introduced at the primary –OH group leaving the secondary –OH unprotected. Some chitosan derivatives are presented based on the reaction nature that led to their functionalization.

11.3.1 Acylated Chitosan

The chitosan $-NH_2$ and –OH groups are opened to ester and amide reactions with an organic acid anhydride and acid chloride, respectively. The reaction temperature and catalyst selection are critical factors in producing acylated chitosan (Hu et al. 2007). Adding lipids or aromatic acyl groups of varying Mws to acylated chitosan often increases solubility in water or an organic solvent. *N*-succinylated chitosan was created by adding succinyl to the chitosan glucosamine unit at the *N*-position. Active groups such as COOH, C_2–NH_2, C_3–OH, and C_6–OH are present in *N*-succinylated chitosan molecules, facilitating their adsorption of divalent copper ions, particularly in contaminated soils. Acylated chitosan provides a sustainably prolonged releasing effect on beneficial compounds and is easy to prepare. It's a novel adjunct component in producing biopackaging due to its low solubility in water (Zhao et al. 2018).

11.3.2 Alkylated Chitosan

Chitosan alkylation may include either amino ($-NH_2$) or hydroxyl (–OH) groups on carbon 3, 6, or both. In contrast to hydroxyl-containing ones, chitosan having amino groups proceeds more rapidly in their reactions, providing superior protection for functional groups of interest (Ma et al. 2008). As a result, *N*-alkylated chitosan derivatives are produced by amino group alkylation to synthesize a series of *N*-alkylated chitosan molecules (Ma et al. 2008). Hemostatic activity testing showed that *N*-alkylated chitosan was functionally superior to unaltered chitosan. Adding the alkyl groups weakens the hydrogen bond between chitosan molecules, making the modified chitosan more promising in improving poultry and seafood shelf life and preservation. This is because the reaction reduces nitrogen concentration, weakens the covalent bonds, and predisposes the molecule to react with other compounds by grafting, making it more biocompatible and biodegradable (Huang et al., 2017).

11.3.3 Carboxyl Alkyl Chitosan Derivatives

Carboxylated chitosan reactions typically proceed through the $-NH_2$ and –OH to yield carboxyl-modified chitosan derivatives. Glyoxylic acid may be used to complete the carboxylation process. Carboxymethyl chitosan (CMC) was made by treating chitosan with monochloroacetic acid under various circumstances. Water solubility is influenced by the circumstances of modification and the degree of carboxymethylation (Mourya and Inamdar 2008). In addition to increasing chitosan's water solubility, carboxylation creates chitosan derivatives with amphiphilic $-NH_2$ and –COOH groups, which are more able to interact with water. These derivatives are used in various cosmetics, food, and agricultural medical industries. This is due to their high-water solubility, high surface activity, ability to form films, high moisture retention, high moisture absorption, and antioxidant, antibacterial, and other biological properties (Wang et al. 2020a).

Chitosan treated with sodium hydroxide (NaOH) produces CMC. This is a water-soluble amphiprotic ether chitosan derivative with –COOH and $-NH_2$ (Figure 11.2). CMC exerts various benefits compared to unmodulated chitosan such as more effective Cu^{2+} complexing and a larger surface area. Furthermore, changing key reaction parameters, such as time, solvent systems, or temperatures, might result in distinct derivatives, resulting in N-CMC, O-CMC, or N, O-CMC products (Chen et al. 2008). At ambient temperature, O-CMC may be produced; at higher temperatures, N, NCMC, or O-CMC can be formed (Joseph et al. 2000). Muzzarelli et al. (1990) examined the bactericidal efficacy of *N*-carboxybutyl chitosan against 298 pathogen strains. They observed that

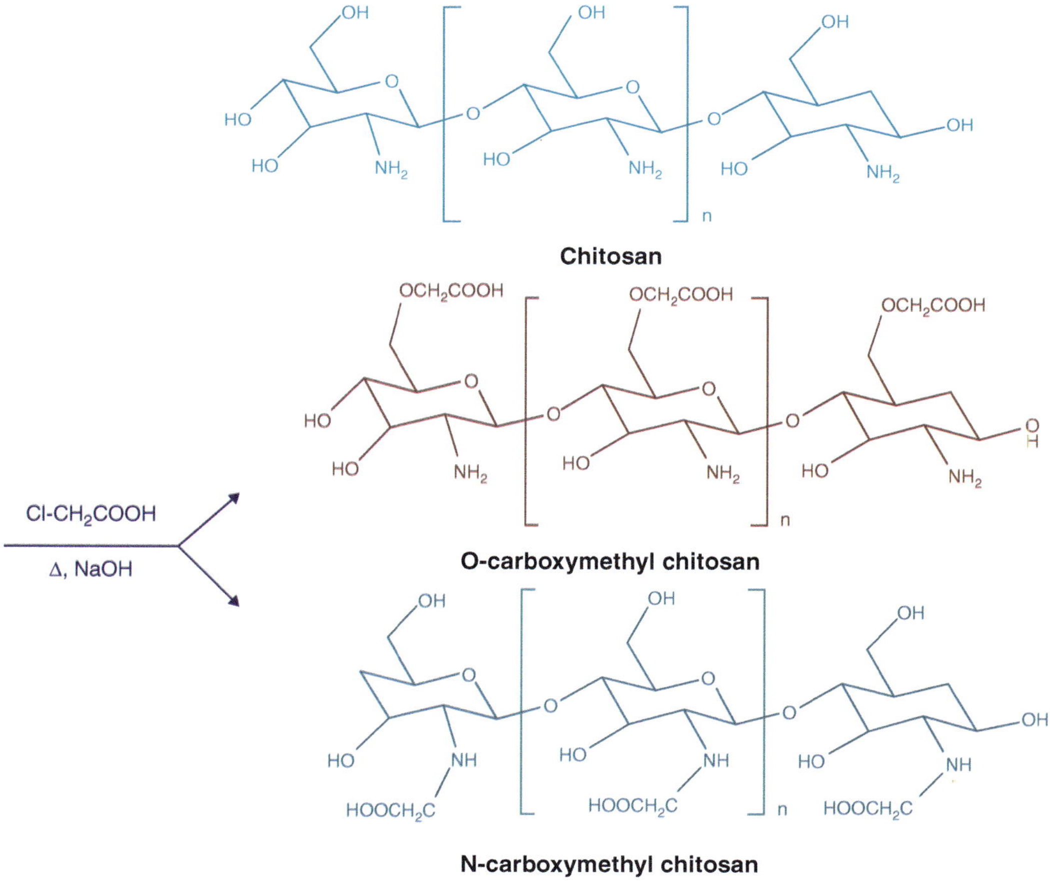

FIGURE 11.2 Schematic illustration of CMC synthetic pathway.

the chemical had antibacterial, antifungal, and inhibiting effects on other pests. These biological attributes enhance their application as antimicrobials and preservatives in food processing.

11.3.4 Esterified and Etherified Chitosan

Some inorganic acid oxygens on the chitosan molecule are involved in the esterification process. Sulfated chitosan may be used instead of heparin or heparin sulfate for various biological purposes (Dimassi et al. 2018). This reaction enhances the molecule's application as an anticoagulant, antiviral agent, strategic binding protein, and catalyst in osteogenic differentiation. Sulfated chitosan has the competent regulatory mechanism as heparin, while esterified chitosan demonstrates high fibre strength.

A comparable alkylating agent is produced in the etherification process of chitosan through its –OH group (alkyne derivatives). Next, a deacetylation procedure is performed on the polymers to produce chitosan ether derivatives. Chitosan ether derivatives are not cytotoxic or capable of causing considerable irritation and fibroblast growth but can cause delayed hypersensitivity and inflammatory responses (Zhao et al. 2018). Hydroxyethyl chitosan is appropriate for medical use due to its strong biocompatibility and biodegradability. They are non-toxic and may be used as a natural textile finishing product due to their softening and bacteriostatic properties. Additionally, hydroxyethyl chitosan inhibits *Escherichia coli* and other common bacteria, making it attractive as a preservative in the cosmetics industry.

11.3.5 Quaternary Chitosan Derivatives

Chitosan synthesized with a permanent positive charge caused by the insertion of a new functional group enhances water solubility and antibacterial activity at neutral pH better than the unmodulated chitosan (Sonia and Sharma 2011). One of the most promising chitosan derivatives is *N,N,N*-trimethyl chitosan (TMC). TMC has a persistent positive charge on the quaternized amino groups, making it soluble in neutral and basic environments (Rúnarsson et al. 2007) (Figure 11.3). Domard et al. (1986) first described the synthesis of TMC, a process now known as the "Domard reaction". In basic circumstances, methyl iodide (MeI) was utilized as a reagent with sodium iodide (NaI) as a catalyst and *N*-methyl pyrrolidone (NMP) as a solvent. Surprisingly, TMC is more active against bacteria than chitosan. Furthermore, due to its promising composition, it may function predictably in enhancing drug absorption and delivery, which supports its applications in medical and food technologies (Benediktsdóttir et al. 2014).

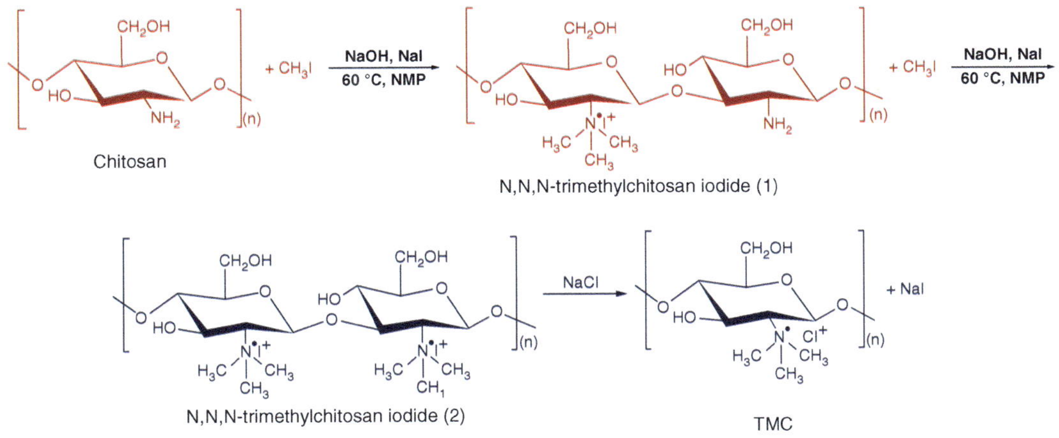

FIGURE 11.3 An illustration of the production stages of trimethyl chitosan (TMC).

The amino, hydroxymethyl (–CH_2OH), and hydroxyl groups in chitosan are active in quaternization processes. Chitosan is often quaternized with methyl iodide, ammonium chloride, and other compounds (Xu et al. 2011; Madera-Santana et al. 2018). Chitosan and quaternary epoxides synthesize the derivatives using different alkyl chains on the epoxides. This produces actionized derivatives with varying degrees of hydrophilicity and hydrophobicity (Mourya and Inamdar 2008), including *N*-2-hydroxypropyl dimethyl ethyl ammonium chloride chitosan (*N*-2-HFCC), *N*-2-hydroxypropyl trimethyl ammonium chloride chitosan (*N*-2-HACC), and *O*-20-hydroxypropyl trimethyl ammonium chloride chitosan (*O*-20-HACC), which were all produced by our team (*O*-20-HACC). Compared to chitosan, *N*-2-HACC demonstrated superior stability and solubility, enhanced antibacterial activity, and decreased toxicity (Jin et al. 2013).

11.3.6 Phosphorylated Chitosan

According to Sakaguchi et al. (1981), phosphorylated chitosan is made by heating chitosan with orthophosphoric acid and urea in *N,N*-dimethyl formamide (DMF). The urea is then added to the reaction media to promote the reaction (Figure 11.4).

Phosphorylated chitosan is also formed by chitosan reaction with phosphorous pentoxide in methanesulfonic acid (Nishi et al. 1984, 1986). The response of chitosan in phosphorous pentoxide (P_2O_5)–methanesulfonic acid is very efficient. The reaction of chitosan with P_2O_5 was found to be water-soluble phosphorylated chitin and chitosan with a high degree of substitution (DS). In this method, methanesulfonic acid acts as a good solvent for chitosan and an efficient catalyst for the esterification reaction (Figure 11.5).

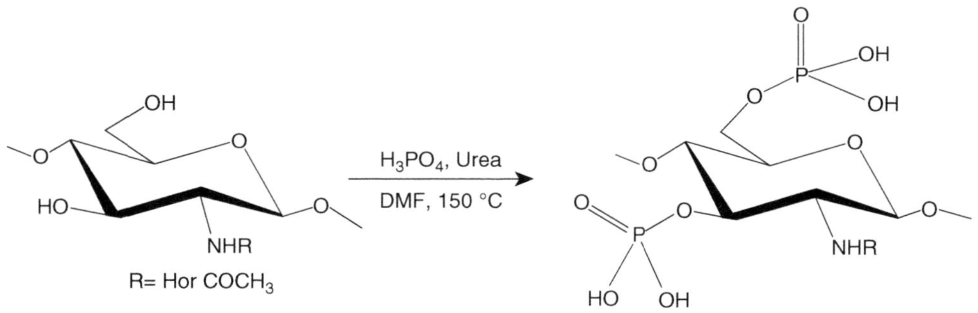

FIGURE 11.4 Synthesis of phosphorylated chitosan by H_3PO_4/urea/DMF method.

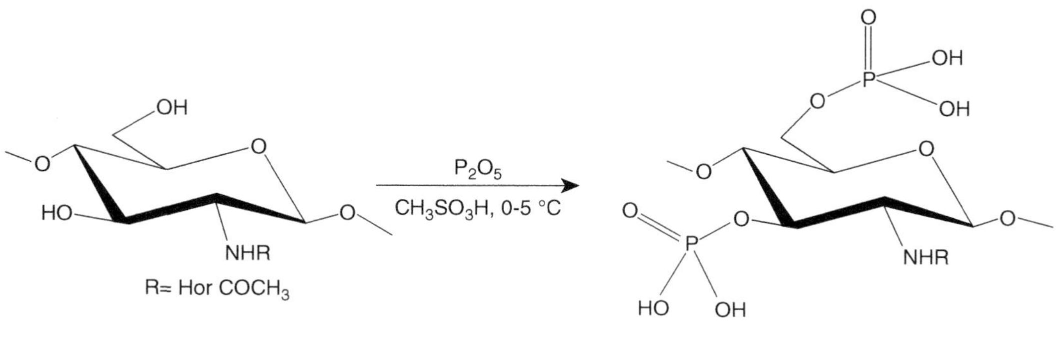

FIGURE 11.5 Synthesis of phosphorylated chitosan by P_2O_5/CH_3SO_3H method.

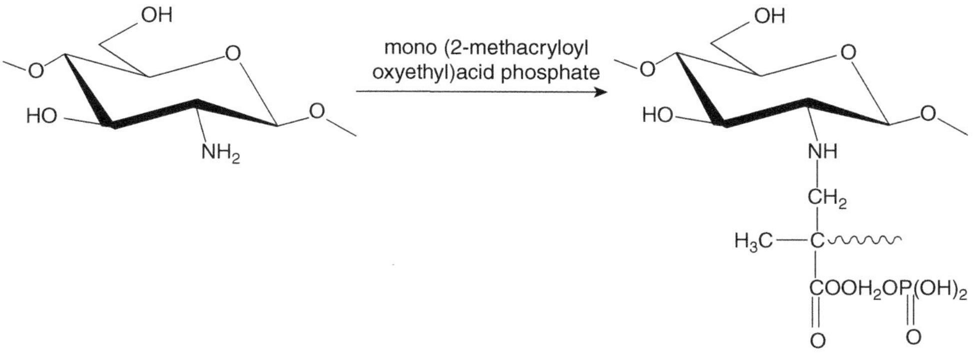

FIGURE 11.6 Chemical illustration of phosphorylated chitosan synthesis by grafting method.

Chitosan may be phosphorylated to synthesize a mono (2-methacryloyl oxyethyl) acid phosphate chitosan complex using the grafting method (Figure 11.6). The phosphorylated chitosan produced has enhanced antibacterial and zwitterionic qualities (Jung et al. 1999).

Goy et al. (2009) and Li et al. (2020) observed that chitosan inhibits bacteria (Gram-negative and Gram-positive) and numerous conidial fungi. Pathogen type, local environmental factors, chitosan source, concentration, and structural properties are crucial determinants of its antimicrobial efficiency (Hosseinnejad and Jafari 2016; Rabea et al. 2003). Chitosan demonstrates better antimicrobial action at pH 6.5 due to the high number of amino groups with positively charged $-NH^{+3}$ in the polymer, which interacted electrostatically with the membrane components from harmful microbes to induce antimicrobial activity (Kong et al. 2010). The antimicrobial activity of chitosan and chitosan derivatives is receiving attention and applied in biotechnology than their antioxidant property. Chemical functionalization of chitosan has been used to improve the antioxidant bioactivity of several chitosan derivatives. The amino and two hydroxyl groups of chitosan monomers are reactive with free radicals, which demonstrates the underlying mechanism of its free radical-scavenging capability (Feng et al. 2008; Wei et al. 2019). Sun et al. (2007) reported that the antioxidant activity of chitosan is substantially a function of its Mw. As Mw decreases, chitosan antioxidant activity increases, indicating that high Mw chitosans might have negligible antioxidant effects, especially against hydroxyl and superoxide radicals. The reason for this observation is linked to shorter chains that are less likely to form intramolecular hydroxyl bonds, leading to more activated amino and hydroxyl groups with less free radical-scavenging roles. Although chitosan and its derivatives are seldom reported as toxic and harmful as demonstrated by their application versatility in food processing and wound therapy, they are also used as dietary fat, in encapsulation or coating for the controlled release and delivery of bioactivity to target sites, or as an excipient (Nagy 2018). Chitosan derivatives have also shown great potential as binders in pharmaceutics and nutraceuticals, particularly in humans for weight reduction and anticholesterolemic effects because they bind and absorb fat (Kaur and Dhillon 2014). To show the weight-loss potential of chitosan, Mhurchu et al. (2004) conducted a randomized controlled experiment on 250 participants. They found that those who took 3 g of chitosan daily for 24 weeks dropped more body weight than individuals who were given a placebo, demonstrating the polymer's clinical importance. The findings also showed that pills formulated from chitosan or its derivatives might help to curb obesity in many parts of the world (Chew and Brownlee 2018).

11.4 TECHNOLOGIES ASSOCIATED WITH CHITOSAN DERIVATIVE APPLICATION

Many synthesis methods exist for modifying chitosan into different uniquely functional derivatives that may be used alone or in combination. These methods include hydrothermal carbonization, solvothermal, carbonization, solvent evaporation/solution cast method, microwave-assisted treatment, pyrolysis, and chemical vapour disposition. Interestingly, these chitosan-based functionalization processes are motivated by the ever-growing need to substitute synthesized harmful industrial chemical compounds with safer, environmentally benign, and reactively stable biomaterials including chitosan. The growing interest in adapting and extending chitosan to various applications also accounted for the rapidly emerging technologies.

11.4.1 Hydrothermal Carbonization (HTC)

A complex pyrolytic process called carbonization is used to raise the elemental carbon content of biological materials and improve their functional or application qualities. It is followed by condensation, isomerization, hydrogenation, and hydrogen-transfer events, and it is a method popular for synthesizing chitosan-based *N*-doped carbonaceous materials. This is one of the methods used for converting chitosan to solid carbonized material. The HTC method is usually conducted at relatively low temperatures and in a neutral or acidic medium, making it a sustainable and reliable process to produce chitosan derivatives destined for packaging or bioplastic production. The use of the HTC method for the modification of chitosan into derivatives underpins its multifunctionality as a versatile biotechnological material with interesting parameters like high recyclability, porosity, electrolyte, and CO_2 properties. Ranging the reaction temperatures and heat-exposure time during the HTC process is a likely sustainable option for optimizing the end function of HTC chitosan and its derivatives. It also theoretically indicates that HTC is expandable and amenable to freeze-dry, microwave, and nanoparticle-assisted technologies (Adolfsson et al. 2020; Wang et al. 2021).

11.4.2 Chemical Vapour Deposition Method

One of the most recent techniques for creating *N*-doped carbon composites based on chitosan is chemical vapour deposition (Kumar et al. 2018). Using this technique, a solid material is deposited from a vapour phase by breaking down a carbon precursor combined with nitrogen gas on a metal catalyst's surface. The process applies to fabricating food-valuable chitosan-based coatings or film patterns with chitosan derivatives' physical structures, elasticity, and porosity, making it biotechnologically versatile in the biomedical, pharmaceutical, and food industries (Chang et al. 2022). The robust functional outcome of the derivatives depends on reaction factors such as substrate material, temperature, solvent volatility, gas mixture composition, gas flow pressure, and materials with desired physicochemical features (Liu et al. 2022).

11.4.3 Solvothermal Method

The solvothermal method is a chemical reaction that takes place in an organic solvent rather than water at a temperature higher than the solvent's boiling point, such as the hydrothermal method, which does not require the use of a solvent (Mruthunjayappa et al. 2022). The reaction between the solid polymer and the organic solvent is catalysed by this condition. Also, the solvothermal method has been used to create carbon-coated *N*-doped TiO_2 nanostructures (CTS-TiO_2) that are photocatalytically active. This chitosan has been utilized as a source of nitrogen and carbon. As the chemical reaction progresses, a suspension of chitosan and titania precursor (titanium(IV) isopropoxide) is formed in a sealed vessel containing ethanol and glacial acetic acid. To produce CTS-TiO_2, the resultant suspension was subjected to solvothermal treatment for 12 hours at 1800°C (Yan et al. 2013).

11.4.4 Graft Copolymerization

There are various techniques adopted to tailor the surface characteristics of chitosan to desired properties. However, graft copolymerization of synthetic and natural polymers with chitosan ranks as one of the most widely used strategies (Kumar et al. 2020). Chitosan and CMC have been grafted with methacrylic acid by using ammonium persulfate (APS) as the initiator in aqueous media. After grafting, the chitosan derivatives exhibited improved solubility in water (Figure 11.6a) (Xie et al. 2002; Sun et al. 2004). The copolymerization of aniline with chitosan in the presence of APS produced films and fibres with effective protonic conductivity (Figure 11.6b) (Yang et al. 1989). Different initiators, such as potassium persulfate (PPS), ceric ammonium nitrate (CAN), thiocarbonation potassium bromate (TCPB), potassium diperiodatocuprate(III) (PDC), 2,2-azobisisobutyronitrile (AIBN), acrylonitrile, 2-hydroxyethyl methacrylate, and ferrous ammonium sulfate (FAS), have been tried as initiators in graft copolymerization (Olteanu 2007; Sanchez-Salvador et al. 2021).

Polyethylene glycol (PEG) is an appropriate graft-forming polymer because of its low toxicity, strong biocompatibility, biodegradability, and solubility in water and organic solvents. Chitosan hydroxyl groups have been used to create PEG grafted with chitosan (Figure 11.6c). Water and aqueous solutions with a broad pH range could dissolve the novel substance (Gorochovceva and Makuška 2004).

11.4.5 Carboxymethylation

The low solubility of chitosan is a setback to its wide applications even though modification techniques have offered means of improving their solubility. Hence, carboxymethylation is one of the modification procedures widely studied, with an underlying mechanism that can entail chitosan dispersion in an aqueous solution of isopropanol and NaOH while being stirred magnetically and kept at room temperature. A mixture of monochloroacetic acid and isopropanol is then added to the resulting suspension (De Abreu and Campana-Filho 2005). Since carboxymethyl chitosan is an amphoteric polymer, the pH affects how soluble it reacts. Even though *O*- and *N*-carboxymethylation may simultaneously occur during the reaction (Figure 11.7), only one of them can be produced under carefully regulated reaction parameters (such as temperature and stoichiometry) (Rinaudo 2006) (Figure 11.8).

11.5 APPLICATION OF CHITOSAN DERIVATIVES IN THE PHARMACEUTICAL AND MEDICAL ECONOMY

Chitosan and its derivatives have been used in food, pharmaceutical, and medicinal applications for several decades. They are used as excipients or carriers in pharmaceutical formulations and delivery systems. Polymers, micelles, liposomes, and nanoparticles are some of the few examples of pharmacological carriers that have garnered considerable interest in recent years (Gindy and Prud'homme 2009). One of the main benefits of their use is to increase the effectiveness and safety of the medication by potentially integrating either hydrophobic or hydrophilic active chemicals depending on the nature of the carrier. Additionally, they may act as ligands to improve drug bioavailability and targeting, as well as increase the stability of the treatments against chemical and enzymatic degradation (Wang et al. 2011). Small active compounds, proteins, genes, vaccines, and oligonucleotides may be adsorbed, encapsulated/coated either electrostatically and/or covalently bonded to the surface or inside the matrix of these carrier systems for controlled release of bioactivity or transport to action sites (Ibraheem et al. 2014). Effective and safe medication delivery has revolutionized polysaccharide-based drug delivery systems in the medical field. As regulators of the pace of bioactivity and delivery of drug actions at a specified location in the body, chitosan-based biomaterials have garnered considerable attention among diverse polymers (Imam et al. 2021). Chitosan-based polymers are cationic polysaccharides studied for their possible use as adjuvant and encapsulation material in human vaccinology due to their attractive biocompatibility and biodegradability (Wu et al. 2020). Even though vaccines may have been linked to some clinical safety concerns, they also exhibit the tendency to efficiently induce immune responses that are highly target-specific and capable of preventing illness. Encapsulating live attenuated pathogens or fully inactivated pathogens using chitosan-based derivatives in vaccinology is attracting attention as a partial solution to

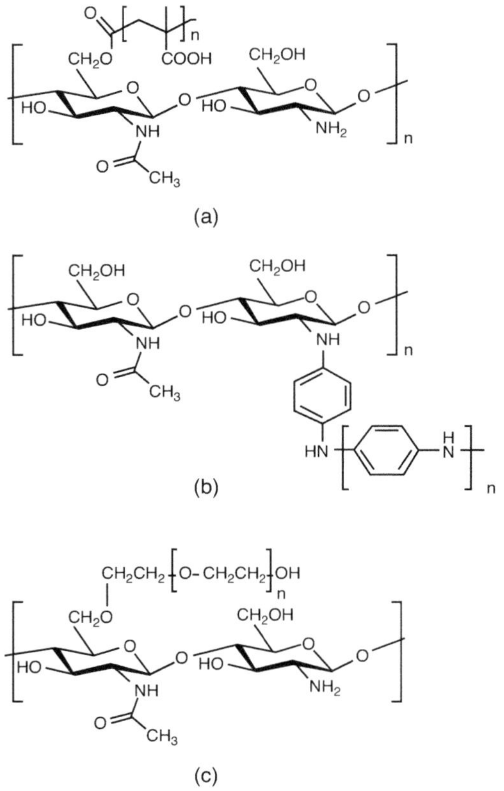

FIGURE 11.7 Chemical structure of graft copolymers: (a) chitosan-PMMA, (b) chitosan-PANI, and (c) chitosan-PEG.

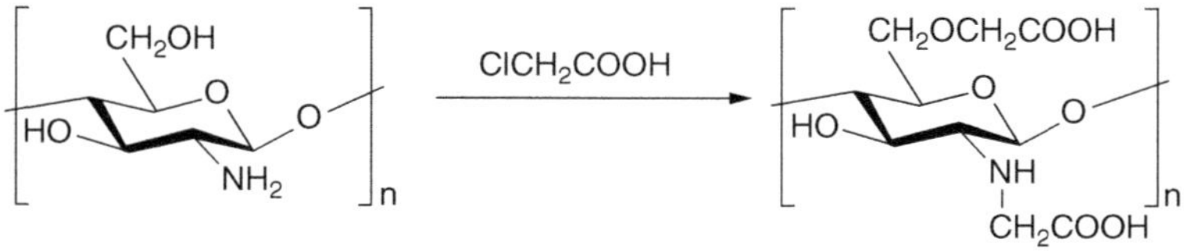

FIGURE 11.8 Illustration of *N*- and *O*-carboxymethylation of chitosan.

promoting the efficiency of vaccines, suppressing virulence reversal, and minimizing attendant clinical risks of drug allergy or reactogenicity (Zuber et al. 2021). Consequently, chitosan's adaptable properties are attractive as a promising biomaterial in the value addition of subunit vaccination formulations and other pharmaceutical applications (Moran et al. 2018).

Furthermore, chitosan and its derivatives, often regarded as medical-grade chitosan, have also been implicated in different biomedical sciences, especially as a fat binder, hemostatic agent, nutraceutical, and wound healing and tissue engineering. Tissue engineering aims to repair, maintain, or increase tissue functions by creating biological replacements for damaged or diseased tissue. This is achieved by an interdisciplinary approach that applies engineering principles to biology and the amenability of chitosan to a wide range of molecular modifications, enhancing seeding cells' adhering ability. For instance, combining chitosan and collagen into a chitosan-collagen scaffold helps improve the latter's cell adhesion property (Becerra et al. 2022). One promising technique for creating effective scaffold materials for tissue regeneration is chitosan conjugation with physiological laminin peptides or protein peptides (Kulkarni et al. 2021). Also, a hydrophilic and biodegradable polymer like polyglutamic acid (PGA) was utilized to alter the chitosan matrix resulting in PGA/chitosan composite biomaterial further supporting chitosan as an excellent candidate for use in tissue engineering (Kim et al. 2008). Mesenchymal stem cells isolated from bone marrow have been the subject of substantial research and have shown encouraging clinical promise (Fathi et al. 2019). Recent studies have demonstrated that the addition of chitosan improves the adhesion, proliferation, and viability of mesenchymal stem cells. The versatile graft copolymerization and biocompatibility properties acknowledged them as a potential biomaterial in tissue regeneration and cell transplantation. It has also opened the application of chitosan in manufacturing tissue analogues including cartilage, nerve, bone, skin, and liver. Non-woven fabrics made of chitin or *N*-carboxybutyl chitosan or polyelectrolyte chitosan complexes have been produced as wound dressings (Dai et al. 2011). It was discovered that these compounds helped regenerate damaged skin.

11.5.1 Therapeutic Applications of Chitosan and Its Derivatives

Recent research showed that this chitosan promotes nerve development and offers new opportunities for boosting nerve regeneration inside tissue-engineered skin by linking neurotrophic factors in the sponge before transplantation (Gingras et al. 2003). Additionally, research revealed that the chitosan/collagen matrix is a suitable substrate for skin stem cells to grow in three dimensions (Shi et al. 2006). For instance, chitosan oligosaccharides have been shown to activate macrophages, and chitosan and chitin are chemoattractants for neutrophils both in vitro and in vivo, which is an essential first step in promoting wound healing (Hoemann and Fong 2017). Chitosan may also hasten healing by encouraging the growth of granulation tissue or replacing damaged epithelium (Jayakumar et al. 2011). Chitosan was found not to inhibit normal human skin fibroblasts but did limit keloid fibroblasts by blocking type 1 collagen production, making it crucial in keloid management (Rasad et al. 2010). Attachment, migration, and proliferation of hSFs (fibroblasts generated from human skin) were greatly enhanced on chitosan membranes treated with argon plasma compared to untreated membranes (Silva et al. 2013). However, the impact of chitosan on fibroblasts and macrophages is controversial, with mixed findings from the available evidence. Several studies have shown that chitosan stimulates fibroblasts and activates macrophages, whereas others have demonstrated that it inhibits fibroblast cell development and does not activate macrophages (Shi et al. 2006). While the reason for the conflicting findings is unclear, it is logical to assume that it was due to differences in the cell population parameters. Water-soluble chitin (WSC) solution had superior tensile strength and healing properties when comparing the potential applications of chitin and chitosan-based biomaterials in biomedicine. This was attributed to WSC's increased hydrophilicity and biodegradability, which improved its compatibility with damaged tissues, increased its activity as a wound-healing facilitator, and showed great potential as a hydrogel for wound dressings (Han 2005).

The prospect of chitosan in gene therapy provides hope for cell-based treatment, which is rapidly advancing, and viral and non-viral gene delivery systems are the two primary categories of vectors adopted in transfection yields. These techniques have generated controversies and are associated with some clinical concerns including carcinogenic as well as immunogenic consequences. Therefore, a growing interest has been ignited in non-viral delivery methods safer alternatives to viral vectors in gene therapy (Lara-Velazquez et al. 2020; Gascón et al. 2013). Chitosan–DNA complexes are better than polygalactosamine–DNA complexes because they are much simpler to produce and show malleable transfection efficiency (Janes et al. 2001). Thus, transfection efficiency can be up-scaled by different chemical, physical, and biological techniques to optimize desirable characteristics of the DNA–chitosan complexes creates stability of key characteristics affected by their chain length and structure.

Conversely, a recent study has investigated chitosan in an experimental mitigation of HIV (Dharshini et al. 2021; Tan et al. 2022). Two types of chitosan oligomers, designated QMW and WMQ-chitosan oligomers (where Q, M, and W represent glutamine, methionine, and tryptophan, respectively), protected C8166 cells against the cytolytic effects of the HIV-1RF strain (Kravanja et al. 2019). Oligomers inhibited HIV-1 reverse protease and transcriptase activities in vitro but had no impact on preventing syncytium formation or lowering the viral load. It was shown that co-culture of HIV-infected and uninfected C8166 cells inhibited syncytium development (Karagozlu et al. 2014). The chitosan oligomers exert their effects on the HIV entry stage, preventing the virus from causing cytopathic effects. Non-toxic and water-soluble molecules called chitooligosaccharides are produced from chitosan by the enzymatic breakdown (Cheung et al. 2015). Furthermore, at concentrations of 3–5 kDa, sulfated chitooligosaccharide III significantly inhibited HIV-1 proliferation, syncytium formation, lytic activity, and p24 antigen generation. The compound interferes with the binding of HIV-1 gp120 to the CD4 cell surface receptor to restrict the virus' entry and virus-cell fusion. This is not the case with unsulfured chitooligosaccharides (Hao et al. 2019). When compared to native chitosan nanoparticles, tenofovir-loaded chitosan-thioglycolic acid-linked nanoparticles have superior biophysical features for mucoadhesion (Meng et al. 2014). The medication lamivudine-PEGylated chitosan has low cytotoxicity and high potency, and it inhibits HIV replication more effectively than lamivudine (Aghasadeghi et al. 2013). Using chitosan as the matrix-forming mucoadhesive polymer, tenofovir disoproxil fumarate dosage forms have been developed (Shailender et al. 2018). Consequently, because of their high cell-targeting effectiveness and outstanding drug loading capacity, chitosan nanoparticles loaded with the anti-HIV medication saquinavir were able to effectively suppress HIV-1 replication in their intended T-cell targets. These results suggest that chitosan nanocarriers might serve as effective delivery system for anti-HIV-1 medicines (Boroumand et al. 2021). For example, the vaginal epithelial cells and *Lactobacillus crispatus* tested negative for cytotoxicity when exposed to tenofovir-loaded chitosan nanoparticles. As the size of the nanoparticles was decreased, their ability to adhere to mucous membranes improved. The most effective outcomes, however, were achieved by a combination of drug encapsulation efficiency and % mucoadhesion for nanoparticles of greater size. Therefore, anti-HIV drug-loaded chitosan nanoparticles of substantial size showed promise for preventing HIV transmission (Dev et al. 2010). To enhance anti-HIV therapy and reduce treatment-related adverse effects, chitosan-*O*-isopropyl-5′-*O*-d4T monophosphate conjugated nano-prodrugs may be used as targeted and sustained polymeric prodrugs (Yang et al. 2010). Zidovudine has been delivered in a regulated manner using chitosan microspheres cross-linked with glutaraldehyde (Nayak et al. 2009).

Novel coronavirus pneumonia is caused mostly by the strain of coronavirus SARS-CoV-2 (COVID-19). Both the direct antiviral action of chitosan and the induction of antiviral immune response mechanisms are effective against viral infections. Electrostatic repulsion between the cationic chitosan and the negatively charged virus causes the membrane of the virus to rupture, resulting in the death of the virus (Safarzadeh et al. 2021). Since *N*-(2-hydroxypropyl)-3-methylchitosan ammonium chloride strongly interacts with the viral S protein, the two molecules combine to create a complex that blocks viral entrance into host cells (Zhang et al. 2022a). The increased antiviral immune response is due to chitosan stimulation of immune cells through the activation of alternative complement system pathways (Li et al. 2013). Chitosan is taken up by macrophages, where it stimulates the release of oxidizing molecules (ROS). To combat viruses, splenocytes produce interferon (IFN) in response to reactive oxygen species (Oberkampf et al. 2018). Hathout and Kassem (2020) remarked that using positively charged polymers like chitosan in protective garments to produce nanofibres with high surface area reduces COVID infectivity. As the positively charged SARS-CoV-2 is repelled by the negatively charged chitosan nanofibres, the viral load in the area surrounding the protective garment decreases (Hathout and Kassem 2020). Hence, the chitosan nanoparticle aerosols made by Bioavanta may be used to deliver any drugs that could be effective against COVID-19. The nanoparticles stick to the airways and slowly degrade, releasing the encapsulated active material and guaranteeing an increased drug concentration (Mittal et al. 2021). Consequently, medication delivery through the intranasal mucosa using chitosan is a highly promising strategy for combating SARS-CoV-2. Taking the COVID-19 medication to the lungs was successful in 9 out of 15 cases. The slow degradation of the nanoparticles encapsulating the active ingredient ensures a high local concentration of the medicine in the airways, where it is most needed. Therefore, intranasal mucosal medication delivery using chitosan is a viable strategy for combating SARS-CoV-2.

11.5.2 Pharmaceutical Applications of Chitosan-Based Materials

Some other biomedical uses of chitosan-based derivatives include the application of amphiphilic copolymers of chitosan as carriers in drug delivery systems. This is due to its tendency to remain stable in hydrophilic and hydrophobic environments and become self-organized in water through micelles or vesicles. The water's supramolecular structure becomes evenly distributed hydrophobic zones. At the same time, soluble chitosan forms nano-sized self-assembled colloidal particles with a hydrophobic core surrounded by a hydrophilic outer shell. This mechanism demonstrates the phenomenon enabling the solubilization of hydrophobic medicine, while the hydrophobic core may act as the nanocontainer for medications that are not easily absorbed by the body. Suffices to say, chitosan derivatives in the form of micelles help to increase the water solubility of a poorly soluble medication, improve bioavailability, decrease toxicity, increase permeability across physiological barriers, and alter drug biodistribution when used as a drug carrier (Torchilin 2007). In contrast to surfactant micelles, chitosan critical micellar concentration and kinetic stability were typically greater (Patist et al. 2001). Some chitosan derivatives have had their hydrophobicity altered to improve the solubility of medications that are notoriously difficult to dissolve. However, chitosan's application in this regard may be limited as the hydrophilic portion of an amphiphilic self-assembling polymer because it is only soluble in aqueous solutions with pH values below its pK_a value (6.5). For this reason, novel amphiphilic chitosan-based polymers have been synthesized using glycol chitosan (Alshamrani et al. 2019). Self-aggregation of these amphiphilic glycol chitosan derivatives and improved stability under physiological circumstances are anticipated to make them useful for enhancing the solubility of poorly soluble medicines. Potential drug delivery carriers for poorly soluble drugs include four major classes of hydrophobically modified chitosan: fatty acid derivatives, carboxymethyl derivatives, steroid derivatives, and aryl and alkyl derivatives. Alternate forms of modified chitosan have also been artificially produced (Riva et al. 2011). Several anticancer medications, such as Mitomycin C, camptothecin, doxorubicin, and paclitaxel, are among the poorly soluble pharmaceuticals contained in amphiphilic chitosan-based nanocarriers (Cho et al. 2008).

11.5.3 Biomedical Uses of Antimicrobial-Grade Chitosan and Chitosan Derivatives

The cationic nature of chitosan has inspired hypothetical suggestions of its antimicrobial mechanism. These implied that a high-molecular-weight chitosan binds to the negatively charged components of a microbial cell wall. In contrast, low-molecular-weight chitosan breaches the cell wall to interact with the DNA, disrupting its transcription process and mRNA production (Badreddine et al. 2008). Chitosan also exerts antimicrobial action by altering the pathogens' membrane permeability to block transport into the cell (Muzzarelli et al. 1990).

In the case of bacteria, Gram-negative bacterial cell walls are more hydrophilic and have a more negative charge on the surface than Gram-positive bacterial cell walls, suggesting that chitosan interacts more with Gram-negative bacterial walls, indicating greater antibacterial activity (Chandrasekaran et al. 2020). In addition, the binding strength of chitosan to the bacterial cell wall was found to vary with the surrounding environment's pH, the Mw of the chitosan, and the degree to which it had been acetylated. Thus, the positive charge in chitosan is enhanced at a low pH environment, resulting in improved attachment affinity to the bacterial cell wall (Jain et al., 2014).

Also, a lower degree of chitosan acetylation is observed by Younes et al. (2014) to increase the antibacterial activity of chitosan. Chitosan has a high mortality rate and a broad range of actions against Gram-positive and Gram-negative bacteria. The antibacterial activity is proportional to the hydrophilicity of the cell wall, indicating a targeted and less harmful effect on mammalian cells (Wang et al. 2019). Most human pathogenic fungi tested were eliminated by a single application of 100 g/mL chitosan, although the toxicity of this concentration was not evident in mammalian HEK293 and COS7 cells (Cheung et al. 2015). Antimicrobial activity against *Staphylococcus aureus* and *E. coli* is significantly enhanced compared to chitosan when ZnO nanoparticles are immobilized on the chitosan matrix by in situ sol–gel conversion of precursor molecules in a single step. The chitosan solution was more effective than the chitosan submicron particles in inhibiting the growth of antibiotic-resistant *Pseudomonas aeruginosa* cells, both in planktonic cultures and in biofilms, which were isolated from patients with long-term diabetic foot infections (Yusof et al. 2019). Furthermore, water-soluble chitosan with low Mw showed anti-infective properties based on data from an in vivo investigation by Park et al. (2008). Gamma-irradiated quaternary ammonium chitosan exhibited excellent antibacterial potency against *E. coli* and *S. aureus*, in addition to swelling ability, mechanical characteristics, and stable water evaporation rate. Likewise, a chitosan-lignosulfonate multilayer modified fibres with chitosan outermost layer exhibited high inhibition of *E. coli* and increased antioxidant activity (Cheung et al. 2015).

The use of chitosan derivatives in preventing oxidative stress, which causes damage to biomolecules including proteins, membrane lipids, and DNA is also paramount in the medical sector (Abd El-Hack et al. 2020; Ngo and Kim 2014). The process has been attributed to the formation of stable radicals through the interaction of unstable free radicals with amino and hydroxyl groups on the pyranose ring (Hajji et al. 2016). Suffice it to say, scavenging free radicals is an area where low-molecular-weight chitosan excels. Prepared from 90%, 75%, and 50% deacetylated chitosan, nine different hetero-chitooligosaccharides were tested for their ability to scavenge free radicals using an electron spin resonance (ESR) spin-trapping technique. Based on the ESR results, hetero-chitooligosaccharides of medium Mw, synthesized from 90% deacetylated chitosan, showed the greatest radical-scavenging potential (Je et al. 2004). Deacetylation values and Mw are linked to the radical-scavenging activity of hetero-chitooligosaccharides. *N*-carboxyethylated derivatives (*N*-CESC) of chitosan were synthesized with varying degrees of carboxyethyl group substitution. Above a pH of 6.5, all three N-CESC samples were easily soluble in water. Based on these results, N-CESC seems to be a viable option for the industrial production of chitosan derivatives with desirable in vitro biochemical properties (Cheung et al. 2015; Tiwari et al. 2022).

Although chitosan has also been implicated in the remediation of cell tumours, correlating its activity to antioxidant properties remains a mystery. Approximately one-sixth of all fatalities globally are attributed to cancer, making it one of the leading threats to global health today. Lymphocytes and bone marrow cells are often destroyed by conventional radiation, chemotherapy, and, more recently, laser technology, which are invasive with devastating systemic consequences. Lung cancer, gastric cancer, and colorectal cancer are a few of the malignancies that may benefit from chitosan nanoparticles' antitumour properties (Zhai et al. 2021). Liu et al. (2017a) demonstrated that chitosan induces apoptosis in tumour cells by increasing caspase-3 expression and reducing Bcl-2 level. CMC increases nitric oxide (NO) levels, stimulates phagocytosis, increases macrophage viability, and induces cytokine synthesis in the tumour microenvironment. Remarkably, CMC is not notably toxic to normal liver cells but has inhibitory effects on BEL-7402 cell development in vitro and sarcoma cells in vivo (Shariatinia 2018). Moreover, chitosan prevents tumour cells from spreading and dividing. Matrix metalloenzymes (MMP) may encode an enzyme that degrades type IV collagen (Konttinen et al. 2011). MMP-2 is implicated in the breakdown of the extracellular matrix during tumour invasion and metastasis. The overexpression of MMP-2 is a hallmark of gastric cancer and is strongly correlated with the disease's ability to metastasize. MMP-2 and CD147 (Cluster of differentiation 147) expression may be downregulated by chitosan, resulting in a dose-dependent reduction of gastric cancer cell metastasis (Zhao et al. 2019). MMP-9 expression is upregulated in tumour tissues, which may release VEGF and promote tumour angiogenesis. Lung metastases have been shown by Shen et al. (2020) to exhibit a greater level of VEGF than initial tumours. It has been demonstrated that tumour invasion and metastasis may be prevented by reducing tumour angiogenesis via VEGF production through MMP-9 expression (Zhao et al. 2019).

11.6 APPLICATION OF CHITOSAN DERIVATIVES IN THE FOOD INDUSTRY

Like in the biomedical and pharmaceutical sectors, chitosan derivates are equally valued in the food industries due to their biocompatibility and hypothetical lack of toxicity. Food products are coated by directly applying a polymeric film layer, such as chitosan derivatives, to their surface. It is expected that these polymeric coatings or films will prolong the shelf life of food products by delaying their disintegration. The capacity of the coating to enhance gas exchange, prevent moisture loss, and microbial growth on food products ultimately affects food products' shelf life and quality. It further delays the deterioration of food products. Furthermore, a variety of natural advantages, including antibacterial, antioxidant, and UV protection, are offered by chitosan derivatives. However, chitosan's greatest flaw, and the reason for its modification, is its incapacity to dissolve in water. It has also been noted that the antibacterial and antioxidant qualities of films made from chitosan derivatives are not very strong (Moalla et al. 2021). Among the methods suggested by researchers to remedy this limitation are modification, fabrication, and grafted copolymerization. By altering different functional groups in the chitosan backbone, several chitosan derivatives have been produced for application as antioxidants, antibacterials, and nutraceuticals in vital biological systems. CMC has been widely used in the clearing of juices like pomegranate juice because of its capacity to interact with both positively and negatively charged molecules, including proline-rich proteins (Rao et al. 2011; Tastan and Baysal 2015). To detect CO_2, a sign of microbial growth, a suspension of chitosan and 2-amino-2-methyl-1-propanol, a substance used to concentrate CO_2 in solution, was created. Because chitosan is soluble in acidic conditions, this solution appears opaque at neutral pH but turns transparent when exposed to CO_2. This acidification results from the conversion of CO_2 to carbonic acid (H_2CO_3). Devices made of chitosan can be used as sensors to monitor changes in the packaging environment that could cause food deterioration during shipping and storage (Jung et al. 2012). The use of this indicator is advised for non-carbonated beverages such as fruit juices, milk, and table wines.

CMC treated with silk peptide (SP) demonstrated exceptional antioxidant activity in tests such as the DPPH, hydroxyl radical, H_2O_2 scavenging activity, and total reducing power studies (Liu et al. 2017b). In a different study, the chitosan derivative that is soluble in water was produced by adding maleic anhydride to the chitosan backbone through an open-chain reaction process. This product's high substitution between water-soluble chitosan derivatives made it more effective in preventing bacterial growth (Feng et al. 2016). Likewise, a water-soluble chitosan derivative made by green synthesis has been proposed for use in the food sector (Hu et al. 2016a). Inadvertent modification of chitosan to improve its reactive amino acid groups and lower its Mw enhanced its antibacterial efficacy (Luo and Wang 2013). To boost the antimicrobial activity of the polymer, chitosan can also be grafted with groups that have antibacterial properties, such as phenolic compounds. When compared to unfunctionalized chitosan, the activity of functionalized chitosan against *E. coli* and *Listeria monocytogenes* was enhanced by the addition of caffeic acid (CA) and gallic acid (GA). *O*-fumaryl ester (OFTMCS) of *N*, *N*,*N*-trimethyl chitosan (TMC) was functionalized with chitosan to improve its antibacterial activity. The antibacterial activity of OFTMCS was found to compare to that exhibited by both TMC and unmodified chitosan against *S. aureus* and *E. coli* (Li et al. 2015).

To create a modified quaternized carboxymethyl chitosan (QCMC)-collagen material for potential use as an antioxidant agent in the food and pharmaceutical industries, QCMC was combined with collagen peptide. Using the carbodiimide technique, fish collagen peptide was first conjugated to the backbone of QCMC, and the antioxidant activity was then measured using a spectrometric approach as a percentage inhibition of H_2O_2. In a different study, its antioxidant activity was enhanced when chitosan underwent additional chemical and enzymatic changes, particularly for the medium-and high-molecular-weight types. These modifications aim to improve the structure of antioxidant molecules to scavenge free radicals and/or make chitosan more soluble (Luo and Wang 2013). Chitosan can be functionalized to increase its antioxidant activity without being poisonous or environmentally harmful by grafting to flavonoids and phenolic acids utilizing oxidative enzymes (Hashemi et al. 2017). For instance, Božič et al. (2012) functionalized chitosan using laccase from *Trametes versicolor* to add CA and GA. In this reaction, the oxidative enzyme converted the phenols into reactive o-quinones, which then interacted with the chitosan's primary amino groups to generate derivatives of phenolic acid that exhibited more potent antioxidant properties than untreated chitosan. It was shown that the derivatives of chitosan at different pH values undergo different functionalization routes and degrees of antioxidant activity because the solubility and protonation of its amino groups are pH-dependent. Chitosan modified at pH 4.5 had the highest level of activity based on the observation of Božič et al. (2012)

Chitosan films made with an essential oil and extract of *Artemisia campestris* showed improved antioxidants and increased UV–Vis light-blocking properties. The films' water resistance and thermal stability were increased up to 300°C by the hydroethanolic extracts and essential oil of *A. campestris* (Moalla et al. 2021). The hydrogen bonding inside the chitosan films may have been enhanced by the reaction of the chitosan derivative with the hydroethanolic extract ions of *A. campestris*. The ability to optically shield UV and visible light is the most important factor in ensuring the thermostability of food. This action prevents food products from UV and visible light exposures, shielding them from oxidation and spoilage by decomposition. Oxidation is prevented in the treated films by polyphenols, well-known antioxidants, and the additive-loaded films' ability to block UV–Vis light. Furthermore, Riaz et al. (2020) examined root extracts from Chinese chives (*Allium tuberosum*). Chinese chives are a good chitosan graft copolymerization material for food packaging because their extracts contain a variety of phenolic components, including GA, isoquercetin, catechin, ferulic acid, p-coumaric acid, and rutin (Riaz et al. 2020). Similarly, the antibacterial properties of the chitosan-extracts combination were enhanced by diallyl disulfide and allyl methyl sulfide. However, when compared to the control films, the chive extract-loaded chitosan films demonstrated superior antioxidant, antibacterial, and biodegradability properties (Huang et al. 2021).

On the other hand, research has investigated the effects of natural colourants made from different plant sources in making food packaging films. These colourants provided additional advantages to the movie in addition to adding colour (Huang et al. 2022). One natural colourant that enhances the mechanical properties of a chitosan/gelatin film is betalains, which are a mixture of nitrogenous compounds comprising betacyanins and betaxanthins. Furthermore, the films' ductility significantly increased, and the betalains-loaded films' antioxidant and UV–Vis light-blocking properties significantly improved. Hu et al. (2020) further recognized the role of chitosan derivatives grafted natural colourants in suppressing foodborne pathogens like *L. monocytogenes*, *S. typhimurium*, *E. coli*, and *S. aureus*. Consequently, more biodegradable antimicrobial-active food packaging films, such as mixes of polyvinyl alcohol and chitosan, are being developed. It was discovered that using *Ocimum tenuiflorum* natural extracts to enhance antibacterial action within a film made from chitosan derivatives significantly increased the films' water resistance (Annu and Ahmed 2021).

It has also been suggested that chitosan-based films that integrate anthocyanins as pH sensors to visualize milk deterioration could make use of a wide range of colours generated by different resonance of anthocyanin structures as the pH changes (Alizadeh-Sani et al. 2020). When the sensor encountered fresh milk, it showed a dark grey colour; when the milk was spoiled, the colour changed to a dark pink colour (Pereira et al. 2015). The shelf life of food produce and products has significantly increased by deploying chitosan-based films and coatings without quality depreciation. Chitosan and its derivatives are anticipated to be the cutting-edge food packaging technologies of the future, with commodification and commercial prospects.

11.7 FUNCTIONAL-USE LIMITATIONS OF CHITOSAN AND ITS DERIVATIVES IN BIOTECHNOLOGY

Chitosan is ubiquitous in application due to its versatile physicochemical properties that can be manipulated or functionalized to meet unique end-use targets using different technologies. Notwithstanding these attractions, chitosan is not infallible irrespective of its applications, forms, and structures. Chitosan derivatives are essential for developing relatively safe and effective materials valuable in drug delivery systems (Wang et al. 2020b). Drug delivery systems are facilitated using chitosan derivatives as hydrogels, beads, or nanoparticles based on the nature of their distinct physicochemical and biological characteristics. However, scientific research now focuses more on discovering newer chitosan modification techniques that can advance chitosan in its diverse derivate forms to economic-scale benefits (Farion et al. 2018).

Also, functionalizing chitosan has expanded biotechnological options for enhancing the diverse but malleable chemical, biological and physical properties like thermostability, membrane permeability, mucoadhesivity, and controlled-release behaviour, significantly supporting their potential use as vaccine carrier systems. Chitosan derivatives have been combined with synthetic or natural ligands, such as mannose, for targeted interactions with cell types to accomplish specific vaccine administration. Despite these benefits, chitosan and its derivatives still face a few obstacles, including uneven reactivity and poor physical stability. Nevertheless, the prospects of chitosan and its derivatives are sufficiently admirable to spur scientists into undertaking more studies on vaccine administration, particularly to meet the growing reservations about the safety of RNA or life-strain vaccine administration (Xing et al. 2018).

Drug delivery systems have been investigated using nanomaterials obtained from chitosan/chitosan derivatives for oral or topical administration and in colon-targeted drug delivery, carcinoma therapy, vaccinology, and gene delivery. However, owing to the possible hazards, the clinical administration of drug vectors linked to chitosan and its derivatives is still restricted (Li et al. 2018). This may be due to the poor understanding of the toxicity of the materials used to form the graft copolymerized carriers. Other numerous productive efforts have been made to understand the role of modified chitosan or derivatives in rapid wound healing, tissue repair, osteogenesis, drug bioavailability, encapsulation efficiency for poorly water-soluble drugs, and cytotoxicity of anticancer drugs without any attention to their chronic side or adverse effects.

In another perspective, chitosan can be copolymerized with certain functional groups to produce derivatives that have improved properties. Despite the inherent prospects in chitosan applications, many of these studies have been unable to successfully translate these laboratory results on the functional use of chitosan derivatives to large-scale commerce or market commodities. Furthermore, regulating the parameters for the synthesis of chitosan derivates to achieve a high level of purity, appropriate DS, appropriate level of reactions, and other properties requires technical sophistication and experience (Bhavsar et al. 2017). While this indicates that some professionalism and technological savvy are crucial to managing the procedures involved in chitosan derivative formation, it could also limit their expanded use in the biomedical and food sectors.

Similarly, despite a few reports and patents in the past 20 years on using chitosan/chitosan derivatives in biomedicine, the procedures for ascertaining the purity or determining the appropriate optimal physiological concentration of their impurity in biological systems remain nascent. Although the implications are unclear, especially in the food processing industries, it can be assumed to limit their applications to tissue repairs, tissue engineering, transplant, and drug delivery systems. This is mainly due to the difficulty in obtaining biopolymers that are sufficiently pure. Additionally, chitosan and its derivatives are widely regarded as non-toxic or safe in human systems even though there is currently a lack of sufficient in vivo investigation and evidence to confirm this perception (Morin-Crini et al. 2019; Wang et al. 2020a). This suggests that more evidence of their in vivo safety or toxicity is necessary for their commodification and use in food, medicine, and pharmaceutics (Varlamov et al. 2020; Shivakumar et al. 2021). For example, the rising controversy on the safety of nanomaterials required further research to prove their safety, validating the applications of chitosan/chitosan derivatives-based nanotherapeutics in biomedicine (Mukhtar et al. 2021). Consequently, only a few chitosan/chitosan derivatives-based nanoparticles have been approved to treat malignancies in clinical settings or clinical trials.

There are various difficulties in the clinical application of antitumour chitosan/chitosan derivatives nanoparticles. The safety of chitosan/chitosan derivatives in vivo has seldom been tested, particularly for those administered via intravenous injection. The chemical nature of chitosan creates a challenge to its modification due to the need to evaluate each new derivative separately. Further studies should focus more on the safety evaluation of intravenous-bound chitosan/chitosan derivative nanoparticles for oncotherapy rather than just developing new derivatives. Also, more research targeting the mechanisms of the antitumour efficacies of chitosan derivatives should be emphasized (Zhang et al. 2019). With more detailed research, it is anticipated that chemical modification techniques will be able to constantly enhance the physical and chemical characteristics of chitosan nanomaterials, making them more suited for usage in medical products and drug delivery systems. Hence, the future applications of chitosan derivatives in biomedicine will be more widespread (Wang et al. 2020a).

After reviewing the literature on chitosan applications and bioeconomic relevance in several critical industrial sectors, other limitations emerged that need to be addressed in future studies. One such issue is the discrepancy in researchers' findings on the characteristics of chitosan or its derivatives. The lack of uniformity in the assay settings, experimental conditions, or standardization makes it difficult to generalize inferences and make meaningful comparisons across studies. For example, most reports on chitosan's antimicrobial activities failed to sufficiently establish the microbes' susceptibility level or the polymer's minimum inhibitory or microbicidal concentration. Although the technique required to verify this is rudimentary, the implication of chitosan antimicrobial action is tied more to lowering microbial load suggesting the need for their standardized inclusion in food products as preservatives (Miot-Sertier et al. 2022; Moller et al. 2022). While existing studies preclude dose-sensitive antimicrobial actions of chitosan, more research is crucial to understanding the molecular intricacies underlying processes and how the process instigates the development of resistance by exposed microbes (Zhang et al. 2022a). In addition, the lack of sufficient empirical data to support the non-toxicity and transgenetic or omic properties of chitosan or its functionalized derivates, especially chitosan fabricated with composite nanoparticles, questions their use as a safe intervention in food and pharmaceutics and suggests the need for them to be regulated like all other antibiotics (Miot-Sertier et al. 2022). While this can be achieved with the cooperation of research institutions, industry, and government regulatory authorities, more safety investigations of nanoparticles made from chitosan need to be undertaken. The cost assessment of the complicated steps required for modifying and functionalizing chitosan for specific end-use has not been sufficiently carried out. This may be limited to their large-scale applications in the industrial economy. Notwithstanding these limitations, chitosan and its derivatives still hold potential future promise for various high-tech applications with a better understanding of their benign, safety dispensation, cost-efficiency, source sustainability, and standards.

11.8 CONCLUSION

Chitosan is a biodegradable polymer material utilized extensively in different biotechnological spheres to offer diverse benefits in pharmaceutics, biomedicine, and food manufacturing. Many chitosan derivatives with superior features such as hydrophilicity, pH sensitivity, and targeting potential have been functionalized to increase chitosan solubility and its range of use. Effective medication excipients may be found in chitosan derivatives. Preparing a drug-loading system using chitosan derivatives, such as micelles, nanoparticles, microspheres, and hydrogels, may improve the stability of pharmaceuticals and food, allowing them to be released gradually, steadily, and at the appropriate dose over time.

Biodegradable materials like chitosan and its derivatives, which are safe for people and the environment, are top priority feedstock in the growing research to find solutions to side-effect problems caused by some medications. Moreover, their over-rated viable properties raise a perception of flawlessly ductile or reliable materials with an extended range of applications. This make-believe is therefore provoking questions about the authenticity of their safety (toxicity and side-effect), ecological intricacies, ease of application, ubiquity, biofunctionality, and lability. Their use as a functional shield, coatings, or encapsulation protects medication and foods from damage by environmental hazards, extends their physiological half-life in the body, and is amenable to composite nanoparticles for multifunctional purposes that may include sensor-induced functions. Among the currently available biological polymers, chitosan derivatives appear to be the most ductile and promising based on extant reports in the body of literature. This relative edge over many other biopolymers is linked to their recently found role in the controlled release of drug bioactivity to target tissues and plastic nature. Suffice it to say chitosan and its derivates are readily stable in in vivo conditions, supporting their optimal functionalization for controlled performance and persistence in releasing bioactive agents or drug targets. It is hypothesized that chitosan derivatives' physical and chemical characteristics permit different enhancements by chemical, biological, and physical modification techniques, making them more user-friendly, and best-suited biotechnology materials for industrial applications. However, further studies are essential to establish their non-toxicity, functional efficiency, and versatility in diverse industrial applications. The potential chitosan derivative applications in biomedicine will grow with time.

REFERENCES

Abd El-Hack, M. E., El-Saadony, M. T., Shafi, M. E., Zabermawi, N. M., Arif, M., Batiha, G. E., Khafaga, A. F., Abd El-Hakim, Y. M., & Al-Sagheer, A. A. (2020). Antimicrobial and antioxidant properties of chitosan and its derivatives and their applications: A review. *International Journal of Biological Macromolecules*, 164, 2726–2744. https://doi.org/10.1016/j.ijbiomac.2020.08.153

Adolfsson, K. H., Yadav, N., & Hakkarainen, M. (2020). Cellulose-derived hydrothermally carbonized materials and their emerging applications. *Current Opinion in Green and Sustainable Chemistry*, 23, 18–24. https://doi.org/10.1016/j.cogsc.2020.03.008

Aghasadeghi, M., Heidari, H., Sadat, M. S., Irani, S., Amini, S., Siadat, D. S., Fazlhashemy, M. E., Zabihollahi, R., Atyabi, S. M., Momen, S. B., Khosraavy, M. S., Davari, M., Mohammadi, T. D., Doroudian, M., & Ardestani, S. M. (2013). Lamivudine-PEGylated chitosan: A novel effective nanosized antiretroviral agent. *Current HIV Research*, 11(4), 309–320.

Alizadeh-Sani, M., Mohammadian, E., Rhim, J.-W., & Jafari, S. M. (2020). pH-sensitive (halochromic) smart packaging films based on natural food colorants for the monitoring of food quality and safety. *Trends in Food Science & Technology*, 105,93–144. https://doi.org/10.1016/j.tifs.2020.08.014

Aljawish, A., Chevalot, I., Jasniewski, J., Scher, J., & Muniglia, L. (2015). Enzymatic synthesis of chitosan derivatives and their potential applications. *Journal of Molecular Catalysis B: Enzymatic*, 112, 25–39.

Alkabli, J. (2022). Progress in preparation of thiolated, crosslinked, and imino-chitosan derivatives targeting specific applications. *European Polymer Journal*, 165, 110998. https://doi.org/10.1016/j.eurpolymj.2022.110998

Al-Remawi, M. (2015). Application of N-hexoyl chitosan derivatives with high degree of substitution in the preparation of super-disintegrating pharmaceutical matrices. *Journal of Drug Delivery Science and Technology*, 29, 31–41. https://doi.org/10.1016/j.jddst.2015.06.001

Alshamrani, M., Sikder, S., Coulibaly, F., Mandal, A., Pal, D., & Mitra, A. K. (2019). Self-assembling topical nanomicellar formulation to improve curcumin absorption across ocular tissues. *AAPS PharmSciTech*, 20(7), 1–16.

Alves, N. M., & Mano, J. F. (2008). Chitosan derivatives obtained by chemical modifications for biomedical and environmental applications. *International Journal of Biological Macromolecules*, 43(5), 401–414.

Annu, A. A., & Ahmed, S. (2021). Eco-friendly natural extract loaded antioxidative chitosan/polyvinylalcohol based active films for food packaging. *Heliyon*, 7, e06550.

Aranaz, I., Mengíbar, M., Harris, R., Paños, I., Miralles, B., Acosta, N., Galed, G., & Heras, Á. (2009). Functional characterization of chitin and chitosan. *Current Chemical Biology*, 3(2), 203–230.

Auta, M., & Hameed, B. (2013). Coalesced chitosan activated carbon composite for batch and fixed-bed adsorption of cationic and anionic dyes. *Colloids and Surfaces B: Biointerfaces*, 105, 199–206.

Ayman, U., Akter, L., Islam, R., Bhakta, S., Rahman, M. S., Islam, M. R., Sultana, N., Sharif, A., Jahan, M. R., Md. Rahman, S., & Haque, Z. (2022). Dietary chitosan oligosaccharides improve health status in broilers for safe poultry meat production. *Annals of Agricultural Sciences*, 67(1), 90–98. https://doi.org/10.1016/j.aoas.2022.05.003

Aytekin, A. O., Morimura, S., & Kida, K. (2011). Synthesis of chitosan–caffeic acid derivatives and evaluation of their antioxidant activities. *Journal of Bioscience and Bioengineering*, 111(2), 212–216.

Badreddine, I., Lafitte, C., Heux, L., Skandalis, N., Spanou, Z., Martinez, Y., Esquerré-Tugayé, M.-T., Bulone, V., Dumas, B., & Bottin, A. (2008). Cell wall chitosaccharides are essential components and exposed patterns of the phytopathogenic oomycete Aphanomyces euteiches. *Eukaryotic Cell*, 7(11), 1980–1993.

Becerra, J., Rodriguez, M., Leal, D., Noris-Suarez, K., & Gonzalez, G. (2022). Chitosan-collagen-hydroxyapatite membranes for tissue engineering. *Journal of Materials Science: Materials in Medicine*, 33(2), 1–16.

Benediktsdóttir, B. E., Baldursson, Ó., & Másson, M. (2014). Challenges in evaluation of chitosan and trimethylated chitosan (TMC) as mucosal permeation enhancers: From synthesis to in vitro application. *Journal of Controlled Release*, 173, 18–31.

Bhavsar, C., Momin, M., Gharat, S., & Omri, A. (2017). Functionalized and graft copolymers of chitosan and its pharmaceutical applications. In *Expert Opinion on Drug Delivery* (Vol. 14, Issue 10). Taylor & Francis. https://doi.org/10.1080/17425247.2017.1241230

Boroumand, H., Badie, F., Mazaheri, S., Seyedi, Z. S., Nahand, J. S., Nejati, M., Baghi, H. B., Abbasi-Kolli, M., Badehnoosh, B., Ghandali, M., Hamblin, M. R., & Mirzaei, H. (2021). Chitosan-based nanoparticles against viral infections. *Frontiers in Cellular and Infection Microbiology*, 11, 643953.

Božič, M., Gorgieva, S., & Kokol, V. (2012). Laccase-mediated functionalization of chitosan by Caffeic and gallic acids for modulating antioxidant and antimicrobial properties. *Carbohydrate Polymers*, 87(4), 2388–2398.

Branca, C., Auditore, L., Loria, D., Trimarchi, M., & Wanderlingh, U. (2013). Radiation synthesis and characterization of poly (ethylene oxide)/chitosan hydrogels. *Journal of Applied Polymer Science*, 127(1), 217–223.

Brasselet, C., Guillaume, P., Dubessay, P., Dols-Lafargue, M., Coulon, J., Vallet-Courbin, A., De Baynast, H., Doco, T., Michaud, P., & Delattre, C. (2019). Modification of chitosan: How generating new functional derivatives? *Applied Sciences*, 9(7), 1321.

Carey, F. A., & Sundberg, R. J. (2007). *Advanced Organic Chemistry: Part B: Reactions and Synthesis* (Vol. 5). Springer.

Çelebi, H., Gurbuz, M., Koparal, S., & Dogan, A. (2013). Development of antibacterial electrospun chitosan/poly (vinyl alcohol) nanofibers containing silver ion-incorporated HAP nanoparticles. *Composite Interfaces*, 20(9), 799–812.

Chandrasekaran, M., Kim, K. D., & Chun, S. C. (2020). Antibacterial activity of chitosan nanoparticles: A review. *Processes*, 8(9), 1173.

Chang, Y.-M., Xiao, J.-Q., Christy, J., Wu, C.-Y., Huang, C.-W., Wu, T.-Y., Chiang, Y.-C., Lin, T.-H., & Chen, H.-Y. (2022). Ice-templated synthesis of multicomponent porous coatings via vapour sublimation and deposition polymerization. *Materials Today Bio*, 16, 100403. https://doi.org/10.1016/j.mtbio.2022.100403

Chen, D., Bai, R., Yong, H., Zong, S., Jin, C., & Liu, J. (2022). Improving the digestive stability and prebiotic effect of carboxymethyl chitosan by grafting with gallic acid: In vitro gastrointestinal digestion and colonic fermentation evaluation. *International Journal of Biological Macromolecules*, 214, 685–696. https://doi.org/10.1016/j.ijbiomac.2022.06.170

Chen, J.-P., Chang, G.-Y., & Chen, J.-K. (2008). Electrospun collagen/chitosan nanofibrous membrane as wound dressing. *Colloids and Surfaces A: Physicochemical and Engineering Aspects*, 313, 183–188.

Chen, W., Li, Y., Yang, S., Yue, L., Jiang, Q., & Xia, W. (2015). Synthesis and antioxidant properties of chitosan and carboxymethyl chitosan-stabilized selenium nanoparticles. *Carbohydrate Polymers*, 132, 574–581.

Cheung, R. C. F., Ng, T. B., Wong, J. H., & Chan, W. Y. (2015). Chitosan: An update on potential biomedical and pharmaceutical applications. *Marine Drugs*, 13(8), 5156–5186.

Chew, K. Y., & Brownlee, I. A. (2018). The impact of supplementation with dietary fibers on weight loss: A systematic review of randomised controlled trials. *Bioactive Carbohydrates and Dietary Fibre*, 14, 9–19.

Cho, K., Wang, X. U., Nie, S., Chen, Z., & Shin, D. M. (2008). Therapeutic nanoparticles for drug delivery in cancer. *Clinical Cancer Research*, 14(5), 1310–1316.

Crini, G. (ed.). (2022). Chitin and chitosan: Production, properties, and applications. In *Chitin and Chitosan*. Academic Press, pp. 149–207. https://doi.org/10.1016/B978-0-323-96119-6.00004-9

Curcio, M., Puoci, F., Iemma, F., Parisi, O. I., Cirillo, G., Spizzirri, U. G., & Picci, N. (2009). Covalent insertion of antioxidant molecules on chitosan by a free radical grafting procedure. *Journal of Agricultural and Food Chemistry*, 57(13), 5933–5938.

Dai, T., Tanaka, M., Huang, Y. Y., & Hamblin, M. R. (2011). Chitosan preparations for wounds and burns: Antimicrobial and wound-healing effects. *Expert Review of Anti-infective Therapy*, 9(7), 857–879.

De Abreu, F. R., & Campana-Filho, S. P. (2005). Preparation and characterization of carboxymethylchitosan. *Polímeros*, 15(2), 79–83. https://doi.org/10.1590/S0104-14282005000200004

Dev, A., Binulal, N. S., Anitha, A., Nair, S. V., Furuike, T., Tamura, H., & Jayakumar, R. (2010). Preparation of poly (lactic acid)/chitosan nanoparticles for anti-HIV drug delivery applications. *Carbohydrate Polymers*, 80(3), 833–838.

Dharshini, K. P., Fang, H., Devi, D. R., Yang, J.-X., Luo, R.-H., Zheng, Y.-T., Brzeziński, M., & Hari, B. N. V. (2021). pH-sensitive chitosan nanoparticles loaded with dolutegravir as milk and food dmixture for paediatric anti-HIV therapy. *Carbohydrate Polymers*, 256, 117440. https://doi.org/10.1016/j.carbpol.2020.117440

Dimassi, S., Tabary, N., Chai, F., Blanchemain, N., & Martel, B. (2018). Sulfonated and sulfated chitosan derivatives for biomedical applications: A review. *Carbohydrate Polymers*, 202, 382–396.

Domard, A., Rinaudo, M., & Terrassin, C. (1986). New method for the quaternization of chitosan. *International Journal of Biological Macromolecules*, 8(2), 105–107.

Elzamly, R. A., Mohamed, H. M., Mohamed, M. I., Zaky, H. T., Harding, D. R. K., & Kandile, N. G. (2021). New sustainable chemically modified chitosan derivatives for different applications: Synthesis and characterization. *Arabian Journal of Chemistry*, 14(8), 103255. https://doi.org/10.1016/j.arabjc.2021.103255

Farion, I. A., Burdukovskii, V. F., Kholkhoev, B. C., Timashev, P. S., & Chailakhyan, R. K. (2018). Functionalization of chitosan with carboxylic acids and derivatives of them: Synthesis issues and prospects of practical use: A review. *Express Polymer Letters*, 12(12), 1081–1105. https://doi.org/10.3144/expresspolymlett.2018.95

Fathi, E., Farahzadi, R., Valipour, B., & Sanaat, Z. (2019). Cytokines secreted from bone marrow derived mesenchymal stem cells promote apoptosis and change cell cycle distribution of K562 cell line as clinical agent in cell transplantation. *PloS One*, 14(4), e0215678.

Feng, T., Du, Y., Li, J., Hu, Y., & Kennedy, J. F. (2008). Enhancement of antioxidant activity of chitosan by irradiation. *Carbohydrate Polymers*, 73(1), 126–132.

Feng, X., Zheng, K., Wang, C., Chu, F., & Chen, Y. (2016). Durable antibacterial cotton fabrics with chitosan based quaternary ammonium salt. *Fibers and Polymers*, 17, 371–379.

Gallo, M., Naviglio, D., Caruso, A. A., & Ferrara, L. (2016). Applications of chitosan as a functional food. In Grumezescu, A. M. (ed.), *Nanotechnology in the Agri-Food Industry, Novel Approaches of Nanotechnology in Food*. Academic Press, pp. 425–464. https://doi.org/10.1016/B978-0-12-804308-0.00013-3

Gascón, A. R., del Pozo-Rodríguez, A., & Solinís, M. Á. (2013). Non-viral delivery systems in gene therapy. In Francisco Martin Molina (ed.), *Gene Therapy-Tools and Potential Applications*. IntechOpen, London. https://doi.org/10.5772/52704

Gindy, M. E., & Prud'homme, R. K. (2009). Multifunctional nanoparticles for imaging, delivery and targeting in cancer therapy. *Expert Opinion on Drug Delivery*, 6(8), 865–878.

Gingras, M., Paradis, I., & Berthod, F. (2003). Nerve regeneration in a collagen–chitosan tissue-engineered skin transplanted on nude mice. *Biomaterials*, 24(9), 1653–1661.

Gorochovceva, N., & Makuška, R. (2004). Synthesis and study of water-soluble chitosan-Opoly (ethylene glycol) graft copolymers. *European Polymer Journal*, 40(4), 685–691. https://doi.org/10.1016/j.eurpolymj.2003.12.005

Goy, R. C., Britto, D. D., & Assis, O. B. (2009). A review of the antimicrobial activity of chitosan. *Polímeros*, 19, 241–247.

Hajji, S., Chaker, A., Jridi, M., Maalej, H., Jellouli, K., Boufi, S., & Nasri, M. (2016). Structural analysis, and antioxidant and antibacterial properties of chitosan-poly (vinyl alcohol) biodegradable films. *Environmental Science and Pollution Research*, 23(15), 15310–15320.

Han, S. S. (2005). Topical formulations of water-soluble chitin as a wound healing assistant. *Fibers and Polymers*, 6(3), 219–223.

Hao, C., Yu, G., He, Y., Xu, C., Zhang, L., & Wang, W. (2019). Marine glycan–based antiviral agents in clinical or preclinical trials. *Reviews in Medical Virology*, 29(3), e2043.

Hashemi Gahruie, H., & Niakousari, M. (2017) Antioxidant, antimicrobial, cell viability and enzymatic inhibitory of antioxidant polymers as biological macromolecules. *International Journal of Biological Macromolecules*, 104, 606–617.

Hathout, R. M., & Kassem, D. H. (2020). Positively charged electroceutical spun chitosan nanofibers can protect health care providers from COVID-19 infection: An opinion. *Frontiers in Bioengineering and Biotechnology*, 8, 885.

Hoemann, C. D., & Fong, D. (2017). Immunological responses to chitosan for biomedical applications. In J. Amber Jennings, Joel D. Bumgardner (eds.), *Chitosan Based Biomaterials Volume* 1. Woodhead Publishing, UK. pp. 45–79.

Hosseinnejad, M., & Jafari, S. M. (2016). Evaluation of different factors affecting antimicrobial properties of chitosan. *International Journal of Biological Macromolecules*, 85, 467–475.

Hu, D., Wang, H., & Wang, L. (2016a). Physical properties and antibacterial activity of quaternized chitosan/carboxymethyl cellulose blend films. *LWT*, 65, 398–405.

Hu, H., Yao, X., Qin, Y., Yong, H., & Liu, J. (2020). Development of multifunctional food packaging by incorporating betalains from vegetable amaranth (Amaranthus tricolor L.) into quaternary ammonium chitosan/fish gelatin blend films. *International Journal of Biological Macromolecules*, 159, 675–684.

Hu, Q., Wang, T., Zhou, M., Xue, J., & Luo, Y. (2016b). In vitro antioxidant-activity evaluation of gallic-acid-grafted chitosan conjugate synthesized by free-radical-induced grafting method. *Journal of Agricultural and Food Chemistry*, 64, 5893–5900.

Hu, Y., Du, Y., Yang, J., Tang, Y., Li, J., & Wang, X. (2007). Self-aggregation and antibacterial activity of N-acylated chitosan. *Polymer*, 48(11), 3098–3106.

Huang, B.-H., Li, S.-Y., Chiou, Y.-J., Chojniak, D., Chou, S.-C., Wong, V. C. M., Chen, S.-Y., & Wu, P.-W. (2021). Electrophoretic fabrication of a robust chitosan/polyethylene glycol/polydopamine composite film for UV-shielding application. *Carbohydrate Polymers*, 273, 118560. https://doi.org/10.1016/j.carbpol.2021.118560

Huang, J., Hu, Z., Li, G., Hu, L., Chen, J., & Hu, Y. (2022). Make your packaging colorful and multifunctional: The molecular interaction and properties characterization of natural colorant-based films and their applications in food industry. *Trends in Food Science and Technology*, 124, 259–277. https://doi.org/10.1016/j.tifs.2022.04.028

Huang, Y., Feng, L., Zhang, Y., He, L., Wang, C., Xu, J., Wu, J., Kirk, T. B., Guo, R., & Xue, W. (2017). Hemostasis mechanism and applications of N-alkylated chitosan sponge. *Polymers for Advanced Technologies*, 28(9), 1107–1114.

Ibraheem, D., Elaissari, A., & Fessi, H. (2014). Gene therapy and DNA delivery systems. *International Journal of Pharmaceutics*, 459(1–2), 70–83.

Imam, S. S., Alshehri, S., Ghoneim, M. M., Zafar, A., Alsaidan, O. A., Alruwaili, N. K., Gilani, S. J., & Rizwanullah, M. (2021). Recent advancement in chitosan-based nanoparticles for improved oral bioavailability and bioactivity of phytochemicals: Challenges and perspectives. *Polymers*, 13(22), 4036.

Jain, A., Duvvuri, L. S., Farah, S., Beyth, N., Domb, A. J., & Khan, W. (2014). Antimicrobial polymers. *Advanced Healthcare Materials*, 3(12), 1969–1985.

Jain, A., Gulbake, A., Shilpi, S., Jain, A., Hurkat, P., & Jain, S. K. (2013). A new horizon in modifications of chitosan: Syntheses and applications. *Critical Reviews in Therapeutic Drug Carrier Systems*, 30(2), 91–181.

Janes, K. A., Calvo, P., & Alonso, M. J. (2001). Polysaccharide colloidal particles as delivery systems for macromolecules. *Advanced Drug Delivery Reviews*, 47(1), 83–97.

Jayakumar, R., Prabaharan, M., Kumar, P. S., Nair, S. V., & Tamura, H. (2011). Biomaterials based on chitin and chitosan in wound dressing applications. *Biotechnology Advances*, 29(3), 322–337.

Je, J. Y., Park, P. J., & Kim, S. K. (2004). Free radical scavenging properties of hetero-chitooligosaccharides using an ESR spectroscopy. *Food and Chemical Toxicology*, 42(3), 381–387.

Jin, Z., Li, W., Cao, H., Zhang, X., Chen, G., Wu, H., Guo, C., Zhang, Y., Kang, H., Wang, Y., & Zhao, K. (2013). Antimicrobial activity and cytotoxicity of N-2-HACC and characterization of nanoparticles with N-2-HACC and CMC as a vaccine carrier. *Chemical Engineering Journal*, 221, 331–341.

Joseph, K., Mattoso, L., Toledo, R., Thomas, S., De Carvalho, L., Pothen, L., Kala, S., & James, B. (2000). Natural fiber reinforced thermoplastic composites. In Elisabete Frollini, Alcides Lopes Leao, Luiz H. C. Mattoso (eds.), *Natural Polymers and Agrofibers Composites*, Embrapa Instrumentacao Agropecuaria, pp. 159–200.

Jung, B. O., Kim, C. H., Choi, K. S., Lee, Y. M., & Kim, J. J. (1999). Preparation of amphiphilic chitosan and their antimicrobial activities. *Journal of Applied Polymer Science*, 72, 1713–1720.

Jung, J., Puligundla, P., & Ko, S. (2012). Proof-of-concept study of chitosan-based carbon dioxide indicator for food packaging applications. *Food Chemistry*, 135, 2170–2174.

Karagozlu, M. Z., Karadeniz, F., & Kim, S. K. (2014). Anti-HIV activities of novel synthetic peptide conjugated chitosan oligomers. *International Journal of Biological Macromolecules*, 66, 260–266.

Kaur, S., & Dhillon, G. S. (2014). The versatile biopolymer chitosan: Potential sources, evaluation of extraction methods and applications. *Critical Reviews in Microbiology*, 40(2), 155–175.

Kazemi, M. S., Mohammadi, Z., Amini, M., Yousefi, M., Tarighi, P., Eftekhari, S., & Tehrani, R. M. (2019). Thiolated chitosan-lauric acid as a new chitosan derivative: Synthesis, characterization and cytotoxicity. *International Journal of Biological Macromolecules*, 136, 823–830. https://doi.org/10.1016/j.ijbiomac.2019.06.132

Kedir, W. M., Abdi, G. F., Goro, M. M., & Tolesa, L. D. (2022). Pharmaceutical and drug delivery applications of chitosan biopolymer and its modified nanocomposite: A review. *Heliyon*, 8(8), e10196. https://doi.org/10.1016/j.heliyon.2022.e10196

Kim, I. Y., Seo, S. J., Moon, H. S., Yoo, M. K., Park, I. Y., Kim, B. C., & Cho, C. S. (2008). Chitosan and its derivatives for tissue engineering applications. *Biotechnology Advances*, 26(1), 1–21.

Kim, J., Kim, Y.-R., Kim, Y., Lim, K. T., Seonwoo, H., Park, S., Cho, S.-P., Hong, B. H., Choung, P.-H., & Chung, T. D. (2013). Graphene-incorporated chitosan substrata for adhesion and differentiation of human mesenchymal stem cells. *Journal of Materials Chemistry, B*, 1(7), 933–938.

Kong, M., Chen, X. G., Xing, K., & Park, H. J. (2010). Antimicrobial properties of chitosan and mode of action: A state-of-the-art review. *International Journal of Food Microbiology*, 144(1), 51–63.

Konttinen, Y. T., Kaivosoja, E., Stegaev, V., Wagner, H. D., Levón, J., Tiainen, V. M., & Mackiewicz, Z. (2011). Extracellular matrix and tissue regeneration. In Steinhoff, G. (ed.), *Regenerative Medicine*. Springer, Dordrecht, pp. 21–80.

Kou, S. G., Peters, L., & Mucalo, M. (2022). Chitosan: A review of molecular structure, bioactivities and interactions with the human body and micro-organisms. *Carbohydrate Polymers*, 282, 119132. https://doi.org/10.1016/j.carbpol.2022.119132

Kravanja, G., Primožič, M., Knez, Ž., & Leitgeb, M. (2019). Chitosan-based (nano) materials for novel biomedical applications. *Molecules*, 24(10), 1960.

Kulkarni, N., Shinde, S. D., Jadhav, G. S., Adsare, D. R., Rao, K., Kachhia, M., Maingle, M., Patil, S. P., Arya, N., & Sahu, B. (2021). Peptide-chitosan engineered scaffolds for biomedical applications. *Bioconjugate Chemistry*, 32(3), 448–465.

Kumar, D., Gihar, S., Shrivash, M. K., Kumar, P., & Kundu, P. P. (2020). A review on the synthesis of graft copolymers of chitosan and their potential applications. *International Journal of Biological Macromolecules*, 163, 2097–2112. https://doi.org/10.1016/j.ijbiomac.2020.09.060

Kumar, D., Kumar, P., & Pandey, J. (2018). Binary grafted chitosan film: Synthesis, characterization, antibacterial activity and prospects for food packaging. *International Journal of Biological Macromolecules*, 115, 341–348.

Lara-Velazquez, M., Alkharboosh, R., Norton, E. S., Ramirez-Loera, C., Freeman, W. D., Guerrero-Cazares, H., Forte, A. J., Quiñones-Hinojosa, A., & Sarabia-Estrada, R. (2020). Chitosan-based non-viral gene and drug delivery systems for brain cancer. *Frontiers in Neurology*, 11, 740.

Lee, D.-S., & Je, J.-Y. (2013). Gallic acid-grafted-chitosan inhibits foodborne pathogens by a membrane damage mechanism. *Journal of Agricultural and Food Chemistry*, 61(26), 6574–6579.

Li, J., Cai, C., Li, J., Li, J., Li, J., Sun, T., Wang, L., Wu, H., & Yu, G. (2018). Chitosan-based nanomaterials for drug delivery. *Molecules*, 23(10), 2661. https://doi.org/10.3390/molecules23102661

Li, Q., Li, Q., Tan, W., Zhang, J., & Guo Z. (2020). Phenolic-containing chitosan quaternary ammonium derivatives and their significantly enhanced antioxidant and antitumor properties. *Carbohydrate Research*, 498, 108169. https://doi.org/10.1016/j.carres.2020.108169

Li, X., Min, M., Du, N., Gu, Y., Hode, T., Naylor, M., Chen, D., Nordquist, R. E., & Chen, W. R. (2013). Chitin, chitosan, and glycated chitosan regulate immune responses: The novel adjuvants for cancer vaccine. *Clinical and Developmental Immunology*, 2013, 387023.

Li, Z., Yang, F., & Yang, R. (2015). Synthesis of chitosan derivative with dualantibacterial functional groups and its antibacterial activity. *Journal of Applied Polymer Science*, 132, 42663.

Liu, L., Xin, Y., Liu, J., Zhang, E., & Li, W. (2017a). Inhibitory effect of chitosan oligosaccharide on human hepatoma cells in vitro. *African Journal of Traditional, Complementary and Alternative Medicines*, 14(4), 272–277.

Liu, M., Min, L., Zhu, C., Rao, Z., Liu, L., Xu, W., Luo, P., & Fan, L. (2017b). Preparation, characterization and antioxidant activity of silk peptides grafted carboxymethyl chitosan. *International Journal of Biological Macromolecules*, 104, 732–738.

Liu, Z., Jiang, X., Li, Z., Zheng, Y., Nie, J.-J., Cui, Z., Liang, Y., Zhu, S., Chen, D., & Wu, S. (2022). Recent progress of photo-excited antibacterial materials via chemical vapor deposition. *Chemical Engineering Journal*, 437(Part 2), 135401. https://doi.org/10.1016/j.cej.2022.135401

Luo, L.-J., Nguyen, D. D., & Lai, J.-Y. (2020). Benzoic acid derivative-modified chitosan-g-poly(N-isopropylacrylamide): Methoxylation effects and pharmacological treatments of Glaucoma-related neurodegeneration. *Journal of Controlled Release*, 317, 246–258. https://doi.org/10.1016/j.jconrel.2019.11.038

Luo, Y., & Wang, Q. (2013). Recent advances of chitosan and its derivatives for novel applications infood science. *Journal of Food Processing & Beverages*, 1(1), 1–13.

Ma, G., Yang, D., Zhou, Y., Xiao, M., Kennedy, J. F., & Nie, J. (2008). Preparation and characterization of water-soluble N-alkylated chitosan. *Carbohydrate Polymers*, 74(1), 121–126.

Ma, J., Xia, W., Zhang, R., Ding, L., Kong, Y., Zhang, H., & Fu, K. (2021). Flocculation of emulsified oily wastewater by using functional grafting modified chitosan: The effect of cationic and hydrophobic structure. *Journal of Hazardous Materials*, 403, 123690. https://doi.org/10.1016/j.jhazmat.2020.123690

Madera-Santana, T. J., Herrera-Méndez, C. H., & Rodríguez-Núñez, J. R. (2018). An overview of the chemical modifications of chitosan and their advantages. *Green Materials*, 6(4), 131–142. https://doi.org/10.1680/jgrma.18.00053

Manocha, S., Dhiman, S., Grewal, A. S., & Guarve, K. (2022). Nanotechnology: An approach to overcome bioavailability challenges of nutraceuticals. *Journal of Drug Delivery Science and Technology*, 72, 103418. https://doi.org/10.1016/j.jddst.2022.103418

Meng, J., Zhang, T., Agrahari, V., Ezoulin, M. J., & Youan, B. B. C. (2014). Comparative biophysical properties of tenofovir-loaded, thiolated and nonthiolated chitosan nanoparticles intended for HIV prevention. *Nanomedicine*, 9(11), 1595–1612.

Mhurchu, C. N., Poppitt, S., McGill, A., Leahy, F., Bennett, D., Lin, R., Ormrod, D., Ward, L., Strik, C., & Rodgers, D. A. (2004). The effect of the dietary supplement, chitosan, on body weight: A randomised controlled trial in 250 overweight and obese adults. *International Journal of Obesity*, 28(9), 1149–1156.

Miot-Sertier, C., Paulin, M., Dutilh, L., Ballestra, P., Albertin, W., Masneuf-Pomarède, I., Coulon, J., Moine, V., Vallet-Courbin, A., Maupeu, J., & Dols-Lafargue M. (2022). Assessment of chitosan antimicrobial effect on wine microbes. *International Journal of Food Microbiology*, 381, 109907. https://doi.org/10.1016/j.ijfoodmicro.2022.109907

Mittal, N., Garg, V., Bhadada, S. K., & Katare, O. P. (2021). Role of novel drug delivery systems in coronavirus disease-2019 (covid-19): Time to act now. *Current Drug Delivery*, 18(3), 289–296.

Moalla, S., Ammar, I., Fauconnier, M.-L., Danthine, S., Blecker, C., Besbes, S., & Attia, H. (2021). Development and characterization of chitosan films carrying Artemisia campestris antioxidants for potential use as active food packaging materials. *International Journal of Biological Macromolecules*, 183, 254–266.

Moine, L., Canali, M. M., Porporatto, C., & Correa, S. G. (2021). Reviewing the biological activity of chitosan in the mucosa: Focus on intestinal immunity. *International Journal of Biological Macromolecules*, 189, 324–334. https://doi.org/10.1016/j.ijbiomac.2021.08.098

Moller, A., Leone, C., Kataria, J., Sidhu, G., Rama, E. N., Kroft, B., Thippareddi, H., & Singh, M. (2022). Effect of a carrageenan/chitosan coating with allyl isothiocyanate on microbial load in chicken breast. *LWT*, 161, 113397. https://doi.org/10.1016/j.lwt.2022.113397

Mondal, A., Dhar, A. K., Banerjee, S., Hasnain, M. S., & Nayak, A. K. (2022). Antimicrobial uses of chitosan. In: Hasnain, M. S., Beg, S., & Nayak, A. K. (eds.), *Chitosan in Biomedical Applications*. Academic Press, pp. 13–36. https://doi.org/10.1016/B978-0-12-821058-1.00009-5

Moran, H. B., Turley, J. L., Andersson, M., & Lavelle, E. C. (2018). Immunomodulatory properties of chitosan polymers. *Biomaterials*, 184, 1–9.

Moreno-Vásquez, M. J., Valenzuela-Buitimea, E. L., Plascencia-Jatomea, M., Encinas-Encinas, J. C., Rodríguez-Félix, F., Sánchez-Valdes, S., Rosas-Burgos, E. C., Ocaño-Higuera, V. M., & Graciano-Verdugo, A. Z. (2017). Functionalization of chitosan by a free radical reaction: Characterization, antioxidant and antibacterial potential. *Carbohydrate Polymers*, 155, 117–127.

Morin-Crini, N., Lichtfouse, E., Torri, G., & Crini, G. (2019). Applications of chitosan in food, pharmaceuticals, medicine, cosmetics, agriculture, textiles, pulp and paper, biotechnology, and environmental chemistry. *Environmental Chemistry Letters*, 17(4), 1667–1692. https://doi.org/10.1007/s10311-019-00904-x

Mourya, V. K., & Inamdar, N. N. (2008). Chitosan-modifications and applications: Opportunities galore. *Reactive and Functional Polymers*, 68(6), 1013–1051.

Mruthunjayappa, M. H., Kotrappanavar, N. S., & Mondal, D. (2022). New prospects on solvothermal carbonisation assisted by organic solvents, ionic liquids and eutectic mixtures: A critical review. *Progress in Materials Science*, 126, 100932. https://doi.org/10.1016/j.pmatsci.2022.100932

Mukhtar, M., Eszter, F., Bartos, C., Zeeshan, M., & Ambrus, R. (2021). Chitosan biopolymer, its derivatives and potential applications in nano-therapeutics: A comprehensive review. *European Polymer Journal*, 160, 110767. https://doi.org/10.1016/j.eurpolymj.2021.110767

Muzzarelli, R., Tarsi, R., Filippini, O., Giovanetti, E., Biagini, G., & Varaldo, P. (1990). Antimicrobial properties of N-carboxybutyl chitosan. *Antimicrobial Agents and Chemotherapy*, 34(10), 2019–2023.

Nagy, V. (2018). *Chitosan-natural antioxidant conjugates: Synthesis, antimicrobial and antioxidant properties.* M.Sc. Thesis, University of Iceland.

Nayak, U. Y., Gopal, S., Mutalik, S., Ranjith, A. K., Reddy, M. S., Gupta, P., & Udupa, N. (2009). Glutaraldehyde cross-linked chitosan microspheres for controlled delivery of zidovudine. *Journal of Microencapsulation*, 26(3), 214–222.

Ngo, D. H., & Kim, S. K. (2014). Antioxidant effects of chitin, chitosan, and their derivatives. *Advances in Food and Nutrition Research*, 73, 15–31.

Nishi, N., Ebina, A., Nishimura, S., Tsutsumi, A., Hasegawa, O., & Tokura, S. (1986). Highly phosphorylated derivatives of chitin, partially deacetylated chitin and chitosan as new functional polymers: Preparation and characterization. *International Journal of Biological Macromolecules*, 8, 311–317.

Nishi, N., Nishimura, S., Ebina, A., Tsutsumi, A., & Tokura, S. (1984). Preparation and characterization of water-soluble chitin phosphate. *International Journal of Biological Macromolecules*, 6, 53–54.

Oberkampf, M., Guillerey, C., Mouriès, J., Rosenbaum, P., Fayolle, C., Bobard, A., Savina, A., Ogier-Denis, E., Enninga, J., Amigorena, S., Leclerc, C., & Dadaglio, G. (2018). Mitochondrial reactive oxygen species regulate the induction of CD8+ T cells by plasmacytoid dendritic cells. *Nature Communications*, 9(1), 1–14.

Olteanu, C. E. (2007). Applications of functionalized chitosan. *Scientific Study and Research: Chemistry and Chemical Engineering, Biotechnology, Food Industry*, 8(3), 227–256.

Omer, A. M., Dey, R., Eltaweil, A. S., Abd El-Monaem, E. M., & Ziora, Z. M. (2022). Insights into recent advances of chitosan-based adsorbents for sustainable removal of heavy metals and anions. *Arabian Journal of Chemistry*, 15(2), 103543. https://doi.org/10.1016/j.arabjc.2021.103543

Panda, P. K., Yang, J.-M., & Chang, Y.-H. (2021). Preparation and characterization of ferulic acid-modified water-soluble chitosan and poly (γ-glutamic acid) polyelectrolyte films through layer-by-layer assembly towards protein adsorption. *International Journal of Biological Macromolecules*, 171, 457–464.

Park, Y. K., Kim, M. H., Park, S. C., Cheong, H. S., Jang, M. K., Nah, J. W., & Hahm, K. S. (2008). Investigation of the antifungal activity and mechanism of action of LMWS-chitosan. *Journal of Microbiology and Biotechnology*, 18(10), 1729–1734.

Patil, M., Rajamani, S. B., Mathad, S. N. M., Patil, A. Y., Hussain, M. A., Alorfii, H. S., Khan, A., Asiri, A. M., Khan, I., & Puttegowda, M. (2022). Microwave-assisted synthesis of poly (acrylamide-co-2-hydroxyethyl methacrylate)/chitosan semi-IPN ZnO nanocomposite membranes for food packaging applications. *Journal of Materials Research and Technology*, https://doi.org/10.1016/j.jmrt.2022.08.079

Patist, A., Oh, S. G., Leung, R., & Shah, D. O. (2001). Kinetics of micellization: Its significance to technological processes. *Colloids and Surfaces A: Physicochemical and Engineering Aspects*, 176(1), 3–16.

Pereira, V. A., de Arruda, I. N. Q., & Stefani, R. (2015). Active chitosan/PVA films with anthocyanins from Brassica oleraceae (Red Cabbage) as time–temperature indicators for application in intelligent food packaging. *Food Hydrocolloids*, 43, 180–188.

Priyanka, D. N., Prashanth, K. V. H., & Tharanathan, R. N. (2022). A review on potential anti-diabetic mechanisms of chitosan and its derivatives. *Carbohydrate Polymer Technologies and Applications*, 3, 100188. https://doi.org/10.1016/j.carpta.2022.100188

Rabea, E. I., Badawy, M. E.-T., Stevens, C. V., Smagghe, G., & Steurbaut, W. (2003). Chitosan as antimicrobial agent: Applications and mode of action. *Biomacromolecules*, 4(6), 1457–1465.

Rao, L., Hayat, K., Lv, Y., Karangwa, E., Xia, S., Jia, C., Zhong, F., & Zhang, X. (2011). Effect of ultrafiltration and fining adsorbents on the clarification of green tea. *Journal of Food Engineering,* 102, 321–326.

Rasad, M. S. B. A., Halim, A. S., Hashim, K., Rashid, A. H. A., Yusof, N., & Shamsuddin, S. (2010). In vitro evaluation of novel chitosan derivatives sheet and paste cytocompatibility on human dermal fibroblasts. *Carbohydrate Polymers*, 79(4), 1094–1100.

Riaz, A., Lagnika, C., Luo, H., Dai, Z., Nie, M., Hashim, M. M., Liu, C., Song, J., & Li, D. (2020). Chitosan-based biodegradable active food packaging film containing Chinese chive (Allium tuberosum) root extract for food application. *International Journal of Biological Macromolecules*, 150, 595–604.

Rinaudo, M. (2006). Chitin and chitosan: Properties and applications. *Progress in Polymer Science*. 31(7), 603–632. https://doi.org/10.1016/j.progpolymsci.2006.06.001

Riva, R., Ragelle, H., Rieux, A. D., Duhem, N., Jérôme, C., & Préat, V. (2011). Chitosan and chitosan derivatives in drug delivery and tissue engineering. In Jayakumar, R., Prabaharan, M., Muzzarelli, R. (eds.), *Chitosan for Biomaterials* II, Springer, Berlin, Heidelberg. pp. 19–44.

Rocha, M., Angélica, M., Coimbra, M. A., & Nunes, C. (2017). Applications of chitosan and their derivatives in beverages: A critical review. *Current Opinion in Food Science*, 15, 61–69. https://doi.org/10.1016/j.cofs.2017.06.008

Rúnarsson, Ö. V., Holappa, J., Nevalainen, T., Hjálmarsdóttir, M., Järvinen, T., Loftsson, T., Einarsson, J. M., Jónsdóttir, S., Valdimarsdóttir, M., & Másson, M. (2007). Antibacterial activity of methylated chitosan and chitooligomer derivatives: Synthesis and structure activity relationships. *European Polymer Journal*, 43(6), 2660–2671.

Safarzadeh, M., Sadeghi, S., Azizi, M., Rastegari-Pouyani, M., Pouriran, R., & Hoseini, M. H. M. (2021). Chitin and chitosan as tools to combat COVID-19: A triple approach. *International Journal of Biological Macromolecules*, 183, 235–244.

Sakaguchi, T., Hirokoshi, T., & Nakajima, A. (1981). Adsorption of uranium by chitin phosphate and chitosan phosphate. *Agricultural and Biological Chemistry*, 45, 2191–2195.

Sanchez-Salvador, J. L., Balea, A., Monte, M. C., Negro, C., & Blanco, A. (2021). Chitosan grafted/cross-linked with biodegradable polymers: A review. *International Journal of Biological Macromolecules*, 178, 325–343. https://doi.org/10.1016/j.ijbiomac.2021.02.200

Schreiber, S. B., Bozell, J. J., Hayes, D. G., & Zivanovic, S. (2013). Introduction of primary antioxidant activity to chitosan for application as a multifunctional food packaging material. *Food Hydrocolloids*, 33(2), 207–214.

Shailender, J., Ravi, P. R., Reddy Sirukuri, M., Dalvi, A., & Keerthi Priya, O. (2018). Chitosan nanoparticles for the oral delivery of tenofovir disoproxil fumarate: Formulation optimization, characterization and ex vivo and in vivo evaluation for uptake mechanism in rats. *Drug Development and Industrial Pharmacy*, 44(7), 1109–1119.

Shariatinia, Z. (2018). Carboxymethyl chitosan: Properties and biomedical applications. *International Journal of Biological Macromolecules*, 120, 1406–1419.

Shen, W. T., Hsu, R. S., Fang, J. H., Hu, P. F., Chiang, C. S., & Hu, S. H. (2020). Marginative delivery-mediated extracellular leakiness and T cell infiltration in lung metastasis by a biomimetic nanoraspberry. *Nano Letters*, 21(3), 1375–1383.

Shi, C., Zhu, Y., Ran, X., Wang, M., Su, Y., & Cheng, T. (2006). Therapeutic potential of chitosan and its derivatives in regenerative medicine. *Journal of Surgical Research*, 133(2), 185–192.

Shivakumar, P., Gupta, M. S., Jayakumar, R., & Gowda, D. V. (2021). Prospection of chitosan and its derivatives in wound healing: Proof of patent analysis (2010–2020). *International Journal of Biological Macromolecules*, 184, 701–712. https://doi.org/10.1016/j.ijbiomac.2021.06.086

Silva, S. S., Popa, E. G., Gomes, M. E., Cerqueira, M., Marques, A. P., Caridade, S. G., Teixeira, P., Sousa, C., Mano, J. F., & Reis, R. L. (2013). An investigation of the potential application of chitosan/aloe-based membranes for regenerative medicine. *Acta Biomaterialia*, 9(6), 6790–6797.

Sonia, T., & Sharma, C. P. (2011). Chitosan and its derivatives for drug delivery perspective. In Jayakumar, R., Prabaharan, M., Muzzarelli, R. (eds.), *Chitosan for Biomaterials* I, Springer, Berlin, Heidelberg. pp. 23–53.

Sun, T., Xie, W., & Xu, P. (2004). Superoxide anion scavenging activity of graft chitosan derivatives. *Carbohydrate Polymers*, 58(4), 374–382. https://doi.org/10.1016/j.carbpol.2004.06.042

Sun, T., Zhou, D., Mao, F., & Zhu, Y. (2007). Preparation of low-molecular-weight carboxymethyl chitosan and their superoxide anion scavenging activity. *European Polymer Journal*, 43(2), 652–656.

Tan, R. S. L., Hassandarvish, P., Chee, C. F., Chan, L. W., & Wong, T. W. (2022). Chitosan and its derivatives as polymeric anti-viral therapeutics and potential anti-SARS-CoV-2 nanomedicine. *Carbohydrate Polymers*, 290, 119500. https://doi.org/10.1016/j.carbpol.2022.119500

Tastan, O., & Baysal, T. (2015). Clarification of pomegranate juice with chitosan: Changes on quality characteristics during storage. *Food Chemistry*, 180, 211–218.

Tiwari, P., Sharma, A., Nayak, V., Verma, R., & Singh, J. (2022). Properties and types of chitosan-based nanomaterials. In Santosh Kumar, Sundararajan V. Madihally (eds.), *Role of Chitosan and Chitosan-Based Nanomaterials in Plant Sciences*. Academic Press, London. pp. 85–117.

Torchilin, V. P. (2007). Micellar nanocarriers: Pharmaceutical perspectives. *Pharmaceutical Research*, 24(1), 1–16.

Tungtong, S., Okonogi, S., Chowwanapoonpohn, S., Phutdhawong, W., & Yotsawimonwat, S. (2012). Solubility, viscosity and rheological properties of water-soluble chitosan derivatives. *Maejo International Journal of Science and Technology*, 6(2), 315.

Varlamov, V. P., Il'ina, A. V., Shagdarova, B. T., Lunkov, A. P., & Mysyakina, I. S. (2020). Chitin/chitosan and its derivatives: Fundamental problems and practical approaches. *Biochemistry (Moscow)*, 85(1), 154–176.

Wang, J., Chen, S., Xu, J.-V., Liu, L.-C., Zhou, J.-C., & Cai, J.-J. (2021). High-surface-area porous carbons produced by the mild KOH activation of a chitosan hydrochar and their CO_2 capture. *New Carbon Materials*, 36(6), 1081–1090. https://doi.org/10.1016/S1872-5805(21)60074-4

Wang, J., Dou, X., Song, J., Lyu, Y., Zhu, X., Xu, L., & Shan, A. (2019). Antimicrobial peptides: Promising alternatives in the post feeding antibiotic era. *Medicinal Research Reviews*, 39(3), 831–859.

Wang, J. J., Zeng, Z. W., Xiao, R. Z., Xie, T., Zhou, G. L., Zhan, X. R., & Wang, S. L. (2011). Recent advances of chitosan nanoparticles as drug carriers. *International Journal of Nanomedicine*, 6, 765.

Ways, T. M., Lau, W. M., & Khutoryanskiy, V. V. (2018). Chitosan and its derivatives for application in mucoadhesive drug delivery systems. *Polymers*, 10(3), 267. https://doi.org/10.3390/polym10030267

Wei, L., Tan, W., Wang, G., Li, Q., Dong, F., & Guo, Z. (2019). The antioxidant and antifungal activity of chitosan derivatives bearing Schiff bases and quaternary ammonium salts. *Carbohydrate Polymers*, 226, 115256.

Woranuch, S., Yoksan, R., & Akashi, M. (2015). Ferulic acid-coupled chitosan: Thermal stability and utilization as an antioxidant for biodegradable active packaging film. *Carbohydrate Polymers*, 115, 744–751.

Wu, D., Zhu, L., Li, Y., Zhang, X., Xu, S., Yang, G., & Delair, T. (2020). Chitosan-based colloidal polyelectrolyte complexes for drug delivery: A review. *Carbohydrate Polymers*, 238, 116126.

Wuts, P. G., & Greene, T. W. (2006). *Greene's Protective Groups in Organic Synthesis*. John Wiley & Sons.

Xia, W., Wei, X.-Y., Xie, Y.-Y., & Zhou, T. (2022). A novel chitosan oligosaccharide derivative: Synthesis, antioxidant and antibacterial properties. *Carbohydrate Polymers*, 291, 119608. https://doi.org/10.1016/j.carbpol.2022.119608

Xie, W., Xu, P., Wang, W., & Liu, Q. (2002). Preparation and antibacterial activity of a watersoluble chitosan derivative. *Carbohydrate Polymers*, 50(1), 35–40. https://doi.org/10.1016/S0144-8617(01)00370-8

Xing, L., Fan, Y. T., Zhou, T. J., Gong, J. H., Cui, L. H., Cho, K. H., & Cho, C. S. (2018). Chemical modification of chitosan for efficient vaccine delivery. *Molecules*, 23(2), 229. https://doi.org/10.3390/molecules23020229

Xu, T., Xin, M., Li, M., Huang, H., Zhou, S., & Liu, J. (2011). Synthesis, characterization, and antibacterial activity of N, O-quaternary ammonium chitosan. *Carbohydrate Research*, 346(15), 2445–2450.

Yan, X. M., Kang, J., Gao, L., Xiong, L., & Mei, P. (2013). Solvothermal synthesis of carbon coated N-doped TiO_2 nanostructures with enhanced visible light catalytic activity. *Applied Surface Science*, 265, 778–783.

Yang, L., Chen, L., Zeng, R., Li, C., Qiao, R., Hu, L., & Li, Z. (2010). Synthesis, nanosizing and in vitro drug release of a novel anti-HIV polymeric prodrug: Chitosan-O-isopropyl-5′-O-d4T monophosphate conjugate. *Bioorganic & Medicinal Chemistry*, 18(1), 117–123.

Yang, S., Tirmizi, S. A., Burns, A., Barney, A. A., & Risen, Jr., W. M. (1989). Chitaline materials: Soluble chitosan-polyaniline copolymers and their conductive doped forms. *Synthetic Metals*, 32(2), 191–200. https://doi.org/10.1016/0379-6779(89)90841-2

Yang, Z., Yang, H., Jiang, Z., Cai, T., Li, H., Li, H., Li, A., & Cheng, R. (2013). Flocculation of both anionic and cationic dyes in aqueous solutions by the amphoteric grafting flocculant carboxymethyl chitosan-graft-polyacrylamide. *Journal of Hazardous Materials*, 254, 36–45.

Yin, S., Ren, L., & Wang, Y. (2013). Argon plasma-induced graft polymerization of PEGMA on chitosan membrane surface for cell adhesion improvement. *Plasma Science and Technology*, 15(10), 1041.

Yoshikawa, S., Takayama, T., & Tsubokawa, N. (1998). Grafting reaction of living polymer cations with amino groups on chitosan powder. *Journal of Applied Polymer Science*, 68(11), 1883–1889.

Younes, I., Sellimi, S., Rinaudo, M., Jellouli, K., & Nasri, M. (2014). Influence of acetylation degree and molecular weight of homogeneous chitosans on antibacterial and antifungal activities. *International Journal of Food Microbiology*, 185, 57–63.

Yusof, N. A. A., Zain, N. M., & Pauzi, N. (2019). Synthesis of ZnO nanoparticles with chitosan as stabilizing agent and their antibacterial properties against Gram-positive and Gram-negative bacteria. *International Journal of Biological Macromolecules*, 124, 1132–1136.

Zhai, X., Li, C., Ren, D., Wang, J., Ma, C., & Abd El-Aty, A. M. (2021). The impact of chitooligosaccharides and their derivatives on the in vitro and in vivo antitumor activity: A comprehensive review. *Carbohydrate Polymers*, 266, 118132.

Zhang, E., Xing, R., Qin, Y., Li, K., & Li, P. (2019). Advances in chitosan-based nanoparticles for oncotherapy. *Carbohydrate Polymers*, 222, 115004. https://doi.org/10.1016/j.carbpol.2019.115004

Zhang, H., & Jin, D. (2012). *Advanced Research on Material Engineering, Chemistry, Bioinformatics II*. Trans Tech Publications Ltd.

Zhang, H., Montemayor, A. M., Wimsatt, S. T., & Tikekar, R. V. (2022a). Effect of combination of UV-A light and chitosan-gallic acid coating on microbial safety and quality of fresh strawberries. *Food Control*, 140, 109106. https://doi.org/10.1016/j.foodcont.2022.109106

Zhang, J., Tan, W., Zhang, Z., Song, Y., Li, Q., Dong, F., & Guo, Z. (2018). Synthesis, characterization, and the antifungal activity of chitosan derivatives containing urea groups. *International Journal of Biological Macromolecules*, 109, 1061–1067.

Zhang, M., Zhang, F., Li, C., An, H., Wan, T., & Zhang, P. (2022c). Application of chitosan and its derivative polymers in clinical medicine and agriculture. *Polymers*, 14(5), 958.

Zhao, D., Yu, S., Sun, B., Gao, S., Guo, S., & Zhao, K. (2018). Biomedical applications of chitosan and its derivative nanoparticles. *Polymers*, 10(4), 462.

Zhao, L., Niu, H., Liu, Y., Wang, L., Zhang, N., Zhang, G., Liu, R., & Han, M. (2019). LOX inhibition downregulates MMP-2 and MMP-9 in gastric cancer tissues and cells. *Journal of Cancer*, 10(26), 6481.

Zou, W., Gu, J., Li, J., Wang, Y., & Chen, S. (2022). Tailorable antibacterial and cytotoxic chitosan derivatives by introducing quaternary ammonium salt and sulfobetaine. *International Journal of Biological Macromolecules*, 218, 992–1001. https://doi.org/10.1016/j.ijbiomac.2022.07.122

Zubareva, A., Shagdarova, B., Varlamov, V., Kashirina, E., & Svirshchevskaya, E. (2017). Penetration and toxicity of chitosan and its derivatives. *European Polymer Journal*, 93, 743–749. https://doi.org/10.1016/j.eurpolymj.2017.04.021

Zuber, P. L., Gruber, M., Kaslow, D. C., Chen, R. T., Giersing, B. K., & Friede, M. H. (2021). Evolving pharmacovigilance requirements with novel vaccines and vaccine components. *BMJ Global Health*, 6(Suppl 2), e003403.

12 Application of Chitosan-Based Composite Coatings Enhanced with Antimicrobial Agents, Including Essential Oils

Ekong Mercy[1] *and Ekpenyong Glory Maurice*[2]
[1]Department of Microbiology, Faculty of Biological Sciences, [111]
[2]University of Cross River State (UNICROSS), Calabar, Nigeria

12.1 INTRODUCTION

Chitosan is a natural polysaccharide obtained from chitin, which is present in the exoskeletons of crustaceans. It can also be found in the exoskeletons of insects and the cell walls of fungi, including yeast. Its types include alkylated chitosan, alginate cellulose chitosan, and chitosan membrane hydrogel (Jayanetti et al., 2023). Preparing chitosan involves various steps, such as preparing the chitin from the biological materials, followed by the deacetylation leading to chitosan. The high demand for effective antimicrobial agents has led to research into novel materials that can combat microbial threats while being safe and environmentally friendly, and the development and use of chitosan-based coatings infused with active agents is a promising technique due to its excellent properties such as the ability to form an edible film on the surfaces of items such as fruits and vegetables. The inhibition of microbial growth and oxygen penetration to plant tissues and, most importantly, its ability to enhance antimicrobial and anti-browning properties, thereby improving the storage and safety of food products, is a valuable feature with potential uses in the food sector (Xing et al., 2011a). Chitosan possesses remarkable bio accessibility, eco-friendliness, surface binding capacity, and non-toxic characteristics. These qualities make it highly beneficial for various applications across multiple sectors, including wound care, drug delivery, food packaging, dietary supplementation, chelation, as well as pharmaceutical and biomaterial uses, and it is currently gaining broader applications in the food industry, environmental protection, and biotechnology (Gallo et al., 2016).

Essential oils (EOs) are natural compounds derived from various aromatic plants (Maurya et al., 2021). They are concentrated in roots, peels, leaves, seeds, fruits, and barks. The synergistic interaction between chitosan and EOs arises from their complementary mechanisms of action. It can be considered a novel approach to vegetable preservation, food preservation, and bioactive packaging as they also have antibacterial, antifungal, and antioxidant properties. The application of EOs in food preservation is often restricted due to their usage, cost, and certain limitations, including their pungent aroma and possible toxicity. Furthermore, adding EOs into chitosan-based composites can impart additional benefits, such as improved stability, controlled release, and extended antimicrobial activity (Cai et al., 2010). Despite the potential advantages of chitosan-based composite when combined with EOs, several challenges remain to be addressed. These include adjusting the composition and properties of the chitosan-based composite to achieve maximum antimicrobial efficacy while ensuring biocompatibility and stability and addressing issues related to scalability and cost-effectiveness.

DOI: 10.1201/9781003355458-12

This study explores the advancements in developing chitosan-based composites combined with EOs, such as essential agents, the mechanism of action, applications across various sectors, and prospects in this rapidly evolving field.

12.2 CHITOSAN

Chitosan is a naturally derived linear polysaccharide obtained by the partial deacetylation of chitin, which is why it is sometimes referred to as deacetylated chitin (Chandy et al., 1990). Chitin itself is a key structural element present in the exoskeletons of insects and crustaceans, such as crabs and shrimps, as well as in the cell walls of fungi. It ranks as the second most abundant natural polysaccharide after cellulose (Cheung et al., 2015). The polymer structure comprises randomly arranged β-(1–4)-linked D-glucosamine and *N*-acetyl-D-glucosamine units. One distinguishing feature of chitosan is its cationic nature, which carries a positive charge, unlike most polysaccharides, which are either neutral or negatively charged in acidic environments. This characteristic enables it to engage in ionic interactions with various negatively charged polymers that are either natural or synthetic (Venkatesan & Kim, 2010). Chitosan is known for its distinctive properties, including biocompatibility, non-toxicity, low risk of allergic reactions, and biodegradability, rendering it appropriate for various applications (Kumar et al., 2004). These attributes are largely determined by the degree of deacetylation, expressed as the molar fraction or percentage of deacetylated units, and by its molecular weight (Aranaz et al., 2009).

12.3 TYPES OF CHITOSAN

12.3.1 Chitosan Based on Structural Modification

12.3.1.1 Alkylated Chitosan

This form of chitosan occurs when an alkyl group is added to chitosan, weakening its intermolecular hydrogen bonds and enhancing its solubility (Kurita & Isogai, 2012). The hydrophobic nature of the alkyl group enhances the solubility of chitosan. However, increasing the size of the alkyl chain reduces solubility, indicating that the solubility of modified forms of chitosan can be controlled by adjusting the chain length (Ma et al., 2008; Yang et al., 2005). *N*-alkylated chitosan can be synthesized through reactions with halogenated alkanes (Palacio et al., 2019) or derived from long-chain fatty acyl groups and higher fatty aldehydes. Its antibacterial and coagulation properties make alkylated chitosan valuable for producing medical gauze (Yang et al., 2005).

12.3.1.2 Acylated Chitosan

This variant of chitosan is the most commonly utilized, and it is produced by the interaction of chitosan with various organic acids and their derivatives, including anhydrides, acyl chlorides, and aromatic or aliphatic acyl groups, which bind to the molecular chain (Cai et al., 2015). The acylation process disrupts hydrogen bonds within and between chitosan molecules, reducing its crystallinity and enhancing its water solubility. Chitosan's molecular structure includes two hydroxyl groups: a primary hydroxyl group located at C6–OH and a secondary hydroxyl group positioned at C3–OH. N-acylated chitosan derivatives offer improved biocompatibility, anticoagulant activity, and enhanced blood compatibility. Additionally, they are non-inflammatory in the human body, making them suitable for use as drug carriers or sustained-release agents in pharmaceutical applications (Medeiros et al., 2018).

12.3.1.3 Thiolated Chitosan

These chitosan derivatives are created by attaching ligands containing –SH groups to the polymer's primary hydroxyl and amino groups and are regarded as highly promising (Christoph et al., 2021).

This alteration boosts its mucoadhesive characteristics, making it stick effectively to mucosal surfaces. Thiolated chitosan is prized in medication delivery for its controlled release properties, enhancing bioavailability and stability. Its applications extend to different biomedical sectors, offering a diverse foundation for creating improved drug delivery systems and other technologies benefiting from the combined features of chitosan and thiol functionality.

12.3.2 Chitosan Based on Molecular Size

12.3.2.1 High-Molecular-Weight Chitosan

This term refers to a form of chitosan with an increased molecular weight, resulting from more repeating units. The molecular weight of a polymer chain is typically measured based on the number of monomer units, such as glucosamine and *N*-acetylglucosamine, present in the polymer chain in the instance of chitosan. The molecular weight exceeds 700 kDa (Román-Doval et al., 2023).

12.3.2.2 Medium-Molecular-Weight Chitosan

This particular chitosan is classified as intermediate in size and degree of deacetylation, falling between the low- and high-molecular-weight variants. It is a versatile substance used across multiple disciplines, such as agriculture, water treatment, and wound healing.

12.3.2.3 Low-Molecular-Weight Chitosan

Chitosan with a low molecular weight has a lesser degree of deacetylation with a smaller molecular size than conventional chitosan. Due to its increased solubility and potential for improved bioavailability, it is regularly used in biomedical applications, drug delivery systems, and as a nutritional supplement.

12.4 SOURCES OF CHITOSAN

12.4.1 Crustaceans

An estimated 20%–30% of crustacean exoskeletons are made up of chitin (Yeul & Rayalu, 2012). Chitosan is commercially produced from crustacean shells through an established industrial process and is available in the market (Tanzina et al., 2022). Challenges in obtaining chitosan from crustaceans include the limited availability of the raw material due to its seasonal and localized nature (Streit et al., 2009). Examples include lobster, shrimps, krill, barnacles, and crayfish.

12.4.2 Insect Cuticles

A total of 525% of insect cuticles comprise chitin, the main chitosan component. Examples include houseflies, silkworms, ants, cockroaches, spiders, beetles, brachiopods, and scorpions,(Abidin et al., 2020).

12.4.3 Fungal Cell Walls

Fungi is utilized in baking, brewing, antibiotic production, organic acid synthesis, and enzyme manufacturing, leading to the generation of fungal biomass wastes. This process and other wastes could be used to extract chitinous polysaccharides. Obtaining chitosan from fungal cell walls is a feasible method to address the challenge of obtaining chitosan from marine sources, which is available seasonally (Huq et al., 2022).

12.5 CHITOSAN SYNTHESIS PROCESS STEPS

The transformation of chitin into chitosan (CH) involves several chemical processes. The primary stages of extracting chitosan are demineralization, deproteinization, and deacetylation.

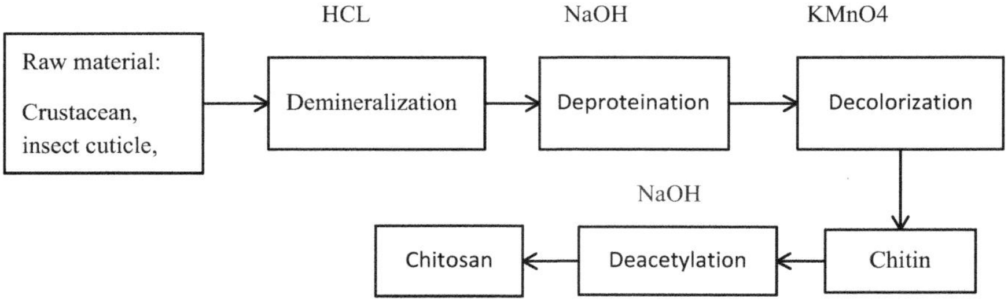

12.5.1 Formulation and Preparation of Chitosan-Based Composite Coating with Essential Oil

Edible biodegradable coatings offer significant advantages to the fruit and vegetable industry by carrying antimicrobial compounds, such as EOs, which help combat pathogens (Vásconez et al., 2009). Common EOs incorporated into chitosan-based coatings include lemon oil, rosemary oil, sunflower oil, lemongrass oil, carvacrol oil, and bergamot oil.

12.5.1.1 Formulation of a Chitosan-Based Coating with Lemon Oil

The process begins by dispersing chitosan in a water-based solution with glacial acetic acid. Once the chitosan has fully dissolved, lemon EO is added, followed by blending. To create the composite chitosan coating, the resulting emulsion undergoes a second homogenization at 165 MPa in a single pass using microfluidization (Perdones et al., 2012).

12.5.1.2 Formulation of a Chitosan-Based Coating with Rosemary Oil

In its powdered form, Chitosan is initially dissolved in an aqueous acetic acid solution at 90°C and stirred for 20 minutes. The mixture is then allowed to cool to room temperature. Separately, a measured amount of clay is stirred into an aqueous acetic acid solution for 24 hours. Afterward, the clay solution is combined with the chitosan solution and mixed for 4 hours. To enhance emulsification, a solution of Tween 80 is added to the mixture and stirred for 30 minutes at 40°C. Finally, the appropriate concentration of rosemary oil is incorporated into the mixture. The resulting solution is degassed under vacuum for 5 minutes and then homogenized to produce the chitosan-oil solution (Abdollahi et al., 2012).

12.5.1.3 Formulation of a Chitosan-Based Coating with Lemongrass Oil

Low-molecular-weight chitosan is used to prepare chitosan solutions with a pH range of 5.6–7.9. Tween 80 and lemongrass EO were then added to chitosan solutions. A homogenizer blends the mixture, effectively incorporating lemongrass EO into the chitosan (Ali et al., 2015).

12.5.1.4 Formulation of a Chitosan-Based Coating with Sunflower Oil

The chitosan-based sunflower oil coating uses blends and films derived from quinoa protein extracts. Quinoa protein extracts are initially dissolved in water, and the pH is adjusted to 8 and 12, respectively. The mixture is stirred and then immersed in distilled water, and the resulting suspensions (Q-8 and Q-12) are centrifuged. The supernatants from the quinoa flour suspensions at pH 8 and 12 were collected. Meanwhile, a second solution is prepared by dissolving various amounts of chitosan in citric acid, sonicating the solution, and allowing it to stand overnight. The final mixture is formed by combining the chitosan solution with Q-8 or Q-12 suspensions. Based on the Q/CH ratio, a Q/CH/SO solution is prepared. The optimal Q/CH ratio can then be mixed with Tween 80 and sunflower oil at different concentrations, with the acidity level of the chitosan solution adjusted to 3.0 (Valenzuela et al., 2013).

12.5.1.5 Formulation of a Chitosan-Based Coating with Carvacrol Oil

A tailored chitosan was stirred in an acetic acid solution for 24 hours. To create the chitosan-oil coating, a carvacrol oil nanoemulsion was then added to the prepared chitosan solution and thoroughly blended (Severino et al., 2015).

12.6 CHITOSAN-BASED COATING

This term describes the protective layer or a chitosan-based film, a naturally derived polymer obtained from chitin. Chitosan is an ideal biopolymer for producing active edible films and coatings because of their non-toxic nature, biocompatibility, biodegradability, and excellent film-forming ability (Dutta et al., 2009). Compared to conventional synthetic packaging, chitosan-based consumable films and coatings have demonstrated significant potential in food preservation. Their key advantage lies in their safety for consumers and the environment (Aiduang et al., 2022). Designed to prolong the storage duration and preserve the food's quality, these coatings consist of a thin, edible, non-toxic layer applied directly to the surface of the food, either with or without later removal. However, their hydrophilic nature makes chitosan films susceptible to moisture. To improve their mechanical, physical, and biological properties, various natural additives are incorporated, such as phenolic compounds, EOs, plant extracts, and other biopolymers (Bigi et al., 2012). Adding active components, such as EOs, further enhances the antibacterial effectiveness of chitosan-based coatings. (Xing et al., 2010).

12.7 TECHNIQUES FOR APPLYING CHITOSAN-BASED COATINGS

12.7.1 Dipping

Dipping and spraying are the most convenient and most often utilized application methods. It is the oldest coating method, dating back to the 12th century. Its application method is exceptionally straightforward: the product is dipped into the coating solution for a predetermined period, drained, and allowed to dry before storage (Skurtys et al., 2001). The dipping technique is best used for irregularly shaped products.

12.7.2 Spraying

The spraying technique is the most widely used approach in the food industry because of its less complicated method. This method involves applying a spray machine to the coating solution by expelling it through the nozzle onto the product surface. Less viscous solutions are recommended over extremely viscous solutions for spraying as the latter can clog the nozzles and are more difficult to spray. Spraying efficiency is influenced by the nozzle size, coating fluidity, and the pressure applied to release the fluid through the nozzles (Andrade et al., 2012). This method of application results in the creation of a uniform layer because of the consistency in the distribution of drop sizes and the overlapping effect. It is primarily utilized on products with a substantial surface area (Embuscado & Huber, 2009).

12.7.3 Solvent Casting

Chitosan is combined with an appropriate solvent (e.g., ionic liquid) and applied onto the substrate, forming a surface-thin film coding on the product's surface. The solvent evaporates after a particular period.

12.7.4 Brushing or Roller Coating

This technique utilizes brushes or rollers to apply chitosan coatings for small-scale applications. This technique enables precise regulation of the thickness of the coating. This technique ensures that the brush or roller is uniformly covered with the chitosan solution for consistent, efficient, and broader coverage.

12.7.5 Spin Coating

This method rotates the coated material, distributing the chitosan solution equally due to the rotational force. It is frequently employed in producing thin films and coatings on flat surfaces.

12.8 ADVANTAGES OF CHITOSAN-BASED COMPOSITE COATING

Antimicrobial Properties: Due to its natural antibacterial properties, chitosan suppresses the proliferation of fungi, bacteria, and other microorganisms, making it valuable in applications such as food packaging and medical devices where microbial contamination poses a concern. Incorporating active components into chitosan-based coatings holds significant potential for enhancing their antibacterial effectiveness and anti-browning capabilities, thereby aiding in preserving fruits and vegetables during storage (Xing et al., 2011a).

Film-Forming Capability: Chitosan forms a flexible and durable film when used as a coating. It effectively reduces moisture and aroma loss, prevents oxygen infiltration into plant tissue, and inhibits microbial growth, enhancing food safety (Mantilla et al., 2013).

Biodegradability and Sustainability: Research has consistently shown that chitosan coatings are environmentally friendly, highly biodegradable, and, in many cases, consumable (Priyadarshi & Rhim, 2020).

Adhesion and Cohesion: Chitosan exhibits strong adhesive properties, allowing it to adhere effectively to various surfaces and enhancing the coating's durability and performance. Its adhesive and cohesive strength can be modified by adjusting the degree of deacetylation (DDA), which influences the crystalline structure of chitosan (Lee et al., 2013).

Biocompatibility: Chitosan is biocompatible, meaning it is generally well-tolerated by living organisms. This renders it appropriate for applications involving interaction with biological systems, such as biomedical devices and drug delivery systems.

12.9 LIMITATIONS OF CHITOSAN-BASED COMPOSITE COATING

Mechanical Strength: Structures composed solely of chitosan as the primary polymer have poor mechanical qualities, such as low tensile strength, fracture stiffness, and rapid degradation (Vaidhyanathan et al., 2021).

Solubility: Chitosan is an essential compound with low acidity and does not dissolve in water or organic solvents. Nevertheless, it can dissolve in a weak acidic solution that is diluted with water and has a pH below 6.5 (Caiqin et al., 2006). This constraint restricts its utilization in specific applications where maintaining stability throughout a broad pH spectrum is crucial.

Sensitivity to Moisture: It is evident that moisture (absorbed water) is crucial in affecting the physical stability of chitosan powder and tablets (Viljoen et al., 2014). This moisture sensitivity can limit its use in applications where exposure to high humidity is a concern.

Processing Challenges: Chitosan's poor solubility makes it difficult to convert into films or coatings. Uniform and effective coatings may necessitate the use of specialized processing procedures.

Cost of Processing: Crustacean shells are a common source of chitosan, yet extracting it can be costly. This could raise the cost of chitosan-based coatings and reduce their ability to compete in some areas.

12.10 APPLICATION OF COMPOSITE-BASED CHITOSAN

12.10.1 Medical and Healthcare Applications

The medical and healthcare sectors utilize chitosan-based composites for various purposes because of their biocompatibility, antibacterial characteristics, and ability to release drugs in a controlled manner.

Wound Healing: Wound healing is a biological process involving the development of granulation and scar tissue. It involves stimulating stem cells and the secretion of cytokines and growth factors upon arrival at the injury location. While the skin can regenerate when the injury is mild, certain cases may lead to delayed healing or an incomplete restoration of structural integrity. Various biomaterials, including chitosan and its derivatives, can be applied to develop skin tissue through wound dressings to address this. An effective wound dressing features a permeable structure that allows oxygen exchange while acting as a protective barrier against external factors (Jones et al., 2020).

Cancer Therapy: In cancer treatment, it is essential to employ safe and capable drug carriers that efficiently deliver therapeutic agents to malignant tissues while minimizing potential harm to healthy cells. Although several drug delivery systems are available, polymeric carriers have gained considerable attention due to their capacity to improve drug targeting and prolong circulation time by reducing drug excretion through urine. Chitosan and chitin have been explored as drug delivery carriers in chemotherapy (Wang et al., 2021).

Drug Delivery Agent: Numerous studies have examined the use of chitosan and chitin as drug delivery agents. Owing to their beneficial properties, these polymers are widely used to fabricate hydrogels for drug delivery. These include bioadhesion, which enhances their ability to adhere to surfaces; a polycationic nature, which facilitates hydrogen formation and ionic bonds; and biocompatibility, which ensures they do not release toxins or trigger immune responses when in contact with living tissues or bodily fluids (Parhi, 2019).

12.10.2 Environmental Applications

Water Treatment: Chitosan-based hydrogels have garnered considerable interest for their essential water purification needs. Its notable attributes include bioavailability, biocompatibility, biodegradability, environmental friendliness, high capacity for adsorbing pollutants, and the ability to absorb water (Chelu et al., 2023). It can function as a flocculant, precipitant, and adsorbent for removing contaminants such as toxic metals, pigments, insecticides, infection-fighting drugs, microbial impurities, and biological pollutants from wastewater (Preeti et al., 2021).

Oil Spill Remediation: The potential of chitosan-based products in oil spill remediation is currently being researched. Chitosan can adsorb hydrophobic compounds, such as oil, hence expressing values in cleaning up oil polluted sites and minimizing environmental harm. Chitosan possesses reactive hydroxyl and amino groups on its main chain, allowing numerous chemical alterations. These modifications give the biopolymer different but specific capabilities.

Biodegradable Films for Agricultural Applications: Chitosan-based films serve as biodegradable crop coverings or crop coatings for agriculture seeds. These films biodegrade organically, minimizing the ecological impact linked to conventional plastic materials. A mixture of pectin and Chitosan, a biodegradable material, was developed specifically for use in food packaging. The composite film shows a lower water vapor transmission rate than the standalone pectic film. Also, composite film exhibits enhanced tensile strength and transparency (Younis & Zhao, 2019).

Soil Amendment: Chitosan helps enhance the ability of polluted soils to break down contaminants. This is done by introducing specific groups of microorganisms (bioaugmentation), supplying nutrients to the existing microorganisms that naturally break down chemicals at the polluted sites (biostimulation), and facilitating the biochemical breakdown of foreign compounds. Chitosan serves as a vehicle for encapsulating microbial communities, enabling their preservation and protection against enzymes in challenging environmental conditions, including pH and temperature (Adamczuk & Jozefaciuk, 2022).

12.10.3 Food Industry Applications

Seafood Preservation: Chitosan has demonstrated high capacity in shell life extension in seafood and aquaculture products, ensuring that they remain fresh by inhibiting the proliferation of bacteria that causes spoilage and preserving the general quality of the product. Over 20 days, applying chitosan coating on grass carp fillets successfully prevents microorganisms' growth and lipids' degradation. Additionally, it reduced the loss of water and maintained the pH levels. In addition, it was discovered that the overall number of viable and psychrotrophic microorganisms was suppressed, leading to a decrease in the production of amines. Applying a chitosan coating can potentially decrease lipid oxidation by generating chelating-free ions. Chitosan-based films can be enhanced with active compounds to improve their preservation capabilities (Kumar et al., 2023).

Meat and Processed Meat Products Preservation: Meat is highly perishable and prone to contamination, with pathogens capable of infiltrating the deeper layers of muscle tissue. Eliminating bacteria through cooking becomes more complicated once bacteria reach these inner layers (Laine et al., 2005). Food oxidation and microbiological deterioration are two significant factors affecting meat quality and longevity. Packing materials are manufactured using antimicrobials and antioxidants in place of chemical food preservatives. Numerous studies have explored the creation of chitosan-based films and their use in processing different types of meat, such as beef, chicken, and pork meat. Additionally, extract from green tea has been incorporated into chitosan films as an active component to extend prolong the storage duration of cooked sausages (Siripatrawan & Noipha, 2012).

Food Quality Monitoring: The quality of food is a critical aspect of the food distribution network. Techniques of analysis, including mass spectrometry and chromatography, used for evaluating food quality are often costly, complex, bulky, and stationary. In contrast, biosensors, known for their high sensitivity, rapid response, compact size, affordability, and portability, have become essential tools in rapid detection. These biosensors are regularly used to detect harmful substances and nutritional components in food (Ye et al., 2009). Notably, chitosan hydrogels can serve as raw materials for producing biosensors and indicators, offering diverse potential applications within the food manufacturing sector, including monitoring food quality. The pH value is one crucial parameter to assess food freshness and decomposition within the supply chain.

12.11 ESSENTIAL OIL-BASED CHITOSAN

There has been a recent surge in research interest in utilizing EOs as bioactive agents of chitosan. These secondary metabolites are derived from various components of aromatic plants (Lv et al., 2011). These substances have antioxidant and/or antibacterial capabilities. They can also engage with the packaging material, decreasing the amount of water vapor that can pass through and affect the mechanical and optical characteristics (Burt, 2004). These inherent and environmentally friendly features have positioned them as the preferred option for numerous active packaging solutions that have been tried in recent years.

Chitosan has been utilized with several EOs to enhance the effectiveness of antioxidants and antimicrobial properties. When coupled with various EOs, it has numerous food model uses and enhances the longevity of perishable foods, including meat and processed meat products, fish and shellfish, grains, and fruits. Furthermore, applying nano-techniques to create chitosan-based coatings and films has opened up new possibilities for preserving food and waste management. These methods have been employed to enclose EOs and enhance chitosan's antioxidant and antimicrobial characteristics (Noori & Hossaeini, 2023).

12.11.1 Essential Oils-Based Chitosan and Their Antimicrobial Properties

Chitosan composites infused with EOs demonstrate improved antibacterial characteristics by harnessing the intrinsic capabilities of chitosan and its bioactive chemicals found in EOs. Research revealed that the chitosan nanoparticles mixed with EOs exhibited a much more significant antibacterial impact than using chitosan or EOs singly (Barrera-Ruiz et al., 2020).

Rosemary Oil (*Rosmarinus officinalis*): Rosemary oil exhibits antimicrobial and antifungal properties, primarily attributed to compounds like rosmarinic acid. This plant, belonging to the *Lamiaceae* family, is native to South Asia and Europe. A variety of studies has been carried out to evaluate the antimicrobial and antioxidant effects of rosemary EO and its potential for extending the longevity of food products (Angioni et al., 2004).

Lemon Oil (*Citrus limon*): This oil, extracted from the *Citrus limon* plant, is generally recognized as safe (GRAS). It includes a diverse combination of volatile compounds such as limonene, beta-pinene, citral, and linalool. These components are known to exhibit antioxidant and antibacterial activities (Kabirou et al., 2022).

Tea Tree Oil (*Melaleuca alternifolia*): Tea tree oil (TTO), derived from *Melaleuca alternifolia*, a plant in the *Myrtaceae* family, is known for its potent antimicrobial, antioxidant, and anti-inflammatory properties. This is because TTO poses several volatile compounds, including p-cymene, α-pinene, α-terpinene, 1,8-cineole, Terpinolene, γ-pinene, α-Terpineol, Limonene, and other trace molecules (Pazyar et al., 2013).

Thyme Oil (*Thymus vulgaris*): Thyme EO has been widely utilized as a flavoring substance in various foods, beverages, and confectionery items. The Food and Drug Administration (FDA) in the United States has classified it as a GRAS food ingredient (Benjemaa et al., 2018). Its antibacterial activity is primarily linked to the presence of carvacrol and thyme, compounds that are also found in other EOs. Additionally, its antioxidant properties are crucial in neutralizing reactive oxygen species and preventing food oxidation (Gonçalves et al., 2017).

12.11.2 Essential Oils-Based Chitosan in Food Conservation and Packaging

Biodegradable ingredients, such as alginate, carrageenan, cellulose, chitosan, pectin, and starch, are essential in developing innovative packaging materials derived from bio-based polymers. Bioactive chitosan has gained widespread popularity as a food wrapping material due to its chemical stability, non-toxic nature, excellent film-forming ability, biocompatibility, antioxidant properties, and effectiveness in inhibiting the growth of yeasts, molds, fungi, and bacteria.

Despite these advantages, chitosan's limited thermal and mechanical stability presents significant challenges when scaling up its application as a biodegradable material for industrial use. Studies have proven that incorporating bioactive EOs into chitosan improves the physical, thermal, and mechanical characteristics of chitosan films while also enhancing their antibacterial and antioxidant activity. As a result, these chitosan-based composites function as efficient materials for proper food wrapping (Shuva et al., 2022).

These active components directly interact with food products, inhibiting microbial growth and preventing chemical reactions contributing to spoilage. This interaction helps maintain the product's integrity, quality, and safety while significantly prolonging its shelf life.

12.11.3 Challenges and Limitations of Essential Oils-Based Chitosan

Although EO-based chitosan composites have certain advantages, they also encounter specific obstacles and constraints that must be considered in their practical use.

EOs Exhibit High Volatility: EOs are volatile and susceptible to gradual evaporation. This leads to a decrease in the effectiveness of antibacterial capabilities and a decline in the aromatic qualities, particularly in situations where the product is stored for long periods.

Compatibility and Stability: Ensuring and sustaining compatibility between chitosan and EOs might pose a challenge ranging from inadequate or incomplete compatibility resulting in separation of phases, decreased stability, and degraded performance of the composite substance.

Chitosan has Low Solubility: Chitosan exhibits restricted solubility at neutral pH settings. This constraint can impact the even dispersion of EOs and their production of chitosan-based composites, especially when water-based treatments are needed.

Cost Considerations: EOs may be quite costly, affecting the overall expense of chitosan-based composites. Due to cost considerations, adoption and usability may be limited, especially in businesses with strict budget limits.

12.12 SYNERGISTIC IMPACTS OF ESSENTIAL OILS AND CHITOSAN

Chitosan-based composite coatings infused with EOs have demonstrated synergistic benefits, especially in food preservation applications. The combined antimicrobial characteristics of chitosan and EOs effectively eliminate the proliferation of bacteria, fungi, and other pathogens, prolonging perishable goods' longevity (Rhim, 2013).

Research conducted by Venkatachalam and Lekjing (2020) found that incorporating thyme EO into chitosan films enhanced their antimicrobial action, showing strong inhibition against *Staphylococcus aureus* and *Salmonella typhimurium*. This antibacterial effect is primarily attributed to thymol, a compound known for its broad-spectrum antimicrobial activity as opposed to Gram-negative and Gram-positive bacteria. Similarly, studies by Mehdizadeh et al. (2012) confirmed that chitosan films infused with EOs exhibited significant antibacterial effects against various pathogenic bacterial strains, unlike oil-free chitosan films, which did not demonstrate such activity.

Baitha et al. (2020) further investigated the effect of EOs on chitosan's antibacterial properties and observed that chitosan, when combined with thyme and clove EOs, showed the highest inhibitory activity against *S. aureus* and *S. typhimurium*. This enhanced antimicrobial performance is due to the bioactive compounds in clove oil, including eugenol and acetate, which possess potent antioxidant and phytochemical properties. Additionally, chitosan's ability to attach to the bacterial cell walls that are negatively charged disrupts membrane permeability and attaches to bacterial DNA, thereby inhibiting replication and eventually leading to cell death.

Moreover, chitosan-EO composites improve the mechanical characteristics of packaging materials while providing a sustainable and natural alternative to synthetic preservatives. The enhanced antimicrobial efficacy results from the water-repellent nature of EOs, which enables them to infiltrate microbial cell membranes more efficiently when embedded in the chitosan matrix. This combination boosts antimicrobial action and strengthens the mechanical and barrier properties of chitosan coatings, making them more suitable for use in food wrapping. Consequently, this approach offers a practical and eco-friendly solution for prolonging the longevity of perishable food products (Noori & Hosseini Marzieh, 2019).

12.13 MECHANISM OF ACTION OF ANTIMICROBIAL PROPERTIES OF CHITOSAN-BASED ESSENTIAL OILS

The positively charged nature of chitosan enables it to interact with the anionic elements of microbial cell membranes, such as lipopolysaccharides in bacteria and phospholipids in fungi. This interaction compromises the membrane's integrity and permeability, causing the leakage of intracellular contents and ultimately leading to cell death. Additionally, chitosan inhibits key enzymes and disrupts essential cellular processes, further enhancing its antimicrobial effectiveness (Kong et al., 2010).

EOs also contribute to antimicrobial activity because of their bioactive compounds, including terpenes, phenolics, and aldehydes. These compounds enhance antimicrobial efficacy by disrupting microbial cell membranes, interfering with metabolic functions, and inhibiting critical enzymes and protein synthesis pathways. By penetrating the lipid bilayer of cell membranes, EOs cause structural damage, resulting in the leakage of cellular components and, eventually, microbial death (Burt, 2004). The antimicrobial mechanism may synergistically enhance their effectiveness in combination (i.e., chitosan and EO). The water-repellent nature of EO compounds allows them to interact more readily with microbial membranes when incorporated into chitosan matrices, enhancing their penetration and disruption of cell structure. Additionally, the positively charged nature of chitosan may facilitate the binding and retention of EO components on the surface of the cell, prolonging their antimicrobial activity (Li et al., 2019).

12.14 COMPATIBILITY AND STABILITY OF CHITOSAN AND ESSENTIAL OILS

Wang et al. (2011) examined the efficacy of chitosan by examining the compatibility and stability of various EOs when combined with chitosan films. The findings indicated that incorporating cinnamon and clove bud EOs enhanced the antimicrobial characteristics of the films, while the specific soil type significantly influenced other physical properties. Thyme EO is particularly notable because of its high concentration of biologically active compounds, including thymol and carvacrol, which exhibit potent bacteriostatic, bactericidal, and antioxidant characteristics.

In research conducted by Altiok and Altiok (2010), chitosan films infused with thyme oil were developed for potential wound-healing applications, with the minimum inhibitory concentration set at 1.2%. The addition of hydrophobic EOs improved the antimicrobial characteristics and the water barrier capabilities of the polymer (Benjakul & Prodpran, 2014). Similarly, Perdones et al. (2016) investigated chitosan systems containing thyme and basil EOs combined with oleic acid, reporting enhanced water vapor permeability across all samples. Rubilar et al. (2013) also found that chitosan films that had grape and carvacrol seed extract exhibited improved barrier properties.

However, the compatibility and stability of chitosan with EOs may vary based on factors such as concentration, pH, and the specific chemical components. Therefore, conducting detailed compatibility and stability assessments is crucial to ensuring optimal functionality and shelf life for the intended applications.

12.15 FUTURE PERSPECTIVES

Chitosan-based composite coatings, enriched with antimicrobial agents like EOs, present promising opportunities across multiple sectors, including food packaging, biomedicine, pharmaceuticals, and agriculture. In the food manufacturing industry, these coatings help prolong the longevity of products by suppressing microbial growth, thereby minimizing the reliance on synthetic preservatives. Their antimicrobial activity and biocompatibility in biomedical applications make them suitable for wound dressings and medication delivery systems.

In agriculture, they may be applied to seeds or plant surfaces to prevent fungal or bacterial infections, promoting healthier crops and reducing the need for chemical pesticides. As research continues, there is potential for further innovation and optimization of these composite coatings for broader applications and increased efficiency.

12.16 REGULATORY CONSIDERATIONS AND SAFETY CONCERNS

The United States Food and Drug Administration (USFDA) designates EOs as GRAS. However, while some chitosan-EO-based products hold regulatory approval with defined permissible values, others do not. According to Section 409 of the Federal Food, Drug, and Cosmetic Act, the GRAS status of a specific EO may be subject to application limits (Code of Federal Regulations [CFR], 2022). Additionally, regulating EOs for human use may change based on the application method and the intended purpose (CFR, 2022). Adherence to good manufacturing practice (GMP) principles is mandatory, regardless of the specific EO used. Additionally, these regulatory guidelines may differ based on the country where the food product is marketed. Compatibility between chitosan, EOs, and several other additives used in the composite coating is confirmed to avoid adverse interactions or unintended consequences. Migration studies are also conducted to assess if any components of the composite coating, including chitosan, migrate into food products and if the components comply with safety regulations regarding migration limits.

The antimicrobial efficacy of the composite coating is also evaluated to ensure that it effectively inhibits the growth of pathogens or spoilage microorganisms without posing risks to human health (Dini et al., 2020).

12.17 DEVELOPMENT OF NOVEL DELIVERY SYSTEM FOR CHITOSAN AND ESSENTIAL OILS

Developing a novel delivery system for chitosan and EOs could involve encapsulating the EOs within chitosan nanoparticles or microspheres. Microencapsulation technology has garnered considerable focus due to its wide range of applications across various sectors, including industry, agriculture, medicine, pharmacy, and biotechnology. In the medical field, numerous formulations utilize microparticles to achieve specific objectives, such as masking unpleasant tastes and odors, controlling drug release, protecting drugs from harsh body fluids like gastric juice, isolating cells, and developing immunoassays.

These microparticles are typically produced with polymers as the shell or matrix materials, with biodegradable and biocompatible polymers being the preferred options. Such systems offer additional benefits by facilitating the incorporation of appropriate drug quantities and enhancing the bioavailability of degradable drugs, as well as improving the permeation of hydrophilic substances across epithelial barriers (Singl & Sinha, 2004).

Furthermore, microparticles provide advantages such as a larger surface area, improved mass transfer estimation, precise kinetic modeling, and controlled drug release into body fluids.

Additionally, optimizing the formulation parameters and incorporating other biocompatible materials could further improve the delivery system's performance and applicability.

12.18 POTENTIAL IMPROVEMENTS AND FUTURE DEVELOPMENT IN THE FIELD

The commercial development of chitosan involves various chemical processes, including deproteinization, decalcification, deacetylation, and decolorization. These processes typically involve exposure to high levels of acidity and alkalinity, which can negatively impact the environment. Thus, determination of a more accurate biological methods for reproduceable results is needed.

The identification of a novel, more blended chitosan is composite chitosan that will improve food quality for commercial purposes.

Nanotechnology-based strategies can potentially enhance the effectiveness of bioactive chitosan films and coatings while supporting the development of stable and scalable hybrid materials (Ahmed, 2015).

In the current research world, more research is essential to optimize chitosan formulations for better stability, efficacy, and targeted delivery of chitosan and EOs.

Interdisciplinary research and collaboration are crucial for advancing the field and unlocking the full potential of chitosan and EOs in various industries, including healthcare, agriculture, and cosmetics.

12.19 CONCLUSION

Chitosan, derived from the deacetylation of chitin, is a biologically sourced compound known for being biodegradable, eco-friendly, biocompatible, and non-toxic, making it a valuable research subject. With its potent antimicrobial properties, chitosan is frequently applied in food preservation, agriculture, and medical treatments. Incorporating antimicrobial agents like EOs into chitosan-based composite coatings presents significant potential across various sectors, including food packaging, biomedicine, pharmaceuticals, and environmental remediation.

These composite coatings exhibit superior antimicrobial activity, extended shelf life, and better mechanical characteristics compared to conventional coatings. Moreover, they offer an eco-friendly and sustainable alternative, helping to reduce the reliance on synthetic chemicals across various industries. However, further research is essential to optimize formulation parameters, assess long-term stability, and explore additional functionalities to maximize the potential benefits of these composite coatings in a large array of uses.

REFERENCES

Xing, X., Li, L., & Zhang.(2012). Effect of TiO nanoparticles on the antibacterial and physical properties of polyethylene-based film: Progress in organic coatings, 73(2), 219–224.

Abdollahi, M., Rezaei, M., & Farzi, M. (2012). A novel active bionanocomposite film incorporating rosemary essential oil and nanoclay into chitosan. *Journal of Food Engineering,* 111, 343–350.

Abidin, N.Z., Kormin, F., Abidin, N.Z., Anuar, N.M., & Abu, Bakar, .M. (2020). The potential of insects as alternative sources of chitin: An overview on the chemical method of extraction from various sources. *International Journal of Molecular Science*, 20, 21–49.

Adamczuk, A., & Jozefaciuk, G. (2022). Impact of chitosan on the mechanical stability of soils. *Molecules*, 27(7), 22–73.

Ahmed, E., M. (2015). Preparation, characterization and application of chitosan: A review. *Journal of Advanced Resolution*, 6(0), 105–121.

Aiduang, W., Chanthaluck, A., Kumla, J., Jatuwong, K., Srinuanpan, S., Waroonkun, T., Oranratmanee, R., Lumyong, S., & Suwannarach, N. (20152022). Amazing fungi for eco-friendly composite materials: A comprehensive review. *Journal of Fungi*, 8, 842.

Ali, A., Noh, N., & Mustafa, M. (2015). Antimicrobial activity of chitosan enriched with lemongrass oil against anthracnose of bell pepper. *Food Packaging and Shelf Life*, 3, 56–61.

Altiok, D., & Altiok, E. (2010) Physical, antibacterial and antioxidant properties of chitosan films incorporated with thyme oil for potential wound healing applications. *Journal of Medical Science*, 21, 223–227.

Andrade, R., Skurtys, O., & Osorio, F. (2012). Atomizing spray systems for application of edible coatings. *Comprehensive Reviews in Food Science and Food Safety*, 11, 323–337.

Angioni, A., Barra, A., & Cereti, E. (2004). Chemical composition, plant genetic differences, antimicrobial and antifungal activity investigation of the essential oil of *Rosmarinus officinalis. Journal of Agricultural and Food Chemistry*, 52, 3530–3535.

Aranaz, I., Mengíbar, M., Harris, R., Paños, I., Miralles, B., Acosta, N., Galed, G., & Heras, Á. (2009). Functional characterization of chitin and chitosan. *Current Chemical Biology*, 3, 203–230.

Baitha, G.E.S., Alkazmi, L.M., Wased, L.G., Beshbishy, A. M., & Rashwan, E.K. (2020). Traditional uses, bioactive chemical constituents, pharmacological and toxicological activities. *Biomolecules*, 10(2), 20–25.

Barrera-Ruiz, D.G., Cuesras-Rosas, G.C., Sancheze-Marines,R.I., Alvarez-Ainza, M.L., Moreno-Ibarra, G.M., Lopez-meneses, A.K, Plascencia-jatomea, M., Cortex-Rocha, M.O(2020). Antimicrobial Activity of Essential oils encapsulated in Chitosan nanoparticles. *Food Science and Technology;* 40, p;568-573.

Benjakul, S., & Prodpran, T. (2014). Structural, morphological and thermal behavior characterizations of fish gelatin film incorporated with basil and citronella essential oils as affected by surfactants. *Food hydrosolHydrosol*, 41, 33–43.

Benjemaa, M., Neves, M., Falleh, H., Isoda, H., Ksouri, R., & Nakajima, M. (2018). Nanoencapsulation of *Thymus capitatus* essential oil: Formulation process, physical stability characterization and antibacterial efficiency monitoring. *Introductory Crop Production,* 113,414–421.

Bigi, F., Haghighi, H., Siesler, H.W., Licciardello, F., & Pulvirenti, A. (2012). Characterization of chitosan-hydroxypropyl methylcellulose blend films enriched with nettle or sage leaf extract for active food packaging applications. *Food Hydrocolloids*, 120, 106–979.

Burt, S. (2004). Essential oils: their Their antibacterial properties and potential applications in foods -: a A review. *International Journal of Food Microbiology*, 94, 223–253.

Cai, J., Dang, Q., Liu, C., Fan, B., Yan, J., Xu, Y., & Li, J. (2015). Preparation and characterization of N-benzoyl-O-acetyl-chitosan. *International Journal of Biological Macromolecules*, 77, 52–58.

Cai, Q., Gu, Z., Chem, Y., Han, W., Fu, T., Song, H., Li, F (2010). Degradation of Chitosan by an electrochemical process. *ELESEVIER*; v 79 (3); p 783-785

Caiqin, Q., Huirong, L., Qi, X., & Yi, L., Juncheng, Z., & Yumin, D. (2006). Water-solubility of chitosan and its antimicrobial activity. *Carbohydrate Polymers*, 63, 367–374.

Chelu, M., Musuc, A., Popa, M., & Calderon, M. (2023). Chitosan hydrogels for water purification applications. *Gels*, 9(8), 664.

Cheung, R.C., Wong, J.H., & Chan, W.Y. (2015). Chitosan: An update on potential biomedical and pharmaceutical applications. *Marine Drugs*, 13(8), 56–86.

Chhandy, C. P. Sharma (1990). Chitosan as a Biomaterial. Biomaterial, Artificial cells and Artificial Organs. *Scientific Research;* 18 (1):1-24.

Christoph, F., Markus, K., & Andreas, B. (2021). Thiolated Chitosan'schitosan's: A multi-talented class of polymers for various applications. *Biomacromolecules*, 22 (1), 24–56.

Code of Federal Regulation; 40C.F.R &50:11 (2022).

Dini, H., Fallah, A. A., Bonyadian, M., Abbasvali, M., & Soleimani, M. (2020). Effect of edible composite film based on chitosan and cumin essential oil-loaded nanoemulsion combined with low-dose gamma irradiation on microbiological safety and quality of beef loins during refrigerated storage. *International Journal of Biological Macromolecules*, 164, 1501–1509.

Dutta, P., Tripati, Shipra T., Mehrotra, G., & DuttaJoydeep, DJ. (2009). Perspectives for chitosan based antimicrobial films in food applications. *Food Chemistry*, 114(4), 1173–1182.

Embuscado, M., & Huber, K. (2009). *Edible Films and Coatings for Food Applications*, Springer, New York. *Food Chemistry,* 114, 1173–1182,

Gallo, M., Naviglio, D., Armone Caruso, A., & Ferrara, L. (2016). Applications of chitosan as a functional food. In A. M. Grumezescu (Ed.), *Novel Approaches of Nanotechnology in Food* (pp. 425–464). Elsevier.

Gonçalves, N.D., Pena, F.D., Sartoratto, A.F., Derlamelina, C.H., Duarte, M.C., & Antunes, A.E. (2017). Encapsulated thyme (*Thymus vulgaris*) essential oil used as a natural preservative in bakery product. *Food Resolution International,* 96, 154–160.

Huq, T., Khan, A., Brown, D., Dhayagude, N., He, Z., & Ni, Y. (2022). "Sources, production and commercial applications of fungal chitosan: a A review.," *Journal of Bioresources and Bioproducts,* 7, 85–98.

Jayanetti, M., Mendis, A., Ekanayake, G., & Vigneswaran, S. (2023). Recent advances in chitosan-based applications: a A review. *Materials*, 16, 20–73.

Jones, M., Kujundzic, M., John, S., & Bismarck, A. (2020). Crab vs. mMushroom: A review of crustacean and fungal chitin in wound treatment. *Mar. ine Drugs*, 18, 64.

Kabirou, O., Qais, A., Hang, Y., Yunfei, X., Yahui, G., Mi, L., Yuhang, D., Kunfeng, L., Yulun, C., & Weirong, Y. (2022). Preparation and characterization of chitosan-based antimicrobial films containing encapsulated lemon essential oil by ionic gelation and cranberry juice, . *Food Chemistry*, 397, 133–781.

Kong, M., X., G., Chen, X., G.,, K., Xing, K., & Park, H.J. (2010). Antimicrobial properties of chitosan and mode of action: a A state-of-the-art review. *International journal Journal of Food Microbiology*, 144(1), 51–63.

Kumar, A., Yadav, S., Pramanik, J., Sivamaruthi, B., Jayeoye, T., Prajapati, B., & Chaiyasut, C. (2023). Chitosan-based composites: Development and perspective in food preservation and biomedical applications. *Polymers*, 15(15), 31–50.

Kumar, M.N., Muzzarelli, R.A., Muzzarelli, C., Sashiwa, H., & Domb, A.J. (2004). Chitosan chemistry and pharmaceutical perspectives. *Chemial Revolutionary Journal*, 104, 6017–6084.

Kurita, Y., & Isogai, A. N. (2012). Alkylations of chitosan promoted with sodium hydrogen carbonate under aqueous conditions. *International Journal of Biology Macromolecules*, 50, 741–746.

Laine, E., Scheftel, J., Boxrud, D., Vought, K., Danila, R., Elfering, K., & Smith, K. (2005). Outbreak of *Escherichia coli* O157:H7 infections associated with non-intact blade-tenderized frozen steaks sold by door-to-door vendors. *Journal of Food Protection*, 68, 1198–1202.

Lee, D., Lim, C., Israelachvili, J., & Hwang, D. (2013). Strong adhesion and cohesion of chitosan in aqueous solutions. *Langmuir*, 29(46), 142–229.

Li, W., Chen, Y., Ye, M., Guo, M., Zhu, Y., Lu, A., &Wang, X. (2019). The antibacterial activity and antibacterial mechanism of chitosan and its application in fish preservation: a A review. *Journal of Aquatic Food Product Technology*, 28(6), 633–655.

Lv, F., Liang, H., Yuan, Q., & Li, C. (2011). In vitro antimicrobial effects and mechanism of action of selected plant essential oil combinations against four food-related microorganisms. *Food Resolution International*, 44, 3057–3064.

Ma, G., Yang, D., Zhou, Y., Xiao, M., Kennedy, J.F., & Nie, J. (2008). Preparation and characterization of water-soluble N-alkylated chitosan. *Carbohydrate Polymers*, 74, 121–126.

Mantilla, M., Castell-Perez, M., Gomes, C., & Moreira, R. (2013). Multilayered antimicrobial edible coating and its effect on quality and shelf-life of fresh-cut pineapple (*Ananas comosus*),"). *LWT—Food Science and Technology*, 51, 37–43.

Maurya, A., Prasad, J., Das, S., & Dwivwdy, A., K. (2021). Essential oils and their application in food safety. *Frontiers in Sustainable Food Systems*, 5, 640–660.

Medeiros, G.L., Carvalho, I. C., & Mansur, H.S. (2018). Amino acid-grafted and N-acylated chitosan thiomers: Construction of 3D bio-scaffolds for potential cartilage repair applications. *International Journal of Biological Macromolecules*, 114, 270–282.

Mehdizadeh, T., Tajik, H., Razavi, Rohani, R.S.M., & Oromiehie, A.R. (2012). Antibacterial, antioxidant and optical properties of edible starch-chitosan composite film containing thymus essential oil. *Resolution Forum International Journal*, 3(3), 167–173.

Noori, S., & Hossaeini, S. (2023). Chitosan-based coatings and films incorporated with essential oils: Applications in food models. *Food Measure,* 17, 4060–4072.

Noori, S.M.A., & Hosseini Marashi, S.M. (2019). Chitosan-based composite films incorporated with tarragon essential oil: Physical, mechanical, structural and antioxidant properties. *Food Hydrocolloids*, 90(19), 246–254.

Palacio, D. A., Urbano, B.F., Palencia, M., & Rivas, B. L. (2019). Preparation of alkylated chitosan-based polyelectrolyte hydrogels: The effect of monomer charge on polymerization. *European Polymers Journal*, 118, 551–560.

Parhi, R., Crini, G., & Lichtfouse, E. (2019). Chitin and chitosan in drug delivery sustainable. In: Crini, G., & Lichtfouse, E. (Eds.), *Agriculture Reviews 36: Chitin and Chitosan: Applications in Food, Agriculture, Pharmacy, Medicine and Wastewater Treatment*. Springer International Publishing, Cham, 175–239.

Pazyar, N., Yaghoobi, R., Bagherani, N., & Kazerouni, A. (2013). A review of applications of tea tree oil in dermatology. *International Journal Dermatology*, 52, 784–790.

Perdones, A., Chiralt, A., & Vargas, M. (2016). Properties of film-forming dispersions and films based on chitosan containing basil or thyme essential oil. *Food hydrosolHydrocolloids*, 57, 271–279.

Perdones, A., Escride, I., Chiralt, A., Vargas, M (2015). Effect of Chitosan-lemon essential oil coating on volatile profile of strawberries during storage. *ScienceDirect*; v 197, p 979-986.

Preeti, P., Anjali, P., Kazunori, N., & Brijesh, K. (2021). Applications of chitosan in environmental remediation: A review. *Chemosphere*, 266, 128–934.

Priyadarshi, R., & Rhim, J. (2020). Chitosan-based biodegradable functional films for food packaging applications. *Innovations of Food Science and Emerging Technologies*, 62, 102–346.

Rhim, JongJ.-Whan. (2013). Antimicrobial and antioxidant activities of Moringa Oleifera essential oil coated onto a polyethylene film. *Food controlControl*, 32(2), 581–587.

Román-Doval, R., Torres, S., Tenorio, A., Gómez, A., & Valencia, A. (2023). Chitosan: Properties and its application in agriculture in context of molecular weight. *Polymers*, 15(13), 28–67.

Rubilar, J., F., Cruz, R.M.S., Silva, H.D., Vicente, A.A., & Vieria, M.C. (2013). Physico-mechanical properties of chitosan films with carvacol and grape seed extract. *Journal of Food Engineering*, 115, 466–474.

Severino, R., Ferrari, G., Vu, K., Donsì, F., Salmieri, S., & Lacroix, M. (2015). Antimicrobial effects of modified chitosan-based coating containing nanoemulsion of essential oils, modified atmosphere packaging and gamma irradiation against *Escherichia coli* O157:H7 and *Salmonella Typhimurium* on green beans. *Food Control*, 50, 215–222.

Shuva, B., Dominic, A., & Azam, A. (2022). Bioactive chitosan and essential oils in sustainable active food packaging: Recent trends, mechanisms, and applications, . *Food Packaging and Shelf Life*, 34, 100–952.

Sinha, V., Single, A.K., Wadhawan, S., Kaushik, R., Kumria, R., Keerti, B., Dhawan, S (2004). Chitosan microsphere as a potential carrier for drugs. *International Journal of Pharmaceutics;* 274 (1-4):1-33.

Siripatrawan, U., & Noipha, S. (2012). Active film from chitosan incorporating green tea extract for shelf-life extension of pork sausages. *Food Hydrocoll*, 27, 102–108.

Skurtys, O., Acevedo, C., Pedreschi, F., Enrione, J., Osorio, F., Aguilera, J.M (2010). Food Hydrocolloid edible films and coatings. *Nova Science*; v.3, p; 41-80.

Streit, F., Koch, F., Laranjeira, M., & Ninow, L. (2009). "Production of fungal chitosan in liquid cultivation using apple pomace as substrate," . *Journal of Microbiology*, 1(5), 20–25.

Tanzina, H., Avik, K., Brown, D., Dhanygude, N,m He, Z., Ni, Y (2022). Sources, Production and Commercial Application of Fungal Chitosan: A Review. *Journal of Bioresources and Bioproducts;* v 1 (2); p 85-98.

Vaidhyanathan, B., Vincent, P., Vadivel, S., Karuppiah, P., ALAl- Dhabi, N., Sadhasivam, D., Vimalraj, S., & Saravanan, S. (2021). Fabrication and investigation of the suitability of chitosan-silver composite scaffolds for bone tissue engineering applications. *Process Biochemical*, 100, 178–187.

Valenzuela, A., Abugoch, L., & Tapia, C. (2013). Quinoa protein-chitosan-sunflower oil edible film: mechanicalMechanical, barrier and structural properties. *Food Science and Technology*, 50, 531–537.

Vásconez, M., Flores, S., Campos, C., Alvarado, J., & Gerschenson, L. (2009). Antimicrobial activity and physical properties of chitosan-tapioca starch based edible films and coatings, . *Food Research International*, 42, 762–769.

Venkata-chalam, K and Lekjing, S (2020). A Chitosan-basededible film with clove essential oil and nisin for improving the quality and sheif life of pork patties in cold storage. *Royal Science of Chemistry;* 10:17777.

Venkatesan, J., & Kim, S. (2010). Chitosan composites for bone tissue engineering- : An overview. *Marine Drugs*, 8, 2252–2266.

Viljoen, J., Steenekamp, J., Marais, A., & Kotzé, A. (2014). Effect of moisture content, temperature and exposure time on the physical stability of chitosan powder and tablets. *Drug Development Pharmeutical*, 40(6), 730–742.

Wang, G., Li, R., Parseh, B., & Du, G. (2021). Prospects and challenges of anticancer agents' delivery via chitosan-based drug carriers to combat breast cancer: A Reviewreview. *Carbohydrate Polymer*, 268, 118–192.

Wang, L., Liu, F., Jiang, Y., Chai, Z., Li, P., Cheng, Y., & Jing, H. (2011). Synergistic antimicrobial activities of natural essential oils with chitosan films. *Journal of agricultural Agricultural and Food Chemistry*, 59, 123–130.

Xing, Y., Li, X., Xu, Q., Yun, J., & Lu, Y. (2010). Antifungal activities of cinnamon oil against Rhizopus nigricans, Aspergillus flavus and Penicillium expansum in vitro and in vivo fruit test. *International Journal of Food Science and Technology*, 45, 1837–1842.

Xing, Y.Q., Xu, Q.Z., Che, Z.X., Li, X., & Li, W. (2011a). Effects of chitosan- oil coating on blue mold disease and the quality of fruits. *Food and functionFunction*, 20(50), 466–474.

Yang, T.C., Chou, C.C., & Li, C.F. (2005). Antibacterial activity of N-alkylated disaccharide chitosan derivatives. *International Journal of Food Microbiology*, 97, 237–245.

Ye, Y., Guo, H., & Sun, X. (2009). Recent progress on cell-based biosensors for analysis of food safety and quality control. *Biosensors and Bioelectronics*, 126, 389–404.

Yeul, V.S., & Rayalu, S.S. (2012). Unprecedented chitin and chitosan: A chemical overview. *Journal Polymer Environment*, 21, 606–614.

Younis, H., & Zhao, G. (2019). Physicochemical properties of the edible films from the blends of high methoxyl apple pectin and chitosan. *International Journal of Biological Macromolecules*, 131, 1057–1066.

13 The Health, Nutraceutical, and Nutritional Benefits of Edible Coatings Based on Chitosan

John Tsado Mathew[1], Abel Inobeme[2], Charles Oluwaseun Adetunji[3], Elijah Yanda Shaba[4], Yakubu Azeh[1], Musah Monday[1], Mary Jumai Jiya[5], Muhammad Aishetu Ibrahim[1], Etsuyankpa Muhammad Bini[6], Ismaila Aliyu Oga[6], Musa Safiyanu Tanko[6], Jibrin Mohammed Ndaejiko[7], Tanko Ezekiel[4], Jonathan Hussaini[1], and Amos Mamman[1]

[1]Department of Chemistry, Ibrahim Badamasi Babangida University Lapai, Niger State, Nigeria

[2]Department of Chemistry, Edo University Uzairue, Auchi, Edo State, Nigeria

[3]Department of Microbiology, Applied Microbiology, Biotechnology and Nanotechnology Laboratory, Edo University Uzairue, Auchi, Edo State, Nigeria

[4]Department of Chemistry, Federal University of Technology, Minna, Niger State, Nigeria

[5]Department of Food Science and Technology, Federal University of Technology, Minna, Niger State, Nigeria

[6]Department of Chemistry, Federal University of Lafia, Nasarawa State, Nigeria

[7]Department of Microbiology, Ibrahim Badamasi Babangida University Lapai, Niger State, Nigeria

13.1 INTRODUCTION

In recent years, there has been a growing emphasis on the development of sustainable and natural food packaging solutions that not only extend shelf life but also offer health and nutritional benefits. Edible coatings have emerged as a promising alternative to traditional packaging materials by providing a protective barrier against moisture loss and microbial decay, while also offering additional functional properties. One such edible coating that has gained significant attention is based on chitosan, a biopolymer derived from chitin found in crustacean exoskeletons (Perez-Vazquez et al., 2023).

Chitosan is known for its unique properties, including biocompatibility, biodegradability, film-forming ability, antimicrobial activity, heavy metal binding ability, antioxidant properties, immunostimulatory effects, and mucoadhesive properties. These characteristics make it an attractive material for the development of edible coatings with potential health, nutraceutical, and nutritional benefits. Chitosan-based edible coatings offer a significant advantage by extending the shelf life of diverse food products. By incorporating chitosan into these coatings, the moisture loss from

DOI: 10.1201/9781003355458-13

fruits and vegetables is effectively reduced during storage and transportation. This prevents wilting and dehydration, thereby maintaining their freshness and quality for longer periods. Chitosan acts as a barrier against microbial contamination by inhibiting the growth of pathogenic microorganisms. Its antimicrobial properties have been demonstrated against bacteria, fungi, and viruses, contributing to improved food safety (Khubiev et al., 2023).

Furthermore, chitosan-based edible coatings have been found to have a positive impact on the nutritional value of fresh produce. During storage, fruits and vegetables undergo natural physiological processes that lead to nutrient degradation. However, studies have shown that chitosan coatings can minimize nutrient loss by creating a protective layer that slows down oxidative reactions and enzymatic degradation. This helps retain important vitamins, minerals, and antioxidant compounds in the coated product (Romanazzi & Moumni, 2022).

In addition to its preservation effects on nutrients, chitosan coatings also offer potential nutraceutical benefits. Nutraceuticals are bioactive compounds found naturally in foods that provide health benefits beyond basic nutrition. Chitosan has been reported to preserve bioactive compounds such as polyphenols and flavonoids in fruits and vegetables. These compounds exhibit antioxidant properties and may contribute to various health benefits, such as reducing inflammation and protecting against chronic diseases (Zabot et al., 2022).

Moreover, chitosan-based edible coatings align with consumer demand for natural packaging solutions. Natural additives are increasingly preferred over synthetic preservatives or additives due to concerns about their potential health risks. Incorporating chitosan as an edible coating allows for the reduction or elimination of synthetic preservatives while still maintaining the quality and safety of the packaged food. Chitosan is derived from natural sources, making it an environmentally friendly option that aligns with sustainable packaging practices (Perez-Vazquez et al., 2023).

The potential health and nutritional benefits of chitosan-based edible coatings extend beyond preserving the quality of fresh produce. Chitosan has been investigated for its antimicrobial, antioxidant, and immunostimulatory properties in various studies. Its antimicrobial activity against pathogenic microorganisms can contribute to improved food safety by reducing the risk of foodborne illnesses. The antioxidant properties of chitosan may help protect against oxidative stress and cellular damage linked to chronic diseases. Furthermore, chitosan has been found to exhibit immunostimulatory effects by stimulating immune cells and enhancing immune responses. This suggests that chitosan-based edible coatings could potentially have positive effects on immune function when consumed as part of a balanced diet (Muñoz-Tebar et al., 2023).

In addition, chitosan-based edible coatings offer a range of potential health, nutraceutical, and nutritional benefits. They provide a protective barrier against moisture loss and microbial decay while helping retain important nutrients in fresh produce during storage and transportation. They have demonstrated antimicrobial activity against pathogenic microorganisms, potential antioxidant effects, and immunostimulatory properties. These benefits align with consumer demands for natural and sustainable packaging solutions (Matloob et al., 2023).

However, it is important to note that further research is needed to fully understand the extent of the health and nutritional benefits of chitosan-based edible coatings. Factors such as coating formulation, application methods, and specific food products may influence the efficacy of chitosan coatings in preserving nutrients and providing nutraceutical benefits. Additionally, studies on the consumer acceptance and sensory aspects of chitosan-coated foods are necessary to ensure widespread adoption and market success (Adiletta et al., 2021).

More so, chitosan-based edible coatings have the potential to revolutionize food packaging by offering not only extended shelf life but also added health and nutritional benefits. These coatings can contribute to improved food safety, nutrient retention, antioxidant effects, and immune support. As research in this area continues to unfold, chitosan-based edible coatings may become a preferred choice for sustainable packaging solutions that prioritize both product quality and consumer well-being (Matloob et al., 2023).

13.2 CHITOSAN: PROPERTIES AND APPLICATIONS

Chitosan, a versatile biopolymer, possesses a multitude of unique properties that render it highly suitable for diverse applications across various industries. In this discussion, we will explore the key characteristics of chitosan and its extensive range of applications.

Biocompatibility: Chitosan demonstrates exceptional biocompatibility, making it highly suitable for various biomedical applications. This unique property allows chitosan to be well-tolerated by living organisms, without causing any harmful effects. As a result, chitosan finds extensive use in areas such as tissue engineering, wound healing, and drug delivery systems within the medical field. Moreover, its resemblance to naturally occurring biomolecules like glycosaminoglycans (GAGs) found in human connective tissues further enhances its biocompatible nature (de Sousa-Victor et al., 2020).

Biodegradability: Chitosan is biodegradable, which means it can be broken down into non-toxic products like glucosamine through enzymatic degradation by lysozymes and chitinases present in many organisms. This property makes chitosan an environmentally friendly material that can be used as an alternative to synthetic polymers with long degradation times. The ability to degrade naturally reduces environmental pollution and waste accumulation (Aranaz et al., 2021).

Film-Forming Ability: Chitosan possesses excellent film-forming properties, making it suitable for the production of thin films and coatings. These films can be applied to various surfaces, including food products, to improve their stability and prolong their shelf life. Chitosan films act as a barrier against moisture loss, preventing dehydration and maintaining the freshness of the coated product. Additionally, chitosan coatings can provide protection against microbial contamination, thus enhancing food safety (Khubiev et al., 2023).

Antimicrobial Activity: One of the notable properties of chitosan is its antimicrobial activity. It has been found to exhibit inhibitory effects against various bacteria, fungi, and viruses. The antimicrobial action of chitosan is attributed to its positive charge at neutral pH, allowing it to interact with negatively charged microbial cell membranes. Chitosan can disrupt the integrity of the cell membrane and cause leakage of cellular contents, ultimately leading to the death of microorganisms. This property makes chitosan a potential natural alternative to synthetic antimicrobials in various applications (Adetunji et al., 2023c, 2023d).

Heavy Metal Binding: Chitosan has a high affinity for heavy metal ions such as lead, mercury, and cadmium. The amino groups present in chitosan form insoluble complexes with these metal ions through coordination bonds or ion exchange reactions. This ability to bind heavy metals makes chitosan suitable for water treatment applications where it can be used as an adsorbent for removing pollutants from contaminated water sources (Mathew et al., 2023a).

Antioxidant Activity: Chitosan exhibits antioxidant properties, which means it can scavenge free radicals and inhibit lipid oxidation. These antioxidant activities are attributed to its structure and high molecular weight. Chitosan has been found to protect cells and tissues from oxidative damage by neutralizing free radicals and preventing lipid peroxidation. This property makes chitosan a potential ingredient in functional foods, nutraceuticals, and dietary supplements (Adetunji et al., 2023a).

Immunostimulatory Effects: Chitosan has been shown to possess immunomodulatory properties, meaning it can enhance the immune response of the body. It stimulates the production of cytokines and activates immune cells such as macrophages and lymphocytes. This immunostimulatory activity of chitosan opens up possibilities for its use in vaccine adjuvants and immunotherapies (Lampe et al., 2022).

Mucoadhesive Properties: Chitosan possesses mucoadhesive properties, allowing it to adhere to mucosal surfaces such as those found in the gastrointestinal tract or nasal cavity. This property enables prolonged contact between drugs or active ingredients and target tissues or cells, enhancing drug absorption across mucosal barriers. Chitosan-based drug delivery systems have been investigated for applications in oral, nasal, ocular, and transdermal drug delivery (Mura et al., 2022).

Given its unique properties, chitosan finds applications in various industries:

Biomedical Industry: In tissue engineering, chitosan plays a crucial role in creating scaffolds that closely resemble the extracellular matrix. These scaffolds facilitate cell growth and tissue regeneration processes. Additionally, chitosan's antimicrobial and hemostatic properties make it an excellent choice for wound healing dressings. Moreover, chitosan-based drug delivery systems enhance drug absorption and enable targeted delivery to specific tissues or cells (Kim et al., 2023).

Food Industry: Chitosan can be used as an edible coating for fruits, vegetables, and other food products. These coatings enhance shelf life by acting as a barrier against moisture loss and microbial decay. Chitosan films can also be applied to food packaging materials to provide antimicrobial protection and extend product freshness (Adetunji et al., 2023b).

Water Treatment: Chitosan's ability to bind heavy metals makes it useful in water treatment applications for removing pollutants from contaminated water sources.

Pharmaceutical Industry: Chitosan-based drug delivery systems are being explored for the controlled release of drugs and targeted therapies. The mucoadhesive properties of chitosan enable efficient drug absorption across mucosal barriers, making it suitable for oral, nasal, ocular, and transdermal drug delivery (Mathew et al., 2023b, 2023c).

Agricultural Industry: Chitosan-based formulations are used as biostimulants and biopesticides in agriculture. They can enhance plant growth, protect against pathogens, and improve crop yield. Chitosan is also used in seed coating to promote germination and seedling establishment (Abimbola et al., 2023).

In addition, chitosan is a versatile biopolymer with a wide range of properties and applications. Its biocompatibility, biodegradability, film-forming ability, antimicrobial activity, heavy metal binding ability, antioxidant properties, immunostimulatory effects, and mucoadhesive properties contribute to its versatility. Chitosan finds applications in biomedical engineering, food science and technology, water treatment, pharmaceuticals, and agriculture. Further research and development are needed to explore its full potential in these areas and optimize its properties for specific applications (Ogundolie et al., 2023).

13.3 OVERVIEW OF CHITOSAN AS A BIOPOLYMER

Chitosan, a biopolymer derived from chitin, has garnered considerable interest across diverse fields owing to its distinctive properties and potential applications. Chitin, the second most abundant polysaccharide in nature after cellulose, serves as the source material for chitosan. Chitosan is obtained by deacetylating chitin through a chemical or enzymatic process, resulting in a polymer with numerous desirable characteristics (Malik et al., 2025).

Chitosan possesses a notable characteristic of being biocompatible, meaning it is compatible with living organisms and does not induce any harmful reactions. This quality renders chitosan suitable for numerous biomedical uses, including wound healing, drug delivery systems, tissue engineering scaffolds, and dental materials. Chitosan's biocompatibility arises from its similarity to naturally occurring biomolecules like GAGs found in human connective tissues. Another important attribute of chitosan is its biodegradability. Due to enzymatic degradation by lysozymes and chitinases present in many organisms, chitosan can be broken down into non-toxic products like glucosamine. This characteristic makes it an environmentally friendly material that can be used as an alternative to synthetic polymers with long degradation times.

Chitosan possesses excellent film-forming properties, which make it suitable for the production of thin films and coatings. These films can be applied to various surfaces, including food products, to improve their stability, extend shelf life, and provide protection against microbial contamination. The film-forming ability of chitosan arises from its ability to form strong intermolecular interactions through hydrogen bonding with other molecules or surfaces. In addition to its film-forming properties, chitosan also exhibits excellent antimicrobial activity. This property has been attributed to several factors, including its positive charge at neutral pH, which allows it to interact with negatively charged microbial cell membranes. Chitosan can disrupt the cell membrane integrity of microorganisms by causing leakage of cellular contents and ultimately leading to their death. This antimicrobial activity makes chitosan a potential natural alternative to synthetic antimicrobials in various applications (Mathew et al., 2025a).

Chitosan's ability to bind with heavy metals has also been extensively studied. Heavy metals such as lead, mercury, and cadmium are known pollutants that can contaminate water sources and have adverse effects on human health. Chitosan's amino groups have a high affinity for heavy metal ions, allowing them to form insoluble complexes that can be easily removed from water through precipitation or filtration processes. Moreover, chitosan exhibits antioxidant properties, which can be attributed to its structure and high molecular weight. Chitosan has been found to scavenge free radicals and inhibit lipid oxidation, thereby protecting cells and tissues from oxidative damage. This antioxidant activity makes chitosan a potential ingredient in functional foods, nutraceuticals, and dietary supplements (Muhammad et al., 2025).

Furthermore, chitosan exhibits immunostimulatory attributes, implying its capability to bolster the body's immune response. Studies have demonstrated that it can stimulate the production of cytokines and activate crucial immune cells like macrophages and lymphocytes. This immunomodulatory activity of chitosan opens up possibilities for its use in vaccine adjuvants and immunotherapies. Furthermore, chitosan has been investigated for its potential hypolipidemic effects. Studies have suggested that chitosan can lower cholesterol levels by interfering with lipid absorption in the digestive tract. The positively charged amino groups of chitosan can bind with negatively charged bile acids, preventing their reabsorption and promoting their excretion from the body (Bodunrinde et al., 2025).

Chitosan's mucoadhesive properties have also been exploited in drug delivery systems. The ability of chitosan to adhere to mucosal surfaces allows for prolonged contact between drugs and target tissues or cells. This property enhances drug absorption across mucosal barriers and improves the bioavailability of drugs. Chitosan-based drug delivery systems have been investigated for applications in oral, nasal, ocular, and transdermal drug delivery (Dawud et al., 2025).

In addition, chitosan is a versatile biopolymer with a wide range of applications due to its biocompatibility, biodegradability, film-forming properties, antimicrobial activity, heavy metal binding ability, antioxidant properties, immunostimulatory effects, hypolipidemic potential, and mucoadhesive properties. Its unique combination of features makes it a promising material for various industries including biomedical engineering, food science and technology, and pharmaceuticals. Further research and development are needed to explore the full potential of chitosan in these areas and to optimize its properties for specific applications (Malik et al., 2025).

Overall, chitosan's characteristics make it an attractive alternative to synthetic materials that are non-biodegradable or have adverse effects on human health or the environment. Its wide availability as a byproduct of the seafood industry adds to its appeal as an environmentally friendly material. However, challenges remain in terms of cost-effectiveness, scale-up production, and standardization of chitosan-based products. Further investigation into these areas will be crucial for realizing the full potential of chitosan as a sustainable biopolymer with various practical applications (Mathew et al., 2025b).

13.4 APPLICATIONS OF CHITOSAN IN VARIOUS INDUSTRIES

Chitosan, a derivative of chitin found in the exoskeletons of crustaceans, has found versatile applications across various industries due to its unique properties. In agriculture, chitosan is employed for crop protection, acting as a biodegradable and eco-friendly solution that enhances plant growth while providing protection against pathogens. The food industry utilizes chitosan in food packaging, leveraging its antimicrobial properties to extend the shelf life of products. Biomedical applications include the use of chitosan in wound dressings, where its hemostatic properties aid in promoting wound healing (Muhammad et al., 2025) (Table 13.1).

TABLE 13.1
Applications of Chitosan in Various Industries

Industry	Application	Benefits	References
Agriculture	Crop protection	Biodegradable, eco-friendly, and safe	Hernández-Soto and Chacón-Cerdas (2021)
Food industry	Food packaging	Antimicrobial properties, extends shelf life	Suvarna et al. (2022)
Biomedical	Wound dressings	Hemostatic, promotes healing	Nguyen et al. (2023)
Water treatment	Water purification	Adsorbs heavy metals, clarifies water	Inobeme et al. (2023c)
Textile	Textile finishing	Antibacterial, odor-resistant	Gulati et al. (2022)
Pharmaceuticals	Drug delivery	Controlled release of drugs	Adetunji et al. (2022) and Inobeme et al. (2023b)
Cosmetics	Skin care products	Anti-aging, moisturizing, and healing properties	Costa et al. (2022)
Environmental	Biodegradable materials	Substitute for plastic in packaging	Inobeme et al. (2023a)
Paper industry	Paper coating	Enhances paper strength and quality	Wu et al. (2020)

In water treatment, chitosan serves as an effective purification agent by adsorbing heavy metals and clarifying water. Textile industries use chitosan for finishing fabrics, imparting antibacterial and odor-resistant properties. In pharmaceuticals, chitosan is valued for drug delivery systems, facilitating controlled release of medications. The cosmetic industry benefits from chitosan's anti-aging, moisturizing, and healing properties in skincare products. In environmental initiatives, chitosan acts as a biodegradable alternative to traditional plastics, particularly in packaging materials. Moreover, chitosan is employed in the paper industry for coating applications, enhancing paper strength and quality. Its diverse applications underscore chitosan's significance as a sustainable and multifunctional material in addressing various industrial challenges (Chimbekujwo et al., 2025).

13.5 ROLE OF EDIBLE COATINGS IN FOOD PRESERVATION

Edible coatings play a crucial role in food preservation by providing a protective barrier to extend the shelf life of various perishable products. These coatings are typically made from natural and edible materials, such as proteins, lipids, carbohydrates, and polysaccharides, and are applied directly onto the surface of food items. The primary objectives of using edible coatings in food preservation include reducing moisture loss, inhibiting microbial growth, and preventing oxidation. This chapter explores the multifaceted role of edible coatings in preserving the freshness and quality of food products (Díaz-Montes & Castro-Muñoz, 2021).

Moisture Retention: One of the key functions of edible coatings is to act as a barrier to moisture. By reducing water vapor transmission, these coatings help to minimize moisture loss from the food product, thereby preventing dehydration. This is particularly important for fruits and vegetables, as moisture loss can lead to wilting, textural changes, and decreased overall quality. Edible coatings create a protective layer that slows down the rate of water evaporation, maintaining the desired moisture content in the food (Aaqil et al., 2024).

Microbial Inhibition: Edible coatings possess antimicrobial properties that inhibit the growth of spoilage microorganisms and pathogens. The physical barrier they create limits the access of microorganisms to the food surface, reducing the risk of contamination. Additionally, some edible coatings incorporate natural antimicrobial agents, such as essential oils or antimicrobial peptides, further enhancing their ability to preserve food by preventing the proliferation of bacteria, molds, and yeasts (Nunes et al., 2023).

Oxidation Prevention: Oxidation is a major cause of deterioration in many food products, leading to flavor changes, discoloration, and nutrient degradation. Edible coatings act as a barrier against oxygen, preventing its contact with the food surface and inhibiting oxidative reactions. Some coatings also contain antioxidants, which can scavenge free radicals and further protect the food from oxidative damage. This is particularly beneficial for products prone to oxidation, such as oils, nuts, and certain fruits (Singh et al., 2022).

Extended Shelf Life: By addressing moisture loss, microbial growth, and oxidation, edible coatings collectively contribute to extending the shelf life of various food items. The protective layer provided by these coatings helps to maintain the freshness, texture, and nutritional quality of the products, allowing them to stay on the market for a more extended period. This not only benefits consumers by reducing food waste but also supports the economic sustainability of food producers and suppliers (Perez-Vazquez et al., 2023).

Enhanced Quality and Appearance: Edible coatings can improve the overall quality and appearance of food products. They can enhance glossiness, color retention, and visual appeal, making the products more attractive to consumers. Additionally, coatings can provide a smooth surface that reduces roughness and maintains the desired texture. This is particularly relevant for fruits and vegetables, as well as other products that are susceptible to surface blemishes and textural changes during storage (Matloob et al., 2023).

Environmental Sustainability: The use of edible coatings aligns with the growing emphasis on sustainable and environmentally friendly food preservation methods. Compared to some traditional packaging materials, edible coatings can be derived from renewable resources and are often biodegradable. This reduces the environmental impact associated with packaging waste and contributes to the development of more sustainable practices in the food industry (Nunes et al., 2023).

In addition, edible coatings play a multifaceted role in food preservation by addressing key factors such as moisture retention, microbial inhibition, and oxidation prevention. Their ability to extend shelf life, enhance quality, and contribute to environmental sustainability makes them a valuable tool in the food industry. As technology advances, further research and innovation in the development of edible coatings are likely to contribute to even more effective and sustainable solutions for preserving a wide range of food products. Health benefits of Chitosan-based edible coatings have been discussed in Perez-Vazquez et al. (2023).

13.6 IMPACT OF CHITOSAN-BASED EDIBLE COATINGS ON FOOD SAFETY

The application of chitosan-based edible coatings has emerged as a promising solution for ensuring food safety. These coatings, derived from the natural polymer chitin found in crustacean shells, offer a multifaceted approach to preserving and enhancing the safety of different food products. With its unique properties, chitosan proves to be a valuable material for creating protective barriers on the surfaces of various food items. This chapter explores the impact of Chitosan-based edible coatings on food safety, covering aspects such as microbial inhibition, oxidation prevention, moisture retention, extended shelf life, enhanced quality, and environmental considerations (Zárate-Moreno et al., 2023).

Microbial Inhibition: One of the notable features of Chitosan-based coatings is their robust antimicrobial activity. Chitosan has been shown to exhibit inhibitory effects against a broad spectrum of microorganisms, including bacteria, molds, and yeasts. When applied as an edible coating, Chitosan forms a protective barrier that hinders the growth and proliferation of these spoilage and pathogenic microorganisms. This microbial inhibition is particularly significant in preventing foodborne illnesses and extending the shelf life of perishable goods. Research has demonstrated that Chitosan disrupts microbial cell membranes, leading to leakage of cellular contents and eventual cell death. This mechanism makes Chitosan an effective natural preservative, offering an alternative to synthetic antimicrobial agents. The use of Chitosan-based coatings has been explored in various food products, ranging from fruits and vegetables to meat and seafood, showcasing its versatility in enhancing the safety of diverse food items (Yan et al., 2021).

Oxidation Prevention: Chitosan's ability to act as a barrier against oxygen contributes to the prevention of oxidative reactions in food products. Oxidation is a key factor in the deterioration of food, leading to changes in color, flavor, and nutritional content. By forming a protective film, Chitosan reduces the contact between the food surface and oxygen, thereby inhibiting the oxidation of lipids and other sensitive compounds. Moreover, Chitosan itself possesses antioxidant properties, scavenging free radicals and further contributing to the preservation of food quality. This dual role of Chitosan-based coatings in preventing oxidation makes them valuable in maintaining the sensory attributes and nutritional integrity of food, ultimately enhancing food safety by preventing the development of rancidity and off-flavors (Nunes et al., 2023).

Moisture Retention: Chitosan-based coatings play a crucial role in moisture retention, especially in fruits, vegetables, and other products prone to dehydration. The formation of a protective film helps reduce water vapor transmission, slowing down the rate of moisture loss from the food. This is vital for preserving the texture, juiciness, and overall quality of the products. The maintenance of optimal moisture content contributes to the prevention of undesirable textural changes, such as wilting in fruits and vegetables, ensuring that the products reach consumers in a safe and appealing condition (Adiletta et al., 2021).

Extended Shelf Life: The combined effects of microbial inhibition, oxidation prevention, and moisture retention contribute to a notable extension of the shelf life of food products with Chitosan-based coatings. By addressing multiple factors that contribute to spoilage, these coatings help reduce food waste and enhance the economic sustainability of the food industry. Consumers benefit from the availability of fresh and safe products for a more extended period, and producers can optimize their supply chains with reduced losses (El-Araby et al., 2024).

Enhanced Quality: Chitosan-based coatings go beyond safety considerations and contribute to the overall quality of food products. These coatings have been shown to preserve the color, texture, and nutritional content of fruits, vegetables, and other perishable items. Improved quality not only enhances the consumer's sensory experience but also reflects positively on the nutritional value of the food. The aesthetic appeal and freshness of products coated with Chitosan make them more attractive to consumers, fostering a positive perception of food safety (Matloob et al., 2023).

Environmental Considerations: In addition to their impact on food safety, Chitosan-based coatings align with the growing demand for environmentally sustainable practices in the food industry. Chitosan is derived from renewable sources, and its biodegradable nature makes it a more eco-friendly alternative to some traditional packaging materials. However, it's essential to consider potential allergies and ensure responsible sourcing practices to mitigate any adverse environmental impacts associated with the extraction of Chitosan (Maliki et al., 2022).

More so, Chitosan-based edible coatings have a substantial impact on food safety by addressing microbial inhibition, oxidation prevention, moisture retention, extended shelf life, and enhanced quality. As research and development in this field continues, Chitosan-based coatings are likely to play an increasingly vital role in shaping the future of sustainable and safe food preservation practices. Their natural origin, coupled with their effectiveness, positions them as a promising solution for the food industry's ongoing efforts to meet the demands of both safety-conscious consumers and environmentally conscious practices (Díaz-Montes & Castro-Muñoz, 2021).

13.7 NUTRACEUTICAL POTENTIAL OF CHITOSAN-BASED EDIBLE COATINGS

Chitosan-based edible coatings have garnered significant attention in recent years not only for their ability to enhance food safety but also for their nutraceutical potential. Nutraceuticals are substances that provide health benefits beyond basic nutrition, and Chitosan, derived from chitin found in crustacean shells, exhibits properties that make it an intriguing candidate for promoting health and well-being. This chapter delves into the nutraceutical potential of Chitosan-based edible coatings, exploring their impact on nutrient preservation, enhanced bioavailability, functional ingredient delivery, and potential support for gastrointestinal health (Adiletta et al., 2021; Inobeme et al., 2024a).

Nutrient Preservation: One of the noteworthy contributions of Chitosan-based coatings to the nutraceutical landscape is their role in nutrient preservation. The coatings act as a protective barrier against oxygen, a major contributor to nutrient degradation. Oxygen exposure can lead to the oxidation of sensitive vitamins and other bioactive compounds, resulting in a loss of nutritional value in food products. Chitosan, with its barrier properties, helps prevent such oxidative reactions, thereby preserving the vitamins and other essential nutrients present in the coated food. The preservation of nutritional content is particularly crucial for fruits, vegetables, and other perishable goods that are susceptible to degradation during storage and transportation. By extending the shelf life of these products, Chitosan-based coatings contribute to ensuring that consumers receive food items with optimal nutritional value (Picos-Corrales et al., 2023).

Enhanced Bioavailability: Chitosan's bioadhesive properties make it an intriguing candidate for improving the bioavailability of certain nutrients. Bioavailability refers to the extent and rate at which a nutrient is absorbed and utilized by the body. Chitosan, due to its ability to adhere to mucosal surfaces in the digestive tract, may enhance the absorption of nutrients. This property could be especially beneficial for fat-soluble vitamins, such as vitamin A, D, E, and K, as well as other lipophilic compounds. By facilitating improved absorption, Chitosan-based coatings may contribute to a more efficient utilization of essential nutrients by the human body, promoting overall health and well-being (Abourehab et al., 2022; Mathew et al., 2024b).

Functional Ingredient Delivery: Chitosan-based coatings serve as effective carriers for functional ingredients with potential health benefits. These functional ingredients may include antioxidants, antimicrobial agents, or bioactive compounds sourced from natural products. By incorporating these ingredients into the coating formulation, Chitosan allows for a controlled release over time. This controlled release is significant for maximizing the efficacy of functional ingredients. For instance, antioxidants released gradually can exert their protective effects on the coated food product, contributing to the reduction of oxidative stress. Additionally, the incorporation of bioactive compounds can transform the coated food into a functional food, offering consumers an option to obtain health-promoting benefits while enjoying their regular diet (Matloob et al., 2023).

Gastrointestinal Health Support: Research has focused on exploring the potential prebiotic effects of chitosan. Prebiotics are substances that stimulate the growth and activity of beneficial bacteria in the gastrointestinal tract. Chitosan, therefore, has been investigated for its ability to promote a healthy gut microbiome. By serving as a substrate for the growth of probiotic microorganisms, Chitosan-based coatings may contribute to maintaining a balanced gut microbiota. Various health benefits are linked to a well-balanced and diverse gut microbiota, such as improved digestion. As interest in gut health continues to rise, Chitosan's potential to support gastrointestinal health adds a valuable dimension to its nutraceutical profile (Roy & Dhaneshwar, 2023).

Chitosan-based edible coatings, originally recognized for their food safety benefits, have emerged as promising contributors to the nutraceutical landscape. The preservation of nutritional content, enhanced bioavailability of nutrients, delivery of functional ingredients, and potential support for gastrointestinal health showcase the diverse ways in which Chitosan can positively impact the health-promoting properties of coated food products. As the demand for functional and health-enhancing foods grows, Chitosan-based coatings offer a natural and versatile solution to meet consumer preferences. Continued research and innovation in the field are likely to unveil further applications and benefits, solidifying Chitosan's place in the evolving landscape of nutraceuticals and functional foods. The integration of Chitosan-based edible coatings into everyday food items represents a harmonious convergence of food safety and nutritional enhancement, addressing both the practical and health-conscious needs of consumers (Wang et al., 2023).

13.8 NUTRITIONAL ADVANTAGES OF CHITOSAN-BASED EDIBLE COATINGS

Chitosan-based edible coatings, primarily recognized for their role in food safety, also offer a spectrum of nutritional advantages that contribute to the overall well-being of consumers. Chitosan, derived from chitin—a natural polymer found in the exoskeletons of crustaceans—exhibits unique properties that make it a valuable material for creating protective coatings on various food items. This chapter explores in depth the nutritional advantages of Chitosan-based edible coatings, encompassing aspects such as the preservation of vitamin content, improved bioavailability, controlled release of bioactive compounds, and the promotion of gastrointestinal health (Thambiliyagodage et al., 2023).

Preservation of Vitamin Content: One of the significant nutritional advantages of Chitosan-based edible coatings is their ability to preserve the vitamin content of coated food items. Vitamins, particularly sensitive to environmental factors like oxygen, light, and temperature, can undergo degradation during storage and transportation. Chitosan acts as a protective barrier against oxidation, preventing the breakdown of essential vitamins such as vitamin C, vitamin A, and various B vitamins. For example, when fruits and vegetables are coated with Chitosan-based films, the coatings create a shield that reduces the exposure of the produce to oxygen. This not only extends the shelf life of these products but also helps maintain their nutritional integrity. Consumers benefit by receiving fruits and vegetables with optimal vitamin content, contributing to their daily nutrient intake and overall health (Nunes et al., 2023).

Improved Bioavailability: Chitosan's bioadhesive properties make it a compelling component for improving the bioavailability of certain nutrients. Bioavailability refers to the proportion of a nutrient that enters the bloodstream and is available for utilization by the body. Chitosan, with its ability to adhere to mucosal surfaces in the digestive tract, can enhance the absorption of nutrients. In particular, Chitosan has shown promise in facilitating the improved bioavailability of fat-soluble vitamins. Coating food items containing these vitamins with Chitosan may lead to better absorption in the gastrointestinal tract. This enhancement in bioavailability is essential for ensuring that the body can efficiently utilize these vital nutrients, contributing to overall well-being (Imam et al., 2021; Mathew et al., 2024a).

Controlled Release of Bioactive Compounds: Chitosan-based coatings serve as effective carriers for bioactive compounds with potential health benefits. These bioactive compounds, such as antioxidants and antimicrobial agents, can be incorporated into the coating formulation. The unique feature of Chitosan is its ability to provide a controlled release of these bioactive compounds over time. For example, when antioxidants are integrated into Chitosan-based coatings, the gradual release of these compounds contributes to prolonged antioxidant activity. This controlled release ensures that the coated food product continues to exhibit health-promoting properties throughout its shelf life. Consumers benefit from sustained exposure to bioactive compounds, potentially reducing oxidative stress and supporting overall health (Thambiliyagodage et al., 2023).

Promotion of Gastrointestinal Health: Chitosan has been studied for its prebiotic effects, making it a potential contributor to gastrointestinal health. Prebiotics are substances that promote the growth and activity of beneficial bacteria in the gut. By serving as a substrate for the growth of probiotic microorganisms, Chitosan-based coatings may contribute to maintaining a balanced gut microbiota. In practical terms, coating food products with Chitosan could introduce its prebiotic effects into the digestive system. This may lead to improved digestion, enhanced nutrient absorption, and overall gastrointestinal well-being. As the understanding of the gut-brain axis and the importance of gut health grows, the potential of Chitosan to positively impact gastrointestinal health adds a valuable dimension to its nutritional profile (Kipkoech et al., 2021; Inobeme et al., 2024b).

Consumer Benefits: The nutritional advantages of Chitosan-based edible coatings translate into tangible benefits for consumers. By choosing products coated with Chitosan, consumers can enjoy fruits, vegetables, and other coated items that not only last longer but also retain their essential nutrients. The improved bioavailability of nutrients contributes to the efficiency of the body's utilization of these substances, promoting overall health. Furthermore, the controlled release of bioactive compounds adds a layer of functional benefits to coated food products. Consumers can access a sustained supply of antioxidants and other health-promoting compounds, potentially supporting their immune system and providing protection against oxidative stress (Adiletta et al., 2021).

In addition, Chitosan-based edible coatings offer a holistic approach to food preservation by combining food safety with notable nutritional advantages. The preservation of vitamin content, improved bioavailability of nutrients, controlled release of bioactive compounds, and potential support for gastrointestinal health showcase the diverse nutritional benefits of incorporating Chitosan into food packaging strategies. As research in this field continues to advance, Chitosan-based edible coatings are likely to play an increasingly vital role in addressing both the nutritional and safety concerns of consumers, contributing to the development of a more health-conscious and sustainable food industry (Muñoz-Tebar et al., 2023).

13.9 MINIMIZATION OF NUTRIENT DEGRADATION DURING STORAGE

Minimization of nutrient degradation during storage is a critical consideration in the food industry, as it directly impacts the nutritional quality of food products available to consumers. Nutrient degradation can occur due to various factors, including exposure to oxygen, light, temperature fluctuations, and moisture. Addressing these challenges is essential to ensure that consumers receive foods with optimal nutritional content. This chapter explores the importance of minimizing nutrient degradation during storage, the key factors influencing degradation, and various strategies employed by the food industry to preserve the nutritional quality of stored food products (Giannakourou & Taoukis, 2021).

Importance of Minimizing Nutrient Degradation: The nutritional quality of food plays a pivotal role in maintaining human health and preventing nutritional deficiencies. Essential vitamins, minerals, and other bioactive compounds are sensitive to external factors, and their degradation during storage can compromise the nutritional value of food items. For consumers, accessing foods with retained nutritional content is not only vital for meeting daily dietary requirements but also contributes to overall well-being and disease prevention. Therefore, minimizing nutrient degradation during storage is crucial for ensuring that the nutritional promises on food labels align with the actual content when consumed (Ofoedu et al., 2021).

Main Factors Influencing Nutrient Degradation: Several factors contribute to nutrient degradation during the storage of food products. Understanding these factors is essential for implementing effective preservation strategies. Exposure to oxygen is a primary cause of nutrient degradation, particularly for sensitive compounds such as vitamins A, C, and E. Oxidation leads to the breakdown of these vitamins and the development of off-flavors and odors. Certain vitamins, such as riboflavin and vitamin A, are sensitive to light. Prolonged exposure to light, especially in transparent packaging, can lead to the degradation of these light-sensitive nutrients. Fluctuations in temperature, especially in conditions where cold storage is required, can affect the stability of vitamins and enzymes. Heat accelerates the degradation process, leading to a loss of nutritional value. Moisture content in food products can contribute to the growth of microorganisms, leading to nutrient degradation. It also affects the texture and quality of the product. Endogenous enzymes present in fruits and vegetables can continue to be active even after harvest, leading to the degradation of certain nutrients, particularly in cut or processed produce (Mieszczakowska-Frąc et al., 2021; Mathew et al., 2024c).

Strategies for Minimizing Nutrient Degradation: The food industry employs various strategies to minimize nutrient degradation during storage, ensuring that consumers receive products with sustained nutritional value. Modified atmosphere packaging involves modifying the composition of the gases within the packaging to slow down oxidation reactions. By reducing oxygen levels and increasing carbon dioxide levels, the shelf life of the product is extended, and nutrient degradation is minimized. Vacuum packaging removes the air from the packaging, reducing exposure to oxygen. This method is particularly effective for preserving the quality of meat and preventing the oxidation of sensitive nutrients. Incorporating natural or synthetic antioxidants into food formulations can help inhibit oxidation reactions. This is commonly used in products containing fats and oils, where oxidation can lead to the degradation of fat-soluble vitamins. Edible coatings, such as those based on Chitosan or other natural polymers, can create a barrier between the food and the external environment. This protective layer helps reduce exposure to oxygen, light, and moisture, thus minimizing nutrient degradation. Maintaining a consistent temperature and humidity level in storage facilities helps minimize nutrient degradation. Cold storage is essential for preserving the nutritional quality of many perishable products. Inactivating enzymes through blanching (brief exposure to high temperatures) is a common practice in the food industry. This step helps prevent enzymatic degradation of nutrients in fruits and vegetables. Packaging that is opaque or provides a barrier to light helps protect light-sensitive nutrients. This is commonly observed in packaging for milk, juices, and certain vitamin supplements (Inobeme et al., 2023d).

As technology advances, the food industry continues to explore innovative solutions to minimize nutrient degradation during storage. Nanotechnology, for instance, is being researched for its potential to create nanoscale delivery systems that protect and enhance the stability of nutrients. Additionally, advancements in smart packaging that can monitor and respond to environmental conditions offer new possibilities for maintaining the nutritional quality of food products throughout their shelf life. Minimizing nutrient degradation during storage is a multifaceted challenge that requires a combination of thoughtful packaging, storage practices, and technological innovations. The food industry's commitment to preserving the nutritional quality of products aligns with the growing awareness among consumers about the link between diet and health. As these efforts continue, consumers can expect to see improvements in the availability of food products that not only taste good but also deliver the promised nutritional benefits, contributing to a healthier and more informed society (Singh et al., 2023).

13.10 CONCLUSION

The application of edible coatings based on Chitosan represents a paradigm shift in enhancing both the health and nutritional attributes of food products. Chitosan, derived from natural sources like crustacean shells, not only contributes to food safety through its antimicrobial properties but also offers a myriad of health and nutraceutical benefits. Through the preservation of essential nutrients, improved bioavailability, and controlled release of bioactive compounds, Chitosan-based coatings contribute to elevated nutritional profiles in coated foods. Moreover, Chitosan's prebiotic effects may positively influence gastrointestinal health, aligning with the growing emphasis on holistic well-being. The eco-friendly nature of Chitosan, being biodegradable and sourced from renewable materials, further reinforces its appeal in the context of sustainable and health-conscious food practices. As research progresses, it becomes increasingly apparent that Chitosan-based edible coatings have the potential to revolutionize the amalgamation of food safety, nutrition, and consumer health. This presents an encouraging opportunity for creating functional and health-promoting foods. Embracing Chitosan in the food industry not only prolongs product shelf life but also harmonizes with the shifting preferences of health-conscious consumers who seek both safety and nutritional value in their diet.

REFERENCES

Aaqil, M., Peng, C., Kamal, A., Nawaz, T., & Gong, J. (2024). Recent Approaches to the Formulation, Uses, and Impact of Edible Coatings on Fresh Peach Fruit. *Foods (Basel, Switzerland)*, *13*(2), 267. https://doi.org/10.3390/foods13020267

Abimbola, O. F., Adetunji, C. O., Dare, B. J., Mathew, J. T., Inobeme, A., Ghazanfar, S., Olaniyan, O. T., Ajiboye, M. D., & Dauda, W. (2023). Packaging of Animal-Based Food Products: The Need for Intelligent Freshness Sensors. In: *Sensing and Artificial Intelligence Solutions for Food Manufacturing*. CRC Press, 89–102. https://doi.org/10.1201/9781003207955-6

Abourehab, M. A. S., Pramanik, S., Abdelgawad, M. A., Abualsoud, B. M., Kadi, A., Ansari, M. J., & Deepak, A. (2022). Recent Advances of Chitosan Formulations in Biomedical Applications. *International Journal of Molecular Sciences*, *23*(18), 10975. https://doi.org/10.3390/ijms231810975

Adetunji, C. O., Bodunrinde, R. E., Inobeme, A., Singh, K. R. B., Mathew, J. T., Olaniyan, O., Ogundolie, F. A., Singh, J., Nayak, V., & Singh, R. P. (2023a). Microbial Community Analysis of Contaminated Soils. In: *Microbial Community Studies in Industrial Wastewater Treatment*. Taylor & Francis eBooks, 83–97. https://doi.org/10.1201/9781003354147-6

Adetunji, C. O., Ogundolie, F. A., Mathew, J. T., Inobeme, A., Olotu, T., Olaniyan, O. T., Ijabadeniyi, O. A., Ajiboye, M. D., Ajayi, O. O., Dauda, W., Ghazanfar, S., & Adetunji, J. B. (2022). Graphene-Based Nanomaterials for Targeted Drug Delivery and Tissue Engineering. In: *Novel Platforms for Drug Delivery Applications*. Elsevier Ltd., 277–288. https://doi.org/10.1016/B978-0-323-91376-8.00014-8

Adetunji, C. O., Olaniyan, O. T, Inobeme, A., Adeyomoye, O., Mathew, J. T. & Akinbo, O. (2023b). Recent Advances in the Characterization and Application of Graphene in the Food Industry. In: *Sensing and Artificial Intelligence Solutions for Food Manufacturing*. CRC Press, 153–164. https://doi.org/10.1201/9781003207955-11

Adetunji, C. O., Olaniyan, O. T., Shariati, M. A., Bodunrinde, R. E., Ajayi, O. O., Adetunji, J. B., Hefft, D. I., Olugbenga, S. M., Rebezov, M., Khayrullin, M., Adeniyi, M. J., Inobeme, A. & Mathew, J. T. (2023c). Pharmacology, Phytochemistry, and Medicinal Usefulness of *Basella alba* L. In T. Pullaiah (Ed.), *Phytochemical Composition and Pharmacy of Medicinal Plants*. Apple Academic Press, Inc. Co-published with CRC Press (Taylor & Francis). Vol. 2, 499–516.

Adetunji, C. O., Olaniyan, O. T., Shariati, M. A., Olugbenga, S. M., Rebezov, M., Ajayi, O. O., Goncharov, A., Bodunrinde, R. E., Adetunji, J. B., Adeniyi, M. J., Hefft, D. I., Inobeme, A., & Mathew, J. T. (2023d). Biological Constituents and Medicinal Attributes of *Spondias mombin* L. In T. Pullaiah (Ed.), *Phytochemical Composition and Pharmacy of Medicinal Plants*. Apple Academic Press, Inc. Co-published with CRC Press (Taylor & Francis). Vol. 2, 517–529.

Adiletta, G., Di Matteo, M., & Petriccione, M. (2021). Multifunctional Role of Chitosan Edible Coatings on Antioxidant Systems in Fruit Crops: A Review. *International Journal of Molecular Sciences*, *22*(5), 2633. https://doi.org/10.3390/ijms22052633

Aranaz, I., Alcántara, A. R., Civera, M. C., Arias, C., Elorza, B., Heras Caballero, A., & Acosta, N. (2021). Chitosan: An Overview of Its Properties and Applications. *Polymers*, *13*(19), 3256. https://doi.org/10.3390/polym13193256

Bodunrinde, R. E., Chimbekujwo, K. I., Oyewole, O. A., Adetunji, C. O., Inobeme, A., Eniola, K. I. T., Yerima, M. B., & Mathew, J. T. (2025). The Use of Nanochitosan for Enhancement in the Quality and Yield of Fruit Crops. In: *Nanochitosan Applications for Enhanced Crop Production and Food Security*. John Wiley & Sons, 111–122. https://doi.org/10.1002/9781394214945.ch5

Chimbekujwo, K. I., Yakubu, J. G., Omoregie, I. P., Oyewole, O. A., Muhammad, W. M., Adetunji, C. O., Eniola, K. I. T., Yerima, M. B., & Mathew, J. T. (2025). Synthesis of Nano-Chitosan Using Agricultural Waste. In: *Nanochitosan Applications for Enhanced Crop Production and Food Security*. John Wiley & Sons, Inc., 39–68. https://doi.org/10.1002/9781394214945.ch2

Costa, E. F., Magalhães, W. V., & Di Stasi, L. C. (2022). Recent Advances in Herbal-Derived Products with Skin Anti-Aging Properties and Cosmetic Applications. *Molecules (Basel, Switzerland)*, *27*(21), 7518. https://doi.org/10.3390/molecules27217518

Dawud, I., Bodunrinde, R. E., Oyewole, O. A., Adetunji, C. O., Inobeme, A., Eniola, K. I. T., Yerima, M. B., & Mathew, J. T. (2025). Application of Nanochitosan for Effective Fruit Production. In: *Nanochitosan Applications for Enhanced Crop Production and Food Security*. John Wiley & Sons, 123–135. https://doi.org/10.1002/9781394214945.ch6

de Sousa-Victor, R., Marcelo da Cunha-Santos, A., Viana de Sousa, B., de Araújo Neves, G., Navarro de Lima-Santana, L., & Rodrigues-Menezes, R. (2020). A Review on Chitosan's Uses as Biomaterial: Tissue Engineering, Drug Delivery Systems and Cancer Treatment. *Materials (Basel, Switzerland)*, *13*(21), 4995. https://doi.org/10.3390/ma13214995

Díaz-Montes, E., & Castro-Muñoz, R. (2021). Edible Films and Coatings as Food-Quality Preservers: An Overview. *Foods (Basel, Switzerland)*, *10*(2), 249. https://doi.org/10.3390/foods10020249

El-Araby, A., Janati, W., Ullah, R., Ercisli, S., & Errachidi, F. (2024). Chitosan, Chitosan Derivatives, and Chitosan-Based Nanocomposites: Eco-Friendly Materials for Advanced Applications (A Review). *Frontiers in Chemistry*, *11*, 1327426. https://doi.org/10.3389/fchem.2023.1327426

Giannakourou, M. C., & Taoukis, P. S. (2021). Effect of Alternative Preservation Steps and Storage on Vitamin C Stability in Fruit and Vegetable Products: Critical Review and Kinetic Modelling Approaches. *Foods (Basel, Switzerland)*, *10*(11), 2630. https://doi.org/10.3390/foods10112630

Gulati, R., Sharma, S., & Sharma, R. K. (2022). Antimicrobial Textile: Recent Developments and Functional Perspective. *Polymer Bulletin (Berlin, Germany)*, *79*(8), 5747–5771. https://doi.org/10.1007/s00289-021-03826-3

Hernández-Soto, A., & Chacón-Cerdas, R. (2021). RNAi Crop Protection Advances. *International Journal of Molecular Sciences*, *22*(22), 12148. https://doi.org/10.3390/ijms222212148

Imam, S. S., Alshehri, S., Ghoneim, M. M., Zafar, A., Alsaidan, O. A., Alruwaili, N. K., Gilani, S. J., & Rizwanullah, M. (2021). Recent Advancement in Chitosan-Based Nanoparticles for Improved Oral Bioavailability and Bioactivity of Phytochemicals: Challenges and Perspectives. *Polymers*, *13*(22), 4036. https://doi.org/10.3390/polym13224036

Inobeme, A., Adetunji, C. O., Ajai, A. I., Inobeme, J., Mathew, J. T., Obar, A., Maliki, M., Nwakife, N., & Eziukwu, C. (2023a). Chemical Nanosensors for Monitoring Environmental Pollution. In: *Advanced Application of Nanotechnology to Industrial Wastewater*. Springer Nature Singapore, 93–103. https://doi.org/10.1007/978-981-99-3292-4_6

Inobeme, A., Adetunji, C. O., Maliki, M., Onyeachu, B. I., Kelani, T., Eziukwu, C. A., Olori, E., Mathew, J. T., & Bamigboye, M. O. (2023b). Strategies to Synthesize, Advantages, and Disadvantages of Pharmaceutical Nanoparticles. In: *Nanotechnology for Drug Delivery and Pharmaceuticals*. Academic Press is an imprint of Elsevier, 371–402. https://doi.org/10.1016/B978-0-323-95325-2.00006-7

Inobeme, A., Adetunji, C. O., Mathew, J. T., Ajai, A. I., Inobeme, J., Bamigboye, M. O., Onyeaku, S., Maliki, M., Eziukwu, C., & Tawa, K. (2023c). Nanotechnology for Bioremediation of Heavy Metals. In: *Microbial Technologies in Industrial Wastewater Treatment*. Springer Nature Singapore, 19–30. https://doi.org/10.1007/978-981-99-2435-6_2

Inobeme, A., Adetunji, C. O., Mathew, J. T., Ajai, A. I., Mann, A., Inobeme, J., Oyedolapo, B., Adekoya, M. A., & Onyeaku, S. (2023d). Diversity, Biodegradation and Bioremediation of Polycyclic Aromatic Hydrocarbons. In: *Microbial Technologies in Industrial Wastewater Treatment*. Springer Nature Singapore, 31–50. https://doi.org/10.1007/978-981-99-2435-6_3

Inobeme, A., Mathew, J. T., Ajai, I. A., Adetunji, C. O., Inobeme, J., Maliki, M., Adekoya, M. A., Shaba, E. Y., Eric, O., Akhor, S., & Eziukwu, C. (2024a). Common Techniques in Food Processing Technologies. In: *Nanobiotechnology for Food Processing and Packaging*. Elsevier Inc., 223–234. https://doi.org/10.1016/B978-0-323-91749-0.00002-2

Inobeme, A., Shahnawaz, M., Adetunji, C. O., Mathew, J. T., Adetuyi, B. O., Popoola, O. A., Olaitan, F. Y., Akinbo, O., Kolawole, O. M., Oyewole, O. A., Eniola, K.I.T., & Yerima, B. M. (2024b). Microplastic as a Pathogenic Vector to Cause Diseases in Marine Biota. In: *Microplastic Pollution*. Springer, *Singapore*, 1, 321–349. https://doi.org/10.1007/978-981-99-8357-5_19

Khubiev, O. M., Egorov, A. R., Kirichuk, A. A., Khrustalev, V. N., Tskhovrebov, A. G., & Kritchenkov, A. S. (2023). Chitosan-Based Antibacterial Films for Biomedical and Food Applications. *International Journal of Molecular Sciences*, *24*(13), 10738. https://doi.org/10.3390/ijms241310738

Kim, Y., Zharkinbekov, Z., Raziyeva, K., Tabyldiyeva, L., Berikova, K., Zhumagul, D., Temirkhanova, K., & Saparov, A. (2023). Chitosan-Based Biomaterials for Tissue Regeneration. *Pharmaceutics*, *15*(3), 807. https://doi.org/10.3390/pharmaceutics15030807

Kipkoech, C., Kinyuru, J. N., Imathiu, S., Meyer-Rochow, V. B., & Roos, N. (2021). In Vitro Study of Cricket Chitosan's Potential as a Prebiotic and a Promoter of Probiotic Microorganisms to Control Pathogenic Bacteria in the Human Gut. *Foods (Basel, Switzerland)*, *10*(10), 2310. https://doi.org/10.3390/foods10102310

Lampe, A. T., Farris, E. J., Brown, D. M., & Pannier, A. K. (2021). High- and low-molecular-weight chitosan act as adjuvants during single-dose influenza A virus protein vaccination through distinct mechanisms. *Biotechnology and Bioengineering*, *118*(3), 1224–1243. https://doi.org/10.1002/bit.27647

Malik, S. I., Chimbekujwo, K. I., Maku, A. M., Oyewole, O. A., Omoregie, I. P., Adetunji, C. O., Eniola, K. I. T., Yerima, M. B., & Mathew, J. T. (2025). The Application of Nanochitosan Biopesticides as a Replacement to Synthetic Pesticides. In: *Nanochitosan Applications for Enhanced Crop Production and Food Security*. John Wiley & Sons, 89–110. https://doi.org/10.1002/9781394214945.ch4

Maliki, S., Sharma, G., Kumar, A., Moral-Zamorano, M., Moradi, O., Baselga, J., Stadler, F. J., & García-Peñas, A. (2022). Chitosan as a Tool for Sustainable Development: A Mini Review. *Polymers*, *14*(7), 1475. https://doi.org/10.3390/polym14071475

Mathew, J. T., Adetunji, C. O., Inobeme, A., Musah, M., Azeh, Y., Etsuyankpa, M. B., Otori, A. A., Chado, I. M., Shaba, E. Y., Akoji, J. N., Mamman, A., & Oyewole, O. A. (2025a). Application of Nanochitosan for the Biodegradation of Agricultural Wastes. In: *Nanochitosan Applications for Enhanced Crop Production and Food Security*. John Wiley & Sons, 361–382. https://doi.org/10.1002/9781394214945.ch16

Mathew, J. T., Adetunji, C. O., Inobeme, A., Musah, M., Azeh, Y., Otori, A. A., Shaba, E. Y., Mamman, A., & Tanko, E. (2023a). Removal of Heavy Metals Using Bio-remedial Techniques. Springer Nature Switzerland AG, 117–130. https://doi.org/10.1007/978-3-031-24086-7_6

Mathew, J. T., Adetunji, C. O., Inobeme, A., Otori, A. A., Musah, M., Azeh, Y., Shaba, E. Y., Muhammad, I. A., Mamman, A., & Oyewole, O. A. (2025b). Relevance of Nanochitosan in Food Sector and Food Packaging Sectors: Current Trends. In: *Nanochitosan Applications for Enhanced Crop Production and Food Security*. John Wiley & Sons, 345–360. https://doi.org/10.1002/9781394214945.ch15

Mathew, J. T., Inobeme, A., Etsuyankpa, M. B., Adetunji, C. O., Shaba, E. Y., Musah, M., Azeh, Y., Jiya, M. J., Tanko, E., Muhammad, A. I., Mamman, A., & Maurice, J. (2024a). Applications of Diatom for Treatment of Industrial Effluents: Waste to Wealth Approach. In: *Marine Biomass: Biorefinery, Bioproducts and Environmental Bioremediation*. Walter de Gruyter GmbH & Co KG, 1–20. https://doi.org/10.1515/9783111353951-001

Mathew, J. T., Inobeme, A., Hussaini, J., Shaba, E. Y., Azeh, Y., Musah, M., Muhammad, E. B., Tanko, M. S., Ismail, H., Adetunji, C. O., Kanwa, A. M., & Mamman, A. (2024b). Properties of Nanomaterials for Corrosion Control. In A. Thakur & A. Kumar (Eds.), *Innovations in Nanomaterials-Based Corrosion Inhibitors*. IGI Global, 74–99. https://doi.org/10.4018/979-8-3693-3088-3.ch004

Mathew, J. T., Inobeme, A., Musah, M., Azeh, Y., Etsuyankpa, M. B., Adetunji, C. O., Shaba, E. Y., Jiya, M. J., Josiah, J. G., Tanko, E., Muhammad, A. I., Mamman, A., & Jemkur, M. (2024c). Utilization of Marine Green Algae for Development of Bioplastics. In: *Marine Biomass: Biorefinery, Bioproducts and Environmental Bioremediation*. Walter de Gruyter GmbH & Co KG, 39–55. https://doi.org/10.1515/9783111353951-003

Mathew, J. T., Musah, M., Azeh, Y., & Muhammed, M. (2023b). Adsorptive Removal of Selected Toxic Metals from Pharmaceutical Wastewater using Fe_3O_4/ZnO Nanocomposite. *Dutse Journal of Pure and Applied Sciences*, *9*(4a), 236–248. https://doi.org/10.4314/dujopas.v9i4a.22

Mathew, J. T., Musah, M., Azeh, Y., & Muhammed, M. (2023c). Kinetic Study of Heavy Metals Removal from Pharmaceutical Wastewater Using Geopolymer/Fe_3O_4 Nanocomposite. *Bima Journal of Science and Technology*, *7*(4), 152–163.

Matloob, A., Ayub, H., Mohsin, M., Ambreen, S., Khan, F. A., Oranab, S., Rahim, M. A., Khalid, W., Nayik, G. A., Ramniwas, S., & Ercisli, S. (2023). A Review on Edible Coatings and Films: Advances, Composition, Production Methods, and Safety Concerns. *ACS Omega*, *8*(32), 28932–28944. https://doi.org/10.1021/acsomega.3c03459

Mieszczakowska-Frąc, M., Celejewska, K., & Płocharski, W. (2021). Impact of Innovative Technologies on the Content of Vitamin C and Its Bioavailability from Processed Fruit and Vegetable Products. *Antioxidants (Basel, Switzerland)*, *10*(1), 54. https://doi.org/10.3390/antiox10010054

Muhammad, W. M., Ibrahim, A. B., Samuel, J. O., Khalifa, A. S., Oyewole, O. A., Adetunji, C. O., Eniola, K. I. T., Yerima, M. B., & Mathew, J. T. (2025). The Role of Nanomaterials in Agriculture as Nanofertilizers. In: *Nanochitosan Applications for Enhanced Crop Production and Food Security*. John Wiley & Sons, Inc., 1–38. https://doi.org/10.1002/9781394214945.ch1

Muñoz-Tebar, N., Pérez-Álvarez, J. A., Fernández-López, J., & Viuda-Martos, M. (2023). Chitosan Edible Films and Coatings with Added Bioactive Compounds: Antibacterial and Antioxidant Properties and Their Application to Food Products: A Review. *Polymers*, *15*(2), 396. https://doi.org/10.3390/polym15020396

Mura, P., Maestrelli, F., Cirri, M., & Mennini, N. (2022). Multiple Roles of Chitosan in Mucosal Drug Delivery: An Updated Review. *Marine Drugs*, *20*(5), 335. https://doi.org/10.3390/md20050335

Nguyen, H. M., Ngoc Le, T. T., Nguyen, A. T., Thien Le, H. N., & Pham, T. T. (2023). Biomedical Materials for Wound Dressing: Recent Advances and Applications. *RSC Advances*, *13*(8), 5509–5528. https://doi.org/10.1039/d2ra07673j

Nunes, C., Silva, M., Farinha, D., Sales, H., Pontes, R., & Nunes, J. (2023). Edible Coatings and Future Trends in Active Food Packaging-Fruits' and Traditional Sausages' Shelf Life Increasing. *Foods (Basel, Switzerland)*, *12*(17), 3308. https://doi.org/10.3390/foods12173308

Ofoedu, C. E., Iwouno, J. O., Ofoedu, E. O., Ogueke, C. C., Igwe, V. S., Agunwah, I. M., Ofoedum, A. F., Chacha, J. S., Muobike, O. P., Agunbiade, A. O., Njoku, N. E., Nwakaudu, A. A., Odimegwu, N. E., Ndukauba, O. E., Ogbonna, C. U., Naibaho, J., Korus, M., & Okpala, C. O. R. (2021). Revisiting Food-Sourced Vitamins for Consumer Diet and Health Needs: A Perspective Review, from Vitamin Classification, Metabolic Functions, Absorption, Utilization, to Balancing Nutritional Requirements. *PeerJ*, *9*, e11940. https://doi.org/10.7717/peerj.11940

Ogundolie, F. A., Adetunji, C. O., Mathew, J. T., Inobeme, A., Ghazanfar, S., Olaniyan, O. T., Ajiboye, M. D., Dauda, W. P., Adetunji, J. B., Akinbo, O., & Adetuyi, B. O. (2023). Intelligent Freshness Indicators in Packaging of Plant-Based Food Products. Academic Press, 323–331. https://doi.org/10.1016/B978-0-323-89864-5.00008-4

Perez-Vazquez, A., Barciela, P., Carpena, M., & Prieto, M. A. (2023). Edible Coatings as a Natural Packaging System to Improve Fruit and Vegetable Shelf Life and Quality. *Foods (Basel, Switzerland)*, *12*(19), 3570. https://doi.org/10.3390/foods12193570

Picos-Corrales, L. A., Morales-Burgos, A. M., Ruelas-Leyva, J. P., Crini, G., García-Armenta, E., Jimenez-Lam, S. A., Ayón-Reyna, L. E., Rocha-Alonzo, F., Calderón-Zamora, L., Osuna-Martínez, U., Calderón-Castro, A., De-Paz-Arroyo, G., & Inzunza-Camacho, L. N. (2023). Chitosan as an Outstanding Polysaccharide Improving Health-Commodities of Humans and Environmental Protection. *Polymers*, *15*(3), 526. https://doi.org/10.3390/polym15030526

Romanazzi, G., & Moumni, M. (2022). Chitosan and Other Edible Coatings to Extend Shelf Life, Manage Postharvest Decay, and Reduce Loss and Waste of Fresh Fruits and Vegetables. *Current Opinion in Biotechnology*, *78*, 102834. https://doi.org/10.1016/j.copbio.2022.102834

Roy, S., & Dhaneshwar, S. (2023). Role of Prebiotics, Probiotics, and Synbiotics in Management of Inflammatory Bowel Disease: Current Perspectives. *World Journal of Gastroenterology*, *29*(14), 2078–2100. https://doi.org/10.3748/wjg.v29.i14.2078

Singh, A. K., Kim, J. Y., & Lee, Y. S. (2022). Phenolic Compounds in Active Packaging and Edible Films/Coatings: Natural Bioactive Molecules and Novel Packaging Ingredients. *Molecules (Basel, Switzerland)*, *27*(21), 7513. https://doi.org/10.3390/molecules27217513

Singh, R., Dutt, S., Sharma, P., Sundramoorthy, A. K., Dubey, A., Singh, A., & Arya, S. (2023). Future of Nanotechnology in Food Industry: Challenges in Processing, Packaging, and Food Safety. *Global Challenges (Hoboken, NJ)*, *7*(4), 2200209. https://doi.org/10.1002/gch2.202200209

Suvarna, V., Nair, A., Mallya, R., Khan, T., & Omri, A. (2022). Antimicrobial Nanomaterials for Food Packaging. *Antibiotics (Basel, Switzerland)*, *11*(6), 729. https://doi.org/10.3390/antibiotics11060729

Thambiliyagodage, C., Jayanetti, M., Mendis, A., Ekanayake, G., Liyanaarachchi, H., & Vigneswaran, S. (2023). Recent Advances in Chitosan-Based Applications: A Review. *Materials (Basel, Switzerland)*, *16*(5), 2073. https://doi.org/10.3390/ma16052073

Wang, Z., Ng, K., Warner, R. D., Stockmann, R., & Fang, Z. (2023). Application of Cellulose- and Chitosan-Based Edible Coatings for Quality and Safety of Deep-Fried Foods. *Comprehensive Reviews in Food Science and Food Safety*, *22*(2), 1418–1437. https://doi.org/10.1111/1541-4337.13116

Wu, M., Xu, R., Liu, C., Li, B., & Long, Z. (2020). Amelioration of Physical Properties and Printability of Paper Coated with N-methylated Chitosan. *Scientific Reports*, *10*(1), 9936. https://doi.org/10.1038/s41598-020-66827-8

Yan, D., Li, Y., Liu, Y., Li, N., Zhang, X., & Yan, C. (2021). Antimicrobial Properties of Chitosan and Chitosan Derivatives in the Treatment of Enteric Infections. *Molecules (Basel, Switzerland)*, *26*(23), 7136. https://doi.org/10.3390/molecules26237136

Zabot, G. L., Schaefer Rodrigues, F., Polano Ody, L., Vinícius Tres, M., Herrera, E., Palacin, H., Córdova-Ramos, J. S., Best, I., & Olivera-Montenegro, L. (2022). Encapsulation of Bioactive Compounds for Food and Agricultural Applications. *Polymers*, *14*(19), 4194. https://doi.org/10.3390/polym14194194

Zárate-Moreno, J. C., Escobar-Sierra, D. M., & Ríos-Estepa, R. (2023). Development and Evaluation of Chitosan-Based Food Coatings for Exotic Fruit Preservation. *Biotech (Basel, Switzerland)*, *12*(1), 20. https://doi.org/10.3390/biotech12010020

14 Antioxidant Prospects of Combining Chitosan and Essential Oil Biofilm Inhibitors in Fruits and Vegetables

Mayowa Jeremiah Adeniyi
Department of Physiology, College of Medicine and Health Sciences, University of Rwanda

14.1 INTRODUCTION

Apart from being important energy sources, foods play critical roles in the maintenance of health and wellness and prevention of diseases. Of foods consumed by humans and mammals, fruits and vegetables have received considerable concerns over the years owing to their high fiber and micronutrient contents. This makes them a palliative nutritional solution for constipation, cardiovascular diseases, obesity and other metabolic disorders. According to the 2023 report of World Health Organization, about 400 g of fruits and vegetables are recommended on daily basis to enhance the overall health and wellness and suppress the risk level of some non-communicable diseases. In United States, 116.21 pounds of fruits were reportedly consumed in year 2022. In the fresh fruits market, it is expected that the volume will increase to 303.70bn kg by 2028 and a volume growth of 4.0% in 2025. However, transportation barriers, costs and affordability, alteration in food environment, inadequate storage facilities alteration in culture, demographic factors and taste and lack of variety and taste, coupled with global climate change and the highly perishable nature of fruits and vegetables, remain to be some of the major constraints to consumption of fruits and vegetables generally and most especially in developing and underdeveloped world.

To overcome the spoilage problems associated with fruits and vegetables, a barrage of policies have been proposed, including shortening of food supply chain, food safety and right to food regulations, nutritional literacy campaigns, enhancement of market access to fruits and vegetables, encouragement of traditional fruits and vegetables and access to finance and price subsidies (Inge, 2021). In low and lower middle income economies, overcoming barriers to consumption of fruits and vegetables involves the availability of production systems for fruits and vegetables in semi-rural areas, inclusive and equitable participation in fruits and vegetables production and the availability of risk capitals (Youri et al., 2021). Other strategies may include farm to farm institution programs, social and behavioral change communication (Kaur, 2023).

Another mechanism of improving the consumption rate of fruits and vegetables is to address their high perishability tendency. Fruits even after harvest are known to be metabolically active, thus generating free radicals that play eventual roles in their ripening and spoilage. Besides primitive preservative approaches such as chilling, freezing, drying and pasteurization, techniques aimed at mitigating spoilage, including electrical, electrothermal, radiation, nanotechnology and chemical methods, have been reported to cause significant improvement in the physicochemical properties (such as flavor, stability and optical and rheological characteristics) as well as nutritional qualities of fruits and vegetables (Sridhar et al., 2021), which collectively indicate the safety and quality of foods.

DOI: 10.1201/9781003355458-14

As part of measures to enhance the safety and quality of foods, plant-derived compounds have been shown to be essential. Chitosan is a derivative from chitin, an integral component of invertebrates, algae and fungi. It has proven to be an indispensable product in biomedicine, pharmacy and pollution control. It is also essential in the preservation of foods including fruits and vegetables owing to its metal-chelating dependent excellent antioxidant capacity. Essential oils, on the other hand, are derived from aromatic plants and are known for their biofilm inhibition due to their remarkable anti-inflammatory and antioxidant prowess. The review was targeted at discussing the possible antioxidant prospects in combining chitosan with essential oils in fruits and vegetables.

14.2 OVERVIEW OF CHITOSAN

Chitosan is a polymer that is derived from chitin (2 acetamido-2-deoxy beta-1,4-D-glucan), an integral part of cell walls of invertebrates including crustaceans and arthropods as well as fungi and algae (Pellis et al., 2022). Chitosan is composed of *N*-acetyl glucosamine and glucosamine and is essential for the regulation of water pollution; it is also applicable in the healthcare, pharmaceuticals and food industries, where it is used as antimicrobial, anti-inflammatory and antioxidant agents for maintaining the stability, improving the shelf life and supporting postharvest storage of many foods including perishable foods like fruits and vegetables. A review by Younes and Rinaudo (2015) identified various ways through which chitin, a precursor for chitosan, can be derived from raw sources such as marine invertebrates. In addition, the role of the chemical structure and solubility in the production process were discussed. In contemporary age, both chitin and chitosan are derived from the exoskeleton of arthropods, shrimps, squid and crabs within the sphere of laboratories. The processes involved in the synthesis of chitin and chitosan include deproteinization, demineralization and deacetylation. Shahbaz et al. (2020) enumerated the various applications of chitosan, a derivative of chitin (the second most abundant polysaccharide), in various fields including cosmetics, biomedical, pharmaceuticals and agriculture and food industries. Apart from the use of biological and chemical methods to convert chitin to chitosan, chitinase-secreting microorganisms can be utilized to convert waste chitin by hydrolysis to mitigate chitin pollution in the environment.

Chitosan is principally synthesized from crab, crayfish and shrimp owing to the availability of crustacean bones as an essential food processing byproduct. Squid pens, a waste byproduct of squid processing, are a new, cheap and renewable chitin and chitosan sources. Shepherd et al. (1997) evaluated the functional properties of squid pen chitosan. Squid chitosan was shown to be colorless with a little ash content and enhanced suspending and thickening properties. When compared with crustacean-derived chitosan, squid-derived chitosan exhibited a low flocculation capacity. Also, when compared with crustacean-derived chitosan, squid-derived chitosan showed more elastic films with enhanced functional properties. Chitosan is non-toxic, renewable, biocompatible, ecofriendly and modifiable. Chitosan is a cationic polysaccharide, exhibiting deacetylation and molecular-weight-dependent mechanical, biological and physicochemical attributes (El-Araby et al., 2024), which make it applicable in waste treatment among others. Owing to their abundance, chitosan may be used as an alternative in fuel cells. Modau et al. (2023) highlighted steps required for developing chitosan and silicon composite membranes for fuel cell usage using 5 g of chitosan and 2%–4% of silica at constant speed for 2 hours. The physical properties of chitosan membrane, such as methanol permeability, proton conductivity, functional groups and water absorption, were also demonstrated.

Chitosan can be modified to nano-sized format known as nanochitosan. Chitosan nanosubstances have a positively charged amine group and thus can combine with negatively charged molecules including proteins and nucleic acids such as deoxyribonucleic acid and ribonucleic acid. Examples of these negatively charged molecules include microbes, bacterial cell wall and viruses. Hence, nanochitosan can be utilized as antimicrobial agent. With the positively charged amine groups, nanochitosan reacts with Gram-negative and Gram-positive bacteria. This combination besides being used in preventing microbial colonization

14.3 OVERVIEW OF THE ESSENTIAL OIL

Microorganisms are known to be capable of developing biofilms as a means of gluing themselves to nutritional cues as well as to shield themselves from hostile environmental and harmful stimuli such as antibiotics. One of the mechanisms that facilitate biofilm formation is quorum sensing, the sharing of information by bacteria with one another regarding cell density. Essential oils are synthetic biofilm-inhibiting oils that are derived from aromatic plants. They exhibit antimicrobial, anti-inflammatory and antioxidant activities. Cáceres et al. (2020) examined anti-biofilm, antimicrobial and anti-quorum sensing impacts of 17 essential oils some microorganisms (*Escherichia coli* and *Staphylococcus epidermidis*). These essential oils were synthesized from plant material through hydrodistillation. To test for antimicrobial effect, microdilution technique was utilized, and the ability of the essential oil to suppress biofilm formation was evaluated. Through quantification of violacein in *Chromobacterium violaceum*, the anti-quorum sensing test was performed. The 17 essential oils were shown to suppress biofilm formation strongly and quorum sensing generation in a dose-dependent pattern.

Antimicrobial resistance is predicated on the tendency of bacteria to form biofilm and biofilm-generating bacteria to form a compact defensive matrix, which makes colonization and adherence to any surfaces possible. Essential oils from *Lippia origanoides* thymol-carvacrol II chemotype plants exhibit a huge potential to suppress diverse bacteria with biofilm-forming activity. Martínez et al. (2023) investigated the effects of essential oils from *L. origanoides* thymol-carvacrol II chemotype plants on the expressions of quorum sensing regulating gene as well as virulence of *Staphylococcus aureus* and *E. coli*. Essential oils from *L. origanoides* thymol-carvacrol II chemotype plants were shown to be effective against the formation of biofilm as well as inhibiting quorum sensing.

Cáceres et al. (2023) assessed the anti-quorum sensing and antimicrobial activities of 15 essential oils against *C. violaceum*. Essential oil derived from *L. origanoides* showed anti-quorum and antimicrobial potentials at 0.15 and 0.37 mg/mL. The anti-biofilm effect of essential oil may be due to suppression of tryptophan metabolisms in the course of violacein synthesis. Guillín et al. (2021) assessed the anti-biofilm and antimicrobial activities of 15 essential oils against *Salmonella typhimurium* and *Salmonella enteritidis*. The essential oil from the *L. origanoides* chemotype thymol-carvacrol II showed the least minimum bactericidal concentration of 0.75 mg/mL and minimum inhibitory concentration of 0.37 mg/mL, but the greatest percentage of biofilm inhibition (confirmed through scanning electron microscope) of greater than 65% on *S. typhimurium* and *S. enteritidis*. Petralito et al. (2024) developed an encapsulated *Lavandula × intermedia* essential oil in lipid-based nanoemulsions through the solvent displacement method. *Lavandula × intermedia* essential oil was doped by incorporating different levels of water-insoluble oil (ripening inhibitor) to maintain and preserve colloidal stability of the formulation. It was observed that the antimicrobial activity of the nanoemulsions reduced as the ripening inhibitor was added. Also in undoped nano *Lavandula × intermedia* essential oil, the antibacterial activity against *Bacillus cereus* and *E. coli* was unaltered up to ten months from the preparation.

Martínez et al. (2021) examined the anti-biofilm, antimotility, antimicrobial and antihemolytic potentials of 15 essential oils from *Thymus vulgaris* and *L. origanoides* thymol-carvacrol II chemotype (LTC II) on *S. aureus* and *E. coli*. Biofilm formation was evaluated using SEM microscopy and crystal violet assay. *T. vulgaris* and *L. origanoides* thymol-carvacrol II chemotype were shown to have a great antimicrobial activity with a maximum inhibitory concentration of 0.45 and 0.75 mg/mL, respectively. As far as *L. origanoides* thymol-carvacrol II chemotype was concerned, the inhibition of biofilm exceeded 70% at subinhibitory concentrations. It was also observed that *L. origanoides* thymol-carvacrol II chemotype and *T. vulgaris* suppressed hemolytic activity of *S. aureus* by 54% and 32%, respectively. There was a decline in the swimming-related mobility capacity in *E. coli* by 55% and 47% with *L. origanoides* thymol-carvacrol II chemotype and *T. vulgaris*, respectively.

14.4 THE ROLE OF OXIDATIVE STRESS IN FRUITS AND VEGETABLES SPOILAGE

Redox reactions occur in every living eukaryotic cell, leading to the formation of free radicals. Interestingly, the cells also contain plethora of antioxidant molecules that serve as regulators of free radical productions. Hence, inadequate production of antioxidant molecules results in free radicals/antioxidant imbalance or oxidative stress. Oxidative stress plays important roles in postharvest spoilage of fruits and vegetables. Therefore, many preservatives work by improving antioxidant strength of postharvest fruits and vegetables. Fruits and vegetables exhibit high respiratory rates and a high tendency of generating free radicals such as superoxide anions and hydrogen peroxides causing spoilage and deteriorations and requiring special storage. They are susceptible to chilling injury (pitting, discoloration and uneven ripening, among others) (Valenzuela et al., 2017). Generation of oxidants by fruits and vegetables is higher during postharvest storage owing to reduction in intrinsic antioxidant molecules (Meitha et al., 2020). Among the mostly reported reactive oxygen species in plants are hydrogen peroxide, singlet oxygen and superoxide anion and hydroxyl radicals. Hydroxyl radicals are generated from hydroperoxides and through the combination of excited environmental oxygen with water. It is also a precursor of hydrogen peroxide and oxygen just like superoxide anion. Unlike superoxide anion, which can be scavenged by superoxide dismutase, hydroxyl radicals are removed by antioxidants such as glutathione, vitamin E, mannitol and melatonin. Production of reactive oxygen species and oxidants by plants can also be influenced by different abiotic stresses. Hydroxyl radical, superoxide radicals, perhydroxy radical, alkoxy radicals, singlet oxygen and hydrogen peroxide are generated in the photosystem I, and the photosystem of chloroplasts is produced in excess due to abiotic stress. Apart from abiotic stress, oxidants and reactive oxygen species play very significant roles as far as growth, gene expression, microbial defense and apoptosis of plants are concerned (Gill and Tuteja, 2010). A number of abiotic stresses affect postharvest storage of crops (Gill et al., 2013). Drought, heavy metals, chilling injury and salinity impair postharvest storage of crops by causing overproduction of oxidants and reactive oxygen species, overwhelming the tripeptide glutathione and glutathione reductase, which are the principal components of ascorbate-glutathione pathway involved in the protection of cells against oxidant damage.

Cadmium is a heavy metal, an environmental hazard and an abiotic stressor (Figure 14.1) derived from soil. It acts as a major threat to food production and storage. A study by Chen et al. (2010) showed that cadmium, when used at a dose of 5 μM, induced toxicity in barley plants through increase in malondialdehyde and hydrogen peroxide. Cadmium exposure decreased catalase and ascorbate peroxidase and caused perturbation in the levels of root monodehydroascorbate reductase, dehydroascorbate reductase and glutathione peroxidase. It caused malondialdehyde accumulation. Rosalie et al. (2018) investigated the impact on oxidative/antioxidant homeostasis and carotenoid

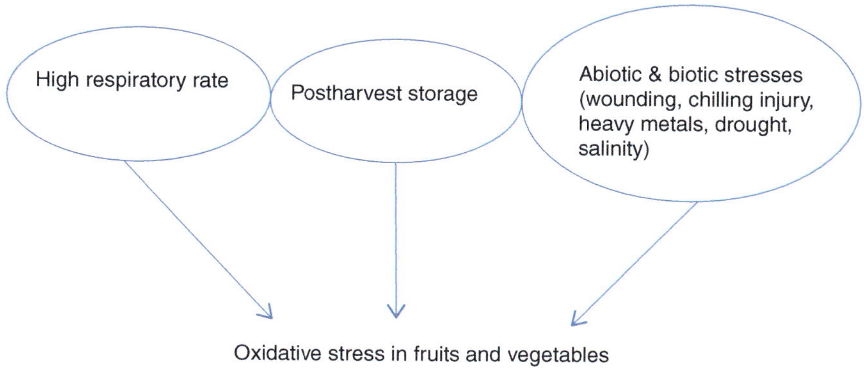

FIGURE 14.1 Sources of oxidative stress in fruits and vegetables. Intrinsic high respiratory rate in fruits and vegetables, together with postharvest storage and abiotic and biotic stresses such as wounding, chilling injury, heavy metals, drought and salinity, causes deranged redox homeostasis in fruits and vegetables.

accumulation in mango fruits for two successive years caused by two week storage at temperatures of 12°C and 7°C and subsequently by fruit ripening at 20°C. Among the parameters evaluated at harvest and post-ripening following cold temperature storage were catalase, superoxide dismutase, ascorbic peroxidase, dehydroascorbate reductase and monodehydroascorbate reductase, hydrogen peroxide and carotenoids levels. During control conditions (a temperature of 20°C), ripening was shown to induce a rise in oxidation, leading to elevated reactive oxygen generation and decline in ascorbate levels. Carotenoid accumulation was exaggerated by ripening, and storage at low temperatures raised cellular respiration crisis severity causing oxidation and oxidative stress in the fruit pulp. Antioxidant responses in younger fruits were shown to reduce when compared to the control, and second harvested fruits exhibited greater antioxidant level required for oxidative stress.

After crops have been harvested, active life ceases and so the intrinsic antioxidant defense system depresses. Hence, postharvest storage and distribution of fruits and vegetables more than often witness elevation in oxidants and reactive oxygen species such as superoxide, hydrogen peroxide, singlet oxygen and hydroxyl radicals (Castillo et al., 2020) owing to shortage of intrinsic antioxidants to protect plant cells from oxidative stress. Mechanical wound occurs when parts of fruit, vegetables or any plants are punctured, cut, crushed, rubbed and struck either deliberatively or accidentally. Mechanic wound induced on fruit is another trigger of oxidative stress in plants and causes breakdown of cell walls, and this exposes membrane lipids to oxidative degradation (Ackah et al., 2022). When mechanical wounds were artificially induced in apples, lipoxygenase and malondialdehyde were shown to increase.

Prolonged low temperature aggravates chilling injury manifestations in fruits and vegetables. Pan et al. (2022) showed that when compared to 1°C, exposure to 6°C caused reduction in superoxide dismutase, catalase, peroxidase, dehydroascorbate reductase, monodehydroascorbate reductase and ascorbate peroxidase and increase in generations of hydrogen peroxide, malondialdehyde and superoxide anions. When *Saccharum officinarum* L. plants were exposed to chilling injury at a temperature of 8°C and 4°C, with a daily lighting duration of 11 hours for a period of 3 days, followed by exposure to temperatures of 25°C and 20°C for a period of 11 hours photoperiod for another 2 days. Zhu et al. (2013) indicated that there was alteration in oxidative markers. Chilling injury caused a rise in superoxide anions, malondialdehyde and hydrogen peroxide. Yu et al. (2023) reported a decline in peroxidase, catalase, ascorbate peroxidase, dehydroascorbate reductase, monodehydroascorbate reductase and glutathione reductase and altered the concentrations of glutathione and ascorbic acid in wounded pear fruits. Wounded pear fruit also elevated production of superoxide anions and hydrogen peroxide.

As far as mechanisms activated by free radicals in plants is concerned, Zhang et al. (2023) showed that hydrogen peroxide-induced cysteine oxidation of NAC WITH TRANS-MEMBRANE MOTIF1-LIKE 1 that occurs in soybean during salt stress enhances its release and transport from endoplasmic reticulum membrane and to the nucleus, respectively. It was also shown that through binding and activation of the promoters of RESPIRATORY BURST OXIDASE HOMOLOG B (GmRbohB) genes, oxidative posttranslational modification on NAC WITH TRANS-MEMBRANE MOTIF1-LIKE 1 residue Cysteine-247 orchestrates downstream amplification of ROS production. This then creates a feed-forward loop to regulate NAC WITH TRANS-MEMBRANE MOTIF1-LIKE 1 activity.

14.5 PLANT-BASED ANTIOXIDANTS

14.5.1 Superoxide Dismutase

During normal and abnormal cellular metabolisms, superoxide anions are generated. To buffer superoxide anions, superoxide dismutase enzymes are produced. Superoxide dismutases catalyze the conversion of superoxide anions to hydrogen peroxide and oxygen. Superoxide dismutases form crucial defense against oxidative stress in the plants (Younus, 2018). Lin et al. (2002) utilized polymerase chain reaction to clone a putative copper/zinc superoxide dismutase from lemon. The analysis of the complementary DNA revealed that it is made up of an open reading frame coding for 152 amino acid residues, with amino acid sequences exhibiting high identity of 65%–84% with Cu/Zn superoxide dismutase sequences from other plant species. Using enzyme activity, the expression of the Cu/Zn superoxide dismutase was ratified via a native gel, which was purified by Ni^{2+}-nitrilotriacetic acid Sepharose superflow. The two active forms, which are 70% monomer and 30% dimer, produced by the purified enzyme are in equilibrium and have a specific activity of 7456 units/mg. The dimer activity was reported to be 65% more than that of the monomer. Ken et al. (2002) reported the presence of a full-length complementary DNA of 794 basepairs that encodes a copper/zinc superoxide dismutase from *Pagrus major* through polymerase chain reaction technique. When the nucleotide sequence of the complementary DNA clone was analyzed, it showed that it contains a full open reading frame coding for 154 amino acid residues with amino acid sequence having high similarity of 53%–91% with the Cu/Zn superoxide dismutase in other plant species. Using enzyme activity that was stained on a native gel and purified through Ni^{2+}-nitrilotriacetic acid Sepharose superflow, the expression of the Cu/Zn superoxide dismutase was confirmed. Ken et al. (2003) reported a whole-length complementary DNA with 803 base pairs that encodes a putative *Epinephelus malabaricus* copper/zinc-superoxide dismutase. This was cloned through polymerase chain reaction method. The analysis of the complementary DNA clone showed that it contains a full open reading frame coding for 154 amino acid residues, and these amino acid sequences exhibited high similarity of 65%–91% with Cu/Zn superoxide dismutase sequences from other species. Using enzyme activity stained on a native gel and purified by Ni^{2+}-nitrilotriacetic acid Sepharose, the expression of the Cu/Zn-SOD was ratified.

In a study by Utami et al. (2023), the crystal structure of lemon superoxide dismutase was examined at 1.86 Å resolution. The lemon copper and zinc superoxide dismutase was shown to resist proteolytic breakdown and active in an acidic environment. The lemon copper and zinc superoxide dismutase exhibited similar structure as copper and zinc superoxide dismutase from plants. Although lemon copper and zinc superoxide dismutase shared similarity with copper and zinc superoxide dismutase from tomato and spinach, it was classified as a cytoplasmic copper and zinc superoxide dismutase. Li et al. (2023) in their study cloned a new copper-zinc superoxide dismutase and characterized it from *Psychropotes verruciaudatus*. In the study, the relative molecular weight of the copper/zinc superoxide dismutase for the organism was 15 kDa, and the optimum temperature for the enzyme is 20°C with a high activity ranging from 0°C to 60°C. *P. verruciaudatus* copper and zinc superoxide dismutase exhibited elevated thermal stability at incubation of 37°C. The enzyme showed a peak activity exceeding 50%. When compared with bovine superoxide dismutase, the enzyme exhibited high stability to gastrointestinal fluid.

14.5.2 Catalase

Catalases are strong metalloenzyme and antioxidants that catalyze the conversion of hydrogen peroxide to water and oxygen in plants including fruits and vegetables. Catalase is an integral component of antioxidant defense and plant development, and it is produced during normal cellular metabolic events. Its activities can also be influenced by exposure to pathogen, ultraviolet radiation and gravity. There is a nexus between intracellular calcium and hydrogen peroxide level within cells in response to abiotic and biotic stresses. Yang and Poovaiah (2002) reported how a calcium-binding protein known as calmodulin, in the presence of calcium, interacts and activates plant catalases, but no effect was reported for calcium/calmodulin as far as fungal, bacterial, human and bovine catalases are concerned. Calcium/calmodulin is capable of down-regulating hydrogen peroxide concentrations in plants by inducing plant catalase activity. Catalase is a powerful antioxidant metalloenzyme located in peroxisomes that also plays a central role in signaling processes under physiological and adverse situations. Unlike animals that possess just a single gene for catalase, plants are multigenic, thus implying the existence of many isoenzymes in plants, which differ from one species of plants to another (Palma et al., 2020). Environmental conditions, tissue distribution, reactive nitrogen species, reactive oxygen species and hydrogen sulfide may influence the expression of plant catalase. Catalase exists as a principal protein that undergoes tyrosine nitration and posttranslational modification in response to reactive nitrogen species in the course of fruit ripening.

Catalases play crucial roles in scavenging H_2O_2 from reactive oxygen species, controlling the growth and development of plants. So far, genome-wide identification and characterization of catalase genes in oil palm have not been reported. In a study by Zhou et al. (2024), five energy and green catalysis (*EgCAT)* genes were extracted via a genome-wide identification technique. Conserved motif analysis and gene structure indicated the conserved form of intron and exon organization in the energy and green catalysis genes. Diverse cis-acting components associated with defense responses, hormone and biotic and abiotic stress were recognized in the promoter regions of energy and green catalysis genes. Through heatmap analysis via transcriptome data, tissue-specific expression of energy and green catalysis genes in five different oil palm tissues were unveiled. These five energy and green catalysis genes were expressed under drought, cold and salt stress.

Retention of active oxygen species plays roles in responses of plant to both abiotic and biotic stressors. In stress situations, active oxygen species may either worsen damage to plants or stimulate diverse defense responses through cell signaling (Dat et al., 2001). Ghorbel et al. (2023) carried out a genome-wide study of catalase gene family involving oat plants to characterize the likely roles of genes that are under the influence of different stressors. The gene structure of AvCAT as well as phylogenetic tree, secondary and tertiary protein structure and physicochemical properties were extrapolated using bioinformatics. In the Saudi oat variety, ten AvCAt genes belonging to three groups were recognized. The identified AvCAT proteins were shown to be situated within the mitochondria, peroxisome and cytoplasm of oat cells. When promoters of these genes were analyzed, diverse cis-elements were discovered as being associated with maturation and development of oat plant as well as response to various external stresses. Through gene analysis, the three different *AvCAT* genes belong to three different subgroups, and they showed detectable modifications in response to salinity and mannitol.

14.5.3 Glutathione Peroxidase

Glutathione peroxidases are a type of peroxidase that catalyzes the conversion of hydrogen peroxide to non-toxic water and oxygen. The action of glutathione peroxidase is similar to catalase only that the former acts predominantly on hydrogen peroxide when it is at high concentrations, while catalase acts when hydrogen peroxide is at low concentration. Glutathione peroxidase is a selenoprotein. In plants, glutathione peroxidase has various isoenzymes that reside in different compartments within the cells. Plant glutathione peroxidase exhibits both thioredoxin peroxidase and glutathione peroxidase activities (Bela et al., 2015). Unlike in animals, plant glutathione peroxidase contains cysteine instead of selenocysteine in its active site. The complex family of glutathione peroxidase is influenced by factors such as type of tissues to biotic and abiotic stress. All the functions of the enzyme in plant are not fully understood. For instance, plants exhibit thiol-related activities, implying the possible involvement of glutathione peroxidase in oxidant/antioxidant homeostasis by maintaining the balance between NADPH and NADP(+) and between thiol and disulfide besides scavenging hydrogen peroxide. The enzyme plays key role in plant development. Plant response to perturbation in photic intensity depends on an enzyme known as NADPH-thioredoxin reductase C. Also, free radicals produced by 2-cysteine peroxiredoxin help in maintaining the oxidant/antioxidant balance in the chloroplast in response to photic challenge. In order to understand the roles of oxidizing signals caused by glutathione peroxidase on chloroplast oxidant/antioxidant homeostasis, Casatejada et al. (2023) created an *Arabidopsis thaliana* double mutant glutathione peroxidase gene known as gpx1gpx7 in the chloroplast. The gene lacks GPX-1 and GPX-7. Glutathione peroxidases are dispensable as far as plant growth is concerned under standard conditions, and this was evidenced by wild type-like phenotype exhibited by the gpx1gpx7 mutant in the chloroplast.

In response to pathogens, plants are known to synthesize large proportion of hydrogen peroxides. Bi et al. (2022) attempted to understand the underlying mechanisms by which plants are able to detect hydrogen peroxide to control immunity, and they realized that endogenous hydrogen peroxide is created when immune activation is detected by the thiol peroxidase through oxidation at Cysteine-51. Stomatal immunity against *Pseudomonas syringae* occurs through this mechanism. Also when cells are immunogenically stimulated, the thiol peroxidase will conjugate through Cysteine-51 with another protein known as type 2C protein phosphatase ABA insensitive 2 resulting in transduction of hydrogen peroxide to type 2C protein phosphatase ABA insensitive 2. This type of oxidation is needed for stomatal immunity in plants.

Glutathione peroxidases were first discovered as a scavenger of active oxygen species. Although non-selenium-dependent glutathione peroxidases containing cysteine have been reported in many organisms, the efficiency of their activity depends on the presence of selenium-containing amino acid known as selenocysteine. Non-selenium-dependent peroxidases are really not true glutathione peroxidases because they reduce active oxygen species through thioredoxin and not glutathione (Herbette et al., 2007). Navrot et al. (2006) concisely highlighted poplar (*Populus trichocarpa*) family of glutathione peroxidase. Generally, glutathione peroxidases, despite being previously thought to be glutathione dependent, do not react with glutathione, but peroxides and thioredoxin are required to restore reduced glutathione. Poplar glutathione peroxidases have far lower specificity with respect to their electron-accepting substrates unlike the chloroplastic glutathione peroxidases. Chloroplastic glutathione peroxidases have high selectivity and specificity to their electron donors with exclusive specificity for thioredoxin of y-type for their reduction. There are six identified genes that code for glutathione peroxidases, and they are found in different poplar organs; two of them are situated in the chloroplast and one colocalizing in mitochondria.

14.6 OTHER PLANT-BASED PROTECTIVE SUBSTANCES

14.6.1 Phenylalanine Ammonia-Lyase

Phenylalanine ammonia-lyase is essential for the metabolism of phenylpropanoid, and it removes ammonia from phenylalanine to produce trans-cinnamic acid. Phenylalanine ammonia-lyase deaminates L-phenylalanine to cause the initiation of phenylpropanoid metabolism. The enzyme is also involved in plants' reaction to abiotic and biotic stresses. Kim and Hwang (2014) examined phenylalanine ammonia lyase in Capsicum annuum (pepper) gene. This was created in the leaves of pepper through infection with avirulent *Xanthomonas campestris* pv. *vesicatoria*. Silenced pepper phenylalanine ammonia lyase gene was more vulnerable to both avirulent and virulent infections of *X. campestris pv. vesicatoria*. The expression of salicylic acid-dependent marker gene in pepper phenylalanine ammonia lyase gene, reactive oxygen species, accumulation of salicylic acid and hypersensitive cell death and phenylalanine ammonia-lyase induction were altered in the Silenced pepper phenylalanine ammonia lyase gene. Phenylalanine is critically involved in the defense against different abiotic and biotic stresses in *Pinus taeda*. Bagal et al. (2012) reported that gymnosperms exhibit phylogenetically diverse set of phenylalanine-lyase genes more than angiosperms. All the five distinct phenylalanine lyase genes, which were identified in *P. taeda*, originated from a group of duplication events that predate and postdate the divergence of gymnosperms and angiosperms.

Since phytophthora root and stem rot of soybean is a destructive disease caused by *Phytophthora sojae*, phenylalanine-lyase plays role in plant responses to pathogenic stress. Zhang et al. (2017b) investigated the likely molecular mechanisms in soybeans in response to *P. sojae*. A soybean phenylalanine-lyase gene which was induced by *Phytophthora sojae* known as *GmPAL2.1* was characterized. *GmPAL2.1* was shown to be involved in the resistance of transgenic soybean to *P. sojae*. GmPAL2.1 plays an active role functioning as a positive modulator in soybean's response to *P. sojae* infection by increasing the levels of daidzein, glyceollins, genistein and salicylic acid. Kao et al. (2002) cloned two phenylalanine ammonia-lyase complementary DNAs that are expressed differentially. These complementary DNAs have unique spatial relationships with biosynthesis of lignin and condensed tannins in the growing root and shoot of aspen. In wounded Leaf, there was slight enhancement of phenylalanine ammonia-lyase and lignin pathway gene expression.

The phenylpropanoid pathway is known to be responsible for the production of condensed tannins, lignin, flavonoids, phenolic glycosides and benzenoids. The pathway is modulated by phenylalanine ammonia-lyase, an enzyme that protects plant from abiotic and biotic stresses and for the healthy development of crop species like willow. de Jong et al. (2015) reported that willow phenylalanine ammonia-1, phenylalanine ammonia-2 and phenylalanine ammonia-3 and poplar genes were orthologous. On the other hand, the expression of willow phenylalanine ammonia-lyase genes was slightly different from that of willow phenylalanine ammonia lyase-2, which was reported to exhibit wider expression than its poplar orthologs.

Reichert et al. (2009) utilized a polymerase chain reaction-based technique to isolate complementary DNA that correspond to four phenylalanine ammonia-lyase genes of tobacco. Using polymerase chain reaction, phenylalanine ammonia-lyase sequences from complementary DNA were amplified, resulting in the production of chimeric clones between phenylalanine ammonia-lyase-1 and phenylalanine ammonia-lyase-4. Lahoucine et al. (2004) examined if isoforms of phenylalanine ammonia-lyase were associated with the immediate next enzyme, cinnamate 4-hydroxylase, in the metabolic pathway of phenylpropanoid. A transgenic tobacco that expresses epitope-tagged versions of phenylalanine ammonia-lyase-1 and phenylalanine ammonia-lyase-2 independently was deployed. When compared with phenylalanine ammonia-lyase-2, phenylalanine ammonia-lyase-1 was shown to exhibit greater affinity for its membrane localization site.

14.6.2 Polyphenol

Polyphenol is an essential organic compound in fruits and vegetables. The major families of polyphenols are lignans, flavonoids, phenols and stilbenes. As antioxidants, polyphenols are scavengers of free radicals and thus confers protection against biotic and abiotic stresses and maintaining the development of plants (Rana et al., 2022). Foods that are rich in polyphenol include teas, soybeans, cocoa, grapes and many others. Evidence from animal studies indicates that polyphenols help in controlling the bioavailability of nitric oxide. Therefore, the compound may influence the control vascular tone and blood pressure (Galleano et al., 2010). Patients with hypertension may receive relief from consumption of polyphenols.

In another development, Leifert and Abeywardena (2008) reported that extracts of grape fruit such as grape juice, grape skin and grape seed are rich sources of anthocyanins, resveratrol and phenolic acids, all of which are capable of suppressing low density lipoprotein and platelet aggregation. By reducing low density lipoprotein, overweight and obese individuals and patients with type II diabetic mellitus may benefit from using this compound.

Keser et al. (2013) studied the antioxidant potentials of the water and ethanol extracts of fresh grape fruit, dried grape fruit, grape seed, pestil, molasses and vinegar. The antioxidant capacity was screened using 2,2-diphenyl-1-picrylhydrazyl, superoxide dismutase, hydrogen peroxide scavenger, metal chelating level, total phenolic level, reducing power and 2,2-azino-is (3-ethylbenzothiazoline-6-sulfonic acid) diammonium salt level. Among the extracts investigated, seed extracts showed the greatest 2,2-azino-is (3-ethylbenzothiazoline-6-sulfonic acid) diammonium salt, 2,2-diphenyl-1-picrylhydrazyl, reducing power and hydrogen peroxide scavenging levels.

Álvarez et al. (2016) evaluated the antioxidant activities of extracts of fruits such as pear, kiwi, green apple, blackberry, raspberry, blueberry and strawberry and vegetables such as green and red pepper and pumpkin. The antioxidant capacity and polyphenol content of different fruits and vegetables were assessed using Folin-Ciocalteu and ferric reducing antioxidant potential techniques, respectively. The polyphenol content was significantly different. The antioxidant capacity was reported to be significantly different in the extracts using both the Folin-Ciocalteu and ferric reducing antioxidant potential techniques.

Basu et al. (2016) assessed the antioxidant activities of extracts of fruits such as jackfruit blueberry, black raspberry, blackberry, strawberry, California table grape and red raspberry in vitro. Techniques that were used for assessing antioxidant activities include ferric reducing antioxidant potential, 2,2-diphenyl-1-picrylhydrazyl, 2,2′-azino-bis (3-ethylbenzothiazoline-6-sulfonic acid) diammonium salt, superoxide anion scavenging, nitric oxide and ferric reducing power. Antioxidant activities were determined using 2,2-diphenyl-1-picrylhydrazyl, ferric reducing antioxidant potential, 2,2′-azino-bis (3-ethylbenzothiazoline-6-sulfonic acid) diammonium salt, nitric oxide, superoxide anion scavenging assays and ferric reducing power. All the fruit extracts showed increasing radical scavenging tendencies, and the radical scavenging activities increased with concentration of the fruit extracts. At 100 µg/mL, black raspberry extract had the peak 2,2-diphenyl-1-picrylhydrazyl activity, red raspberry extract showed the highest ferric reducing power, and jackfruit extract had the greatest 2,2′-azino-bis (3-ethylbenzothiazoline-6-sulfonic acid) diammonium salt, nitric oxide and radical scavenging activities.

Gu et al. (2019) measured the antioxidant properties of polyphenols such as 1,1-diphenyl-2-picrylhydrazyl, total flavonoid content, 2,2′-azinobis-(3-ethylbenzo-thiazoline-6-sulfonic acid), total phenolic content, ferric ion reducing antioxidant power and tannin level in cherry, papaya, apple, mango, blueberry, strawberry, raspberry, black carrot, garlic, dark grapes, onion, ginger, plum, peach, apricot and pear. Mango was shown to have the peak phenolic acid, overall antioxidant level and flavonoid content among all the fruits under study.

Proteggente et al. (2002) evaluated phenolic content and the antioxidant capacity of varieties of fruits and vegetables, which are indigenous and locally available in the United Kingdom market. The antioxidant capacities of methanolic and aqueous extracts of the fruits were evaluated comparatively through Ferric Reducing Ability of Plasma, Oxygen Radical Absorbance Capacity and Trolox Equivalent Antioxidant Capacity assays. Anthocyanin-enriched fruit and vegetables such as red plum, strawberry and raspberry exhibited peak antioxidant activities. This was followed by flavanone-enriched foods such as grapefruit and orange and flavonols-enriched foods such as leek, onion, green cabbage and spinach. Hydroxycinnamate-enriched fruit such as tomato, pear, apple and peach showed lower antioxidant levels.

14.7 COMBINED EFFECT OF CHITOSAN AND ESSENTIAL OIL ON ANTIOXIDANT STRENGTH

Chitosan exhibits high antioxidant potentials due to its strong metal chelating ability. Essential oils possess anti-inflammatory and antioxidant activities. Therefore, combining chitosan and essential oil demonstrates superior antioxidant potentials than either chitosan or essential oils in fruits and vegetables. A study by Zhang et al. (2022) demonstrated that combining chitosan, a precursor of nanochitosan with essential oil, is capable of enhancing the shelf life of citrus fruits which at 20°C ± 2°C storage temperature and relative humidity of 75% ± 5% for a period of 60 days. There was enhancement of the antioxidant activities improving the levels of peroxidase, catalase and superoxide dismutase in citrus fruits treated with lemon essential oil. However, administration of 1.0% chitosan and 1.0% lemon essential oil caused a peak level of vitamin C, total soluble solids, titratable acid and catalase, peroxidase and superoxide dismutase. Rasha et al. (2023) investigated the combination effects of chitosan and extracts of curcumin, garlic and beetroot on lipid peroxidation levels and other parameters of tuna files following 14 days of refrigeration. The results were compared with uncoated tuna files. The combination of chitosan and extracts of curcumin, garlic and beetroot was shown to reduce lipid peroxidation, thus improving the quality of foods.

When chitosan coating was combined with cinnamon oil, for the preservation of jujube fruits for 60 days at 4°C, Xing et al. (2015) reported that the levels of superoxide dismutase, peroxidase and polyphenol oxidase at the end of storage were 13.40, 14.53 and 63.6 U/g, respectively, following the combination of chitosan coating and cinnamon oil. Also, the levels of malondialdehyde and phenolics were 19.43 μmol/g and 34.51 mg/g, respectively. The study indicated that combining chitosan coating with cinnamon oil enhanced jujube fruits preservation. During storage at 4°C and 8°C, Popescu et al. (2022) evaluated the microbiological and physicochemical properties of apple slices and organic strawberries with the aim of determining the effect of ascorbic or acetic acid and sea buckthorn or grape seed essential oils containing medium- and high-molecular-weight chitosan-based coating. There was preservation of antioxidant and phenolic contents of foods with these chitosan-based coatings when compared with the control, resulting in improvements in shelf life.

Li et al. (2019) examined the possibility that the combination of chitosan with citrus essential oil will be effective and efficient in preserving pacific mackerel. Combining chitosan with citrus essential oil coating inhibits oxidative stress and reactive oxygen species than chitosan alone. When stored at −3°C for 12 days, pacific mackerel that was coated with the combination of chitosan and citrus essential oil exhibited low levels of thiobarbituric acid reactive species, an index of lipid peroxidation and food corruption. In a study by Cao et al. (2020), chlorogenic acid was combined with chitosan in an attempt to maintain and improve the quality of snakehead fish fillets. In the study, fish fillets were immersed in 2% chitosan, 0.5% chlorogenic acid in 2% chitosan, 0.5% chlorogenic acid in 2% chitosan, 0.2% chlorogenic acid in 2% chitosan or 1.0% chlorogenic acid in 2% chitosan and then stored at 2°C±0.5°C. There was a significant increase in antioxidant values when chitosan and chlorogenic acids were combined. Lipid oxidation was also reduced in fish fillet coated with 2% chitosan, 0.5% chlorogenic acid in 2% chitosan and 1.0% chlorogenic acid in 2% chitosan. This suggests that chitosan and chlorogenic acid are efficient in fish fillet preservation.

Karakosta et al. (2022) studied the effect of combining chitosan coating with laurel essential oil on the storage life of water buffalo meat, which was stored at 4°C under aerobic packaging conditions and for a period of 18 days. The level of thiobarbituric acid was found to be around 2.12 mg malondialdehyde/kg throughout storage. This suggests that combining chitosan coating with laurel essential oils improved the storage life of water buffalo meat. Brenda et al. (2019) assessed the effect of combining chitosan coating with clove essential oil on the storage of frozen tambaqui (*Colossoma macropomum*) fillets. The fillets were packaged individually with the aid of polyethylene plastic film and stored at −18°C. Lipid oxidation and other analyses were done for three months. When frozen tambaqui were treated with 2% chitosan and chitosan+clove essential oil clove essential oil at 0.16% and 0.08%, respectively, there was effective delay in lipid oxidation (less than 0.83 mg eq. kg/MDA). The study suggests that owing to the inhibition of chemical deterioration, combining chitosan coating with clove essential oil could act as a natural way of preserving frozen tambaqui as it can inhibit the chemical.

Carrapiso et al. (2023) examined the extent to which combining chitosan coating and olive leaf extract was effective in slowing down the spoilage of refrigerated pork burgers kept under 40% oxygen and 60% carbon dioxide environmental atmospheric situation with no additives. Combining chitosan coating and olive leaf extract was shown to inhibit changes in lipid oxidation better than chitosan. Shiow et al. (2013) reported that in strawberries stored at either 5°C or 10°C, the extent of decay when compared with control was low, while the shelf life was prolonged when the fruits were treated by immersion with chitosan solutions ranging from 0.5 g/100 mL, 1.0 g/100 mL and 1.5 g/100 mL for a period of 5 minutes at a temperature of 20°C. When strawberries were treated with chitosan, the levels of antioxidant enzyme activity, such as glutathione-peroxidase, catalase, guaiacol peroxidase, dehydroascorbate reductase and monodehydroascorbate reductase, were increased. There was also an increase in phenolics, anthocyanins and flavonoids such as ellagic acid glucoside, ellagic acid, *p*-coumaroyl glucose, quercetin 3-glucuronide, quercetin 3-glucoside, kaempferol 3-glucoside, cyanidin 3-glucoside, kaempferol 3-glucuronide, pelargonidin 3-glucoside, cyanidin 3-glucoside-succinate and pelargonidin 3-glucoside-succinate. 2,2-Di (4-tert-octylphenyl)-1-picrylhydrazyl, peroxyl radicals, oxygen radical absorbance capacity and hydroxyl radical were significantly different from the untreated fruits. Chitosan when compared with control also attenuated reduction in reduced glutathione level, ascorbic acid and β-1,3-glucanase activities.

Zhang et al. (2022) assessed the possibility of utilizing chitosan and lemon essential oil for improving the storage life of citrus stored for 60 days at 20°C±2°C and 75%±5% relative humidity. The fruits were treated with 1.0% chitosan and 1.0% lemon essential oil. Other agents added included 1.0% calcium chloride and 0.03% natamycin. Combining chitosan with essential oil and others regulated malondialdehyde level and elevated activities of catalase, superoxide dismutase and peroxidase in citrus. Gao et al. (2018) investigate the effect of combining chitosan with cinnamaldehyde on fresh mandarin fruit cv. Ponkan at different concentrations such as chitosan, 0.5% cinnamaldehyde+chitosan, 1.0% cinnamaldehyde+chitosan and 1.5% cinnamaldehyde+chitosan. The effects of these formulations on total antioxidant capacity and malondialdehyde of fruits stored for more than 100 days at 20°C were investigated. There was an increase in the antioxidant capacity and decline in the accumulation of malondialdehyde in fruits treated with chitosan and cinnamaldehyde especially at 0.5% cinnamaldehyde+chitosan.

To mitigate the impact of reactive oxygen species orchestrated by fruit wounds, Ackah et al. (2022) treated apple fruits (cv. Fuji) that were artificially wounded with 2.5% w/v of chitosan and stored them at 21°C–25°C with relative humidity of 81%–85% for 7 days. The results were compared with non-wounded fruits. Chitosan treatment was shown to induce the activities of superoxide dismutase, peroxidase and catalase. Indices of lipid peroxidation such as malondialdehyde and lipoxygenase were reduced following chitosan treatment in wounded fruit when compared with untreated wounded fruits. Reduced glutathione, glutathione, glutathione reductase, monodehydroascorbate reductase, ascorbate peroxide and dehydroascorbate reductase were all elevated at fruit wounds, following chitosan treatment when compared with the untreated wounded fruits. Eissa (2008) investigated the impact of different formulations of chitosan coating on enzyme activities in fresh cut mushroom. Freshly cut mushrooms were treated with 5, 10 and 20g of chitosan per liter, stored at 4°C and immersed in polyethylene bag for 15 days. The utilization of chitosan coating slowed down morphological changes caused by lowered activities of peroxidase, catalase, laccase, phenylalanine ammonia lyase, peroxidase and phenol.

Zhang et al. (2017a) utilized chitosan, alginate and pullulan for postharvest preservation of strawberry fruit stored at 4°C and reported the impacts of these polysaccharides on antioxidant enzymes. Treatment with chitosan, alginate and pullulan was shown to improve the activities of catalase, peroxidase, superoxide dismutase and ascorbate peroxidase, thus assuaging lipid peroxidation and membrane deterioration. Yu et al. (2023) investigated the effect of combining chitosan and chitooligosaccharide on the healing of wounded pear fruits. Pear fruit was artificially wounded and was treated with 1 g/L of chitosan and chitooligosaccharide. The combination of chitosan and chitooligosaccharide enhanced the levels of nicotinamide adenine dinucleotide phosphate oxidase and superoxide dismutase and enhanced O_2^- and H_2O_2 production at wound sites when compared to chitosan alone. The combination also enhanced activities of peroxidase, catalase, glutathione reductase, monodehydroascorbate reductase, ascorbate peroxidase and dehydroascorbate reductase as well as increased ascorbic acid and glutathione concentrations when compared with chitosan-treated pear fruits. Ghosh et al. (2021) studied the effect of combining chitosan and putrescine on reactive oxygen species in relation to the shelf life of Capsicum fruit. Combining chitosan and putrescine caused a significant decline in the concentrations of oxygen free radicals and hydrogen peroxide. The combination of chitosan and putrescine improved the postharvest shelf life of chili fruit when compared to chitosan alone by improving catalase, superoxide dismutase, guaiacol peroxidases, glutathione peroxidase, ascorbate peroxidase, glutathione reductase and NADPH oxidase.

14.8 CONCLUSION AND FUTURE PERSPECTIVES

The available evidence indicates that combining chitosan with essential oil is capable of conferring stronger antioxidant protection on fruits and vegetables. Specifically, the combined utilization of chitosan and essential oils causes greater antioxidant benefits in citrus by maintaining malondialdehyde at non-high level while increasing the activities of peroxidase, superoxide dismutase, polyphenol oxidase and catalase. The combination of chitosan and essential oil may influence catalase activities by modulating calcium-calmodulin signaling, a redox signaling. In view of the importance of fruits and vegetables to the health and wellness of mammals, policy attention should be diverted toward biological means of boosting antioxidant strength of foods such as the utilization of chitosan and essential oil combination. This will mitigate spoilage and postharvest loss of fruits and vegetables while also improving shelf life, quality and texture of foods.

CONFLICT OF INTEREST

None declared.

REFERENCES

Achnine L, Blancaflor EB, Rasmussen S, Dixon RA. Colocalization of l-phenylalanine ammonia-lyase and cinnamate 4-hydroxylase for metabolic channeling in phenylpropanoid biosynthesis. *Plant Cell*, 2004; *16*(11):3098–3109.

Ackah S, Bi Y, Xue S, Yakubu S, Han Y, Zong Y, Atuna RA, Prusky D. Post-harvest chitosan treatment suppresses oxidative stress by regulating reactive oxygen species metabolism in wounded apples. *Front Plant Sci,* 2022; *13*:959762.

Álvarez R, Araya H, Navarro-Lisboa R, Lopez de Dicastillo C. Evaluation of polyphenol content and antioxidant capacity of fruits and vegetables using a modified enzymatic extraction. *Food Technol Biotechnol*, 2016; *54*(4):462–467.

Bagal UR, Leebens-Mack JH, Lorenz WW, Dean JFD. The phenylalanine ammonia lyase (PAL) gene family shows a gymnosperm-specific lineage. *BMC Genomics*, 2012; *13*(Suppl 3):S1.

Basu P, Maier C. *In vitro* antioxidant activities and polyphenol contents of seven commercially available fruits. *Pharmacognosy Res*, 2016; *8*(4):258–264.

Bela K, Horváth E, Gallé Á, Szabados L, Tari I, Csiszár J. Plant glutathione peroxidases: emerging role of the antioxidant enzymes in plant development and stress responses. *J Plant Physiol*, 2015; *176*:192–201.

Bi G, Hu M, Fu L, Zhang X, Zuo J, Li J, Yang J, Zhou JM. The cytosolic thiol peroxidase PRXIIB is an intracellular sensor for H_2O_2 that regulates plant immunity through a redox relay. *Nat Plants*, 2022 Oct; *8*(10):1160–1175.

Brouwer ID, Harris J. Overcoming the global obstacles to fruit and vegetable consumption to build healthy diets. Research program on agriculture for nutrition and health, 2021.

Cáceres M, Hidalgo W, Stashenko E, Torres R, Ortiz C. Essential oils of aromatic plants with antibacterial, anti-biofilm and anti-quorum sensing activities against pathogenic bacteria. *Antibiotics (Basel)*, 2020 Mar 30; *9*(4):147.

Cáceres M, Hidalgo W, Stashenko EE, Torres R, Ortiz C. Metabolomic analysis of the effect of *Lippia origanoides* essential oil on the inhibition of quorum sensing in *Chromobacterium violaceum*. *Antibiotics (Basel)*, 2023 Apr 26; *12*(5):814.

Cao X, Islam MN, Chitrakar B, Duan Z, Xu W, Zhong S. Effect of combined chlorogenic acid and chitosan coating on antioxidant, antimicrobial, and sensory properties of snakehead fish in cold storage. *Food Sci Nutr*, 2020 Jan 8; *8*(2):973–981.

Carrapiso AI, Pimienta M, Martín L, Cardenia V, Andrés AI. Effect of a chitosan coating enriched with an olive leaf extract on the characteristics of pork burgers. *Foods*, 2023 Oct 12; *12*(20):3757.

Casatejada A, Puerto-Galán L, Pérez-Ruiz JM, Cejudo FJ. The contribution of glutathione peroxidases to chloroplast redox homeostasis in Arabidopsis. *Redox Biol*, 2023 Jul; *63*:102731.

Castillo A, Meitha K, Pramesti Y, Suhandono S. Reactive Oxygen Species and Antioxidants in Postharvest Vegetables and Fruits. SP-8817778, VL-2020, 2020.

Chen F, Wang F, Wu F, Mao W, Zhang G, Zhou M. Modulation of exogenous glutathione in antioxidant defense system against Cd stress in the two barley genotypes differing in Cd tolerance. *Plant Physiol Biochem*, 2010 Aug; *48*(8):663–672.

Chitin SU. Characteristic, sources, and biomedical application. *Curr Pharm Biotechnol*, 2020; *21*(14):1433–1443.

Dat JF, Inzé D, Van Breusegem F. Catalase-deficient tobacco plants: tools for in planta studies on the role of hydrogen peroxide. *Redox Rep*, 2001; *6*(1):37–42.

de Jong F, Hanley SJ, Beale MH, Karp A. Characterisation of the willow phenylalanine ammonia-lyase (PAL) gene family reveals expression differences compared with poplar. *Phytochemistry*, 2015 Sep; *117*:90–97.

Dijkxhoorn Y, de Steenhuijsen Piters B, Brouwer I, Hengsdijk H, Tichar T. Enhancing fruit and vegetable consumption in low- and middle income countries through a food systems approach. Wageningen Economic Research, 2021.

Eissa HAA. Effect of chitosan coating on shelf-life and quality of fresh-cut mushroom. *Polish J Food Nutr Sci*, 2008; *58*(1):95–105.

El-Araby A, Janati W, Ullah R, Ercisli S, Errachidi F. Chitosan, chitosan derivatives, and chitosan-based nanocomposites: eco-friendly materials for advanced applications (a review). *Front Chem*, 2024 Jan 4; *11*:1327426.

Elsabagh R, Ibrahim SS, Elsayed M, et al. Chitosan edible coating: a potential control of toxic biogenic amines and enhancing the quality and shelf life of chilled tuna filets. *Front Sustain Food Syst*, 2023; 7. doi:10.3389/fsufs.2023.1177010

Galleano M, Pechanova O, Fraga CG. Hypertension, nitric oxide, oxidants, and dietary plant polyphenols. *Curr Pharm Biotechnol*, 2010 Dec; *11*(8):837–848.

Gao Y, Kan C, Chen M, Chen C, Chen Y, Fu Y, Wan C, Chen J. Effects of chitosan-based coatings enriched with cinnamaldehyde on mandarin fruit cv. Ponkan during room-temperature storage. *Coatings*, 2018; *8*(10):372.

Ghorbel M, Zribi I, Chihaoui M, Alghamidi A, Mseddi K, Brini F. Genome-wide investigation and expression analysis of the catalase gene family in oat plants (*Avena sativa* L.). *Plants (Basel)*, 2023 Oct 26; *12*(21):3694.

Ghosh A, Saha I, Debnath SC, Hasanuzzaman M, Adak MK. Chitosan and putrescine modulate reactive oxygen species metabolism and physiological responses during chili fruit ripening. *Plant Physiol Biochem*, 2021 Jun; *163*:55–67.

Gill SS, Anjum NA, Hasanuzzaman M, Gill R, Trivedi DK, Ahmad I, Pereira E, Tuteja N. Glutathione and glutathione reductase: a boon in disguise for plant abiotic stress defense operations. *Plant Physiol Biochem*, 2013 Sep; *70*:204–212.

Gill SS, Tuteja N. Reactive oxygen species and antioxidant machinery in abiotic stress tolerance in crop plants. *Plant Physiol Biochem*, 2010 Dec; *48*(12):909–930.

Gu C, Howell K, Dunshea FR, Suleria HAR. LC-ESI-QTOF/MS characterisation of phenolic acids and flavonoids in polyphenol-rich fruits and vegetables and their potential antioxidant activities. *Antioxidants (Basel)*, 2019 Sep 17; *8*(9):405.

Guillín Y, Cáceres M, Torres R, Stashenko E, Ortiz C. Effect of essential oils on the inhibition of biofilm and quorum sensing in *Salmonella enteritidis* 13076 and *Salmonella typhimurium* 14028. *Antibiotics (Basel)*, 2021 Oct 1; *10*(10):1191.

Herbette S, Roeckel-Drevet P, Drevet JR. Seleno-independent glutathione peroxidases. More than simple antioxidant scavengers. *FEBS J*, 2007 May; *274*(9):2163–2180.

Kao YY, Harding SA, Tsai CJ. Differential expression of two distinct phenylalanine ammonia-lyase genes in condensed tannin-accumulating and lignifying cells of quaking aspen. *Plant Physiol*, 2002 Oct; *130*(2):796–807.

Karakosta LK, Vatavali KA, Kosma IS, Badeka AV, Kontominas MG. Combined effect of chitosan coating and laurel essential oil (*Laurus nobilis*) on the microbiological, chemical, and sensory attributes of water buffalo meat. *Foods*, 2022; *11*(11):1664.

Kaur S. Barriers to consumption of fruits and vegetables and strategies to overcome them in low- and middle-income countries: a narrative review. *Nutr Res Rev*, 2023; *36*(2):420–447. doi:10.1017/S0954422422000166

Ken CF, Cheng YF, Chang CF, Lin CT. Copper/zinc-superoxide dismutase from Epinephelus malabaricus cDNA and enzyme property. *J Agric Food Chem*, 2003 Sep 10; *51*(19):5688–5694.

Ken CF, Weng DF, Duan KJ, Lin CT. Characterization of copper/zinc-superoxide dismutase from Pagrus major cDNA and enzyme stability. *J Agric Food Chem*, 2002 Feb 13; *50*(4):784–789.

Keser S, Celik S, Turkoglu S. Total phenolic contents and free-radical scavenging activities of grape (Vitis vinifera L.) and grape products. *Int J Food Sci Nutr*, 2013 Mar; *64*(2):210–216.

Kim DS, Hwang BK. An important role of the pepper phenylalanine ammonia-lyase gene (PAL1) in salicylic acid-dependent signalling of the defence response to microbial pathogens. *J Exp Bot*, 2014 Jun; *65*(9):2295–2306.

Leifert WR, Abeywardena MY. Cardioprotective actions of grape polyphenols. *Nutr Res*, 2008 Nov; *28*(11):729–737.

Li Y, Chen Z, Zhang P, Gao F, Wang J, Lin L, Zhang H. Characterization of a novel superoxide dismutase from a deep-sea sea cucumber (*Psychoropotes verruciaudatus*). *Antioxidants (Basel)*, 2023 Jun 7; *12*(6):1227.

Li Y, Wu C, Wu T, Yuan C, Hu Y. Antioxidant and antibacterial properties of coating with chitosan-citrus essential oil and effect on the quality of Pacific mackerel during chilled storage. *Food Sci Nutr*, 2019 Feb 11; *7*(3):1131–1143. doi:10.3389/fmicb.2021.768414

Lin MW, Lin MT, Lin CT. Copper/zinc-superoxide dismutase from lemon cDNA and enzyme stability. *J Agric Food Chem*, 2002 Dec 4; *50*(25):7264–7270.

Martínez A, Manrique-Moreno M, Klaiss-Luna MC, Stashenko E, Zafra G, Ortiz C. Effect of essential oils on growth inhibition, biofilm formation and membrane integrity of *Escherichia coli* and *Staphylococcus aureus*. *Antibiotics (Basel)*, 2021 Nov 30; *10*(12):1474.

Martínez A, Stashenko EE, Sáez RT, Zafra G, Ortiz C. Effect of essential oil from *Lippia origanoides* on the transcriptional expression of genes related to quorum sensing, biofilm formation, and virulence of *Escherichia coli* and *Staphylococcus aureus*. *Antibiotics (Basel)*, 2023 May 3; *12*(5):845.

Meitha K, Pramesti Y, Suhandono S. Reactive oxygen species and antioxidants in postharvest vegetables and fruits. *Int J Food Sci*, 2020 Dec 10; *2020*:8817778.

Modau L, Sigwadi R, Mokrani T, Nemavhola F. Chitosan membranes for direct methanol fuel cell applications. *Membranes (Basel)*, 2023 Oct 20; *13*(10):838.

Navrot N, Collin V, Gualberto J, Gelhaye E, Hirasawa M, Rey P, Knaff DB, Issakidis E, Jacquot JP, Rouhier N. Plant glutathione peroxidases are functional peroxiredoxins distributed in several subcellular compartments and regulated during biotic and abiotic stresses. *Plant Physiol*, 2006 Dec; *142*(4):1364–1379.

Nwe N, Furuike T, Tamura H. Isolation and characterization of chitin and chitosan from marine origin. *Adv Food Nutr Res*, 2014; *72*:1–15.

Palma JM, Mateos RM, López-Jaramillo J, Rodríguez-Ruiz M, González-Gordo S, Lechuga- Sancho AM, Corpas FJ. Plant catalases as NO and H_2S targets. *Redox Biol*, 2020 Jul; *34*:101525.

Pan Y, Jiang Y, Huang Q, Zhu Y, Nie Y, Yuan R, Zhang Z. Abnormal chilling injury of postharvest papaya is associated with the antioxidant response. *J Food Biochem*. 2022 Oct; *46*(10):e14272.

Pellis A, Guebitz GM, Nyanhongo GS. Chitosan: sources, processing and modification techniques. *Gels*. 2022 Jun 21; *8*(7):393.

Petralito S, Garzoli S, Ovidi E, et al. Long-term stability of *Lavandula x intermedia* essential oil nanoemulsions: can the addition of the ripening inhibitor impact the biocidal activity of the nanoformulations? *Pharmaceutics*, 2024 Jan 14; *16*(1):108.

Popescu P-A, Palade LM, Nicolae I-C, Popa EE, Miteluţ AC, Drăghici MC, Matei F, Popa ME. Chitosan-based edible coatings containing essential oils to preserve the shelf life and postharvest quality parameters of organic strawberries and apples during cold storage. *Foods*, 2022; *11*(21):3317.

Proteggente AR, Pannala AS, Paganga G, Van Buren L, Wagner E, Wiseman S, Van De Put F, Dacombe C, Rice-Evans CA. The antioxidant activity of regularly consumed fruit and vegetables reflects their phenolic and vitamin C composition. *Free Radic Res*, 2002 Feb; *36*(2):217–233.

Rana A, Samtiya M, Dhewa T, Mishra V, Aluko RE. Health benefits of polyphenols: a concise review. *J Food Biochem*, 2022 Oct; *46*(10):e14264.

Reichert AI, He XZ, Dixon RA. Phenylalanine ammonia-lyase (PAL) from tobacco (Nicotiana tabacum): characterization of the four tobacco PAL genes and active heterotetrameric enzymes. *Biochem J*, 2009 Nov 11; *424*(2):233–242.

Rosalie R, Léchaudel M, Dhuique-Mayer C, Dufossé L, Joas J. Antioxidant and enzymatic responses to oxidative stress induced by cold temperature storage and ripening in mango (Mangifera indica L. cv. 'Cogshall') in relation to carotenoid content. *J Plant Physiol*, 2018 May-Jun; *224–225*:75–85. doi:10.1016/j.jplph.2018.03.011. Epub 2018 Mar 26. PMID: 29605751.

Shepherd R, Reader S, Falshaw A. Chitosan functional properties. *Glycoconj J*, 1997 Jun; *14*(4):535–542.

Sridhar A, Ponnuchamy M, Kumar PS, Kapoor A. Food preservation techniques and nanotechnology for increased shelf life of fruits, vegetables, beverages and spices: a review. *Environ Chem Lett*, 2021; *19*(2):1715–1735.

Statistica. Fresh fruits – Worldwide. 2024; Accessed on 14/03/2024. https://www.statista.com/outlook/cmo/food/fruits-nuts/fresh-fruits/worldwide

Utami RA, Yoshida H, Kartadinata LH, Abdillah VA, Faratilla CR, Retnoningrum DS, Ismaya WT. Direct relationship between dimeric form and activity in the acidic copper-zinc superoxide dismutase from lemon. *Acta Crystallogr F Struct Biol Commun*, 2023 Dec 1; *79*(Pt 12):301–307.

Valenzuela JL, Manzano S, Palma F, Carvajal F, Garrido D, Jamilena M. Oxidative stress associated with chilling injury in immature fruit: postharvest technological and biotechnological solutions. *Int J Mol Sci*, 2017 Jul 8; *18*(7):1467.

Vieira BB, Mafra JF, da Rocha Bispo AS, Ferreira Ma, de Lima Silva F, Rodrigues AVN, Evangelista-Barreto NS. Combination of chitosan coating and clove essential oil reduces lipid oxidation and microbial growth in frozen stored tambaqui (Colossoma macropomum) fillets. *LWT*, 2019; *116*:108546.

Wang SY, Gao H. Effect of chitosan-based edible coating on antioxidants, antioxidant enzyme system, and postharvest fruit quality of strawberries (Fragaria x aranassa Duch.). *LWT Food Sci Technol*, 2013; *52*(2):71–79.

World Health Organization. Increasing fruit and vegetable consumption to reduce the risk of noncommunicable diseases. Intervention .| Last updated: 9 August 2023; Accessed on 3/18/2024. https://www.who.int/tools/elena/interventions/fruit-vegetables-ncds

Xing Y, Lin H, Cao D, Xu Q, Han W, Wang R, Che Z, Li X. Effect of chitosan coating with cinnamon oil on the quality and physiological attributes of China jujube fruits. *Biomed Res Int*, 2015; *2015*:835151.

Yang T, Poovaiah BW. Hydrogen peroxide homeostasis: activation of plant catalase by calcium/calmodulin. *Proc Natl Acad Sci U S A*, 2002 Mar 19; *99*(6):4097–4102.

Younes I, Rinaudo M. Chitin and chitosan preparation from marine sources. Structure, properties and applications. *Mar Drugs*, 2015 Mar 2; *13*(3):1133–1174.

Younus H. Therapeutic potentials of superoxide dismutase. *Int J Health Sci (Qassim)*, 2018 May-Jun; *12*(3):88–93.

Yu L, Zhang X, Zhang F, Tang Y, Gong D, Oyom W, Li Y, Prusky D, Romanazzi G, Bi Y. Chitosan and chitooligosaccharide regulated reactive oxygen species homeostasis at wounds of pear fruit during healing. *Int J Biol Macromol*, 2023 Jun 15; *240*:124395.

Zhang W, Lin M, Feng X, Yao Z, Wang T, Xu C. Effect of lemon essential oil-enriched coating on the postharvest storage quality of citrus fruits. *Food Sci Technol Campinas*, 2022; *42*:e125421.

Zhang C, Li L, Sun J, et al. Effects of polysaccharide-based edible coatings on quality and antioxidant enzyme system of strawberry during cold storage. SP-9746174. VL-2017, 2017a.

Zhang C, Wang X, Zhang F, et al. Phenylalanine ammonia-lyase2.1 contributes to the soybean
response towards Phytophthora sojae infection. *Sci Rep*, 2017b; 7:7242.

Zhang W, Zhi W, Qiao H, et al. H_2O_2-dependent oxidation of the transcription factor GmNTL1 promotes salt tolerance in soybean. *Plant Cell*, 2023 Dec 21; *36*(1):112–135.

Zhou L, John Martin JJ, Li R, et al. Catalase (CAT) gene family in oil palm (*Elaeis guineensis* Jacq.): genome-wide identification, analysis, and expression profile in response to abiotic stress. *Int J Mol Sci*, 2024 Jan 25; *25*(3):1480.

Zhu JJ, Li YR, Liao JX. Involvement of anthocyanins in the resistance to chilling-induced oxidative stress in Saccharum officinarum L. leaves. *Plant Physiol Biochem*, 2013 Dec; *73*:427–433.

15 Chitosan and Its Modification
A Promising Biomaterial for Food and Medical Technology

John Tsado Mathew[1], *Jonathan Hussaini*[1], *Abel Inobeme*[2], *Elijah Yanda Shaba*[3], *Mary Jummai Jiya*[4], *Tanko Ezekiel*[3], *Yakubu Azeh*[1], *Musah Monday*[1], *Muhammad Aishetu Ibrahim*[1], *Etsuyankpa Muhammad Bini*[5], *Ismaila Aliyu Oga*[5], *Musa Safiyanu Tanko*[5], *Muhammed Jibrin Ndejiko*[6], *Charles Oluwaseun Adetunji*[7], *and Amos Mamman*[1]

[1]Department of Chemistry, Ibrahim Badamasi Babangida University Lapai, Niger State, Nigeria

[2]Department of Chemistry, Edo University Uzairue, Auchi, Edo State, Nigeria

[3]Department of Chemistry, Federal University of Technology, Minna, Niger State, Nigeria

[4]Department of Food Science and Technology, Federal University of Technology, Minna, Niger State, Nigeria

[5]Department of Chemistry, Federal University of Lafia, Nasarawa State, Nigeria

[6]Department of Microbiology, Ibrahim Badamasi Babangida University Lapai, Niger State, Nigeria

[7]Department of Microbiology, Applied Microbiology, Biotechnology and Nanotechnology Laboratory, Edo University Uzairue, Auchi, Edo State, Nigeria

15.1 INTRODUCTION

Chitosan, a versatile biomaterial derived from chitin, has gained significant attention in recent years due to its unique properties and wide-ranging applications in medical and food technology. Chitin, primarily found in the exoskeletons of crustaceans and insects, undergoes deacetylation to form chitosan. This natural polymer is characterized by its biocompatibility, biodegradability, antimicrobial activity, and film-forming ability, making it a promising candidate for various biomedical and food-related applications (Desai et al., 2023).

The medical field has witnessed a growing interest in chitosan and its derivatives owing to their ability to interact with biological systems without triggering adverse reactions. The biocompatibility of chitosan makes it an attractive material for drug delivery systems, wound healing, and tissue engineering. Chitosan's cationic nature allows for efficient binding with negatively charged cell membranes, facilitating controlled drug release. Additionally, its hemostatic properties make it valuable in wound dressings, where it can promote blood clotting and accelerate the healing process. The biodegradability of chitosan further enhances its appeal for use in implantable medical devices, reducing the risk of long-term adverse effects (de Sousa et al., 2020; Malik et al., 2025).

DOI: 10.1201/9781003355458-15

Furthermore, chitosan's antimicrobial activity has been exploited in medical applications to combat infections. The polymer's ability to disrupt the cell membranes of bacteria and fungi makes it an effective agent in the development of antimicrobial coatings for medical devices and surfaces. This feature is particularly crucial in the prevention of nosocomial infections, addressing a significant challenge in healthcare settings. In the realm of food technology, chitosan finds application as a natural and sustainable ingredient with various functional properties. Its film-forming ability makes it an excellent candidate for food packaging materials, providing an eco-friendly alternative to synthetic polymers. Chitosan films can enhance the shelf life of perishable food products by acting as a barrier to oxygen and moisture, thereby reducing the risk of spoilage and deterioration (Confederat et al., 2021; Chimbekujwo et al., 2025).

Chitosan's positive impact on food preservation extends beyond packaging. Its antimicrobial properties can be harnessed to develop edible coatings for fruits and vegetables, offering a natural and safe method to inhibit microbial growth and extend the freshness of produce. This application aligns with the increasing consumer demand for minimally processed and preservative-free food options. Moreover, chitosan derivatives, such as *N,O*-carboxymethyl chitosan and quaternized chitosan, exhibit enhanced properties that further expand their applications. These derivatives often possess improved solubility, stability, and bioactivity, making them suitable for more specialized roles in both medical and food technology (Muñoz-Tebar et al., 2023).

Furthermore, the unique characteristics of chitosan and its derivatives position them as promising biomaterials for diverse applications in medical and food technology. The biocompatibility, biodegradability, antimicrobial activity, and film-forming ability of chitosan contribute to its versatility and desirability in these fields. As research continues to uncover novel properties and applications, chitosan is poised to play a crucial role in advancing both medical and food-related technologies, offering sustainable and effective solutions to contemporary challenges (Thambiliyagodage et al., 2023).

15.2 SIGNIFICANCE OF CHITOSAN IN BIOMATERIALS

The significance of chitosan as a biomaterial stems from its inherent properties that address crucial requirements in biomedical and food-related applications. Its biocompatibility ensures that chitosan is well-tolerated by living organisms, minimizing adverse reactions when employed in medical interventions or consumed in food products. Additionally, the biodegradability of chitosan contributes to its environmental friendliness, as it undergoes natural degradation processes without leaving harmful residues (Table 15.1).

TABLE 15.1
Significance of Chitosan in Biomaterials

Aspect	Significance	Examples of Applications	References
Biocompatibility	Chitosan's ability to interact harmoniously with biological systems, minimizing adverse reactions	Wound healing, tissue engineering, implants, and drug delivery systems	Trucillo (2024)
Biodegradability	Natural degradation of chitosan without leaving harmful residues	Environmental friendliness and reduced ecological impact	Maliki et al. (2022) and Mathew et al. (2025a)
Antimicrobial activity	Inherent ability to inhibit the growth of microorganisms	Food preservation, coatings for perishable products	Rollini et al. (2020) and Muhammad et al. (2025)
Film-forming capabilities	Formation of films and coatings for various applications	Edible coatings for fruits and vegetables, drug delivery systems	Matloob et al. (2023)

In the context of medical applications, chitosan has demonstrated immense potential in wound healing, owing to its hemostatic properties and ability to promote tissue regeneration. The controlled release of therapeutic agents facilitated by chitosan-based drug delivery systems further enhances its role in medical interventions. Meanwhile, in food technology, chitosan's antimicrobial activity has been harnessed to extend the shelf life of perishable products, and its film-forming capabilities enable the development of edible coatings for fruits and vegetables, providing protective barriers against external factors (Kim et al., 2023).

15.3 CHITOSAN: PROPERTIES AND EXTRACTION

Chitosan, derived from chitin through a deacetylation process, is a natural polysaccharide that has gained significant attention for its unique properties and versatile applications in various fields, including biomedical and food technology. Understanding the properties and extraction methods of chitosan is crucial for harnessing its full potential in these applications (Muthu et al., 2021).

Structural and Chemical Properties: Chitosan exhibits a distinctive chemical structure that sets it apart from chitin. While chitin is composed of acetylated units of *N*-acetylglucosamine, chitosan is obtained by partially or fully deacetylating chitin, resulting in a polymer comprising glucosamine and *N*-acetylglucosamine units. The degree of deacetylation influences chitosan's properties, with higher degrees often associated with enhanced solubility and reactivity. The amino and hydroxyl functional groups in the chitosan structure contribute to its unique properties. The amino groups can be protonated, imparting cationic characteristics to chitosan, which is particularly relevant in applications such as drug delivery and water treatment. The presence of hydroxyl groups allows for hydrogen bonding, leading to chitosan's film-forming capabilities, a property exploited in various biomedical and food technology applications (Aranaz et al., 2021; Mathew et al., 2025a).

Sources of Chitosan: Chitosan is primarily extracted from the exoskeletons of crustaceans, such as crabs and shrimp, and from the cell walls of fungi. The source significantly influences the quality and properties of the chitosan. Crustacean-derived chitosan is often favored due to its typically higher degree of deacetylation, which enhances solubility and functional properties. This results in superior film-forming ability, antimicrobial activity, and overall performance in various applications. In contrast, fungal-derived chitosan, while also valuable, may exhibit differences in these attributes, often leading to variations in its effectiveness and suitability for specific uses. Therefore, the choice of source is crucial in determining the quality and application potential of the chitosan product (Pellis et al., 2022).

Extraction Methods: Several extraction methods are employed to obtain chitosan from its natural sources. Commonly used methods, including alkaline hydrolysis, enzymatic hydrolysis, and chemical treatments, are outlined below:

Alkaline Hydrolysis: Alkaline hydrolysis is a key method for extracting chitosan from chitin, involving the treatment of chitin with alkaline substances like sodium hydroxide. This process breaks the acetyl groups from the chitin polymer, converting it into chitosan. Following the alkaline treatment, the chitosan is washed to remove residual chemicals and then neutralized to obtain a purified product. This method is favored for its simplicity and effectiveness in achieving the deacetylation needed to produce chitosan. The resulting chitosan is then used in various applications, benefiting from the efficiency and straightforward nature of alkaline hydrolysis in the extraction process (Novikov et al., 2023).

Enzymatic Hydrolysis: Enzymatic methods for chitosan extraction use chitinase enzymes to selectively deacetylate chitin, offering a precise and controlled approach. This method is preferred due to its specificity and the mild reaction conditions it employs, which help preserve the integrity of the chitosan structure. By carefully controlling the enzymatic process, it is possible to achieve chitosan with specific degrees of deacetylation tailored for various applications. Enzymatic hydrolysis minimizes unwanted side reactions and degradation, resulting in high-purity chitosan that maintains desirable properties for use in pharmaceuticals, food packaging, and other fields.

This approach is advantageous for applications requiring precise control over chitosan's chemical and physical characteristics (Kaczmarek et al., 2019; Azeh et al., 2023).

Chemical Treatments: Chemical treatments involve the use of acids, such as hydrochloric acid, to break down the acetyl groups in chitin. The process requires careful control of reaction conditions to prevent excessive degradation of chitosan. Chemical treatments are effective but may result in lower-molecular-weight chitosan. The choice of extraction method depends on factors such as the desired properties of the chitosan, the intended application, and economic considerations. Each method has its advantages and limitations, and researchers continue to explore innovative approaches to improve the efficiency and sustainability of chitosan extraction (Ssekatawa et al., 2021).

Properties and Applications: The properties of chitosan make it a versatile biomaterial with diverse applications in medical and food technology.

Biocompatibility: Chitosan's biocompatibility is a critical factor for its success in medical applications. It is well-tolerated by living organisms, which minimizes the risk of adverse reactions and enhances its suitability for use in various medical contexts. This property is especially important in applications such as wound healing, where chitosan supports tissue repair and reduces infection risk. In tissue engineering, its biocompatibility facilitates cell growth and integration, promoting the development of functional tissue. For implants, chitosan's compatibility with biological tissues helps ensure successful integration and reduces the likelihood of rejection. Its low toxicity and favorable interaction with the body make chitosan a valuable material for the development of medical devices and therapies (Inobeme et al., 2023a).

Biodegradability: Chitosan's biodegradability significantly enhances its environmentally friendly profile. As a naturally occurring polymer, chitosan breaks down through biological processes into non-toxic byproducts such as carbon dioxide, water, and other natural substances. This degradation occurs without the need for complex disposal methods or significant environmental management. By minimizing the accumulation of waste and reducing long-term ecological impact, chitosan supports sustainable practices and reduces the burden on waste management systems. Its ability to degrade in natural environments makes it a preferable alternative to conventional synthetic polymers, which can persist in ecosystems for extended periods and contribute to pollution. The environmentally benign nature of chitosan underscores its potential as a key component in sustainable materials, contributing to a reduction in waste and fostering a more circular approach to material use (de Sousa et al., 2020; Dawud et al., 2025).

Antimicrobial Activity: Chitosan's cationic nature, resulting from the protonation of its amino groups, imparts inherent antimicrobial properties. This positive charge allows chitosan to interact with negatively charged microbial cell membranes, disrupting their function and inhibiting growth. In food technology, this antimicrobial feature is harnessed for food preservation and packaging materials. Chitosan-based coatings and films can effectively reduce microbial contamination, extend the shelf life of perishable goods, and maintain food safety. By preventing spoilage and protecting against pathogens, chitosan enhances the overall quality and longevity of food products. Its natural origin and safety profile further support its use in food-related applications, offering a sustainable alternative to synthetic preservatives while contributing to improved food preservation techniques (Adetunji et al., 2023a).

Film-Forming Capabilities: Chitosan's film-forming capabilities are highly valuable in various applications, including drug delivery systems, wound dressings, and edible coatings for fruits and vegetables. In drug delivery, chitosan films offer controlled release of therapeutic agents, ensuring a steady and targeted release over time. For wound dressings, chitosan films provide a protective barrier that supports healing while maintaining a moist environment conducive to tissue repair. In food technology, chitosan-based edible coatings act as a barrier against moisture loss and microbial contamination, helping to preserve freshness and extend shelf life. The versatility of chitosan films in these applications stems from their ability to form strong, flexible layers that are both protective and functional, making them ideal for diverse uses across medical and food industries (Mathew et al., 2023a).

In addition, chitosan's properties and extraction methods make it a valuable biomaterial with wide-ranging applications. Its biocompatibility, biodegradability, antimicrobial activity, and film-forming capabilities position chitosan as a material of choice in medical and food technology. Ongoing research continues to explore new extraction techniques, improve properties, and uncover novel applications, further establishing chitosan as a cornerstone in biomaterial science (Thambiliyagodage et al., 2023).

13.4 MEDICAL APPLICATIONS OF CHITOSAN

Chitosan, a versatile biomaterial derived from chitin, has demonstrated remarkable potential in various medical applications. Its unique properties, including biocompatibility, biodegradability, and antimicrobial activity, make it particularly suitable for wound healing, tissue engineering, drug delivery, and hemostasis (Baharlouei and Rahman, 2022).

Wound Healing: Chitosan's use in wound healing is one of its most prominent medical applications. The biocompatibility of chitosan promotes favorable interactions with the wound site, while its antimicrobial properties help prevent infections. Chitosan-based wound dressings create a conducive environment for tissue regeneration by absorbing excess exudate, maintaining a moist environment, and facilitating gas exchange. The hemostatic properties of chitosan are crucial in controlling bleeding, making it effective for various types of wounds, including surgical incisions, lacerations, and traumatic injuries. Chitosan's ability to accelerate the clotting process contributes to faster wound closure and reduced risk of complications. Furthermore, chitosan-based scaffolds can be tailored to mimic the extracellular matrix, providing structural support and promoting cell adhesion and proliferation. This makes them valuable in promoting tissue regeneration and wound closure, particularly in chronic wounds or large tissue defects (Matica et al., 2019; Mathew et al., 2025b).

Tissue Engineering: Chitosan plays a pivotal role in tissue engineering, where the goal is to create functional tissues or organs for transplantation. Its biocompatibility and biodegradability make it an ideal scaffold material for tissue regeneration. Chitosan scaffolds can be engineered to match the mechanical properties of specific tissues and provide a supportive structure for cell growth. Chitosan's positive charge facilitates interactions with negatively charged cell surfaces, promoting cell adhesion and proliferation. Moreover, chitosan can be modified to enhance its mechanical strength, porosity, and degradation rate, allowing for customization based on the targeted tissue or organ. In tissue engineering applications, chitosan has been explored for various tissues, including bone, cartilage, skin, and nerve. Chitosan-based constructs show promise in overcoming the limitations associated with traditional tissue engineering materials, contributing to advancements in regenerative medicine (Islam et al., 2020).

Drug Delivery Systems: Chitosan's ability to form nanoparticles, films, and gels makes it an attractive material for drug delivery systems. Nanoparticles composed of chitosan can encapsulate therapeutic agents, protecting them from degradation and facilitating controlled release. The controlled release allows for sustained therapeutic effects, reducing the frequency of administration and improving patient compliance. Chitosan-based drug delivery systems have been explored for various applications, including cancer therapy, antimicrobial treatment, and anti-inflammatory drugs. The cationic nature of chitosan facilitates interactions with negatively charged cell membranes, promoting cellular uptake of drug-loaded nanoparticles. The mucoadhesive properties of chitosan are particularly advantageous for drug delivery to mucosal surfaces, such as the gastrointestinal tract and respiratory system. Chitosan-based formulations have been investigated for targeted drug delivery to specific tissues, improving the therapeutic efficacy while minimizing side effects (Mikušová and Mikuš, 2021).

Hemostasis: Chitosan's hemostatic properties have found applications in controlling bleeding during surgical procedures, trauma, and emergency situations. Chitosan dressings or hemostatic agents can be applied directly to the bleeding site, promoting rapid clot formation and minimizing blood loss. The mechanism of chitosan-induced hemostasis involves its interaction with blood components, leading to platelet activation and aggregation. Additionally, chitosan's positive charge may enhance interactions with red blood cells and promote the formation of a stable blood clot. Chitosan-based hemostatic products have been used in military and healthcare settings to address hemorrhage and control bleeding in situations where rapid intervention is critical. The biocompatibility of chitosan minimizes the risk of adverse reactions, making it a valuable tool in emergency medicine (Wang et al., 2019; Bodunrinde et al., 2025).

However, the medical applications of chitosan highlight its versatility and potential in advancing healthcare. From wound healing and tissue engineering to drug delivery and hemostasis, chitosan's properties make it a valuable biomaterial in the development of innovative solutions for various medical challenges. Ongoing research continues to explore new formulations, delivery methods, and applications, further expanding the role of chitosan in the field of medicine (Adetunji et al., 2023b).

15.5 CHITOSAN IN FOOD TECHNOLOGY

Chitosan, a natural polysaccharide derived from chitin, has gained considerable attention in the realm of food technology due to its unique properties and versatile applications. From extending the shelf life of perishable products to developing innovative food packaging materials, chitosan offers a range of benefits in enhancing food quality, safety, and sustainability (Stefanowska et al., 2023).

Antimicrobial Properties: One of the key attributes that make chitosan valuable in food technology is its inherent antimicrobial activity. Chitosan exhibits antimicrobial effects against a broad spectrum of microorganisms, including bacteria, fungi, and yeasts. This property can be harnessed to inhibit the growth of spoilage microorganisms and pathogens, thereby extending the shelf life of fresh produce, seafood, and other perishable foods. Chitosan's antimicrobial action is attributed to its cationic nature, which allows it to interact with negatively charged microbial cell membranes, leading to membrane disruption and cell death. This makes chitosan a natural and effective alternative to chemical preservatives in food preservation (Yan et al., 2021).

Edible Coatings: Chitosan's film-forming capabilities make it an excellent candidate for developing edible coatings for fruits, vegetables, and other food products. These coatings act as a protective barrier, reducing moisture loss, gas exchange, and microbial contamination. Edible chitosan coatings can contribute to the preservation of food quality, texture, and appearance during storage and transportation. In addition to its protective role, chitosan coatings can serve as carriers for bioactive compounds, such as antioxidants or antimicrobial agents, further enhancing the overall quality and safety of the coated food products. The ability to tailor the properties of chitosan films allows for customization based on the specific requirements of different food items (Ogundolie et al., 2023).

Antioxidant Activity: Chitosan possesses inherent antioxidant properties that can contribute to the preservation of food products by inhibiting oxidative reactions. Oxidation is a major factor leading to the deterioration of food quality, causing changes in flavor, color, and nutritional content. Chitosan's antioxidant activity helps mitigate these undesirable effects, maintaining the sensory and nutritional attributes of the food. The antioxidant potential of chitosan arises from its ability to scavenge free radicals and chelate metal ions that catalyze oxidative reactions. Incorporating chitosan into food formulations can, therefore, enhance the oxidative stability of oils, fats, and other susceptible components, contributing to the overall quality and longevity of food products (Demircan and Velioglu, 2024).

Low-Fat Product Development: Chitosan's lipid-binding properties make it a promising ingredient in the development of low-fat food products. By binding with lipids, chitosan can reduce the absorption and assimilation of dietary fats in the digestive system. This makes it a potential tool for formulating low-fat versions of various food items, contributing to the development of healthier food options. The use of chitosan in low-fat product development aligns with the growing consumer demand for healthier food choices and reduced fat intake. Chitosan's role in fat reduction can be explored in diverse food categories, including dairy products, meat analogs, and baked goods, providing a sustainable solution to address health-conscious consumer preferences (Liu et al., 2021).

Metal Ion Removal in Food Processing: Chitosan's chelating properties make it effective in metal ion removal, a valuable application in food processing. Metal ions can be present in food ingredients or water used in food production, and their presence may lead to undesirable changes in color, flavor, and stability. Chitosan can be employed as a chelating agent to sequester and remove metal ions, improving the overall quality of the final food product. Chitosan's metal ion removal capabilities are particularly relevant in industries where water quality is crucial, such as brewing, dairy processing, and beverage production. By acting as a natural chelating agent, chitosan contributes to the purification and quality enhancement of food processing streams (Thambiliyagodage et al., 2023).

In addition, chitosan's diverse applications in food technology highlight its potential to address multiple challenges in the food industry. From extending shelf life through antimicrobial activity to contributing to low-fat product development and improving oxidative stability, chitosan offers a range of solutions to enhance food quality, safety, and sustainability. Ongoing research in this field continues to explore innovative ways to leverage the unique properties of chitosan for the benefit of the food industry and consumers alike (Kumar et al., 2023).

15.6 CHITOSAN IN TISSUE ENGINEERING

Chitosan, a biocompatible and biodegradable polysaccharide derived from chitin, has emerged as a valuable biomaterial in tissue engineering. Its biocompatibility ensures harmonious interactions with biological systems, making it well-tolerated by living tissues. The ability of chitosan to undergo natural degradation aligns with tissue regeneration processes, avoiding the need for subsequent removal (Table 15.2).

TABLE 15.2
Chitosan in Tissue Engineering

Aspect	Significance	Examples of Applications	References
Biocompatibility and biodegradability	Chitosan's compatibility with biological systems, promoting favorable interactions with cells and tissues. Its ability to undergo natural degradation, aligning with tissue regeneration processes	Scaffolds for tissue engineering, implants for tissue regeneration	Kim et al. (2023)
Scaffold and structural support	Chitosan's role in providing a three-dimensional scaffold that mimics the extracellular matrix, offering structural support for cell attachment, proliferation, and tissue regeneration	Bone tissue engineering, cartilage regeneration, skin substitutes	Krishani et al. (2023)
Cell adhesion and proliferation	Chitosan's positive charge promoting interactions with negatively charged cell surfaces, enhancing cell adhesion and proliferation	Enhancing cell growth in scaffolds, promoting tissue regeneration	Islam et al. (2020)
Customization for specific tissues	Chitosan's versatility in modification to tailor mechanical properties, porosity, and degradation rate, allowing customization based on the targeted tissue or organ	Engineering chitosan scaffolds for bone, cartilage, skin, nerve regeneration	Liu et al. (2023)

In tissue engineering, chitosan plays a crucial role in providing three-dimensional scaffolds that mimic the extracellular matrix, offering structural support for cell attachment, proliferation, and tissue regeneration. The positive charge of chitosan promotes interactions with negatively charged cell surfaces, enhancing cell adhesion and proliferation. This makes it particularly effective in promoting cell growth within scaffolds, contributing to the regeneration of various tissues such as bone, cartilage, skin, and nerves (Adetunji et al., 2022).

Furthermore, chitosan's versatility allows for the customization of mechanical properties, porosity, and degradation rate, enabling tailored scaffolds for specific tissues or organs. The applications of chitosan in tissue engineering hold significant promise for advancing regenerative medicine and addressing challenges associated with tissue repair and replacement. Ongoing research continues to explore innovative ways to optimize chitosan-based materials for enhanced tissue regeneration and functional restoration (Wang et al., 2023).

15.7 CHITOSAN'S ROLE IN METAL ION REMOVAL

Chitosan, a natural polysaccharide derived from chitin, has demonstrated significant effectiveness in metal ion removal, making it a promising material for various applications in environmental remediation and food processing. The chelating properties of chitosan, arising from its amino and hydroxyl functional groups, enable it to bind with metal ions and facilitate their removal from aqueous solutions (Fatullayeva et al., 2022).

Chelating Mechanism: Chitosan's chelating mechanism involves the formation of coordinate bonds between the amino and hydroxyl groups of chitosan and metal ions present in a solution. The lone pairs of electrons on the amino and hydroxyl groups can coordinate with metal ions, leading to the formation of stable complexes. The resulting chitosan–metal complexes can be easily separated from the solution, effectively removing metal ions (Aranaz et al., 2021).

Metal Ion Removal in Environmental Remediation: Chitosan's ability to remove metal ions from water has garnered attention in environmental remediation efforts. Contamination of water sources with heavy metals, such as lead, cadmium, and copper, poses significant threats to ecosystems and human health. Chitosan-based materials, such as chitosan beads or membranes, have been employed to selectively adsorb and remove these metal ions from water. The use of chitosan in water treatment is advantageous due to its biodegradability, low toxicity, and high efficiency in metal ion removal. The chitosan-based adsorbents offer a sustainable solution for mitigating metal pollution, addressing concerns related to water quality and ecosystem health (Mathew et al., 2023a,b).

Metal Detoxification in Food Processing: Chitosan's metal ion removal capabilities also find applications in food processing, where metal contaminants can adversely affect the quality and safety of food products. Chitosan can act as a natural chelating agent to bind with undesirable metal ions present in food ingredients or processing water, thereby preventing their incorporation into the final food product. In addition to its metal ion removal function, chitosan has been explored as a food preservative due to its antimicrobial properties. By inhibiting the growth of spoilage microorganisms, chitosan contributes to extending the shelf life of food products, providing a dual benefit in terms of food safety and quality (Inobeme et al., 2023a,b).

Chitosan-Based Adsorbents: Chitosan can be modified and incorporated into various forms of adsorbents, enhancing its metal ion removal efficiency. Chitosan beads, membranes, and nanoparticles have been extensively studied for their adsorption capacities and selectivity toward different metal ions. The porous structure of chitosan-based materials, coupled with the availability of functional groups, facilitates efficient metal ion binding. The modification of chitosan through processes such as crosslinking or grafting further enhances its adsorption properties. These modifications can tailor chitosan-based adsorbents to selectively target specific metal ions, making them versatile tools in the removal of a wide range of contaminants from diverse matrices (Benettayeb et al., 2023).

Future Perspectives and Challenges: While chitosan presents promising prospects in metal ion removal, certain challenges and considerations remain. The regeneration and reusability of chitosan-based adsorbents, the influence of solution pH on metal ion adsorption, and the potential interference of co-existing ions are aspects that require careful evaluation and optimization. Moreover, the scalability and economic feasibility of chitosan-based metal ion removal processes need to be addressed for widespread practical applications. Research efforts are ongoing to develop innovative techniques and materials that enhance the efficiency and cost-effectiveness of chitosan in metal ion removal processes (Gamage et al., 2023; Mathew et al., 2023c).

Chitosan's role in metal ion removal underscores its potential contributions to environmental sustainability and food safety. The chelating properties of chitosan make it an effective and eco-friendly solution for mitigating metal pollution in water sources and preventing metal contamination in food processing. As research in this field advances, chitosan-based materials are likely to play an increasingly vital role in addressing metal-related challenges, contributing to a cleaner environment and safer food supply chains. The ongoing exploration of chitosan's capabilities in metal ion removal reflects its continued relevance in addressing contemporary environmental and food safety concerns (Shankar et al., 2023).

15.8 CONCLUSION

The diverse properties of chitosan and its derivatives position them as promising biomaterials with substantial applications in both medical and food technology. The biocompatibility, biodegradability, and antimicrobial activity of chitosan make it a valuable component in wound healing, tissue engineering, and drug delivery systems, addressing critical challenges in the medical field. Chitosan's role extends to food technology, where its antimicrobial properties, film-forming capabilities, and lipid-binding attributes contribute to food preservation, packaging, and the development of healthier, low-fat food products.

The ability of chitosan to serve as a scaffold for tissue engineering provides a platform for regenerative medicine, offering solutions for various tissue regeneration applications. Moreover, its chelating properties make chitosan effective in metal ion removal, supporting environmental remediation efforts and ensuring food safety. As research continues to unravel the full potential of chitosan and its derivatives, their applications are poised to advance, addressing evolving challenges in the healthcare and food industries. The sustainability, versatility, and biocompatibility of chitosan underscore its significance as a biomaterial, paving the way for innovative developments that hold promise for the future of medical and food technologies. The ongoing exploration of chitosan's multifaceted capabilities invites further collaboration between scientific disciplines, fostering the development of novel solutions and contributing to the progress of biomaterial science.

REFERENCES

Adetunji, C. O., Mathew, J. T., Inobeme, A., Taiwo, O., Olaniyan, O., Shakira, G., & Wadazani, D. (2023a). Application of biosensors for detection and monitoring of water quality. In: *Organic and Inorganic Materials Based Sensors*. John Wiley & Sons, Inc., 1065–1077. https://doi.org/10.1002/9783527834266.ch46

Adetunji, C. O., Ogundolie, F. A., Mathew, J. T., Inobeme, A., Olotu, T., Olaniyan, O. T., Ijabadeniyi, O. A., Ajiboye, M. D., Ajayi, O. O., Dauda, W., Ghazanfar, S., & Adetunji, J. B. (2022). Graphene-based nanomaterials for targeted drug delivery and tissue engineering. In: *Novel Platforms for Drug Delivery Applications*. Elsevier Ltd., 277–288. https://doi.org/10.1016/B978-0-323-91376-8.00014-8

Adetunji, C. O., Olaniyan, O. T., Shariati, M. A., Olugbenga, S. M., Osemwegie, O. O., Rebezov, M., Ajayi, O. O., Makhmudov, F., Bodunrinde, R. E., Adetunji, J. B., Adeniyi, M. J., Hefft, D. I., Inobeme, A., & Mathew, J. T. (2023b). Medical attributes of *Breynia disticha*. In: J. R. Forst & G. Forst. eds. *Phytochemical Composition and Pharmacy of Medicinal Plants*. Apple Academic Press, Inc. Co-published with CRC Press (Taylor & Francis). Vol. 2, 531–538.

Aranaz, I., Alcántara, A. R., Civera, M. C., Arias, C., Elorza, B., Heras Caballero, A., & Acosta, N. (2021). Chitosan: an overview of its properties and applications. *Polymers*, *13*(19), 3256. https://doi.org/10.3390/polym13193256

Azeh, Y., Musah, M., Mathew, J. T., Badeggi, U. M., Estuyankpa, M. B., Muhammad, Z. D., Oluwaseyi, D. S., Umar, M. T., Abdulkarim, M. A., Muhammad, A. I., Ibrahim, A. A., Ismail, H., Muhammad, K. T., Umar, G., & Ahmed, T. (2023). In situ synthesis of polyaniline nanohybrid and formulation of polyaniline/carboxymethyl cellulose/ethylene glycol nanocomposite: study of its conducting and antibacterial properties. *Journal of Materials and Polymer Science*, *3*(4), 1–9. https://doi.org/10.47485/2832-9384.1041

Baharlouei, P., & Rahman, A. (2022). Chitin and chitosan: prospective biomedical applications in drug delivery, cancer treatment, and wound healing. *Marine Drugs*, *20*(7), 460. https://doi.org/10.3390/md20070460

Benettayeb, A., Seihoub, F. Z., Pal, P., Ghosh, S., Usman, M., Chia, C. H., Usman, M., & Sillanpää, M. (2023). Chitosan nanoparticles as potential nano-sorbent for removal of toxic environmental pollutants. *Nanomaterials (*Basel, Switzerland), *13*(3), 447. https://doi.org/10.3390/nano13030447

Bodunrinde, R. E., Chimbekujwo, K. I., Oyewole, O. A., Adetunji, C. O., Inobeme, A., Eniola, K. I. T., Yerima, M. B., & Mathew, J. T. (2025). The use of nanochitosan for enhancement in the quality and yield of fruit crops. In: *Nanochitosan Applications for Enhanced Crop Production and Food Security*. John Wiley & Sons, 111–122. https://doi.org/10.1002/9781394214945.ch5

Chimbekujwo, K. I., Yakubu, J. G., Omoregie, I. P., Oyewole, O. A., Muhammad, W. M., Adetunji, C. O., Eniola, K. I. T., Yerima, M. B., & Mathew, J. T. (2025). Synthesis of nano-chitosan using agricultural waste. In: *Nanochitosan Applications for Enhanced Crop Production and Food Security*. John Wiley & Sons, Inc., 39–68. https://doi.org/10.1002/9781394214945.ch2

Confederat, L. G., Tuchilus, C. G., Dragan, M., Sha'at, M., & Dragostin, O. M. (2021). Preparation and antimicrobial activity of chitosan and its derivatives: a concise review. *Molecules (*Basel, Switzerland), *26*(12), 3694. https://doi.org/10.3390/molecules26123694

Dawud, I., Bodunrinde, R. E., Oyewole, O. A., Adetunji, C. O., Inobeme, A., Eniola, K. I. T., Yerima, M. B., & Mathew, J. T. (2025). Application of Nanochitosan for Effective Fruit Production. In: *Nanochitosan Applications for Enhanced Crop Production and Food Security*. John Wiley & Sons, 123–135. https://doi.org/10.1002/9781394214945.ch6

de Sousa, V. R., Marcelo da Cunha Santos, A., Viana de Sousa, B., de Araújo Neves, G., Navarro de Lima Santana, L., & Rodrigues Menezes, R. (2020). A review on chitosan's uses as biomaterial: tissue engineering, drug delivery systems and cancer treatment. *Materials (*Basel, Switzerland), *13*(21), 4995. https://doi.org/10.3390/ma13214995

Demircan, B., & Velioglu, Y. S. (2024). Control of browning, enzyme activity, and quality in stored fresh-cut fruit salads through chitosan coating enriched with bergamot juice powder. *Foods (*Basel, Switzerland), *13*(1), 147. https://doi.org/10.3390/foods13010147

Desai, N., Rana, D., Salave, S., Gupta, R., Patel, P., Karunakaran, B., Sharma, A., Giri, J., Benival, D., & Kommineni, N. (2023). Chitosan: a potential biopolymer in drug delivery and biomedical applications. *Pharmaceutics*, *15*(4), 1313. https://doi.org/10.3390/pharmaceutics15041313

Fatullayeva, S., Tagiyev, D., Zeynalov, N., Mammadova, S., & Aliyeva, E. (2022). Recent advances of chitosan-based polymers in biomedical applications and environmental protection. *Journal of Polymer Research*, *29*(7), 259. https://doi.org/10.1007/s10965-022-03121-3

Gamage, A., Jayasinghe, N., Thiviya, P., Wasana, M. L. D., Merah, O., Madhujith, T., & Koduru, J. R. (2023). Recent application prospects of chitosan based composites for the metal contaminated wastewater treatment. *Polymers*, *15*(6), 1453. https://doi.org/10.3390/polym15061453

Inobeme, A., Adetunji, C. O., Mathew, J. T., Ajai, A. I., Inobeme, J., Bamigboye, M. O., Onyeaku, S., Maliki, M., Eziukwu, C., & Tawa, K. (2023a). Nanotechnology for bioremediation of heavy metals. In: *Microbial Technologies in Industrial Wastewater Treatment*. Springer Nature Singapore, 19–30. https://doi.org/10.1007/978-981-99-2435-6_2

Inobeme, A., Adetunji, C. O., Mathew, J. T., Okonkwo, S., Bamigboye, M. O., Ajai, A. I., Afoso, E., & Inobeme, J. (2023b). Advanced nanotechnology for the degradation of persistent organic pollutants. In: Shah, M. ed. *Microbial Degradation and Detoxification of Pollutants*. Berlin, Boston: De Gruyter, 51–72. https://doi.org/10.1515/9783110743623-003

Islam, M. M., Shahruzzaman, M., Biswas, S., Nurus Sakib, M., & Rashid, T. U. (2020). Chitosan based bioactive materials in tissue engineering applications: a review. *Bioactive Materials*, *5*(1), 164–183. https://doi.org/10.1016/j.bioactmat.2020.01.012

Kaczmarek, M. B., Struszczyk-Swita, K., Li, X., Szczęsna-Antczak, M., & Daroch, M. (2019). Enzymatic modifications of chitin, chitosan, and chitooligosaccharides. *Frontiers in Nioengineering and Biotechnology*, *7*, 243. https://doi.org/10.3389/fbioe.2019.00243

Kim, Y., Zharkinbekov, Z., Raziyeva, K., Tabyldiyeva, L., Berikova, K., Zhumagul, D., Temirkhanova, K., & Saparov, A. (2023). Chitosan-Based Biomaterials for Tissue Regeneration. *Pharmaceutics*, *15*(3), 807. https://doi.org/10.3390/pharmaceutics15030807

Krishani, M., Shin, W. Y., Suhaimi, H., & Sambudi, N. S. (2023). Development of scaffolds from Bio-Based Natural Materials for Tissue Regeneration Applications: A Review. *Gels*, *9*, 100. https://doi.org/10.3390/gels9020100

Kumar, A., Yadav, S., Pramanik, J., Sivamaruthi, B. S., Jayeoye, T. J., Prajapati, B. G., & Chaiyasut, C. (2023). Chitosan-based composites: development and perspective in food preservation and biomedical applications. *Polymers*, *15*(15), 3150. https://doi.org/10.3390/polym15153150

Liu, S. H., Chen, R. Y., & Chiang, M. T. (2021). Effects and mechanisms of chitosan and chitosanoligosaccharide on hepatic lipogenesis and lipid peroxidation, adipose lipolysis, and intestinal lipid absorption in rats with high-fat diet-induced obesity. *International Journal of Molecular Sciences*, *22*(3), 1139. https://doi.org/10.3390/ijms22031139

Liu, S., Yu, J. M., Gan, Y. C., Qiu, X. Z., Gao, Z. C., Wang, H., Chen, S. X., Xiong, Y., Liu, G. H., Lin, S. E., McCarthy, A., John, J. V., Wei, D. X., & Hou, H. H. (2023). Biomimetic natural biomaterials for tissue engineering and regenerative medicine: new biosynthesis methods, recent advances, and emerging applications. *Military Medical Research*, *10*(1), 16. https://doi.org/10.1186/s40779-023-00448-w

Malik, S. I., Chimbekujwo, K. I., Maku, A. M., Oyewole, O. A., Omoregie, I. P., Adetunji, C. O., Eniola, K. I. T., Yerima, M. B., & Mathew, J. T. (2025). The application of nanochitosan biopesticides as a replacement to synthetic pesticides. In: *Nanochitosan Applications for Enhanced Crop Production and Food Security*. John Wiley & Sons, 89–110. https://doi.org/10.1002/9781394214945.ch4

Maliki, S., Sharma, G., Kumar, A., Moral-Zamorano, M., Moradi, O., Baselga, J., Stadler, F. J., & García-Peñas, A. (2022). Chitosan as a tool for sustainable development: a mini review. *Polymers*, *14*(7), 1475. https://doi.org/10.3390/polym14071475

Mathew, J. T., Adetunji, C. O., Inobeme, A., Musah, M., Azeh, Y., Etsuyankpa, M. B., Otori, A. A., Chado, I. M., Shaba, E. Y., Akoji, J. N., Mamman, A., & Oyewole, O. A. (2025a). Application of nanochitosan for the biodegradation of agricultural wastes. In: *Nanochitosan Applications for Enhanced Crop Production and Food Security*. John Wiley & Sons, 361–382. https://doi.org/10.1002/9781394214945.ch16

Mathew, J. T., Adetunji, C. O., Inobeme, A., Musah, M., Azeh, Y., Otori, A. A., Shaba, E. Y., Mamman, A. & Tanko, E. (2023a). Removal of heavy metals using bio-remedial techniques. In: Shah M. P. ed. *Modern Approaches in Waste Bioremediation*, 117–130. https://doi.org/10.1007/978-3-031-24086-7_6

Mathew, J. T., Adetunji, C. O., Inobeme, A., Otori, A. A., Musah, M., Azeh, Y., Shaba, E. Y., Muhammad, I. A., Mamman, A., & Oyewole, O. A. (2025b). Relevance of nanochitosan in food sector and food packaging sectors: current trends. In: *Nanochitosan Applications for Enhanced Crop Production and Food Security*. John Wiley & Sons, 345–360. https://doi.org/10.1002/9781394214945.ch15

Mathew, J. T., Musah, M., Azeh, Y., & Muhammed, M. (2023b). Adsorptive removal of selected toxic metals from pharmaceutical wastewater using Fe_3O_4/ZnO nanocomposite. *Dutse Journal of Pure and Applied Sciences*, *9*(4a), 236–248. https://doi.org/10.4314/dujopas.v9i4a.22

Mathew, J. T., Musah, M., Azeh, Y., & Muhammed, M. (2023c). Kinetic study of heavy metals removal from pharmaceutical wastewater using geopolymer/Fe_3O_4 nanocomposite. *Bima Journal of Science and Technology*, *7*(4), 152–163. https://doi.org/10.56892/bima.v7i4.539

Matica, M. A., Aachmann, F. L., Tøndervik, A., Sletta, H., & Ostafe, V. (2019). Chitosan as a Wound Dressing Starting Material: Antimicrobial Properties and Mode of Action. *International Journal of Molecular Sciences*, *20*(23), 5889. https://doi.org/10.3390/ijms20235889

Matloob, A., Ayub, H., Mohsin, M., Ambreen, S., Khan, F. A., Oranab, S., Rahim, M. A., Khalid, W., Nayik, G. A., Ramniwas, S., & Ercisli, S. (2023). A review on edible coatings and films: advances, composition, production methods, and safety concerns. *ACS Omega*, *8*(32), 28932–28944. https://doi.org/10.1021/acsomega.3c03459

Mikušová, V., & Mikuš, P. (2021). Advances in chitosan-based nanoparticles for drug delivery. *International Journal of Molecular Sciences*, *22*(17), 9652. https://doi.org/10.3390/ijms22179652

Muhammad, W. M., Ibrahim, A. B., Samuel, J. O., Khalifa, A. S., Oyewole, O. A., Adetunji, C. O., Eniola, K. I. T., Yerima, M. B., & Mathew, J. T. (2025). The role of nanomaterials in agriculture as nanofertilizers. In: *Nanochitosan Applications for Enhanced Crop Production and Food Security*. John Wiley & Sons, Inc., 1–38. https://doi.org/10.1002/9781394214945.ch1

Muñoz-Tebar, N., Pérez-Álvarez, J. A., Fernández-López, J., & Viuda-Martos, M. (2023). Chitosan edible films and coatings with added bioactive compounds: antibacterial and antioxidant properties and their application to food products: a review. *Polymers*, *15*(2), 396. https://doi.org/10.3390/polym15020396

Muthu, M., Gopal, J., Chun, S., Devadoss, A. J. P., Hasan, N., & Sivanesan, I. (2021). Crustacean waste-derived chitosan: antioxidant properties and future perspective. Antioxidants (Basel, Switzerland), *10*(2), 228. https://doi.org/10.3390/antiox10020228

Novikov, V. Y., Derkach, S. R., Konovalova, I. N., Dolgopyatova, N. V., & Kuchina, Y. A. (2023). Mechanism of heterogeneous alkaline deacetylation of chitin: a review. *Polymers*, *15*(7), 1729. https://doi.org/10.3390/polym15071729

Ogundolie, F. A., Adetunji, C. O., Mathew, J. T., Inobeme, A., Ghazanfar, S., Olaniyan, O. T., Ajiboye, M. D., Dauda, W. P., Adetunji, J. B., Akinbo, O., & Adetuyi, B. O. (2023). Intelligent freshness indicators in packaging of plant-based food products. Academic Press, 323–331. https://doi.org/10.1016/B978-0-323-89864-5.00008-4

Pellis, A., Guebitz, G. M., & Nyanhongo, G. S. (2022). Chitosan: sources, processing and modification techniques. *Gels (*Basel, Switzerland), *8*(7), 393. https://doi.org/10.3390/gels8070393

Rollini, M., Musatti, A., & Cavicchioli, D. (2020). From cheese whey permeate to Sakacin-A/bacterial cellulose nanocrystal conjugates for antimicrobial food packaging applications: a circular economy case study. *Scientific Reports*, *10*, 21358. https://doi.org/10.1038/s41598-020-78430-y

Shankar, S., Joshi, S., & Srivastava, R. K. (2023). A review on heavy metal biosorption utilizing modified chitosan. *Environmental monitoring and assessment*, *195*(11), 1350. https://doi.org/10.1007/s10661-023-11963-7

Ssekatawa, K., Byarugaba, D. K., & Wampande, E. M. (2021). Isolation and characterization of chitosan from Ugandan edible mushrooms, Nile perch scales and banana weevils for biomedical applications. *Scientific Reports*, *11*, 4116. https://doi.org/10.1038/s41598-021-81880-7

Stefanowska, K., Woźniak, M., Dobrucka, R., & Ratajczak, I. (2023). Chitosan with natural additives as a potential food packaging. *Materials (*Basel, Switzerland), *16*(4), 1579. https://doi.org/10.3390/ma16041579

Thambiliyagodage, C., Jayanetti, M., Mendis, A., Ekanayake, G., Liyanaarachchi, H., & Vigneswaran, S. (2023). Recent advances in chitosan-based applications: a review. *Materials (*Basel, Switzerland), *16*(5), 2073. https://doi.org/10.3390/ma16052073

Trucillo, P. (2024). Biomaterials for drug delivery and human applications. *Materials (*Basel, Switzerland), *17*(2), 456. https://doi.org/10.3390/ma17020456

Wang, Q., Wang, X., & Feng, Y. (2023). Chitosan hydrogel as tissue engineering scaffolds for vascular regeneration applications. *Gels (*Basel, Switzerland), *9*(5), 373. https://doi.org/10.3390/gels9050373

Wang, Y. W., Liu, C. C., Cherng, J. H., Lin, C. S., Chang, S. J., Hong, Z. J., Liu, C. C., Chiu, Y. K., Hsu, S. D., & Chang, A. H. (2019). Biological effects of chitosan-based dressing on hemostasis mechanism. *Polymers*, *11*(11), 1906. https://doi.org/10.3390/polym11111906

Yan, D., Li, Y., Liu, Y., Li, N., Zhang, X., & Yan, C. (2021). Antimicrobial properties of chitosan and chitosan derivatives in the treatment of enteric infections. *Molecules (*Basel, Switzerland), *26*(23), 7136. https://doi.org/10.3390/molecules26237136

Index

Note: **Bold** page numbers refer to tables and *italic* page numbers refer to figures.

A

Abiotic stress mitigator, 22–23
Abscisic acid (ABA), 22
Abundant biopolymer, 61
Acetylation levels, 21
Active packaging, 2, 6, 46, 72, **72**, 165, 213
Active pharmaceutical ingredients (APIs), 56
Acylated chitosan, 179, 206
Additives, natural, 43
Agricultural and crop protection applications, of chitosan-based formulations, 167–168
Alginate, *51*, 51–52, 139
Aliphatic polyesters, 139
Alkylated chitosan, 179, 206
Allyl isothiocyanate (AITC), 92
Amino acids, oxidation of, 87
Ammonium persulfate (APS), 185
AMR, *See* Antimicrobial resistance (AMR)
Antibacterial activity of chitosan, 25, 26
Antibacterial agents, 4
Antibacterial qualities of chitosan, 2, 80, *90*
Anti-browning mechanism of chitosan, 27
Antifungal activities of chitosan, 20, 26, 27, 87, 167
Antifungal efficacy of CSNPs, 91
Antifungal property of chitosan, 21
Antimalarial drugs, 56
Antimicrobial action in coatings, mechanisms of, 160, **161**
Antimicrobial activity of chitosan, **109**, 183
Antimicrobial agents
  in composite coatings, 160, **161**
  organic, 160
Antimicrobial drugs, 4
Antimicrobial effectiveness of chitosan, 89
Antimicrobial efficacy of silver nanoparticles, 160
Antimicrobial-grade chitosan and chitosan derivatives, biomedical uses of, 189–191
Antimicrobial properties, 4–5
  of chitosan coatings, 107
  of chitosan composites, 159
  enhancing preservation, 67
Antimicrobial resistance (AMR), 45, 239
Antimicrobials, natural, 4
Antinematode, antimicrobial and, 20–22
Antioxidant
  activity of chitosan, 27, 124
  conventional, 4
  plant-based, 241–244
  properties of chitosan, 222
  qualities of chitosan coatings, 2
  strength, combined effect of chitosan and essential oil on, 246–248
Anti-TB drugs, 57
Antiviral activity of chitosan, 27, 188
Antiviral immune response mechanisms, 188
APIs, *See* Active pharmaceutical ingredients (APIs)
Aquatic ecosystems, 146
Aquatic environments, biodegradability in, 145–147
Aspergillus flavus, 91
AvCAT, gene structure of, 243

B

Bacillus subtilis, 21, 88
Bacteria, gram-positive, 84
Bacterial growth
  extracellular mechanism of chitosan inhibition of, 85
  intracellular mechanism of chitosan inhibition of, *86*, 86–87
Bakery products, 65, 67, 103, 106, 107
Barrier properties, 3, 63–65, 68, 70, 71, 138–140, 143, 148, 158, 159, 162–163
  enhanced, 72, **72**, 128, 170
  exceptional oxygen-barrier properties, 17
  gas, 165
  material, 170
Bifidobacteria, 83
Bioactive compounds, 18
Bioactive ingredients, chitosan in encapsulation and delivery of, 18
Biocompatibility, chitosan, 103, 138, 210, 223, **254**, 256, **259**
Biodegradability
  in aquatic environments, 145–147
  chitosan, 68, 103, 118, 135, 137, 143, 223, 256
  of chitosan-based packaging, 134, 140–141
  factors influencing, 143–145, **144–145**
  renewable source and, 68
Biodegradable films, for agricultural applications, 212
Biodegradable packaging, 4
  materials, 138–140
  motivation for, 136–137
Biodegradable polymers, 139, 194
Biodegradation of chitosan-based packaging, 135
Biofertilizer enhancer, 167
Biological control agents, susceptibility of, 20
Biomaterials, in food coatings, 50
Biomedical implants, 162
Biopolymer
  abundant, 61
  chitosan as, 224–225
  natural, 4
  for packaging, chitosan as promising, 137–138
Biopreservation, 13, 45, 128
Biosensors, 10, 212
Biostimulant, for plant growth, 168

C

Cadmium, 240
Caffeic acid (CA), 191
Calcium, 242

Callose, 27
Calmodulin, 242
Candida spp., 91
Canning, 43
Carbon dioxide, 105
Carboxyl alkyl chitosan derivatives, 180–181
Carboxymethylation, 185
Carboxymethyl chitosan (CMC), 180
β-Carotene, **19**
Carvacrol oil, formulation of chitosan-based coating with, 209
Catalases, 242–243
Catechin and quercetin, **19**
Cationic behavior of chitosan, 13, 189
Cellulose, *14*, *50*, 50–51
  -based materials, 139
  chemical reactions, *15*
Cell walls, fungal, 207
Chemical vapour deposition, 184
Chilling damage, 44
Chitin, 2, *14*, 20
  biological and chemical mechanisms of, *80*
  chemical reactions, *15*
  deacetylation of, 80–82, 157
Chitinase, 21, 27, 81, 177, 223
Chitin-containing raw materials, chitosan extraction from, 141
Chitin into chitosan, transformation of, 208
Chitinolytic microbes, 22
Chitosan, 1, 2, 10–12, *14*, 52, *53*, 206
  action in plant growth, 28
  analogs and chemical differences, 14, *15*
  antibacterial activity, 25, 26, 82
  antibacterial properties, *90*
  antibacterial qualities, 80
  anti-browning mechanism of, 27
  antifungal potential of, 20
  antimicrobial activity, 183, 223
  antimicrobial advantage, 69
  antimicrobial effectiveness of, 89
  antioxidant activity, 27, 124, 223
  antiviral activity, 27, 188
  based on molecular size, 207
  based on structural modification, 206–207
  biocompatibility, 103, 138, 210, 223, **254**, 256, **259**
  biodegradability, 68, 103, 118, 135, 137, 143, 223, 256
  biological and chemical mechanisms of, *80*
  in biomaterials, significance, **254**, 254–255
  as biopolymer, 224–225
  biotechnological uses of, 14
  cationic behavior of, 13, 189
  chemical functionalization of, 178, 183
  chemical reactions, *15*
  and chitosan derivatives, 178–179
  customizable biopolymer properties, 68
  degree of DA of, 89
  as eco-friendly material, 68–69
  economic value of, 13
  effectiveness of, 85
  electrostatic flexibility in polymers, 83
  as emulsifying agent, 16
  in encapsulation and delivery of bioactive ingredients, 18
  enzyme inhibitory activity in food products, 27–28
  extraction and processing of, 68
  extraction from chitin-containing raw materials, 141
  fabrication and modification of, 12
  film-forming characteristics of, 117, 159, 223
  in food and agriculture, 14–16, 258–259
  in gene therapy, 187
  heavy metal binding, 223
  high hydrogen-donating capacity of, 27
  HMW, 84, 85, *86*, 87, 88, 91
  immunostimulatory effects, 223
  intermolecular cross-linking of, 13
  in meat preservation, 123–124
  mechanism of antifungal activities of, 26
  mechanisms of action of, 108–109, **109**
  medical applications of, 257–258
  minimum inhibitory concentration (MIC) of, 89
  modification methods, *178*
  mold growth by, 87–88
  molecular structure
    and film-forming ability, 62–63
    and impact on film formation, 63–64
  molecular weight of, 88, 169
  mucoadhesive properties, 223, 225
  nanoparticles, 17
  overview, 101, 238
  as packaging and film material, 16–17
  polymer, bioeconomic applications, 176
  as promising biopolymer for packaging, 137–138
  properties, 12–14, *13*
    and applications, 223–224
    and characteristics, 117–118
    and extraction, 255–257
  role in metal ion removal, 260–261
  role in sustaining food texture, 28
  sources of, 207
    crustaceans, 207
    fungal cell walls, 207
    insect cuticles, 207
  synthesis, 208–209
  in tissue engineering, **259**, 259–260
  versatility, 2, 68, 70, 101, 104
Chitosan and derivatives, in biotechnology, 193–194
Chitosan and essential oils
  compatibility and stability of, 215
  development of novel delivery system for, 216
Chitosan applications, *11*, 18, **226**
  in agriculture, 19–20, 224
    antimicrobial and antinematode, 20–22
    contaminated soil remediation, 23–24
    plant defense mechanism, 20
    plant growth stimulant and abiotic stress mitigator, 22–23
    postharvest treatment, 22
    seed coating/priming, 24
  in biomedical industry, 224
  in food industry, 224
  in food preservation, 103–104, **104**
  in pharmaceutical industry, 224
  in poultry industry, 122–123
  in water treatment, 224
Chitosanases, 81
Chitosan-based coating, 1–2, 4–7, 209
  with carvacrol oil, 209
  enhanced with antimicrobial agents

future perspectives, 215–216
potential improvements and future development, 216–217
regulatory considerations and safety concerns, 216
with essential oil, 208
with lemongrass oil, 208
with lemon oil, 208
mechanisms of action, 121–122
mechanisms of microbial inhibition by, 84
with rosemary oil, 208
with sunflower oil, 208–209
techniques for applying
brushing or roller coating, 210
dipping, 209
solvent casting, 210
spin coating, 210
spraying, 209
Chitosan-based composite coating
advantages of, 210
fundamentals of, 158–160
limitations of, 210–211
performance of, 159
synergistic combination of, 170
Chitosan-based derivatives, biomedical uses of, 189
Chitosan-based edible coatings, 80, 222
on food safety, 227–228
nutraceutical potential of, 228–229
nutritional advantages of, 230–231
Chitosan-based edible films and coatings
engineering properties of, 72
future prospects and developments in, 71, **72**
Chitosan-based essential oils, mechanism of action of antimicrobial properties of, 215
Chitosan-based films, 67
mechanical properties of, 62
mechanical strength of, 64
solubility and water vapor permeability of, 62
Chitosan-based in development and application, existing challenges of, 168–170
Chitosan-based materials
pharmaceutical applications of, 189
uniqueness of, 29
Chitosan-based matrices, integration strategies with, 161–162
Chitosan-based packaging
biodegradability of, 134, 135, 140–141
materials, preparation of, 141–143
properties and applications of, 140, **140**, **147–148**, 147–149
utilization of, 134
Chitosan-based systems, **19**
Chitosan coatings
antibacterial and antioxidant qualities of, 2
antimicrobial properties of, 107
Chitosan composites, antimicrobial properties of, 159
Chitosan concentration, tailoring mechanical strength through, 64–65
Chitosan derivatives, 177
antimicrobial-grade chitosan and, biomedical uses of, 189–191
application, technologies associated with, 183–184
carboxymethylation, 185
chemical vapour deposition method, 184
graft copolymerization, 184–185
hydrothermal carbonization, 184
solvothermal method, 184
chemical nature and end-use function of, 179
in food industry, 191–192
in pharmaceutical and medical economy, 185–187, *186*
quaternary, 181–182
therapeutic applications of, 187–189
Chitosan edible coating
applications and limitations of, 105–106
application techniques of, *85*
benefits, limitations, and future directions, 93–95
coating formulation of, 82–83
effectiveness of, 89
factors influencing efficacy of, 88–89
future directions and opportunities for, 95
inhibition of bacterial growth by, 84–85
methods and properties of, 83–84
with preservation methods, **94**
production, attributes, and antimicrobial influence of, 81
properties and applications, 101–102, **102**
as viable preservation alternative, 90–93, *93*
Chitosan edible coating and MAP
challenges and opportunities, 110
synergistic effects of, 106–108, **107**
Chitosan films, 192
Chitosan inhibition of bacterial growth
extracellular mechanism of, 85
intracellular mechanism of, *86*, 86–87
Chitosan-nanobased material, 17–18
Chitosan nanoparticles (CSNPs), 91
Chitosan-sensitive fungi, 87–88
Cholecalciferol (vitamin D), 18
Cholesterol, 15, 225
Chronic illnesses, 41–42
Circular economy principles, 70, 135, 137, 138, 149, 150
Cold plasma technology, 128
Composite-based chitosan, application of
environmental applications, 211–212
food industry applications, 212
medical and healthcare applications, 211
Composite coatings, antimicrobial agents in, 160, **161**
Consumer awareness, 94, 117, 136
Contaminated soil remediation, 23–24
Conventional antioxidants, 4
Conventional packaging materials, 17
Conventional plastics, 146
Cooling systems, evaporative, 42, 45
Corn gluten flour, 53
Corn-zein protein, 53
Coronavirus SARS-CoV-2 (COVID-19), 188
Crop biofertilization techniques, 22
Crop residue management, 167–168
Crustaceans, 207
Cutting-edge technologies, 49

D

Dairy products, 3, 5, 43, 80, 93, 103–105, 109, 128, 259
Deacetylation (DA), 12, 62, 142
of chitin, 61–63, 80–82, 157
Decalcification, 82
Decomposition, microbial, 4
Degree of deacetylation (DD), 177

Demineralization, 12, 141, 208
Depolymerization, 12
Deproteinization, 141, 238
Dipping method, food coating, 54
DPPH approach, 18
Drug delivery, 161
  agent, 211
  systems, 193
Drug-resistant TB (DR-TB), 57
Drugs, antimicrobial, 4
Dry products, low water vapor permeability for, 67

E

Eco-friendly material, chitosan as, 68–69
Ecosystem, aquatic, 146
Edible coating, 2–4, 40–41, 49
  foods
    in malaria treatment, 55–56
    in tuberculosis treatment, 56–57
  properties of, 50, 102–103
Edible fruit coating, therapeutic prospects of, 54–55
Electrostatic coating method, 54
Electrostatic interactions, 25
Emulsifying agent, chitosan as, 16
Encapsulation, 18, **19**
Encapsulation efficiency (EE), of tripolyphosphate (TPP)-chitosomes, 18
Energy-efficient production, 70
Environmental remediation, 162
Enzymatic activity, 39, 43, 87
Enzymatic browning, 102, 105
Enzyme inhibitory activity in food products, 27–28
EOs, *See* Essential oils (EOs)
Epigallocatechin gallate (EGCG), **19**
EPR, *See* Extended producer responsibility (EPR)
Escherichia coli, 15, 88, 92, 181, 190–192, 239
Essential oil-based chitosan, 212–213
  and antimicrobial properties, 213
Essential oil nano emulsions (Eon), 92
Essential oils-based chitosan
  challenges and limitations of, 214
  in food conservation and packaging, 213–214
  synergistic impacts of, 214–215
Essential oils (EOs), 4–5, 205
  formulation and preparation of chitosan-based composite coating with, 208
  overview, 239
Esterified chitosan, 181
Etherified chitosan, 181
Ethylene, 105
Eugenol, **19**
Evaporative coolant structure, 39
Evaporative cooling systems, 41, 42, 45
Extended producer responsibility (EPR), 136
Extracellular polymeric substances (EPS), 91

F

Fabricated chitosan complex, 10
FAO, *See* Food and Agriculture Organization (FAO)
Fertilizer, NPK, 23
Fibers, natural, 147
Film-forming properties of chitosan, 63, 159
Fish and meat, enhancing shelf life in multi-component products of, 126
Fishery product shelf life
  challenges in, **119**, 119–121
  chitosan's role in extending, 124–125, **125**
Flavonoids, 55
Flavour loss, 103, 106–108
Food and agriculture, chitosan in, 14
Food and Agriculture Organization (FAO), 5, 79
Foodborne illnesses, 79, 91, 103, 106, 107, 109, 120, 158, 222, 227
Food coating, 49, 54
  applications of, 49
  biomaterials utilized in, 50
  dipping method, 54
  electrostatic coating method, 54
  polysaccharides employed in, *50*, 50–52, *51*
  protein, 53
  spraying method, 54
Food industry, chitosan derivatives in, 191–192
Food loss, 79
Food matrix, 16, 26, 28, 105, 108, 177
Food oxidation, 212
Food packaging
  innovative technologies in, 157
  meeting functional and aesthetic requirements in, 65, **66**
Food packaging industry, 57
  applications of chitosan-based in, 165–166
Food preservation, 3–4, 42–43
  chitosan's applications in, 103–104, **104**
  chitosan's significance in, 117
  conventional methods of, 1
  emerging technologies in, 128
  gas composition and effects on, 104–105
  role of edible coatings in, 226–227
  versatile biopolymer for, 117–118
  water vapor permeability and impact on, 66–68
Food processing, 49
  chitosan in, 15–16
  conditions, 16, 18
  metal detoxification in, 260
  synthetic pesticides in, 15
Food products
  chitosan enzyme inhibitory activity in, 27–28
  water vapor on, 66–67
Food Quality Monitoring, 212
Food quality, preservation of, 106
Food safety, impact of chitosan-based edible coatings on, 227–228
Food spoilage, 25
  and loss, 79
Food system, hypothetical chitosan mechanism of action in, 25–26
Food technology, chitosan in, 258–259
Food waste, edible films reducing, 68–69, 71
Fossil fuels, renewable resource and reduced dependency on, 70
Freeze-drying, 43
Fresh fruit, 41
Fresh product preservation, 39
Fruits and vegetables, 41–42, 49
Fruits and vegetables spoilage, oxidative stress in, *240*, 240–241

Fungi, 82, 87
cell walls, 207
chitosan-sensitive, 87–88

G

Gallic acid (GA), 191
Gas composition
control, **109**
and effects on food preservation, 104–105
Generally recognized as safe (GRAS), 83, 127, 138, 213
Gene therapy, chitosan in, 187
Global food security, 5
Glucanase, 20, 27
Glucans, 20, 27
Glutathione peroxidases, 243–244
Good Manufacturing Practices (GMP), 127
Graft copolymerization, 184–185
Gram-negative bacteria, 5
Gram-positive bacteria, 5, 84
GRAS, *See* Generally recognized as safe (GRAS)
Green tea (GT), 18

H

Hazard Analysis and Critical Control Points (HACCP), 121
High-molecular-weight chitosan, 207
High-Mw chitosan forms, 25
High-pressure processing (HPP), 4, 43, 45, 126, 128
HTC, *See* Hydrothermal carbonization (HTC)
Hugh-molecular-weight (HMW) chitosan, 84, 85, *86*, 87, 88, 91
'Hurdle technique', 11
Hydrothermal carbonization (HTC), 184
Hydroxyl radicals, 240
Hypothetical chitosan mechanism, of action in food systems, 25–26

I

Immunostimulatory effects, 221–223, 225
Inhibition of yeast, 87–88
Innovative processing techniques, 126
Innovative technologies, in food packaging, 157
Insect cuticles, 207
Ion removal, chitosan role in metal, 260–261

L

Lactic acid bacteria (LAB), 45
Lactobacilli, 83
Lactobacillus plantarum (LAB 12), **19**
Lactobacillus rhamnosus GGLMG 18243, **19**
Lemongrass oil, formulation of chitosan-based coating with, 208
Lemon oil (*Citrus limon*), 213
formulation of chitosan-based coating with, 208
Lifecycle assessments (LCAs), 135
Lipid-based coatings, 40, 52
Lipid oxidation, 2, 25, 27, 52, 53, 91, 92, 102, 105, 117, 119, 121, 122, 247
and enzymatic browning, 102
and enzymatic reactions, 119
inhibition, 123, 223, 247
prevention, 121, 122, 126
Lipopolysaccharides (LPS), 84, *86*
Low-molecular-weight (LMW) chitosan, 84–88, *86*, 207

M

Malaria, edible coating foods in treatment of, 55–56
MAP, *See* Modified atmosphere packaging (MAP)
Maritime food industry, 81
Matrix-forming polymer, 18
Meat, merits of edible coating on, *55*
Meat preservation, chitosan in, 123–124
Meat product shelf life, challenges in, **119**, 119–121
Mechanisms of action of chitosan, 108–109, **109**
Medical and healthcare devices, applications of chitosan-based in, 166–167
Medium-molecular-weight chitosan, 207
Metal detoxification, in food processing, 260
Metal ion removal, chitosan role in, 260–261
Metal oxides, 160
Microbes, chitinolytic, 22
Microbial activity management, 4
Microbial contamination, 117
Microbial control, 44–45
Microbial decomposition, 4
Microbial deterioration, 39
Microbial growth and oxygen penetration, 205
Microbial spoilage, 42, 66, 79, 93, 103, 104, 118, 119, 122
Microbiological deterioration, 212
Microbiological safety, 103, 122
Microcapsules, 163
Microencapsulation, 216
Microorganisms, 82, 239
Microparticles, 216
Minimum inhibitory concentration (MIC) of chitosan, 89
Modified atmosphere packaging (MAP), 4, 6, 44, 100, 128
Moisture loss, 43–44
Multi-component products, enhancing shelf life in, 126

N

Nano compounds, 163, 164
and role in coatings, 162–164, **164**
Nano emulsion, 49
Nanoparticles
chitosan, 17
silver, 5, 21, 45, 83, 158–160, 163
antimicrobial efficacy of, 160
zinc oxide, 158, 159
Nanotechnology, 158
in packaging, 128
Natural additives, 43, 222
Natural antimicrobials, 4
Natural biopolymers, 4
Natural fibres, 147
Nitrogen, 105
Non-thermal preservation methods, 5
NPK fertilizer, 23
Nutraceuticals, 15, 177, 222
encapsulation systems, 18
potential of chitosan-based edible coatings, 228–229
Nutrient degradation, minimization of, 231–232
Nutrient release mechanism, 23

O

Oil spill remediation, 211
Organic antimicrobial agents, 160
Outer membrane protein A (Oompa), 26
Oxidation, of amino acids, 87
Oxidation, lipid, 2, 25, 27, 52, 53, 91, 92, 102, 105, 117, 119, 121, 122, 247
  and enzymatic browning, 102
  and enzymatic reactions, 119
  inhibition, 123, 223, 247
  prevention, 121, 122, 126
Oxidative processes, 117
Oxidative stress
  in fruits and vegetables spoilage, *240*, 240–241
  sources of, *240*
Oxidized lipids, 27
Oxygen, 105
Oxygen species, retention of active, 243
Ozone treatment, 129

P

Packaging
  active packaging, 2, 6, 46, 72, **72**, 213
    with chitosan derivatives, 165
  chitosan-based packaging
    biodegradability of, 134, 135, 140–141
    biodegradation studies of, 135
    potential applications of, **147–148**, 147–149
    properties and applications of, 140, **140**
    utilization of, 134
  conventional packaging materials, 17
  food packaging, 57
    applications of chitosan-based in, 165–166
    innovative technologies in, 157
    meeting functional and aesthetic requirements in, 65, **66**
  innovative technologies in food, 157
  modified atmosphere packaging (MAP), 4, 100
  nanotechnology in, 128
  plastic, 2
  sustainable packaging solutions, 70–71
  technology, 6–7
Pasteurization, 43
Pathogenic microorganisms, 25
PCMs, *See* Phase-change materials (PCMs)
PEF, *See* Pulsed electric fields (PEF)
Perishable food products, preservation of, 116
Petrochemicals, reduced dependency on, 68
PHA, *See* Polyhydroxyalkanoates (PHA)
Pharmaceutical and medical economy, chitosan derivatives in, 185–187, *186*
Phase-change materials (PCMs), 46
Phenolics, 55
Phenylalanine ammonia-lyase, 244–245
Phosphorylated chitosan, *182*, 182–183, *183*
Photostability, UV resistance and, 163
Physical barrier, **109**
PLA, *See* Polylactic acid (PLA)
Plant-based antioxidants, 241–244
Plant-based protective substances, 244
Plant disease management, 167
Plant growth
  bio stimulant for, 168
  chitosan action in, 28
  stimulant, 22–23
Plants' natural defense mechanisms, 20
Plant viral infections, 21
Plasmodium parasites, 55–56
Plastics
  conventional, 146
  packaging, 2
  pollution, 70
Polybutylene adipate terephthalate (PBAT), 139
Polydispersity, molecular weight and, 169
Polyesters, aliphatic, 139
Polyethylene glycol (PEG), 185
Polyethylene (PE), 146
Polyhydroxyalkanoates (PHA), 139
Polyhydroxy urethanes (PHUs), 139
Polylactic acid (PLA), 138–139, 146
Polymers
  chitosan electrostatic flexibility in, 83
  starch-based, 139
Polyphenol, 245–246
Polyphenol oxidase, 17, 26, 27, 105, 246
Polyphenols, encapsulation of, **19**
Polypropylene (PP), 146
Polysaccharide-based coatings, 3
Polysaccharide-based edible coatings, 4
Polysaccharide, employed in food coatings, *50*, 50–52, *51*
Polystyrene (PS), 146
Polyunsaturated fatty acids (PUFA), 87–88
Postharvest losses, 5
Poultry industry, chitosan application in, 122–123
Poultry product shelf life, challenges in, **119**, 119–121
Preservation
  antimicrobial properties enhancing, 67
  food, 3–4, 42–43, 106
  fresh product, 39
  of perishable food products, 116
Probiotics, 128
  microorganisms, encapsulation, **19**
  strains, 55
Protein food coatings, 53
Pulsed electric fields (PEF), 4, 43, 128
Pythium aphanidermatum, 20

Q

Quaternary chitosan derivatives, 181–182
Quaternized carboxymethyl chitosan (QCMC), 191

R

Reactive oxygen species (ROS), 86–88
Redox reactions, 240
Renewable source, and biodegradability, 68
Rifampicin-resistant TB (RR-TB), 57
Rosemary oil, formulation of chitosan-based coating with, 208
Rosemary oil (*Rosmarinus officinalis*), 213

S

SCFAs, *See* Short-chain fatty acids (SCFAs)
Seafood, 3–6, *11*, 15, 53, 125

industry, 62, 68, 70, 118, 124, 137, 141, 225
preservation, 124, 165, 212
Seed priming, 24
Seed treatment, 167
Sensory attributes, 102–104, 110, 117, 119, 122, 123
of chitosan-based films, 62
of foods, 128, 228
of meat products, 124
preservation of tuna, 125
Shelf life
fishery product
challenges in, **119**, 119–121
chitosan's role in extending, 124–125, **125**
meat product, challenges, **119**, 119–121
multi-component products, 126
poultry product, challenges in, **119**, 119–121
Short-chain fatty acids (SCFAs), 83
Silk peptide (SP), 191
Silver nanoparticles (NPs), 5, 21, 45, 83, 158–160, 163
antimicrobial efficacy of, 160
Small-scale farmers, 40, 41
Smart packaging, 4–6, 72, 95, 128, 150, 165, 232
Soil amendment, 212
Soil fertilization, 11
Soil remediation, contaminated, 23–24
Solvothermal method, 184
Spoilage
contamination and, 67
crop, 22
detection of, 72
fish, 17
fish fillet, 52
food, 25, 27, 79
fruits and vegetables, oxidative stress in, *240*, 240–241
growth of spoilage microorganisms, 67, 69, 79–95, 101–106, 118, 121, 124, 129, 162, 165, 168, 226, 258, 260
microbial, 42, 66, 79, 93, 103, 104, 118, 119, 122
minimizing, 62
prevention, 65, 69, 71, 119, 122
reducing, 159
Spraying method, food coating, 54
Staphylococcus aureus, 15, 84, 190, 214
Starch, 52
Starch-based polymers, 139
Sunflower oil, formulation of chitosan-based coating with, 208–209
Superoxide dismutase, 241–242
Surface coating material, 18
Surface of food, 49
Sustainable packaging solutions, aligning with demand for, 70–71
Sustainable storage solutions, 45–46
Synthetic pesticides, in food processing, 15

T

TBAR (Thio barbituric Acid Reactive Substances) values, 91
Tea tree oil (*Melaleuca alternifolia*), 213
Thiolate chitosan, 206–207
Thyme oil (*Thymus vulgaris*), 213
Time-temperature indicators (TTIs), 6
Tissue engineering, 162
chitosan in, **259**, 259–260
Titanium dioxide, 128, 163
TPP, *See* Tripolyphosphate (TPP)
Trimethyl chitosan (TMC), *181*
Tripolyphosphate (TPP), 18
Tripolyphosphate (TPP)-chitosomes, 18
encapsulation efficiency (EE) of, 18
TTIs, *See* Time-temperature indicators (TTIs)
Tuberculosis (TB), effects of edible coating foods in treatment of, 56–57
Turmeric plants, 20

U

Ultraviolet (UV)
resistance and photostability, 163
therapy, 4
United States Food and Drug Administration (USFDA), 216
US Food and Drug Administration, 15

V

Vacuum packing, 44
Vapor pressure differential (VPD), 44
Vegetable food coatings, therapeutic prospects of, 54–55
Vegetables, fruits and, 41–42, 49
Versatile biopolymer, for food preservation, 117–118
Vitamin C, 42
Vitamin, encapsulation of, **19**

W

Wall teichoic acids (WTAs), negative charges of, *86*
Waste management, 135–137, 145, 147, 150, 256
Water retention, soil amendment for, 168
Water-soluble chitin (WSC), 187
Water vapor, on food products, 66–67
Water vapor permeability, 67
for fresh produce, 67
and impact on food preservation, 66–68
for perishable goods, 67
World Health Organization, 237
Wound healing, 161

Y

Yeast, inhibition of, 87–88

Z

Zinc oxide nanoparticles, 158, 159